Springer Proceedings in Complexity

Springer Proceedings in Complexity publishes proceedings from scholarly meetings on all topics relating to the interdisciplinary studies of complex systems science. Springer welcomes book ideas from authors. The series is indexed in Scopus. Proposals must include the following:

- name, place and date of the scientific meeting
- a link to the committees (local organization, international advisors etc.)
- scientific description of the meeting
- list of invited/plenary speakers
- an estimate of the planned proceedings book parameters (number of pages/articles, requested number of bulk copies, submission deadline)

Submit your proposals to: christoph.baumann@springer.com

More information about this series at http://www.springer.com/series/11637

Clemens Mensink · Wanmin Gong ·
Amir Hakami
Editors

Air Pollution Modeling and its Application XXVI

Springer

Editors
Clemens Mensink
VITO NV
Mol, Belgium

Amir Hakami
Department of Civil
and Environmental Engineering
Carleton University
Ottawa, ON, Canada

Wanmin Gong
Environment and Climate Change Canada
Science and Technology Branch
Toronto, ON, Canada

ISSN 2213-8684 ISSN 2213-8692 (electronic)
Springer Proceedings in Complexity
ISBN 978-3-030-22054-9 ISBN 978-3-030-22055-6 (eBook)
https://doi.org/10.1007/978-3-030-22055-6

This Springer imprint is published by the registered company Springer Nature Switzerland AG
The registered company address is: Gewerbestrasse 11, 6330 Cham, Switzerland

History of the International Technical Meeting (ITM) on Air Pollution Modeling and Its Application

Pilot Studies

1969–1974	Air Pollution Pilot Study (*Pilot Country*: *USA*)
1975–1979	Air Pollution Assessment Methodology and Modeling (*Pilot Country*: *Germany*)
1980–1984	Air Pollution Control Strategies and Impact Modeling (*Pilot Country*: *Germany*)

Pilot Follow-Up Meetings

Pilot Country—USA (*R. A. McCormick, L. E. Niemeyer*)

February 1971	Eindhoven, The Netherlands	First Conference on Low Pollution Power Systems Development
July 1971	Paris, France	Second Meeting of the Expert Panel on Air Pollution Modeling

NATO/CCMS International Technical Meetings (ITM) on Air Pollution Modeling and Its Application

Subsequent meetings were supported by the NATO Committee for Challenges to Modern Society and were designated NATO/CCMS International Technical Meetings (ITM) on Air Pollution Modeling and its Application.

October	1972	Paris, France	3rd	ITM
May	1973	Oberursel, Federal Republic of Germany	4th	ITM
June	1974	Roskilde, Denmark	5th	ITM

Pilot Country—Germany (Erich Weber)

September	1975	Frankfurt, Federal Republic of Germany	6th	ITM
September	1976	Airlie House, USA	7th	ITM
September	1977	Louvain-la-Neuve, Belgium	8th	ITM
August	1978	Toronto, Canada	9th	ITM
October	1979	Rome, Italy	10th	ITM

Pilot Country—Belgium (Chris De Wispelaere)

November	1980	Amsterdam, The Netherlands	11th	ITM
September	1981	Menlo Park, California, USA	12th	ITM
September	1982	Ile des Embiez, France	13th	ITM
September	1983	Copenhagen, Denmark	14th	ITM
April	1985	St. Louis, Missouri, USA	15th	ITM

Pilot Country—The Netherlands (Han van Dop)

April	1987	Lindau, Federal Republic of Germany	16th	ITM
September	1988	Cambridge, UK	17th	ITM
May	1990	Vancouver, BC, Canada	18th	ITM
September	1991	Ierapetra, Greece	19th	ITM

Pilot Country—Denmark (Sven-Erik Gryning)

November	1993	Valencia, Spain	20th	ITM
November	1995	Baltimore, Maryland, USA	21st	ITM
May	1997	Clermont-Ferrand, France	22nd	ITM
September	1998	Varna, Bulgaria	23rd	ITM
May	2000	Boulder, Colorado, USA	24th	ITM

Pilot Country—Portugal (Carlos Borrego)

September	2001	Louvain-la-Neuve, Belgium	25th	ITM
May	2003	Istanbul, Turkey	26th	ITM
October	2004	Banff, Canada	27th	ITM
May	2006	Leipzig, Germany	28th	ITM

NATO/SPS International Technical Meetings (ITM) on Air Pollution Modeling and Its Application

In 2007, NATO's Committee for Challenges to Modern Society was disbanded and replaced by NATO's Committee on Science for Peace and Security (NATO/SPS), which continued its support for the ITM.

September	2007	Aveiro, Portugal	29th	ITM

Pilot Country—Canada (Douw Steyn)

May	2009	San Francisco, California, USA	30th	ITM
September	2010	Torino, Italy	31st	ITM
May	2012	Utrecht, The Netherlands	32nd	ITM

International Technical Meetings (ITM) on Air Pollution Modeling and Its Application

In 2012, the NATO Committee on Science for Peace and Security refocused its mandate, and the ITM became independent of NATO/SPS support.

September	2013	Miami, USA	33rd	ITM
May	2015	Montpellier, France	34th	ITM

Pilot Country—Belgium (Clemens Mensink)

October	2016	Chania (Crete), Greece	35th	ITM
May	2018	Ottawa, Canada	36th	ITM

List of Participants

Sebnem Aksoyoglu, Paul Scherrer Institute
Stefano Alessandrin, National Center for Atmospheric Research
Stavros Antonopoulos, Environment and Climate Change Canada
Calvin Arter, University of North Carolina at Chapel Hill
Saravanan Arunachalam, University of North Carolina at Chapel Hill
Marina Astitha, University of Connecticut
Alexander Baklanov, World Meteorological Organization
Sabine Banzhaf, Freie Universität Berlin
Jerzy Bartnicki, Norwegian Meteorological Institute
Carlos Borrego, University of Aveiro
Véronique Bouchet, Environment Canada
Richard Burnett, Health Canada
María Allué Camacho, Agencia Estatal de Meteorología (AEMET)
Hana Chaloupecká, Institute of Thermomechanics of the CAS
Robyn Chatwin-Davies, Carleton University
Nadine Chaumerliac, LaMP/CNRS
Yilin Chen, Georgia Institute of Technology
Jack Chen, Environment and Climate Change Canada
Pieter De Meutter, Belgian Nuclear Research Centre
Laurent Deguillaume, LaMP/OPGC/CNRS
Andy Delcloo, Royal Meteorological Institute of Belgium
Denis Dionne, EGS Ecosupport
Congtru Doan, MOECC
Anthony Dore, Centre for Ecology and Hydrology
Annie Duhamel, Environment Canada
Zita Ferenczi, Hungarian Meteorological Service
Melanie Fillingham, Carleton University/Golder Associates
Arlene Fiore, LDEO/Columbia
Dennis Fudge, Saskatchewan Ministry of Environment
Stefano Galmarini, European Commission—EU Science Hub
Valerie Garcia, U.S. Environmental Protection Agency

Fernando Garcia Menendez, North Carolina State University
Martin Gauthier, RWDI
Camilla Geels, Aarhus University
Angele Genereux, Carleton University
Wanmin Gong, Environment and Climate Change Canada
Cristina Guerreiro, NILU—Norwegian Institute for Air Research
Nitsa Haikin, TAU/NRCN
Amir Hakami, Carleton University
Tomas Halenka, Charles University
Yvonne Hall, Ontario Ministry of Environment and Climate Change
Steven Hanna, Hanna Consultants
Risto Hänninen, Finnish Meteorological Institute
Lucas Henneman, Harvard Chan School of Public Health
Christian Hogrefe, U.S. Environmental Protection Agency
Li Huang, Ministry of Environment and Climate Change Strategy
Li Huang, Ontario Ministry of Environment and Climate Change
Ulas Im, Aarhus University
Cesunica Ivey, University of California, Riverside
Peter Jackson, University of Northern British Columbia
Jukka-Pekka Jalkanen, Finnish Meteorological Institute
Jianhui Jiang, Paul Scherrer Institute
Oriol Jorba Casellas, Barcelona Supercomputing Center
George Kallos, National and Kapodistrian University of Athens
Maria Kanakidou, University of Crete
Daiwen Kang, U.S. Environmental Protection Agency
Eleni Karnezi, Carnegie Mellon University
Ari Karppinen, Finnish Meteorological Institute
Niko Karvosenoja, Finnish Environment Institute (SYKE)
Anke Kelker, Environment Canada
Janya Kelly, Golder Associates Ltd.
Pavel Kishcha (Kichtcha), Tel Aviv University
Rostislav Kouznetsov, Finnish Meteorological Institute
Maciej Kryza, Wroclaw University
Kaarle Kupiainen, Finnish Environment Institute (SYKE)
Nana-Owusua Kwamena, Canadian Nuclear Safety Commission
Pius Lee, Department of Commerce—NOAA/ARL
Wouter Lefebvre, VITO
Joana Leitao, IASS—Potsdam
Fabian Lenartz, ISSeP
Jinliang Liu, Ontario Ministry of Environment and Climate Change
Deborah Luecken, U.S. Environmental Protection Agency
Huiying Luo, University of Connecticut
Bino Maiheu, VITO
Paul Makar, Environment and Climate Change Canada
Randall Martin, Dalhousie University

Isabel Martinez Marco, Agencia Estatal de Meteorología (AEMET)
Rohit Mathur, U.S. Environmental Protection Agency
Volker Matthias, Helmholtz-Zentrum Geesthacht
Jeffery McQueen, NOAA/NWS/NCEP/EMC
Richard Menard, Environment and Climate Change Canada
Clemens Mensink, VITO
Ana Miranda, University of Aveiro
Luisa Molina, Molina Center for Energy and the Environment
Mike Moran, Environment and Climate Change Canada
Heather Morrison, Environment and Climate Change Canada
Nicolas Moussiopoulos, Aristotle University of Thessaloniki
Jennifer Moutinho, Georgia Institute of Technology
Rodrigo Munoz-Alpizar, Environment and Climate Change Canada
Sergey Napelenok, U.S. Environmental Protection Agency
David Niemi, Environment Canada
Talat Odman, Georgia Institute of Technology
Yasar Burak Oztaner, Carleton University
Amanda Pappin, NSERC
Georgina Paull, Environment Canada and Climate Change
Julien Poirier, WSP Canada Inc.
Marje Prank, Cornell University
Yu Qian, Georgia Institute of Technology
Jacinthe Racine, Environment and Climate Change Canada
Martin Otto Paul Ramacher, Helmholtz-Zentrum Geesthacht
Shuzhan Ren, Environment and Climate Change Canada
Alain Robichaud, Environment and Climate Change Canada
Vera Rodrigues, University of Aveiro
Armistead Russell, Georgia Institute of Technology
Rebecca Saari, University of Waterloo
Pablo Saide, UCLA
Abby Salb, Ontario Ministry of Environment and Climate Change
Roberto San Jose, UPM NIF Q2818015F
Verica Savic-Jovcic, Environment and Climate Change Canada
K. Heinke Schluenzen, Universitaet Hamburg
Kirill Semeniuk, Environment Canada
Joana Soares, Environment and Climate Change Canada
Marjan Soltan Zadeh, Carleton University
Vanisa Surapipith, National Astronomic Research Institute of Thailand
Jianguo Tan, Shanghai Meteorological Service
Camille Taylor, Golder Associates Ltd.
Renske Timmermans, TNO
Silvia Trini Castelli, CNR National Research Council
Andreas Uppstu, Finnish Meteorological Institute
Eric van der Swaluw, National Institute for Public Health and the Environment
Julius Vira, Cornell University

Sina Voshtani, Carleton University
Benjamin Weinstien, BC Ministry of Environment
E. Charles White, University of Toronto
Ralf Wolke, Leibniz Institute for Tropospheric Research
David Wong, U.S. Environmental Protection Agency
Greg Yarwood, Ramboll
Calin Zaganescu, Environment and Climate Change Canada
Xuesong Zhang, University of Toronto
Shunliu Zhao, Carleton University
Sergej Zilitinkevich, Finnish Meteorological Institute

Organizing Committee

Members of the Scientific Committee for the 36th International Technical Meeting on Air Pollution Modeling and its Application

Clemens Mensink, Belgium
Ekaterina Batchvarova, Bulgaria
Wanmin Gong, Canada
Ari Karppinen, Finland
Laurent Deguillaume, France
Volker Matthias, Germany
Maria Kanakidou, Greece
Silvia Trini Castelli, Italy
Hilde Fagerli, Norway
Ana Isabel Miranda, Portugal
Oriol Jorba, Spain
Renske Timmermans, The Netherlands
Ulas Im, Turkey
Tony Dore, UK
Rohit Mathur, USA

Honorary Life Members

Chris De Wispelaere, Belgium
Douw Steyn, Canada
Sven-Erik Gryning, Denmark
Werner Klug, Germany

Carlos Borego, Portugal
Han van Dop, The Netherlands
S. T. Rao, USA
Frank Schiermeier, USA

Preface

Worldwide air pollution remains a threat to human health and the environment. Despite the efforts over the last decades, especially in the USA and Europe, many of the air pollutant concentrations are still too high, leading to persistent air quality problems. "Air pollution is still responsible for more than 400.000 premature deaths in Europe each year," states the European Environment Agency.

Since 1969, scientists and policy makers from all over the world meet to discuss these problems at the International Technical Meeting on Air Pollution Modeling and its Application (ITM), using air quality models to study the atmospheric transport, chemistry, impacts and sources of air pollution, its effect on human health, and its interaction with climate change. Their results, new insights, and discussions are very much appreciated by regulators and policy makers and help them to underpin national and international abatement programs and protocols.

This volume contains the abstracts and papers presented at the 36th ITM, held in Ottawa, Canada, from May 14–18, 2018. The 36th ITM was organized by Carleton University and Environment and Climate Change Canada (Host Country) and VITO in Belgium (Pilot Country) in conjunction with the annual WMO-GURME meeting. Key topics presented at this ITM correspond to the sections in this book and include: regional and intercontinental modeling; local- and urban-scale modeling; emission modeling and processing; data assimilation and air quality forecasting; model assessment and verification; aerosols in the atmosphere; modeling air pollution in a changing climate; and air quality effects on human health and ecology.

The ITM was attended by 131 participants representing 22 countries. Keynotes were presented by Randall Martin, Dalhousie University (air quality effects on human health), Arlene Fiore, Columbia University (modeling air pollution in a changing climate), Luisa T. Molina, Massachusetts Institute of Technology (local- and urban-scale modeling), and Richard Burnett, Health Canada and University of Washington (air quality effects on human health).

On behalf of the ITM Scientific Committee and as organizers and editors, we would like to thank all the participants who contributed to the success of the meeting and its high scientific level. We especially recognize the organizational and support efforts of the chairpersons and rapporteurs. Special thanks to the sponsoring institutions: Environment and Climate Change Canada, Carleton University (Canada), VITO (Belgium), the Federal Public Planning Service Science Policy of Belgium, the World Meteorological Organization (WMO), and Ramboll, who gave a special grant to award prizes to Early Career Researchers for the best paper or poster presentations.

The next meeting will be held from September 23 to 27, 2019, in Hamburg, Germany.

Mol, Belgium — Clemens Mensink, Scientific Committee Chair

Toronto, Canada — Wanmin Gong

Ottawa, Canada — Amir Hakami, Local Conference Organizers

Contents

Part IX Special Sessions

Contributors

Jan Aben RIVM, Bilthoven, The Netherlands

Ha Ai School of Civil and Environmental Engineering, Georgia Institute of Technology, Atlanta, GA, USA

Bruce Ainslie Environment and Climate Change Canada, Ottawa, Canada

Ayodeji Akingunola Air Quality Modelling and Integration Section, Air Quality Research Division, Science and Technology Branch, Environment and Climate Change Canada, Toronto, ON, Canada

Yayne-Abeba Aklilu Environmental Monitoring and Science Division, Alberta Environment and Parks, Edmonton, AL, Canada

Sebnem Aksoyoglu Laboratory of Atmospheric Chemistry (LAC), Paul Scherrer Institute (PSI), Villigen, Switzerland

J. Al Qahtani Saudi Aramco, Dhahran, Saudi Arabia

I. Alexiou Saudi Aramco, Dhahran, Saudi Arabia

Victor Almanza Molina Center for Energy and the Environment, La Jolla, CA, USA

Pinhas Alpert Department of Geophysics, Tel Aviv University, Tel-Aviv, Israel

Ummugulsum Alyuz Eurasia Institute of Earth Sciences, Istanbul Technical University, Istanbul, Turkey

D. G. Amanatidis Environmental Chemical Processes Laboratory, Department of Chemistry, University of Crete, Heraklion, Greece

Fulvio Amato Institute of Environmental Assessment and Water Research, IDAEA-CSIC, Barcelona, Spain

Mikael Skou Andersen Department of Environmental Science, Aarhus University, Roskilde, Denmark

Stavros Antonopoulos Meteorological Service of Canada (MSC), Environment and Climate Change Canada (ECCC), Montreal, QC, Canada

Wyat Appel National Exposure Research Laboratory, U.S. Environmental Protection Agency, Research Triangle Park, Durham, NC, USA

Jan Arndt Helmholtz-Zentrum Geesthacht, Institute of Coastal Research, Geesthacht, Germany

Calvin Arter Department of Environmental Sciences and Engineering, Institute for the Environment, University of North Carolina at Chapel Hill, Chapel Hill, NC, USA

Sarav Arunachalam Institute for the Environment, University of North Carolina at Chapel Hill, Chapel Hill, NC, USA

Ana Ascenso CESAM & Department of Environment and Planning, University of Aveiro, Aveiro, Portugal

Marina Astitha Civil and Environmental Engineering, University of Connecticut, Storrs, CT, USA

Armin Aulinger Department of Chemical Transport Modeling, Helmholtz-Zentrum Geesthacht, Institute of Coastal Research, Geesthacht, Germany

Roni Avissar University of Miami, Florida, USA

Jonathan Baik Meteorological Service of Canada, ECCC, Vancouver, BC, Canada

Barry Baker NOAA/Air Resources Laboratory, College Park, MD, USA; UMD/Cooperative Institute for Climate and Satellites, College Park, MD, USA; Center for Spatial Information Science and Systems, George Mason University, Fairfax, VA, USA

Alessandra Balzarini Ricerca sul Sistema Energetico (RSE SpA), Milan, Italy

Sabine Banzhaf Institute of Meteorology, Freie Universität Berlin, Berlin, Germany

Rocio Baro Department of Physics, University of Murcia, Physics of the Earth, Campus de Espinardo, Murcia, Spain

Jerzy Bartnicki Norwegian Meteorological Institute, Oslo, Norway

Stephen Beagley Air Quality Research Division, Science and Technology Branch, Environment and Climate Change Canada, Toronto, ON, Canada

William Bealey Centre for Ecology & Hydrology, Edinburgh, Scotland, UK

M. Beauchemin High Impact Weather National Laboratory, Meteorological Service of Canada, Environment and Climate Change Canada, Montreal, Canada

Stephane Belair Meteorological Research Division, Science and Technology Branch, Environment and Climate Change Canada, Dorval, QC, Canada

Michal Belda Department of Atmospheric Physics, Faculty of Mathematics and Physics, Charles University, Prague, Czech Republic

Roberto Bellasio Enviroware srl, Concorezzo, MB, Italy

Jaime Benavides Earth Sciences Department, Barcelona Supercomputing Center (BSC), Barcelona, Spain

Roberto Bianconi Enviroware srl, Concorezzo, MB, Italy

Johannes Bieser Helmholtz-Zentrum Geesthacht, Institute of Coastal Research, Geesthacht, Germany;
Department of Chemical Transport Modeling, Helmholtz-Zentrum Geesthacht, Institute of Coastal Research, Geesthacht, Germany

Andrea Bisignano CNR-Institute of Atmospheric Sciences and Climate, Turin, Italy

Daniel Blanco-Ward CESAM & Department of Environment and Planning, University of Aveiro, Aveiro, Portugal

Bob Bornstein SJSU, San Jose, CA, USA

Carlos Borrego CESAM & Department of Environment and Planning, University of Aveiro, Aveiro, Portugal

Adam E. Bourassa Department of Physics and Engineering Physics, University of Saskatchewan, Saskatoon, SK, Canada

László Bozó Hungarian Meteorological Service, Budapest, Hungary

Christine Braban Centre for Ecology & Hydrology, Edinburgh, Scotland, UK

Jørgen Brandt Department of Environmental Science, Aarhus University, Roskilde, Denmark

Nicolas Bruffaerts Mycology and Aerobiology Unit, Sciensano, Brussels, Belgium

Johan Camps Belgian Nuclear Research Institute, Mol, Belgium

Edward Carnell Centre for Ecology & Hydrology, Edinburgh, Scotland, UK

Hana Chaloupecká Institute of Thermomechanics, Czech Academy of Sciences, Prague, Czech Republic;
Faculty of Mathematics and Physics, Charles University, Prague, Czech Republic

Joseph Chang RAND Corporation, Arlington, VA, USA

Christina Feng Chang Civil and Environmental Engineering, University of Connecticut, Storrs, CT, USA

Chris Charron Ontario Ministry of the Environment and Climate Change, Toronto, ON, Canada

N. Chaumerliac Université Clermont Auvergne, CNRS LaMP, Clermont-Ferrand, France

Jack Chen Air Quality Research Division, Science and Technology Branch, Environment and Climate Change Canada, Ottawa, ON, Canada

Y. Chen Leibniz Institute for Tropospheric Research (TROPOS), Leipzig, Germany;
Lancaster Environment Centre, Lancaster University, Lancaster, UK

Philip Cheung Air Quality Research Division, Science and Technology Branch, Environment and Climate Change Canada, Toronto, ON, Canada

Eleftherios Chourdakis Laboratory of Heat Transfer and Environmental Engineering, Department of Mechanical Engineering, Aristotle University, Thessaloniki, Greece

Jesper Heile Christensen Department of Environmental Science, Aarhus University, Roskilde, Denmark

Giancarlo Ciarelli Laboratoire Inter-Universitaire Des Systèmes Atmosphériques (LISA), Institut Pierre Simon Laplace, UMR CNRS 7583, Université Paris Est Créteil et Université Paris Diderot, Créteil, France

Anthony Ciccone Golder Associates Ltd, Mississauga, ON, Canada

Cathy Clerbaux UPMC Université Paris 6, Université Versailles St-Quentin, LATMOS-IPSL, CNRS/INSU, Paris, France;
Spectroscopie de l'Atmosphére, Service de Chimie Quantique et Photophysique, Université Libre de Bruxelles, Brussels, Belgium

Sílvia Coelho CESAM & Department of Environment and Planning, University of Aveiro, Aveiro, Portugal

Amanda Cole Measurements and Analysis Research Section, Air Quality Research Division, Environment and Climate Change Canada, Downsview, Canada

Augustin Colette INERIS, Institut National de l'Environnement Industriel et des Risques, Verneuil-en-Halatte, France

Ellen Cooter Computational Exposure Division, National Exposure Research Laboratory, U.S. Environmental Protection Agency, Washington, DC, NC, USA

Christophe Corbel Deepki, Paris, France

Guido Cosemans VITO, Mol, Belgium

Gabriele Curci Department of Physical and Chemical Sciences, Center of Excellence CETEMPS, University of L'Aquila, L'Aquila, Italy

Arlindo da Silva NASA Goddard Space Flight Center, Greenbelt, MD, USA

N. Daskalakis LAMOS, Institute of Environmental Physics (IUP), University of Bremen, Bremen, Germany

Didier Davignon Air Quality Modelling Applications Section, ECCC, Montreal, QC, Canada

Pieter De Meutter Royal Meteorological Institute of Belgium, Brussels, Belgium;
Department of Physics and Astronomy, Ghent University, Ghent, Belgium;
SCK CEN, Belgian Nuclear Research Centre, Mol, Belgium

Wilco de Vries RIVM, Bilthoven, The Netherlands

D. A. Degenstein Department of Physics and Engineering Physics, University of Saskatchewan, Saskatoon, SK, Canada

Bart Degraeuwe VITO, Mol, Belgium;
Air and Climate Unit, Directorate C—Energy, Transport and Climate, European Commission—Joint Research Centre, Brussels, Belgium

Jean DeGranpré Air Quality Research Division, Environment and Climate Change Canada, Toronto, ON, Canada

L. Deguillaume Université Clermont Auvergne, CNRS LaMP, Clermont-Ferrand, France

Andy Delcloo Royal Meteorological Institute of Belgium, Brussels, Belgium;
Royal Meteorological Institute of Belgium, Brussels, Belgium;
Department of Physics and Astronomy, Ghent University, Ghent, Belgium

A.-M. Delort Université Clermont Auvergne, CNRS, SIGMA Clermont, ICCF, Clermont-Ferrand, France

Bruce Denby Norwegian Meteorological Institute (MET), Oslo, Norway

Hugo A. C. Denier van der Gon Department of Climate, Air and Sustainability, TNO, Utrecht, The Netherlands

Martin Deshaies-Jacques Air Quality Research Division, Environment and Climate Change Canada, Dorval, Canada

Felix Deutsch VITO, Mol, Belgium

Congtru Doan Ontario Ministry of the Environment and Climate Change, Toronto, ON, Canada

Anthony Dore Centre for Ecology & Hydrology, Edinburgh, Scotland, UK

Ulrike Dragosits Centre for Ecology & Hydrology, Edinburgh, Scotland, UK

Guy Driesen VITO, Mol, Belgium

Sebastien Dujardin Department of Geography, University of Namur, Namur, Belgium

James East Department of Civil, Construction and Environmental Engineering, North Carolina State University, Raleigh, NC, USA

Imad El Haddad Laboratory of Atmospheric Chemistry (LAC), Paul Scherrer Institute (PSI), Villigen, Switzerland

Chris Emery Bay Area Air Quality Management District, San Francisco, CA, USA

Martin Eriksson Shipping and Marine Technology, Chalmers University of Technology, Gothenburg, Sweden

Hilde Fagerli EMEP MSC-W, Norwegian Meteorological Institute, Oslo, Norway

Georgios Fanourgakis Environmental Chemical Processes Laboratory, Department of Chemistry, University of Crete, Heraklion, Greece

Aidan Farrow Centre for Atmospheric and Instrumentation Research (CAIR), University of Hertfordshire, Hatfield, UK

Zita Ferenczi Hungarian Meteorological Service, Budapest, Hungary

Melanie Fillingham Golder Associates Ltd., Ottawa, ON, Canada

Vivian Fischer Helmholtz-Zentrum Geesthacht, Geesthacht, Germany

Johannes Flemming European Centre for Medium Range Weather Forecast (ECMWF), Reading, UK

Kristen Foley National Exposure Research Laboratory, U.S. Environmental Protection Agency, Research Triangle Park, Durham, NC, USA

Renate Forkel Karlsruher Institut für Technologie, Institute of Meteorology and Climate Research, Atmospheric Environmental Research, Garmisch-Partenkirchen, Germany

Hosein Foroutan Department of Biomedical Engineering and Mechanics, Virginia Tech, Blacksburg, VA, USA

V. Fortin Environmental Numerical Weather Prediction Research, Meteorological Research Division, Environment and Climate Change Canada, Dorval, Canada

Andrea Fraser Ricardo Energy & Environment, Harwell, Oxon, UK

Erik Fridell Swedish Environmental Research Institute, Göteborg, Sweden

Dennis Fudge Saskatchewan Ministry of Environment, Regina, Canada

G. Galanis Department of Physics, National and Kapodistrian University of Athens, Athens, Greece;
Naval Academy of Greece, Section of Mathematics, Piraeus, Greece

Stefano Galmarini European Commission Joint Research Centre (JRC), Ispra, Italy

Carla Gama CESAM & Department of Environment and Planning, University of Aveiro, Aveiro, Portugal

Agustin Garcia Center for Atmospheric Sciences, National University of Mexico, Mexico City, Mexico

Valerie Garcia Computational Exposure Division, National Exposure Research Laboratory, U.S. Environmental Protection Agency, Washington, DC, NC, USA

Fernando Garcia-Menendez Department of Civil, Construction and Environmental Engineering, North Carolina State University, Raleigh, NC, USA

Carlos Pérez García-Pando Earth Sciences Department, Barcelona Supercomputing Center (BSC), Barcelona, Spain

Camilla Geels Department of Environmental Science, Aarhus University, Roskilde, Denmark

Robert C. Gilliam Computational Exposure Division, National Exposure Laboratory, US EPA, Ann Arbor, MI, USA

Wanmin Gong Air Quality Research Division, Science and Technology Branch, Environment and Climate Change Canada, Toronto, ON, Canada

Scott L. Goodrick Southern Research Station, US Forest Service, Athens, GA, USA

Sylvie Gravel Air Quality Research Division, ECCC, Montreal, QC, Canada; Air Quality Research Division, Environment and Climate Change Canada, Toronto, ON, Canada

Alex Guenther Department of Earth System Science, University of California, Irvine, CA, USA

Marc Guevara Earth Sciences Department, Barcelona Supercomputing Center (BSC), Barcelona, Spain

Jakub Guzikowski Department of Climatology and Atmosphere Protection, University of Wrocław, Institute of Geography and Regional Development, Wrocław, Poland

Nitsa Haikin Department of Geosciences, Tel Aviv University, Tel Aviv, Israel; Department of Physics, NRCN, Beer Sheva, Israel

Tomas Halenka Department of Atmospheric Physics, Faculty of Mathematics and Physics, Charles University, Prague, Czech Republic

Jane Hall Centre for Ecology & Hydrology, Bangor, Wales, UK

Yvonne Hall Ontario Ministry of the Environment and Climate Change, Toronto, ON, Canada

Rafiq Hamdi Royal Meteorological Institute of Belgium, Brussels, Belgium; Department of Physics and Astronomy, Ghent University, Ghent, Belgium

Steven Hanna Hanna Consultants, Kennebunkport, ME, USA

Otto Hänninen National Institute for Health and Welfare (THL), Kuopio, Finland

Risto Hänninen Finnish Meteorological Institute, Helsinki, Finland

Kaj Mantzius Hansen Department of Environmental Science, Aarhus University, Roskilde, Denmark

Marijke Hendrickx Mycology and Aerobiology Unit, Sciensano, Brussels, Belgium

C. Hendriks Department Climate, Air and Sustainability, TNO, Utrecht, The Netherlands

Lucas R. F. Henneman Harvard TH Chan School of Public Health, Boston, MA, USA

Jerold A. Herwehe Computational Exposure Division, National Exposure Laboratory, US EPA, Ann Arbor, MI, USA

Christian Hogrefe Atmospheric and Environmental Systems Modeling Division, Center for Environmental Measurement and Modeling, U.S. Environmental Protection Agency, Research Triangle Park, NC, USA;
National Exposure Research Laboratory, Computational Exposure Division, Office of Research and Development, U.S. Environmental Protection Agency, Research Triangle Park, NC, USA

Emese Homolya Hungarian Meteorological Service, Budapest, Hungary

Hans Hooyberghs VITO, Mol, Belgium

Yongtao Hu School of Civil and Environmental Engineering, Georgia Institute of Technology, Atlanta, GA, USA

Ling Huang Ramboll, Novato, CA, USA

Peter Huszar Department of Atmospheric Physics, Faculty of Mathematics and Physics, Charles University, Prague, Czech Republic

Virginie Hutsemékers AwAC, Jambes, Belgium

Ulas Im Department of Environmental Science, Aarhus University, Roskilde, Denmark

Irena Ivanova Air Quality Research Division, Environment and Climate Change Canada, Toronto, ON, Canada

Peter L. Jackson Environmental Science and Engineering Programs, University of Northern British Columbia, Prince George, BC, Canada

Jukka-Pekka Jalkanen Atmospheric Composition Research, Finnish Meteorological Institute, Helsinki, Finland

Zbyněk Jaňour Institute of Thermomechanics, Czech Academy of Sciences, Prague, Czech Republic

Stijn Janssen VITO, Mol, Belgium

Jianhui Jiang Laboratory of Atmospheric Chemistry (LAC), Paul Scherrer Institute (PSI), Villigen, Switzerland

Xiaoyan Jiang Department of Earth System Science, University of California, Irvine, CA, USA

Zhe Jiang School of Earth and Space Sciences, University of Science and Technology of China, Hefei, Anhui, China

Pedro Jimenez-Guerrero Department of Physics, University of Murcia, Physics of the Earth, Campus de Espinardo, Murcia, Spain

Lasse Johansson Atmospheric Composition Research, Finnish Meteorological Institute, Helsinki, Finland

M. Joly Université Clermont Auvergne, CNRS, SIGMA Clermont, ICCF, Clermont-Ferrand, France

Dylan Jones Department of Physics, University of Toronto, Toronto, ON, Canada

Oriol Jorba Earth Sciences Department, Barcelona Supercomputing Center (BSC), Barcelona, Spain

Klára Jurčáková Institute of Thermomechanics, Czech Academy of Sciences, Prague, Czech Republic

Nikos Kalivitis Environmental Chemical Processes Laboratory, Department of Chemistry, University of Crete, Heraklion, Greece;
National Observatory of Athens, Athens, Greece

George Kallos Department of Physics, National and Kapodistrian University of Athens, Athens, Greece;
University of Athens, Athens, Greece

Maria Kanakidou Environmental Chemical Processes Laboratory, Department of Chemistry, University of Crete, Heraklion, Greece

Farah Kanani-Sühring Institute of Meteorology and Climatology, Leibniz University Hannover, Hannover, Germany

Daiwen Kang National Exposure Research Laboratory, U.S. Environmental Protection Agency, Research Triangle Park, Durham, NC, USA

Leena Kangas Finnish Meteorological Institute, Helsinki, Finland

Matthias Karl Department of Chemical Transport Modeling, Helmholtz-Zentrum Geesthacht, Institute of Coastal Research, Geesthacht, Germany

Jan Karlicky Department of Atmospheric Physics, Faculty of Mathematics and Physics, Charles University, Prague, Czech Republic

Eleni Karnezi Department of Chemical Engineering, Carnegie Mellon University, Pittsburgh, PA, USA

Ari Karppinen Finnish Meteorological Institute (FMI), Helsinki, Finland

Niko Karvosenoja Finnish Environment Institute (SYKE), Helsinki, Finland

Mari Kauhaniemi Finnish Meteorological Institute, Helsinki, Finland

Martin Keller Department of Physics, University of Toronto, Toronto, ON, Canada

Radka Kellnerová Institute of Thermomechanics, Czech Academy of Sciences, Prague, Czech Republic

Janya Kelly Golder Associates Ltd, Mississauga, ON, Canada

Sue Kemball-Cook Ramboll, Novato, CA, USA

Klaus Ketelsen Independent Software Consultant, Hannover/Berlin, Germany

Basit Khan Karlsruher Institut für Technologie, Institute of Meteorology and Climate Research, Atmospheric Environmental Research, Garmisch-Partenkirchen, Germany

Hyuncheol Kim NOAA/Air Resources Laboratory, College Park, MD, USA; UMD/Cooperative Institute for Climate and Satellites, College Park, MD, USA; Center for Spatial Information Science and Systems, George Mason University, Fairfax, VA, USA

Ioannis Kioutsioukis Department of Physics, University of Patras, University Campus, Rio, Patras, Greece

I. Kirchner Freie Universität Berlin (FUB), Berlin, Germany

Pavel Kishcha Department of Geophysics, Tel Aviv University, Tel-Aviv, Israel

Nutthida Kitwiroon Environmental Research Group, Kings' College London, London, UK

Bonyoung Koo Bay Area Air Quality Management District, San Francisco, CA, USA

Rostislav Kouznetsov Finnish Meteorological Institute, Helsinki, Finland

R. Kranenburg Department Climate, Air and Sustainability, TNO, Utrecht, The Netherlands

Roy Wichink Kruit RIVM, Bilthoven, The Netherlands

Maciej Kryza Department of Climatology and Atmosphere Protection, Institute of Geography and Regional Development, Wrocław University, Wrocław, Poland

Jaakko Kukkonen Finnish Meteorological Institute (FMI), Helsinki, Finland

Kaarle Kupiainen Finnish Environment Institute (SYKE), Helsinki, Finland; Ministry of the Environment, Helsinki, Finland

Mona Kurppa University of Helsinki, Helsinki, Finland

Paul Kushner Department of Physics, University of Toronto, Toronto, ON, Canada

Hugo Landry Air Quality Modelling Applications Section, ECCC, Montreal, QC, Canada

Pius Lee NOAA/Air Resources Laboratory, College Park, MD, USA; Center for Spatial Information Science and Systems, George Mason University, Fairfax, VA, USA

Filip Lefebre VITO, Mol, Belgium

Wouter Lefebvre VITO, Mol, Belgium

Wenfang Lei Molina Center for Energy and the Environment, La Jolla, CA, USA

Fabian Lenartz ISSeP, Liège, Belgium

Sylvie Leroyer Meteorological Research Division, Science and Technology Branch, Environment and Climate Change Canada, Dorval, QC, Canada

S. M. Li Air Quality Research Division, Science and Technology Branch, Environment and Climate Change Canada, Toronto, ON, Canada

Donghai Liang Rollins School of Public Health, Emory University, Atlanta, GA, USA

Ciao-Kai Liang Department of Environmental Sciences and Engineering, University of North Carolina at Chapel Hill, Chapel Hill, NC, USA

Neng-Huei Lin Department of Atmospheric Sciences, National Central University, Taoyuan, Taiwan

Po-Hsiung Lin National Taiwan University, Taipei, Taiwan

Tang-Huang Lin Department of Atmospheric Sciences, National Central University, Taoyuan, Taiwan

Cong Liu School of Energy and Environment, Southeast University, Nanjing, China

Gin-Rong Liu Department of Atmospheric Sciences, National Central University, Taoyuan, Taiwan

Jinliang Liu Ontario Ministry of the Environment and Climate Change, Toronto, ON, Canada

Peng Liu NRC Research Associate, Computational Exposure Division, National Exposure Research Laboratory, U.S. Environmental Protection Agency, Research Triangle Park, NC, USA

Diogo Lopes CESAM & Department of Environment and Planning, University of Aveiro, Aveiro, Portugal

Myriam Lopes CESAM & Department of Environment and Planning, University of Aveiro, Aveiro, Portugal

Deborah Luecken U.S. Environmental Protection Agency, Research Triangle Park, NC, USA

Alexandru Lupu Modelling and Integration Research Section, Air Quality Research Division, Environment and Climate Change Canada, Downsview, Canada;
Air Quality Research Division, Environment and Climate Change Canada, Toronto, ON, Canada

Bino Maiheu VITO, Mol, Belgium

Jackson Mak Golder Associates Ltd., Ottawa, ON, Canada

Paul Makar Air Quality Research Division, Science and Technology Branch, Environment and Climate Change Canada, Toronto, ON, Canada;
Air Quality Modelling and Integration Section, Air Quality Research Division, Science and Technology Branch, Environment and Climate Change Canada, Toronto, ON, Canada

Piero Malguzzi CNR-Institute of Atmospheric Sciences and Climate, Bologna, Italy

Ilja Maljutenko Tallinn University of Technology, Tallinn, Estonia

Björn Maronga Institute of Meteorology and Climatology, Leibniz University Hannover, Hannover, Germany;
Geophysical Institute, University of Bergen, Bergen, Norway

Rohit Mathur Atmospheric and Environmental Systems Modeling Division, Center for Environmental Measurement and Modeling, U.S. Environmental Protection Agency, Research Triangle Park, NC, USA;
Computational Exposure Division, National Exposure Research Laboratory, U.S. Environmental Protection Agency, RTP, Washington, D.C., USA

Volker Matthias Helmholtz-Zentrum Geesthacht, Institute of Coastal Research, Geesthacht, Germany

Matthias Mauder Karlsruher Institut für Technologie, Institute of Meteorology and Climate Research, Atmospheric Environmental Research, Garmisch-Partenkirchen, Germany

Jamie McEvoy Golder Associates Ltd., Ottawa, ON, Canada

Marcelo Mena-Carrasco Climate Change Group, World Bank, Washington, D. C., USA

Richard Ménard Air Quality Research Division, Environment and Climate Change Canada, Dorval, Canada

Sylvain Ménard Air Quality Modelling Applications Section, ECCC, Montreal, QC, Canada

Clemens Mensink VITO, Mol, Belgium

Swen Metzger ResearchConcepts io GmbH, Freiburg im Breisgau, Germany

Ana Isabel Miranda CESAM & Department of Environment and Planning, University of Aveiro, Aveiro, Portugal

Luisa T. Molina Molina Center for Energy and the Environment, La Jolla, CA, USA

Erwan Monier Joint Program on the Science and Policy of Global Change, Massachusetts Institute of Technology, Cambridge, MA, USA

Alexandra Monteiro CESAM & Department of Environment and Planning, University of Aveiro, Aveiro, Portugal

Jacques Montpetit Meteorological Service of Canada (MSC), Environment and Climate Change Canada (ECCC), Montreal, QC, Canada

Michael Moran Modelling and Integration Research Section, Air Quality Research Division, Environment and Climate Change Canada, Downsview, Canada;
Air Quality Research Division, Science and Technology Branch, Environment and Climate Change Canada, Toronto, ON, Canada

C. Mouchel-Vallon Université Clermont Auvergne, CNRS LaMP, Clermont-Ferrand, France;
NCAR, National Center for Atmospheric Research, Boulder, CO, USA

Nicolas Moussiopoulos Laboratory of Heat Transfer and Environmental Engineering, Department of Mechanical Engineering, Aristotle University, Thessaloniki, Greece

Jennifer L. Moutinho School of Civil and Environmental Engineering, Georgia Institute of Technology, Atlanta, GA, USA

James A. Mulholland School of Civil and Environmental Engineering, Georgia Institute of Technology, Atlanta, GA, USA

Rodrigo Munoz-Alpizar Air Quality Modelling and Application, Meteorological Service of Canada, Environment and Climate Change Canada, Dorval, QC, Canada

Benjamin N. Murphy National Exposure Research Laboratory, U.S. Environmental Protection Agency, Research Triangle Park, Durham, NC, USA

S. Myriokefalitakis IERSD, National Observatory of Athens, Penteli, Greece;
LAMOS, Institute of Environmental Physics (IUP), University of Bremen, Bremen, Germany

Sergey Napelenok Atmospheric Modelling and Analysis Division, National Exposure Research Laboratory, U.S. Environmental Protection Agency, Research Triangle Park, NC, USA

Athanasios Nenes Laboratory of Atmospheric Processes and their Impacts, School of Architecture, Civil and Environmental Engineering, École Polytechnique Federale de Lausanne, Lausanne, Switzerland;
Institute of Chemical Engineering Sciences, Foundation for Research and Technology (ICEHT-FORTH), Hellas, Patras, Greece

Brayden Nilson Environmental Science and Engineering Programs, University of Northern British Columbia, Prince George, BC, Canada

Uarporn Nopmongcol Ramboll Environ, Novato, CA, USA

Tereza Novakova Department of Atmospheric Physics, Faculty of Mathematics and Physics, Charles University, Prague, Czech Republic

Catherine Nowakowski Civil and Environmental Engineering, University of Connecticut, Storrs, CT, USA

Vincenzo Obiso Earth Sciences Department, Barcelona Supercomputing Center (BSC), Barcelona, Spain

M. Talat Odman School of Civil and Environmental Engineering, Georgia Institute of Technology, Atlanta, GA, USA

Emmanouil Oikonomakis Laboratory of Atmospheric Chemistry (LAC), Paul Scherrer Institute (PSI), Villigen, Switzerland

Balbir Pabla Air Quality Research Division, Science and Technology Branch, Environment and Climate Change Canada, Toronto, ON, Canada

Li Pan NOAA/Air Resources Laboratory, College Park, MD, USA;
UMD/Cooperative Institute for Climate and Satellites, College Park, MD, USA;
Center for Spatial Information Science and Systems, George Mason University, Fairfax, VA, USA

Spyros N. Pandis Department of Chemical Engineering, Carnegie Mellon University, Pittsburgh, PA, USA;
Department of Chemical Engineering, University of Patras, Patra, Greece

Carlos Pérez-García Pando Earth Sciences Department, Barcelona Supercomputing Center (BSC), Barcelona, Spain

Marco Pandolfi Department of Chemical Engineering, Carnegie Mellon University, Pittsburgh, PA, USA;
Department of Chemical Engineering, University of Patras, Patra, Greece;
Institute of Environmental Assessment and Water Research, (IDAEA-CSIC), Barcelona, Spain

P. Patlakas Department of Physics, National and Kapodistrian University of Athens, Athens, Greece

Ville-Veikko Paunu Finnish Environment Institute (SYKE), Helsinki, Finland

Radenko Pavlovic Air Quality Policy-Issue Response Section, Environment and Climate Change Canada, Montreal, QC, Canada;
Air Quality Modelling Applications Section, ECCC, Montreal, QC, Canada

María Teresa Pay Earth Sciences Department, Barcelona Supercomputing Center (BSC), Barcelona, Spain

Wim Peelaerts VITO, Mol, Belgium

Diane Pendlebury Air Quality Research Division, Environment and Climate Change Canada, Toronto, ON, Canada

H. Perroux Université Clermont Auvergne, CNRS LaMP, Clermont-Ferrand, France

Oliver Pescott Centre for Ecology & Hydrology, Wallingford, UK

Roger Pielke Sr. University of Colorado, Colorado State University, Colorado, USA

Bret D. Pienkosz Department of Civil, Construction and Environmental Engineering, North Carolina State University, Raleigh, NC, USA

Guido Pirovano Ricerca sul Sistema Energetico (RSE SpA), Milan, Italy

Jonathan E. Pleim Computational Exposure Division, National Exposure Laboratory, US EPA, Ann Arbor, MI, USA

David Plummer Canadian Centre for Climate Modelling and Analysis, Environment and Climate Change Canada, Toronto, ON, Canada

George Pouliot National Exposure Research Laboratory, U.S. Environmental Protection Agency, Research Triangle Park, Durham, NC, USA;
Computational Exposure Division, National Exposure Laboratory, US EPA, Ann Arbor, MI, USA

Luca Pozzoli European Commission, Joint Research Centre (JRC), Ispra, VA, Italy

Marje Prank Finnish Meteorological Institute, Helsinki, Finland;
Atmospheric Composition Research Unit, Finnish Meteorological Institute, Helsinki, Finland;
Department of Earth and Atmospheric Sciences, Cornell University, Ithaca, USA

André S. H. Prévôt Laboratory of Atmospheric Chemistry (LAC), Paul Scherrer Institute (PSI), Villigen, Switzerland

Yu Qian School of Civil and Environmental Engineering, Georgia Institute of Technology, Atlanta, GA, USA

Markus Quante Helmholtz-Zentrum Geesthacht, Institute of Coastal Research, Geesthacht, Germany

Xavier Querol Institute of Environmental Assessment and Water Research, IDAEA-CSIC, Barcelona, Spain

Siegfried Raasch Institute of Meteorology and Climatology, Leibniz University Hannover, Hannover, Germany

Sandra Rafael CESAM & Department of Environment and Planning, University of Aveiro, Aveiro, Portugal

Martin Otto Paul Ramacher Department of Chemical Transport Modeling, Helmholtz-Zentrum Geesthacht, Institute of Coastal Research, Geesthacht, Germany;
Helmholtz-Zentrum Geesthacht, Institute of Coastal Research, Geesthacht, Germany

Limei Ran National Exposure Research Laboratory, U.S. Environmental Protection Agency, Research Triangle Park, Durham, NC, USA

Shuzhan Ren Air Quality Research Division, Science and Technology Branch, Environment and Climate Change Canada, Toronto, ON, Canada

Alain Robichaud Modelling and Integration Research Section, Air Quality Research Division, Environment and Climate Change Canada, Dorval, Canada

Alfredo Rocha CESAM & Department of Physics, University of Aveiro, Aveiro, Portugal

Vera Rodrigues CESAM & Department of Environment and Planning, University of Aveiro, Aveiro, Portugal

Rebecca Rose Ricardo Energy & Environment, Harwell, Oxon, UK

Shawn Roselle National Exposure Research Laboratory, U.S. Environmental Protection Agency, Research Triangle Park, Durham, NC, USA

Eva Roth University of Southern Denmark, Esbjerg, Denmark

Ed Rowe Centre for Ecology & Hydrology, Bangor, Wales, UK

G. Roy High Impact Weather National Laboratory, Meteorological Service of Canada, Environment and Climate Change Canada, Montreal, Canada

Armistead G. Russell School of Civil and Environmental Engineering, Georgia Institute of Technology, Atlanta, GA, USA

O. Russell Bullock Jr. Computational Exposure Division, National Exposure Laboratory, US EPA, Ann Arbor, MI, USA

Emmanuele Russo Institute of Meteorology, Freie Universität Berlin, Berlin, Germany

Christian Saad Meteorological Service of Canada (MSC), Environment and Climate Change Canada (ECCC), Montreal, QC, Canada

Pablo Saide Department of Atmospheric and Oceanic Sciences and Institute of the Environment and Sustainability, University of California, Los Angeles, CA, USA

Abby Salb Ontario Ministry of the Environment and Climate Change, Toronto, ON, Canada

Jeremy Sarnat Rollins School of Public Health, Emory University, Atlanta, GA, USA

Mourad Sassi Air Quality Policy-Issue Response Section and Air Quality Modelling Applications Section, Environment and Climate Change Canada, Montreal, QC, Canada

Verica Savic-Jovcic Air Quality Research Division, Environment and Climate Change Canada, Toronto, ON, Canada

Mikko Savolahti Finnish Environment Institute (SYKE), Helsinki, Finland

Rick Saylor NOAA/ARL/Atmospheric Turbulence and Diffusion Division, Oak Ridge, TN, USA;
Center for Spatial Information Science and Systems, George Mason University, Fairfax, VA, USA

M. Schaap Department Climate, Air and Sustainability, TNO, Utrecht, The Netherlands;
Freie Universität Berlin (FUB), Berlin, Germany

W. Schröder Leibniz Institute for Tropospheric Research (TROPOS), Leipzig, Germany

Volker Schunicht Anjoda Technical Solutions, Prince George, BC, Canada

Tejas Shah Ramboll, Novato, CA, USA

M. Shaw Measurements and Analysis Research Section, Air Quality Research Division, Environment and Climate Change Canada, Downsview, Canada

Huizhong Shen Georgia Institute of Technology School of Civil and Environmental Engineering, Atlanta, GA, USA

Carlos Silveira CESAM & Department of Environment and Planning, University of Aveiro, Aveiro, Portugal

Janet Simkin Newcastle University, Newcastle upon Tyne, UK

Nele Smeets VITO, Mol, Belgium

Michelle Snyder Institute for the Environment, University of North Carolina at Chapel Hill, Chapel Hill, NC, USA

Joana Soares Air Quality Modelling and Integration Section, Air Quality Research Division, Environment and Climate Change, Toronto, ON, Canada

Mikhail Sofiev Atmospheric Composition Research, Finnish Meteorological Institute, Helsinki, Finland

Viktoria Sofieva Finnish Meteorological Institute, Helsinki, Finland

Ranjeet Sokhi Centre for Atmospheric and Instrumentation Research (CAIR), University of Hertfordshire, Hatfield, UK

Efisio Solazzo European Commission Joint Research Centre (JRC), Ispra, Italy

Albert Soret Earth Sciences Department, Barcelona Supercomputing Center (BSC), Barcelona, Spain

Sandra Sorte CESAM & Department of Environment and Planning, University of Aveiro, Aveiro, Portugal

John Spagnol Environmental Science and Engineering Programs, University of Northern British Columbia, Prince George, BC, Canada

G. Spindler Leibniz Institute for Tropospheric Research (TROPOS), Leipzig, Germany

Boris Starobinets Department of Geophysics, Tel Aviv University, Tel-Aviv, Israel

C. Stathopoulos Department of Physics, National and Kapodistrian University of Athens, Athens, Greece

Amy Stephens Centre for Ecology & Hydrology, Edinburgh, Scotland, UK

Gabriele Stiller Karlsruhe Institute of Technology, Karlsruhe, Germany

Ana Stojiljkovic Finnish Environment Institute (SYKE), Helsinki, Finland

Craig Stroud Air Quality Research Division, Science and Technology Branch, Environment and Climate Change Canada, Toronto, ON, Canada

Mark Sutton Centre for Ecology & Hydrology, Edinburgh, Scotland, UK

Chunling Tang Computational Exposure Division, National Exposure Research Laboratory, U.S. Environmental Protection Agency, Washington, DC, USA

Sim Tang Centre for Ecology & Hydrology, Edinburgh, Scotland, UK

Youhua Tang NOAA/Air Resources Laboratory, College Park, MD, USA;
UMD/Cooperative Institute for Climate and Satellites, College Park, MD, USA;
Center for Spatial Information Science and Systems, George Mason University, Fairfax, VA, USA

Camille Taylor Golder Associates Ltd., Ottawa, ON, Canada

Andrew Teakles Meteorological Service of Canada, ECCC, Dartmouth, NS, Canada

Piet Termonia Royal Meteorological Institute of Belgium, Brussels, Belgium;
Department of Physics and Astronomy, Ghent University, Ghent, Belgium

M. Thürkow Freie Universität Berlin (FUB), Berlin, Germany

R. Timmermans Department Climate, Air and Sustainability, TNO, Utrecht, The Netherlands

Samuel Tomlinson Centre for Ecology & Hydrology, Edinburgh, Scotland, UK

Daniel Tong NOAA/Air Resources Laboratory, College Park, MD, USA;
UMD/Cooperative Institute for Climate and Satellites, College Park, MD, USA;
NOAA/ARL/Atmospheric Turbulence and Diffusion Division, Oak Ridge, TN, USA;
Center for Spatial Information Science and Systems, George Mason University, Fairfax, VA, USA

Silvia Trini Castelli CNR-Institute of Atmospheric Sciences and Climate, Turin, Italy

George Tsegas Laboratory of Heat Transfer and Environmental Engineering, Department of Mechanical Engineering, Aristotle University, Thessaloniki, Greece

Paolo Tuccella Department of Physical and Chemical Sciences, University of L'Aquila, L'Aquila, Italy;
Center of Excellence CETEMPS, University of L'Aquila, L'Aquila, Italy

Alper Unal Eurasia Institute of Earth Sciences, Istanbul Technical University, Istanbul, Turkey

Andreas Uppstu Finnish Meteorological Institute, Helsinki, Finland;
Atmospheric Composition Research, Finnish Meteorological Institute, Helsinki, Finland

Ambarish Vaidyanathan Centers for Disease Control and Prevention, Atlanta, GA, USA

Karen van de Vel VITO, Mol, Belgium

Eric van der Swaluw RIVM, Bilthoven, The Netherlands

Stijn Van Looy VITO, Mol, Belgium

D. van Pinxteren Leibniz-Institut Für Troposphärenforschung (TROPOS), Leipzig, Germany

Addo van Pul RIVM, Bilthoven, The Netherlands

Marlies Vanhulsel VITO, Mol, Belgium

Jean Vankerkom VITO, Mol, Belgium

Andreas Veira German Meteorological Service, Offenbach, Germany

Pradeepa Vennam Bay Area Air Quality Management District, San Francisco, CA, USA

Willem W. Verstraeten Royal Meteorological Institute of Belgium, Brussels, Belgium

R. Vet Measurements and Analysis Research Section, Air Quality Research Division, Environment and Climate Change Canada, Downsview, Canada

Peter Viaene VITO, Mol, Belgium

Carolina Viceto CESAM & Department of Physics, University of Aveiro, Aveiro, Portugal

Massimo Vieno Centre for Ecology and Hydrology, Penicuik, UK

V. Vinatier Université Clermont Auvergne, CNRS, SIGMA Clermont, ICCF, Clermont-Ferrand, France

Julius Vira Atmospheric Composition Research, Finnish Meteorological Institute, Helsinki, Finland;
Cornell University, Ithaca, NY, USA

Marta Garcia Vivanco INERIS, Institut National de l'Environnement Industriel et des Risques, Verneuil-en-Halatte, France;
CIEMAT, Madrid, Spain

Penny Vlahos Marine Sciences, University of Connecticut, Groton, CT, USA

Stijn Vranckx VITO, Mol, Belgium

Sheng-Hsiang Wang Department of Atmospheric Sciences, National Central University, Taoyuan, Taiwan

Małgorzata Werner Department of Climatology and Atmosphere Protection, Institute of Geography and Regional Development, University of Wrocław, Wrocław, Poland

Jason West Department of Environmental Sciences and Engineering, University of North Carolina at Chapel Hill, Chapel Hill, NC, USA

E. Charles White Department of Physics, University of Toronto, Toronto, ON, Canada

A. Wiedensohler Leibniz Institute for Tropospheric Research (TROPOS), Leipzig, Germany

James Wilkinson Terra-Technologies, Eugene, OR, USA

Gary Wilson Bay Area Air Quality Management District, San Francisco, CA, USA

Peter Wind EMEP MSC-W, Norwegian Meteorological Institute, Oslo, Norway

N. Wirgot Université Clermont Auvergne, CNRS, SIGMA Clermont, ICCF, Clermont-Ferrand, France

R. Wolke Leibniz Institute for Tropospheric Research (TROPOS), Leipzig, Germany

David Wong National Exposure Research Laboratory, U.S. Environmental Protection Agency, Research Triangle Park, Durham, NC, USA;
Computational Exposure Division, National Exposure Laboratory, US EPA, Ann Arbor, MI, USA

Jia Xing School of Environment, Tsinghua University, Beijing, China

Greg Yarwood Bay Area Air Quality Management District, San Francisco, CA, USA;
Ramboll Environ, Novato, CA, USA

Erik Ytreberg Shipping and Marine Technology, Chalmers University of Technology, Gothenburg, Sweden

Miguel Zavala Molina Center for Energy and the Environment, La Jolla, CA, USA

Junhua Zhang Air Quality Research Division, Science and Technology Branch, Environment and Climate Change Canada, Toronto, ON, Canada

Xuesong Zhang Department of Physics, University of Toronto, Toronto, ON, Canada

Yuqiang Zhang Computational Exposure Division, National Exposure Research Laboratory, U.S. Environmental Protection Agency, RTP, NC, Washington, D.C., USA

Qiong Zheng Air Quality Research Division, Science and Technology Branch, Environment and Climate Change Canada, Toronto, ON, Canada

Part I
Regional and Intercontinental Modeling

Chapter 1
Establishing the Origin of Particulate Matter in Eastern Germany Using an Improved Regional Modelling Framework

R. Timmermans, R. Kranenburg, C. Hendriks, M. Thürkow, I. Kirchner, D. van Pinxteren and M. Schaap

Abstract In Eastern Germany winter episodes with PM10 exceedances of EU limit values are often connected to continental air masses, combining polluted air from Eastern Europe with air pollution from local urban sources. The EU air quality legislation requires the analysis of the contribution of such cross-boundary transport to exceedances for development of effective policy measures. To this end we have performed a source apportionment study to establish the main sources of particulate matter during high PM episodes in Eastern Germany. We have run the LOTOS-EUROS model with its labelling based source apportionment tool and found that the contribution from cross boundary transport becomes more dominant during the episodes. The results from the model are currently compared to a measurement based source attribution (PMF) and a more detailed evaluation of episodes with exceedances is being performed. The modelled concentrations have been evaluated against PM composition observations which revealed that especially the modelling of carbonaceous aerosol is challenging and lead to a large underestimation of modelled PM10 levels in winter. The use of an updated bottom-up inventory in combination with temperature dependent temporal variability for residential combustion emissions leads to an increase of carbonaceous aerosols and reduction of PM biases. Another improvement was realised through an update of the deposition routine over snow that leads to a strong reduction of the underestimation during cold PM episodes with snow conditions.

R. Timmermans (✉) · R. Kranenburg · C. Hendriks · M. Schaap
Department Climate, Air and Sustainability, TNO, Utrecht, The Netherlands
e-mail: renske.timmermans@tno.nl

M. Thürkow · I. Kirchner · M. Schaap
Freie Universität Berlin (FUB), Berlin, Germany

D. van Pinxteren
Leibniz-Institut Für Troposphärenforschung (TROPOS), Leipzig, Germany

C. Mensink et al. (eds.), *Air Pollution Modeling and its Application XXVI*, Springer Proceedings in Complexity,
https://doi.org/10.1007/978-3-030-22055-6_1

1.1 Introduction

In the 2008 Air Quality Directive (2008/50/EG), the EU established limit values for the PM10 concentration in ambient air. The daily limit of 50 μg/m^3 should not be exceeded on more than 35 days a year, and the maximum allowed annual average is 40 μg/m^3. The daily limit value is still exceeded on too many days for a number of observation stations in Germany. Although the number of exceedance days has reduced in the last few years, there are still episodes with very high concentrations of PM. High PM observations in Eastern Germany are often correlated to transport of air masses from south-easterly direction. The pollution in these air masses mixes with local pollution, adding up to levels exceeding the limit values for PM. This south-easterly transport occurs often with a high pressure system causing stable conditions over Eastern Europe, which in winter means low temperatures in the area. When Eastern Germany is more fed with westerly flows, this is usually during weather conditions causing high atmospheric mixing, which leads to dilution of pollution.

The air quality guidelines state that for regions with exceedances associated with long-range transport, the contribution of transboundary transport of PM to the exceedance of limit values should be assessed and discussed between neighboring countries. Unfortunately, separating national and foreign contributions as well as source sectors using observation-based methods is hardly possible. This requires a dedicated analysis using receptor-oriented approaches such as Positive Matrix Factorisation (PMF), ideally in combination with chemistry transport modelling aimed at identifying the origin of air pollution.

1.2 Method

1.2.1 LOTOS-EUROS Model and Source Apportionment

In this study we have performed a source apportionment study using the LOTOS-EUROS model and its source apportionment tool. LOTOS-EUROS is an open source 3D chemistry transport model. The grid model simulates air pollution concentrations in the lower troposphere on a regular Eulerian grid with variable resolution over Europe [1]. The model is used for a wide range of applications supporting scientific research, regulatory programmes and air quality forecasts. The vertical layer structure makes the model very efficient in use.

For the source apportionment a so-called labelling system is used to track the impact of emission categories within a LOTOS-EUROS simulation. Besides species concentrations the contributions of pre-defined source categories are calculated. The labelling routine is implemented for primary, inert aerosol tracers as well as chemically active tracers containing a C, N or S atom, as these are conserved and traceable. For details and validation of this source apportionment module we refer to [2].

1.2.2 Model Improvements

From comparisons with groundbased observations it was found that the model largely underestimates the PM concentrations during episodes with exceedances. The underestimation was especially visible in the comparison of carbonaceous aerosols. To this end a few model improvements were realised.

A new scientific based emission database for residential combustion has been implemented. This database includes emission factors from the GAINS model for residential fossil fuel combustion. And includes the impact of condensable material [3]. The revised residential wood combustion emissions are higher than those officially reported by a factor of 2–3.

The temporal variability of the residential PM emissions was further improved by using the heating degree days concept. In this concept the emissions vary on a daily basis based on ambient temperature. Hence, during cold spells the emissions are increased compared to periods with relatively warm winter weather.

The deposition routine over snowy surfaces was evaluated. Stability and deposition factors were updated using settings appropriate for snowy conditions, leading to a decrease of deposition under these conditions.

1.3 Results

1.3.1 PM Modeling

Figure 1.1 shows a PM10 timeseries for the first quarter of 2017 for the station of Melpitz. It can be seen that the model underestimates the observed concentrations

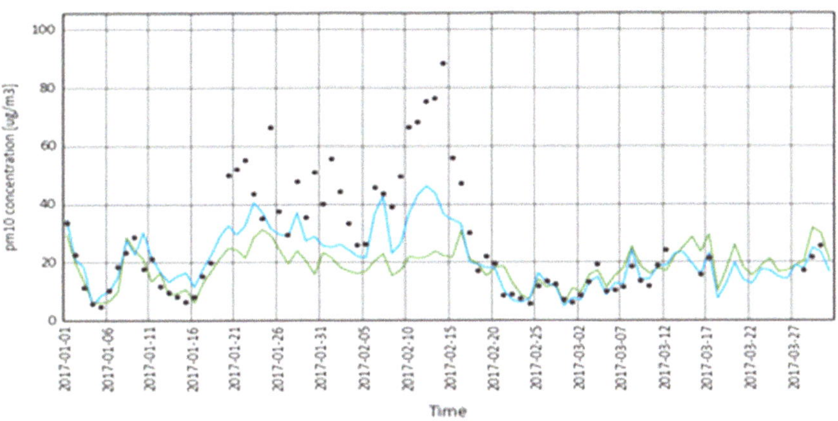

Fig. 1.1 PM10 concentration [$\mu g/m^3$] for the station of Melpitz for January–March 2017 from the LOTOS-EUROS model before (green) and after improvements (blue) and from the measurements (black dots)

during the episode from ~ 20 January to halfway February. The implemented model improvements lead to a considerable improvement of the model performance. Nevertheless a strong underestimation remains present. There are several possible reasons for this. For example, the amount of condensables in the improved emission database is based on standard temperatures, during very cold conditions, the amount can be much larger. Another reason could be the meteorological data driving the model. The meteorological conditions during stable cold conditions are often difficult to represent, the (vertical) mixing may be too high.

1.3.2 Source Apportionment

Figure 1.2 shows the contribution from local German sources versus sources in Poland and other countries to the modeled PM10 concentrations for the first quarter of 2017. It can be seen that for the days with highest PM concentrations the contribution from Poland and other countries increases while the local contribution decreases. This is consistent with previous studies showing a correlation between high PM observations and long-range transport in Eastern Germany.

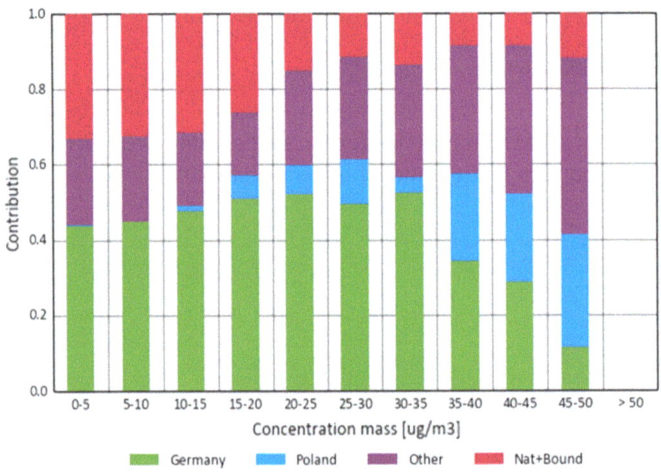

Fig. 1.2 Source contribution from Germany (green), Poland (blue) other countries (purple) and natural/boundary conditions (red) as function of PM concentration at Melpitz for January–March 2017

1.4 Conclusion

The model performance during PM episodes has been improved through the use of improved emissions and improved timing in combination with updated deposition numbers over snow. Additional investigation is needed to further close the gap with observations during cold stable episodes with PM exceedances.

The source apportionment shows an increased contribution from neighbouring countries (and even further away) during high PM concentrations. The main sector contributing during the analysed events is residential combustion (not shown here). Currently, the results from the model are being compared to a measurement based source attribution (PMF) and a more detailed evaluation of episodes is being performed.

Questions and Answers

QUESTIONER: W. Lefebre

QUESTION: Do you have a clue of the uncertainty of the source contribution (especially on the regions)?

ANSWER: We do not have any numbers on the uncertainty. The comparison of the model results with PMF results are aimed at getting more insight into the uncertainty of the contributions from different source sectors. Unfortunately PMF does not give you any information on the geographical origin. However there are also trajectory based analyses that will be used to check the confidence in the contributions from different regions.

QUESTIONER: Anthony Dore

QUESTION: You mentioned that sea salt concentrations were overestimated in the boundary conditions. Sea salt emissions are highly sensitive to wind speed. Was the overestimate caused by emissions being too high during extreme wind events?

ANSWER: The high sea-salt concentrations in the boundary conditions originate from a model version of the global -IFS model known to produce too high sea salt concentrations. This was indeed most prominent during extreme wind events, since the sea salt emissions are proportional to the wind speed^3. This problem is believed to be solved for newer versions of the model.

Acknowledgements The work presented here was funded through the Umweltbundesamt (UBA) Project number: FKZ 3716 51 203 0.

References

1. A.M.M. Manders, P.J.H. Builtjes, L. Curier, H.A.C. Denier van der Gon, C. Hendriks, S. Jonkers, R. Kranenburg, J.J.P. Kuenen, A.J. Segers, R.M.A. Timmermans, A.J.H. Visschedijk, R.J. Wichink Kruit, W.A.J. van Pul, F.J. Sauter, E. van der Swaluw, D.P.J. Swart, J. Douros, H. Eskes, E. van Meijgaard, B. van Ulft, P. van Velthoven, S. Banzhaf, A.C. Mues, R. Stern, G. Fu, S. Lu, A. Heemink, N. van Velzen, M. Schaap, Curriculum vitae of the LOTOS–EUROS (v2.0) chemistry transport model. Geosci. Model Dev. **10**, 4145–4173 (2017). https://doi.org/10.5194/gmd-10-4145-2017

2. R. Kranenburg, A.J. Segers, C. Hendriks, M. Schaap, Source apportionment using LOTOS-EUROS: module description and evaluation. Geosci. Model Dev. **6**, 721–733 (2013). https://doi.org/10.5194/gmd-6-721-2013
3. H.A.C. Denier van der Gon, R. Bergström, C. Fountoukis, C. Johansson, S.N. Pandis, D. Simpson, A.J.H. Visschedijk, Particulate emissions from residential wood combustion in Europe – revised estimates and an evaluation. Atmos. Chem. Phys. **2015**(15), 6503–6519 (2015). https://doi.org/10.5194/acp-15-6503-2015

Chapter 2
Ozone in the Eastern United States: Production Efficiency Variability Over Time and Between Sources

Lucas R. F. Henneman, Huizhong Shen, Cong Liu, Yongtao Hu, James A. Mulholland and Armistead G. Russell

Abstract The eastern United States has seen dramatic air pollution emissions reductions since the turn of the century. These emissions reductions have in turn been linked to widespread reductions in ozone (O_3)—between 2000 and 2016, the US EPA reports a reduction in 4th highest mean daily annual 8-hr O_3 of 15% (from 82.3 to 69.6 ppb) across 206 sites nationwide. Reductions, however, have not been spatially uniform or linear with emissions reductions, and therefore motivate an investigation into spatial and source-specific O_3 production efficiency (OPE). OPE is a measure of the number of O_3 molecules produced per emitted NO_X ($NO_X = NO + NO_2$) molecule. We assess OPE using both model-based and empirical approaches. We modelled July OPE in 2001 and 2011 using CMAQ-DDM version 5.0 with a 12 km resolution over the eastern US. CMAQ-modelled OPE is taken as a ratio of electricity generating unit and mobile source sensitivities, and controls for differences in O_3 and NO_Z deposition rates. Measurements were taken from the SEARCH network, which reports sub-daily observations of many gaseous and particulate species along with meteorological measurements at eight sites in the southeastern US. Using measurement data, we stratified days based on their emissions-independent photochemical state, and estimated OPE using a spline model to assess the relationship between O_3 and NO_X reaction products (denoted NO_Z). Both approaches yield an increase in OPE with decreasing NO_Z, indicating an increasing effectiveness at lowering O_3 for subsequent NO_X emissions reductions. Electricity generating unit OPEs are low near individual sources, but generally higher than on-road mobile source OPEs throughout the domain, suggesting that further utility NO_X emissions reductions will reduce regional O_3 concentrations more efficiently than mobile source NO_X emissions reductions.

L. R. F. Henneman (✉)
Harvard TH Chan School of Public Health, 655 Huntington Avenue Building 2, 4th Floor, Boston, MA 02115, USA
e-mail: henneman@hsph.harvard.edu

H. Shen · Y. Hu · J. A. Mulholland · A. G. Russell
Georgia Institute of Technology School of Civil and Environmental Engineering, 790 Atlantic Drive, Atlanta, GA 30332-0355, USA

C. Liu
School of Energy and Environment, Southeast University, Nanjing, China

C. Mensink et al. (eds.), *Air Pollution Modeling and its Application XXVI*, Springer Proceedings in Complexity,
https://doi.org/10.1007/978-3-030-22055-6_2

2.1 Introduction

Air quality policies have long targeted the highest O_3 concentrations, and regulations in place since mid-1990s have largely been successful at reducing these concentrations [1]. Recently established health-based standards, however, have prescribed increasingly lower O_3 concentrations, which presumably call for further emissions reductions.

The effectiveness of future emissions reductions, however, depends on the continued responsiveness of O_3 concentrations to emissions changes. OPE—the number of O_3 molecules formed per NO_X molecule emitted—has been used previously to assess the potential for reducing O_3 with NO_X emissions reductions. A high OPE, for example, suggests greater effectiveness at reducing O_3 with future emissions cuts. We endeavour to investigate how O_3 will continue to respond to emissions reductions at increasingly low NO_Z concentrations.

2.2 Method

2.2.1 *Chemical Transport Model-Based OPE*

We assess changing concentrations of gaseous air pollution species throughout the eastern United States using the Community Multiscale Air Quality model with the Decoupled Direct Method (CMAQ–DDM) version 5.0 with a 12 km resolution [2]. Meteorological fields for years 2001 and 2011 were processed using the Weather Research Forecast (WRF) model version 3.6.1 Emissions were processed using EPA SMOKE platforms for the 2002 and 2011 National Emissions Inventories. CMAQ modelling was evaluated in detail by Henneman et al. [3].

We model concentrations and sensitivities to power plant and mobile emissions sources in four July scenarios: one each in years 2001 (2001_{BASE}) and 2011 (2011_{BASE}), and two hypothetical scenarios with 2011 meteorology—the first with 50% NO_X emissions reductions ($2011_{50\%NO_X}$) and the second with 90% emissions reductions ($2011_{10\%NO_X}$). Concentrations and sensitivities used in the OPE calculations below were taken as the average from 2 to 3 pm local time to match previous studies.

OPEs were calculated with multiple approaches. The first, a brute force method, assumes linear changes in OPE:

$$OPE^{BF} = \frac{\Delta O_3}{\Delta NO_Z} \tag{2.1}$$

where Δ denotes changing concentrations for the same locations and time of O_3 and NO_Z (the sum of NO_X reaction products: peroxyacetyl nitrate, peroxynitric acid, organic nitrate, nitric acid, nitrous acid, nitrate radical, aerosol nitrate, and dinitrogen pentoxide as N).

A second estimate of OPE applies ratios of O_3 sensitivities (S [*ppb*]) to mobile and electricity generating unit (EGU) emissions:

$$OPE^{NODEP} = \frac{S_{O_3,MOB} + S_{O_3,EGU}}{S_{NO_Z,MOB} + S_{NO_Z,EGU}} \tag{2.2}$$

This equation directly estimates the ratio of O_3 attributable to major emissions sources to NO_Z attributable to the same sources. It does, however, omit effects of differing O_3 and NO_Z deposition rates, which we rectify by adding source-specific deposition sensitivities (e.g., $S^{DEP}_{O_3,MOB}$):

$$OPE = \frac{\left(S_{O_3,MOB} + S^{DEP}_{O_3,MOB}\right) + \left(S_{O_3,EGU} + S^{DEP}_{O_3,EGU}\right)}{\left(S_{NO_Z,MOB} + S^{DEP}_{NO_Z,MOB}\right) + \left(S_{NO_Z,EGU} + S^{DEP}_{NO_Z,EGU}\right)} \tag{2.3}$$

where $S^{DEP}_{i,j}$ [*ppb*] is the sensitivity of species *i* deposition to emissions from source *j* and must be calculated from CMAQ output $S^{DEP*}_{i,j}$[kg km^2]. This is achieved by dividing $S^{DEP*}_{i,j}$ by total deposition (DEP^*_i [kg]) and multiplying by concentration [*ppb*].

The form of Eq. 2.3 can be further refined to calculate source-specific OPE:

$$OPE_{MOB} = \frac{\left(S_{O_3,MOB} + S^{DEP}_{O_3,MOB}\right)}{\left(S_{NO_Z,MOB} + S^{DEP}_{NO_Z,MOB}\right)} \tag{2.4}$$

$$OPE_{EGU} = \frac{\left(S_{O_3,EGU} + S^{DEP}_{O_3,EGU}\right)}{\left(S_{NO_Z,EGU} + S^{DEP}_{NO_Z,EGU}\right)} \tag{2.5}$$

2.2.2 Observation-Based OPE

Long-term detailed observation records from the Southeastern Aerosol Research and Characterization (SEARCH) network were used to estimate OPE at eight urban and rural sites. OPE was calculated by regressing average 2–3 pm O_3 concentrations against NO_Z concentrations at each site. To assess only air parcels with similar histories, only days with photochemical state (PS*, a measure of the atmosphere's emissions-independent reactivity described in detail by Henneman et al. [4]) in the 80th percentile were included in the regressions. Fits were assessed using a flexible spline model to capture increasing OPEs at lower NO_Z concentrations.

2.3 Results

Comparisons between brute-force OPEs (which do not account for differences in O_3 and NO_Z deposition rates) calculated using differences between three scenarios ($2011_{BASE} - 2011_{50\% NO_X}$ and $2011_{50\% NO_X} - 2011_{10\% NO_X}$) and OPE^{NODEP} yield a slope near one, establishing credibility in the sensitivities approach for estimating OPE.

Nationwide OPE corrected for differing deposition rates is slightly higher than OPE^{NODEP} (Table 2.1), but the difference decreases as NO_X emissions decrease. Spatially, OPE is elevated in urban areas and near highways, and decreases with increasing latitude (Fig. 2.1). OPE increase with decreasing NO_X emissions; however, the rate of increase in OPE decreases as NO_X emissions approach zero.

Average domain-wide mobile (EGU) OPEs are slightly lower (higher) than total OPEs in 2001 and 2011 (Table 2.1). Spatially, OPE_{MOB} generally mirrors the distribution of total OPE, whereas OPE_{EGU} is elevated across the domain except for areas near electricity generating units.

At SEARCH sites, empirically-derived OPE behaves similarly to CMAQ-simulated OPE, i.e., OPEs increase with decreasing NO_Z concentration (Fig. 2.2). The empirical estimates find slightly lower OPEs than CMAQ, with a more pronounced difference in rural areas at high NO_Z concentrations. Both the CMAQ-estimated and empirical OPEs, however, reach location-specific maxima as NO_Z concentrations approach zero.

2.4 Discussion

Increasing OPEs with decreasing NO_X emissions suggests that further emissions reductions will be increasingly effective at reducing the highest levels of O_3. As OPE approaches site-specific maxima, however, the increasing benefits of decreasing NO_X emissions wanes. These site-specific maxima reflect site-specific meteorological conditions and differing availability of other atmospheric species. That OPE_{EGU} is greater than OPE_{MOB} in most areas suggests that further NO_X emissions reductions from EGUs will be more efficient at reducing O_3 than mobile NO_X emissions reductions on a per mass emitted basis. These effects, however, vary by location.

In future work, there is potential to investigate OPE on days that have low photochemical state. Current policies have been effective at reducing the highest O_3 concentrations, but wintertime and night time O_3 concentrations have increased as near-source NO_X inhibition of O_3 formation has decreased with deceasing emissions.

Questions and Answers

QUESTIONER: Randall Martin

QUESTION: It is interesting that the OPE for EGUs exceeded the OPE for more diffuse traffic sources. Could this reflect stack heights, or numerical diffusion of EGU emissions within the grid box, or another explanation?

Table 2.1 Information on the modelling runs. Emissions totals reflect the entire domain. Concentrations and OPEs are averaged over the polygon superimposed on the inland region of each of the plots in Fig. 2.1 to exclude boundary effects and high OPEs over water. Reprinted with permission from Henneman et al. [5]. Copyright (2017) American Chemical Society

	July NO_X Emissions, 10^3 tons			Mean July Concentrations, ppb (std. dev.)			Mean July OPE (std. dev.)			
	ALL	MOB	EGU	O_3	NO_Z	NO_X	OPE	OPE^{NODEP}	OPE_{MOB}	OPE_{EGU}
2001	1,364	773	432	54.7 (4.7)	2.7 (0.9)	1.6 (2.7)	11.2 (2.9)	10.9 (3.1)	10.7 (3.0)	12.7 (3.1)
2011	679	364	169	48.1 (5.5)	1.9 (0.6)	1.1 (1.3)	14.0 (3.3)	13.8 (3.3)	13.8 (3.5)	15.3 (3.6)
2011 50% NOx	339	182	85	39.5 (4.7)	1.1 (0.4)	0.5 (0.6)	15.7 (2.3)	15.7 (2.4)		
2011 10% NOx	68	36	17	28.6 (3.6)	0.3 (0.1)	0.1 (0.1)	16.4 (1.7)	16.4 (1.7)		

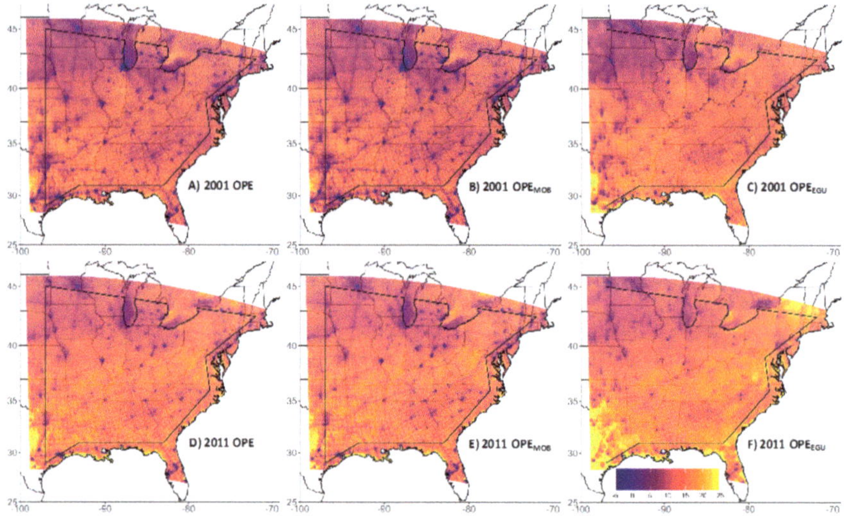

Fig. 2.1 CMAQ-modelled mean daily July deposition-corrected OPE (2–3 p.m.) values for 2001 (A–C) and 2011 (D–F), including total OPE (A&D), mobile OPE (B&E), and EGU OPE (C&F). Reprinted with permission from Henneman et al. [5]. Copyright (2017) American Chemical Society

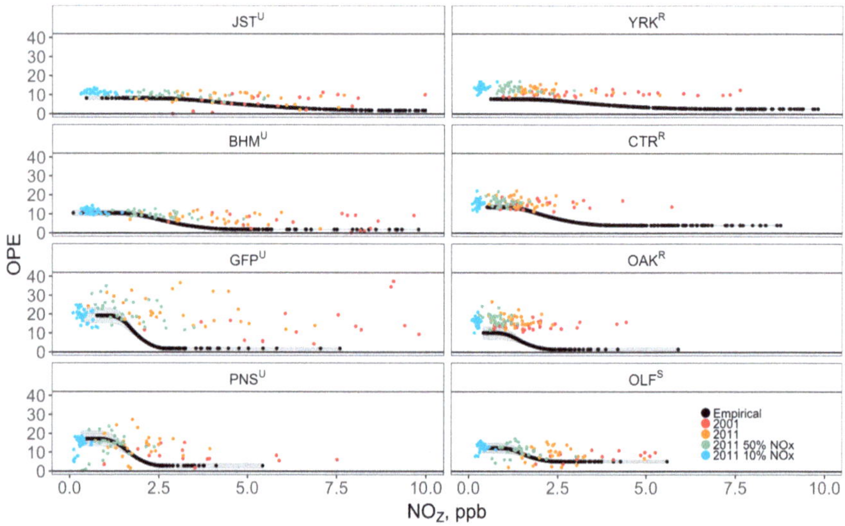

Fig. 2.2 OPE at each SEARCH site estimated by the empirical spline and CMAQ deposition-corrected sensitivities approaches. Empirical values represent all days with PS* > 80th percentile, and CMAQ values are days in July. Grey shading is the 95% confidence interval around the spline model. Reprinted with permission from Henneman et al. [5]. Copyright (2017) American Chemical Society

ANSWER: EGU OPE's are higher than mobile sources primarily because they are emitted in areas with lower NO_X concentrations and higher VOC concentrations. Lower NO_X concentrations lead to slower conversion of NO_X to NO_Z, and higher VOC concentrations mean faster cycling of each NO_X molecule between NO and NO_2.

Acknowledgements This material is based upon work supported by Health Effects Institute and the National Science Foundation Graduate Research Fellowship under Grant No. DGE-1148903. Lucas Henneman's travel to the 36th ITM was supported by Atmosphere Journal's (ISSN 2073-4433) 2018 Travel Award for Young Investigators.

References

1. H. Simon, A. Reff, B. Wells, J. Xing, N. Frank, Ozone trends across the United States over a period of decreasing NOx and VOC emissions. Environ. Sci. Technol. **49**(1), 186–195 (2014)
2. D. Byun, K.L. Schere, Review of the governing equations, computational algorithms, and other components of the models-3 Community Multiscale Air Quality (CMAQ) modeling system. Appl. Mech. Rev. **59**(2), 51 (2006)
3. L.R. Henneman, C. Liu, Y. Hu, J.A. Mulholland, A.G. Russell, Air quality modeling for accountability research: operational, dynamic, and diagnostic evaluation. Atmos. Environ. **2017**(166), 551–565 (2017)
4. L. R. Henneman, H. H. Chang, D. Lavoue, J. A. Mulholland, A. G. Russell, Accountability assessment of regulatory impacts on ozone and PM2.5 concentrations using statistical and deterministic pollutant sensitivities. Air Qual. Atmos. Heal. **10**(6), 695–711 (2017)
5. L. Henneman, C. Liu, H. Shen, Y. Hu, J. A. Mulholland, A. G. Russell, responses in ozone and its production efficiency attributable to recent and future emissions changes in the Eastern United States. Environ. Sci. Technol. (2017). acs.est.7b04109

Chapter 3
Unravelling the Origin of High Ozone Concentrations in Southwestern Europe

María Teresa Pay, Carlos Pérez-García Pando, Marc Guevara, Oriol Jorba, Sergey Napelenok and Xavier Querol

Abstract This study aims to quantify the contributions to surface ozone (O_3) concentration in the Iberian Peninsula (IP) from the main NO_x emission sectors in the region along with the imported O_3 during a 10-day episode. The work applies an Integrated Source Apportionment Method within the CALIOPE air quality system at 4-km resolution. This study finds that the imported O_3 is overall the larger contribution to its surface concentration. Contributions from local/regional sources are decisive in the O_3 peaks downwind of main nitrogen oxides (NO_x) hotspots in the IP under stagnant conditions.

3.1 Introduction

Southwestern EU experiences severe O_3 episodes in summer. Several studies have advanced our knowledge on the dynamics of O_3 episodes based on a comprehensive understanding of the circulation patterns [2]. However, there is a lack of studies identifying the sources responsible for the high O_3 over the region, which is crucial to design adequate mitigation measures. Chemical Transport Models (CTMs) quantify the impact of sector-specific pollution at the surface of a receptor region. The most widely used source apportionment approach in CTMs is the brute force, but it is susceptible to numerical instability in non-linear regimens. Recently, CTM versions include methods that tag multiple pollutants by source all the way through the pollutant's lifetime. This integrated source apportionment method allows identifying main sources contributing to high O_3 concentrations under the actual atmospheric

M. T. Pay (✉) · C. Pérez-García Pando · M. Guevara · O. Jorba
Earth Sciences Department, Barcelona Supercomputing Center (BSC), Barcelona, Spain
e-mail: maria.pay@bsc.es

S. Napelenok
Atmospheric Modelling and Analysis Division, National Exposure Research Laboratory, U.S. Environmental Protection Agency, Research Triangle Park, NC, USA

X. Querol
Institute of Environmental Assessment and Water Research, IDAEA-CSIC, Barcelona, Spain

C. Mensink et al. (eds.), *Air Pollution Modeling and its Application XXVI*, Springer Proceedings in Complexity,
https://doi.org/10.1007/978-3-030-22055-6_3

conditions, which is a preliminary step towards designing refined and efficient emission abatement scenarios. The Integrated Source Apportionment Method (ISAM) within the Community Multiscale Air Quality (CMAQ) model has demonstrated quite promising performance for the O_3 tagging, exhibiting less noise in locations where brute force results are demonstrably inaccurate [1].

3.2 Methodology

3.2.1 Air Quality Model

Our analysis focuses on the 10-day period from July 21st to 31st, 2012, which is representative of typical summer synoptic conditions in the region according to studies on circulation type classification [4]. We use the ISAM within the CALIOPE air quality forecast system (www.bsc.es/caliope) at 4-km horizontal resolution over the IP. CALIOPE first runs over Europe at 12-km resolution (EU domain) and then over the IP at 4-km (IP domain). The WRF-ARWv3.6 model provides meteorological fields. The EU domain uses meteorological initial and boundary conditions from the Final Analyses provided by the National Centers of Environmental Prediction. Boundary conditions for reactive gases and aerosols come from the global MOZART-4/GEOS-5 model. The system is configured with 38 sigma layers up to 50 hPa. HERMESv2.0 emission model provides disaggregated emissions to the CALIOPE system based on local information and state-of-the-art bottom-up approaches for the most polluting sectors. The MEGANv2.0.4 estimates the Volatile Organic Compounds (VOC) and NO_x emissions from vegetation.

3.2.2 Ozone Source Apportionment

The ISAM tags both O_3 and its precursor emissions from each source sector and calculates all the O_3 concentrations ensuring mass conservation within one simulation. Each tagged species undertakes normal physical processes without perturbing the actual conditions. ISAM uses the ratio H_2O_2/HNO_3 to determine whether O_3 is NO_x- or VOC-sensitive (above or below 0.35, respectively). In this study, we quantified the contributions to O_3 in the IP from the main NO_x emission sectors (Fig. 3.1), which include the energy production, the manufacturing industries, the road transport, and the non-road transport. We also quantified the contribution from the chemical boundary and initial conditions. Hereinafter, we name the O_3 from boundary conditions as the imported O_3 to the IP domain, which includes the O_3 produced in Europe and the O_3 globally transported provided by the MOZART-4/GEOS-5 model l model.

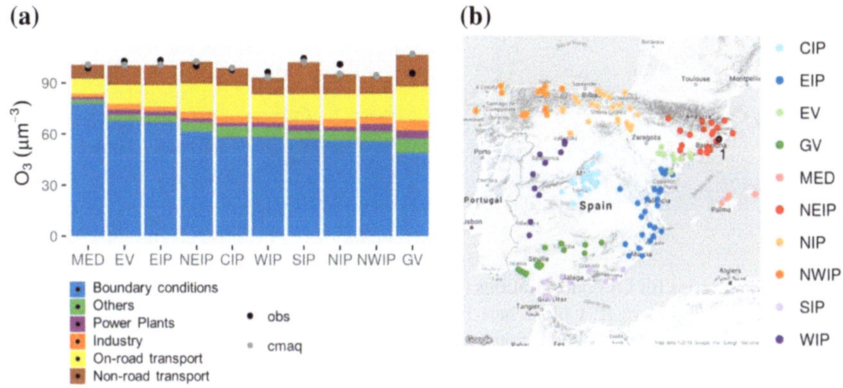

Fig. 3.1 **a** Daily mean contribution of tagged O_3 during exceedances of the observed 120 μgm^{-3} for MDA8 O_3 averaged by (**b**) the identified receptor regions. Black and grey dots represent observed and modelled daily mean O_3 concentration during exceedances of 120 μgm^{-3} of the MDA8. Regions are the central IP (CIP), the eastern IP (EIP), the Ebro Valley (EV), the Guadalquivir Valley (GV), the Mediterranean Sea (MED), the northeastern IP (NEIP), the northern IP (NIP), the northwestern IP (NWIP), the southern IP (SIP) and the western IP (WIP)

3.3 Results

3.3.1 Statistical Evaluation

CALIOPE has been evaluated in detail elsewhere [3]. The model slightly overestimates for hourly and maximum daily 8-hour averaged (MDA8) O_3 concentration (+12 μgm^{-3} and +6 μgm^{-3}, respectively). The r is above 0.6 in more than 50% of the stations and above 0.7 in 25% of the stations. The mean bias for hourly and MDA8 O_3 concentrations are lower at rural background stations (± 4 μgm^{-3}) than at industrial, traffic and urban stations (between +6 and +16 μgm^{-3}) at the 50% of the stations. CALIOPE underestimates the hourly NO_2 concentrations ranging from a mean bias of -7 μgm^{-3} at traffic stations to -2 μgm^{-3} at rural background stations. These biases partly explain the high overestimation of the hourly and MDA8 O_3 concentration at traffic and urban stations, as well as the systematic overestimation of hourly O_3 concentration at nighttime.

3.3.2 Source-Sector Contribution During Peak Episodes

The sector contribution is grouped into ten O_3 receptor sub-regions (Fig. 3.1). In all sub-regions, the external contribution accounted for more than 45% of the O_3 under exceedances of the 120 μgm^{-3} threshold for the MDA8. Our analysis indicates that the imported O_3 is typically controlled by the downward mixing of O_3 upper

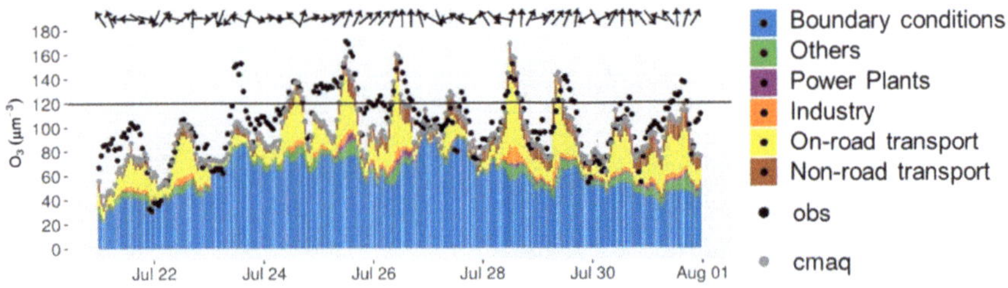

Fig. 3.2 Time series for O_3 concentrations (in μgm^{-3}) in the study episode at a rural background station in the northeastern IP. Colour bars indicate the ozone tags. Black and grey dots indicate the observed and modelled concentrations, respectively. Black horizontal lines represent O_3 target value. Black arrows indicate wind vectors. Station 1 in Fig. 3.1

layers transported from beyond the IP. The sub-regions involving the biggest cities in Spain (central IP and northeastern IP) show the highest road transport contribution to O_3 (up to 40% in a daily peak in events). Meanwhile, in industrial regions (north and northwestern IP and Guadalquivir Valley) the power plants and industrial sectors contribute to O_3 up to 11%. Overall, the non-road traffic is a contributor as significant as the road transport in all sub-regions (10–19%).

As an example, Fig. 3.2 shows source apportionment time series for O_3 at one rural background stations in the northeastern IP located downwind the Barcelona Metropolitan area (station 1). Modelled O_3 peaks (>120 μgm^{-3}) show a high temporal correlation with observations, which suggests that the model reproduces the main transport paths, photochemical processes, and relative contributions from different sources. Imported O_3 is one of the main contributors to hourly O_3 concentrations (from 40 to 100 μgm^{-3}), but during hourly peaks, the on-road transport contribution sharply increases up to 80 μgm^{-3}. The on-road transport O_3 concentrations arriving at rural areas in the northeastern IP mainly come from the Barcelona Metropolitan Area as a result of the afternoon sea breezes. However, under specific meteorological patterns, these winds also transport along precursors from other cities located in the NW Mediterranean Basin.

3.4 Conclusions

The present work assessed the origin of surface O_3 over the IP. Imported O_3 is the main contributor to the surface daily mean O_3 concentration. Besides, during high events, imported O_3 are added to relevant local and regional anthropogenic contributions, which are decisive in the O peaks downwind of main NO_x hotspots in IP. The on-road transport emissions are a significant contributor in rural areas downwind the big cities in Spain (up to 16–18% to the daily mean O_3 concentration under exceedances of the target value for human health protection), contributing up to 70 μgm^{-3} on an hourly basis downwind the Barcelona Metropolitan Area. The non-road transport sector is a contributor to O_3 as significant as the on-road transport over inland regions (10–19% of the daily mean O_3 concentration in episodes).

Questions

Questioner: Mike Moran

Question: What are your future works with the ISAM tool?

Answer: This integrated source apportionment approach has demonstrated several advantages. First, it allows identifying the main sources contributing to high O_3 levels under actual atmospheric conditions, which is a preliminary step towards designing refined and efficient emission abatement scenarios. Second, as we show below, it supports enhanced model evaluation and therefore potential model improvements by identifying problems in emission estimates (sectors or regions) or chemical boundary conditions. Due to the potential of the tool, we plan further detailed and longer O_3 source apportionment studies targeting other nonattainment regions in Europe, which will be necessary prior to define local mitigation measurements that complement national and European-wide strategies to reach the European O_3 objectives.

Acknowledgements This study was supported by the Spanish Ministry of Economy and Competitiveness and FEDER funds under the project PAISA (CGL2016-75725-R) and the "Red Española de Supercomputación" (AECT-2017-1-0008). The views expressed in this article are those of the authors and do not necessarily represent the views or policies of the U.S. Environmental Protection Agency.

References

1. R.H.F. Kwok, K.R. Baker, S.L. Napelenok, G.S. Tonnesen, Photochemical grid model implementation and application of VOC, NO_x, and O_3 source apportionment. Geosci. Model Dev. **8**, 99–114 (2015)
2. M.M. Millán, R. Salvador, E. Mantilla, G. Kallos, Photooxidant dynamics in the Mediterranean basin in summer: results from European research projects. J. Geophys. Res. **102**, 8811–8823 (1997)
3. M.T. Pay, F. Martínez, M. Guevara, J.M. Baldasano, Air quality forecasts at kilometer scale grid over Spanish complex terrains. Geosci. Model Dev. **7**, 1979–1999 (2014)
4. V. Valverde, M.T. Pay, J.M. Baldasano, Circulation-type classification derived on a climatic basis to study air quality dynamics over the Iberian Peninsula. Int. J. Climatol. **35**(8) (2014)

Chapter 4
Sensitivity of Ambient Atmospheric Formaldehyde to VOC and NO_x Emissions: Implications for Predicting Multi-pollutant Benefits of Emission Reductions

Deborah Luecken and Sergey Napelenok

Abstract This study uses a photochemical Air Quality Model applied across the continental US to identify source categories and chemical species (hydrocarbons and nitrogen oxides) that have the largest impact on concentrations of ambient formaldehyde. We contrast the sensitivities of formaldehyde to those of ozone. Although reactions of organic radicals with nitrogen oxide can produce high yields of formaldehyde, the concentrations are more sensitive to hydrocarbons. Biogenic sources of hydrocarbons contribute the most to formaldehyde sensitivity in July, with contributions from isoprene, other alkenes and direct emissions. These results indicate that different strategies may be needed to reduce ambient ozone and formaldehyde concentrations.

4.1 Introduction

Formaldehyde (HCHO) is a ubiquitous trace chemical in the troposphere which plays an important role in atmospheric photochemistry because it reacts quickly and provides a major source of new radicals which drives ozone (O_3) production. HCHO also is important per se because it can adversely impact human health: inhalation exposure causes upper airway irritation and it is a probable human carcinogen [7]. In national studies of risk from 187 Hazardous Air Pollutants (HAPs), HCHO contributes over half of total cancer risk, and 9% of noncancer risk, making it a national risk driver in 99% of US census tracts [5].

HCHO is emitted from many sources, with fuel combustion activities being predominant anthropogenic sources. HCHO also is produced in the atmosphere from emissions of almost every organic hydrocarbon. Unraveling the role of these two

D. Luecken (✉) · S. Napelenok
U.S. Environmental Protection Agency, Research Triangle Park, NC, USA
e-mail: luecken.deborah@epa.gov

S. Napelenok
e-mail: napelenok.sergey@epa.gov

C. Mensink et al. (eds.), *Air Pollution Modeling and its Application XXVI*,
Springer Proceedings in Complexity, https://doi.org/10.1007/978-3-030-22055-6_4

routes complicates efforts to identify and quantify sources of HCHO, and hence to develop actions to reduce HCHO concentrations. The goal of this study is to use sensitivity tools within an Air Quality Model, CMAQ-DDM, to determine which sources and hydrocarbons have the largest impacts on ambient HCHO concentrations.

4.2 Modeling Methods

This study applied the CMAQ version 5.02 air quality model, configured as in the 2011 NATA ([6], using 12×12 km^2 grids over the continental US and parts of Canada and Mexico, and 24 vertical layers. Simulations were performed for two entire months covering January and July, 2011. Chemistry was modeled with the CB05tu mechanism [8] with additional HAPs. More details are given in Luecken et al. [3].

Sensitivity coefficients were calculated with DDM-3D [2] within CMAQ [4]. Sensitivity is defined as the response of a model variable (concentrations of HCHO and O_3) due to a change in a specified model parameter (nitrogen oxides (NO_x) and hydrocarbon emissions from anthropogenic emission categories and natural emissions, and emissions of eight hydrocarbon model species groupings). Emissions categories and VOC groupings are shown in Table 4.1. Emissions of hydrocarbons and NO_x from each source sector were calculated separately; hydrocarbons were calculated separately for anthropogenic and biogenic sources.

Table 4.1 Sensitivities studied with DDM, by source category (left side) and hydrocarbons (right side)

Source categories tracked	Hydrocarbons tracked
Mobile sources (onroad, nonroad, marine, rail)	Model species PAR + ethane (PAR + ETHA)
Oil and gas (point and non-point)	Methanol + ethanol (MEOH + ETOH)
EGU and other point sources	Formaldehyde (FORM)
Fires (wild, prescribed and agricultural)	Acetaldehyde and larger aldehydes (ALD2 + ALDX)
Residential wood combustion	Alkenes (model species ETHE + OLE + IOLE)
Total biogenic hydrocarbons, biogenic NO_x	Isoprene (ISOP)
Boundary conditions	Terpenes (TERP)
	Aromatics (model species TOL + XYL)
	Methane (from constant background)

4.3 Results

The sensitivities of HCHO and O_3 to source category in July are shown in Figs. 4.1 and 4.2, for both overall emissions and anthropogenic emissions only. HCHO is most sensitive to hydrocarbon emissions from biogenic sources, and more sensitive to hydrocarbons than NO_x, partially because biogenic emissions of reactive hydrocarbons are large in summer. When sensitivities are analyzed for anthropogenic sources

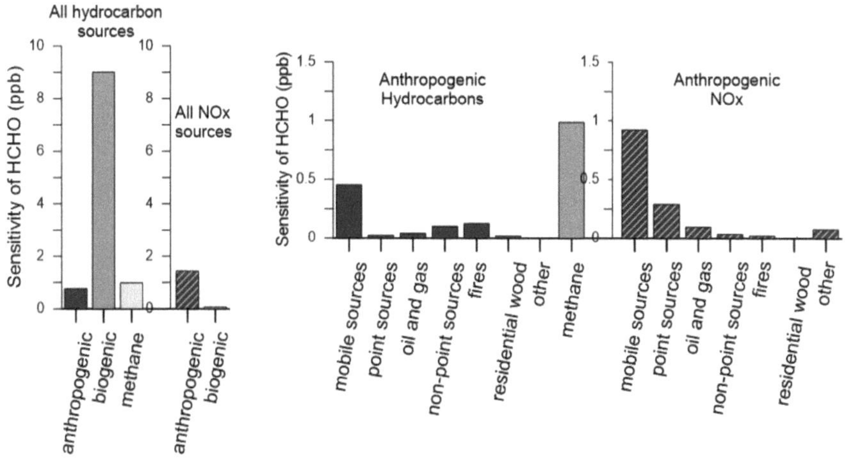

Fig. 4.1 Sensitivity of HCHO to emissions of hydrocarbons and NO_x from different source sectors, including biogenic and anthropogenic sources (left) and anthropogenic sources only (right)

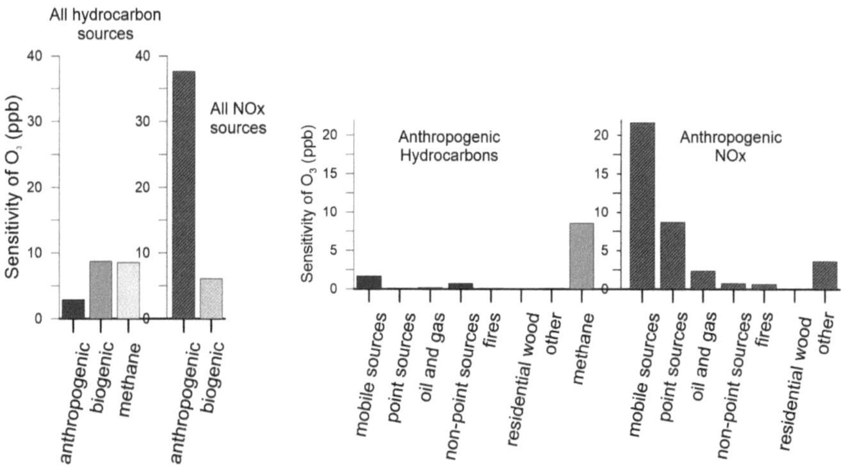

Fig. 4.2 Sensitivity of O_3 to emissions of hydrocarbons and NO_x from different source sectors, including biogenic and anthropogenic sources (left) and anthropogenic sources only (right)

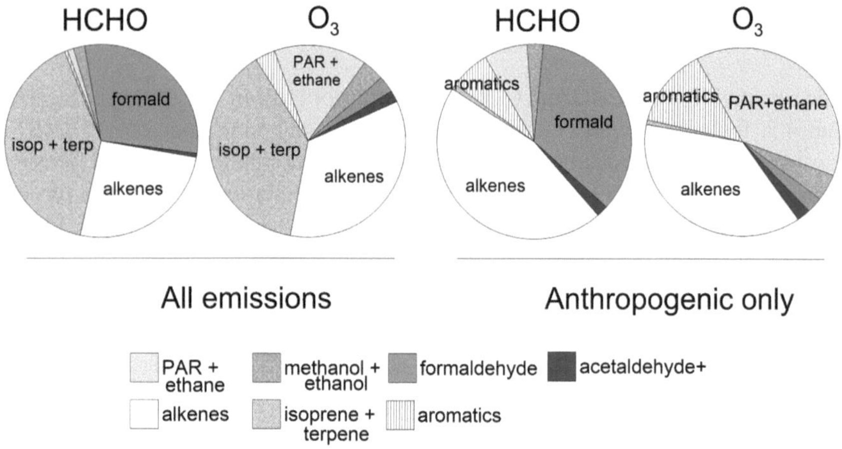

Fig. 4.3 Sensitivity of HCHO and O_3 to hydrocarbon classes for total emissions (left) and for anthropogenic emissions only (right)

only—those which regulations might control—HCHO is sensitive to both hydrocarbon and NO_x emissions, with large contributions from mobile and point sources. In contrast, O_3 is highly sensitive to NO_x sources (Fig. 4.2), with some impact from biogenic hydrocarbons. Methane, which is relatively long-lived, contributes to both HCHO and O_3.

4.3.1 Sensitivity to Hydrocarbons

The sensitivities for both HCHO and O_3 to individual organic hydrocarbon classes are shown in Fig. 4.3 for overall emissions and when only anthropogenic emissions are considered. The sensitivity to the overall emissions shows the signature of biogenic sources, where isoprene plus terpene account for 1/3 of the overall sensitivity. However, biogenic sources of alkenes and direct biogenic emissions of HCHO also impact HCHO concentrations. When only anthropogenic emissions are considered, both HCHO and O_3 have the largest sensitivity to alkenes (ethene, propene, butene and others), but the remainder of the sensitivity is different between HCHO and O_3.

4.4 Conclusion

This study predicts that biogenic sources of hydrocarbons have the largest impact on HCHO concentrations, with isoprene contributing about 1/3 of the total biogenic sensitivity of HCHO and O_3 concentrations, but other alkenyl species, and direct HCHO emissions, are equally important. Satellite column measurements of HCHO are used to evaluate and improve isoprene inventories [1], so it is important to accurately characterize how HCHO is formed from isoprene versus from other biogenic VOCs. Overall, these results predict that HCHO has much smaller sensitivity to NO_x than hydrocarbons, indicating that HCHO formation is not limited by NO_x, although O_3 can be. When only anthropogenic emissions are considered, HCHO is found to be sensitive to both NO_x and hydrocarbons, with mobile sources contributing substantially. Condensed chemical mechanisms are used to develop emission control strategies, so it is important to confirm these sensitivities using detailed mechanisms and dynamic evaluations.

The goal of multipollutant control strategies is to develop emission reductions targeting one important pollutant, but provide co-benefits in reducing other harmful pollutants. While both O_3 and HCHO are formed under similar conditions (high actinic flux, availability of reactive hydrocarbons and NO_x), they show strikingly different sensitivities. NO_x reduction strategies will have the most impact overall on O_3. Hydrocarbon-focused strategies would be the most optimal for reducing HCHO, although NO_x controls are predicted to have some small positive impact in summer.

Questioner: Paul Makar

Question: To what extent does the gas-phase mechanism employed conserve carbon number, and how much of the missing HCHO could be due to losses of carbon on the way down the reaction sequence towards HCHO?

Answer: The mechanism used is CB05 with aromatic updates. While it does not conserve carbon, the carbon that the mechanism drops would not participate strongly in ozone formation, i.e. relatively non-reactive (such as CO_2) or produced in small yields. Some of these could eventually form a small amount of HCHO further downwind, but it would not likely account for the 30% under prediction.

Disclaimer Although this work was reviewed by EPA and approved for publication, it may not necessarily reflect official agency policy.

References

1. M. Bauwens, T. Stavrakou, J.F. Müller, I. De Smedt, M. Van Roozendael, G.R. van der Werf, C. Wiedinmyer, J.W. Kaiser, K. Sindelarova, A. Guenther, Nine years of global hydrocarbon emissions based on source inversion of OMI formaldehyde observations. Atmos. Chem. Phys. **16**, 10133–10158 (2016)
2. A. Hakami, M.T. Odman, A.G. Russell, High-order, direct sensitivity analysis of multidimensional air quality models. Environ. Sci. Technol. **37**, 2442–2452 (2003)
3. D.J. Luecken, S.L. Napelenok, M. Strum, R. Scheffe, S. Phillips, Sensitivity of ambient atmospheric formaldehyde and ozone to precursor species and source types across the U.S. Environ. Sci. Tech. **52**(8), 4668–4675 (2017). https://doi.org/10.1021/acs.est.7b05509
4. S.L. Napelenok, D.S. Cohan, M.T. Odman, S. Tonse, Extension and evaluation of sensitivity analysis capabilities in a photochemical model. Environ. Mod. Softw. **23**, 994–999 (2008)

5. R.D. Scheffe, M. Strum, S.B. Phillips, J. Thurman, A. Eyth, S. Fudge, M. Morris, T. Palma, R. Cook, Hybrid modeling approach to estimate exposures of Hazardous Air Pollutants (HAPs) for the National Air Toxics Assessment (NATA). Environ. Sci. Technol. **50**, 12356–12364 (2016)
6. U.S. EPA (2015) Technical support document, EPA's 2011 National-scale air toxics assessment. U.S. Environmental Protection Agency, Office of air quality, planning, and standards, Research Triangle Park, NC. https://www.epa.gov/sites/production/files/2015-12/documents/2011-nata-tsd.pdf
7. U.S. National Library of Medicine (2017) Hazardous Substances Data Bank (HSDB): formaldehyde. https://toxnet.nlm.nih.gov/newtoxnet/hsdb.htm
8. G.Z. Whitten, G. Heo, Y. Kimura, E. McDonald-Buller, D.T. Allen, W.P.L. Carter, G. Yarwood, A new condensed toluene mechanism for Carbon Bond: CB05-TU. Atmos. Env. **44**, 5346–5355 (2010)

Chapter 5
Effects of Using Two Different Biogenic Emission Models on Ozone and Particles in Europe

Jianhui Jiang, Sebnem Aksoyoglu, Giancarlo Ciarelli, Emmanouil Oikonomakis and André S. H. Prévôt

Abstract In this paper, we discuss the importance of biogenic volatile organic compound (BVOC) emissions used in air quality simulations and how the model results are affected by the choice of the BVOC emission model. The European air quality in 2011 was simulated using CAMx regional air quality model with two different BVOC emission models: PSI-model and MEGAN. Especially isoprene and monoterpene emissions calculated by the two models differed significantly both in amounts and their spatial distribution. In general, MEGAN produced much higher isoprene emissions while PSI-model generated more monoterpene emissions. The difference in emissions between the two models was shown to be as high as a factor of 3 in summer. The choice of the BVOC emission model had significant consequences especially on the formation of organic aerosols as well as on ozone and inorganic aerosols. Using MEGAN led to relatively higher ozone concentrations in summer while much more SOA (secondary organic aerosol) was formed when PSI-model was applied. Our results suggest that the amount and spatial distribution of BVOC emissions might affect the oxidant concentrations (OH and nitrate radicals, ozone) leading to significant differences in SOA, ozone, particulate nitrate and sulfate concentrations calculated by different BVOC emission models.

J. Jiang · S. Aksoyoglu (✉) · E. Oikonomakis · A. S. H. Prévôt
Laboratory of Atmospheric Chemistry (LAC), Paul Scherrer Institute, 5232 Villigen PSI, Switzerland
e-mail: sebnem.aksoyoglu@psi.ch

J. Jiang
e-mail: jianhui.jiang@psi.ch

E. Oikonomakis
e-mail: emmanouil.oikonomakis@psi.ch

A. S. H. Prévôt
e-mail: andre.prevot@psi.ch

G. Ciarelli
Laboratoire Inter-Universitaire des Systèmes Atmosphériques (LISA), UMR CNRS 7583, Université Paris Est Créteil et Université Paris Diderot, Institut Pierre Simon Laplace, Créteil, France
e-mail: giancarlo.ciarelli@lisa.u-pec.fr

C. Mensink et al. (eds.), *Air Pollution Modeling and its Application XXVI*, Springer Proceedings in Complexity,
https://doi.org/10.1007/978-3-030-22055-6_5

5.1 Introduction

The BVOC (biogenic volatile organic compounds) emissions have a significant influence on atmospheric reactions leading to formation of secondary pollutants such as ozone and secondary particles in many areas, especially in summer. In many intercomparison studies, modelers share the same anthropogenic emissions, but the biogenic emissions usually differ [3], making the comparison of results from different models difficult. Since the uncertainties in BVOC emission estimates are very high [8, 9], it is important to know the range of variability in simulated pollutant concentrations while using different BVOC emissions. In this study, we investigated the effects of using two different BVOC emission models on ozone as well as inorganic and organic particles in Europe during summer and winter periods in 2011.

5.2 Methods

We simulated the European air quality in 2011 using the regional air quality model CAMx (Comprehensive Air quality Model with extensions, v6.30, www.camx.com) and the meteorological model WRF-ARW (Weather Research and Forecasting Model, v3.7.1; [10]. Our model domain covered Europe with a horizontal resolution of $0.25° \times 0.125°$ and there were fourteen terrain-following layers. The gas-phase mechanism was CB6r2 [5]. We used the fine/coarse option to calculate the concentrations of fine particles ($PM_{2.5}$) and the VBS scheme for organic aerosols. Initial and boundary conditions were obtained from the global model MOZART [6]. Anthropogenic emissions were based on the TNO-MACC-III inventory [7]. Two different biogenic emission models were used in this study to estimate BVOC emissions (isoprene, monoterpenes and sesquiterpenes) as well as NO emissions from soil. The first model was the PSI-model which was first developed by Andreani-Aksoyoglu and Keller [1] and updated later by Oderbolz et al. [8]. It calculates BVOC emissions using temperature and photosynthetically active radiation (PAR) from WRF, land cover data from USGS and the vegetation inventory GlobCover (http://due.esrin.esa.int/page_globcover.php). The second model was the widely used MEGAN v2.1 [4] that covers 147 individual compounds within 19 categories. MEGAN also used the same meteorological data from WRF. The land use data was the Community Land Model version 4 (CLM4) and the leaf area index (LAI) was used to simulate changes of vegetation during the year. The ozone and organic aerosol (OA) concentrations calculated in two simulations using PSI-model and MEGAN, respectively, were compared with ACSM/AMS (Aerosol Chemical Speciation Monitor/Aerosol Mass Spectrometer) measurements at 8 European sites. Modelled ozone concentrations were compared with measurements at 537 rural AIRBASE (European Air Quality Database v7) stations.

5.3 Results and Discussion

We focussed on summer since the emissions in winter were much lower, especially for isoprene. The comparison of biogenic emissions from the two BVOC models suggests that MEGAN model generates more isoprene, but much less monoterpene emissions than the PSI-model in Europe (Fig. 5.1) while the difference in sesquiterpene emissions was relatively small (<5%). In spite of three-times higher isoprene emissions in MEGAN, summer ozone was only slightly higher (<10%) than ozone calculated by the PSI-model (Fig. 5.2, left panel). On the other hand, higher monoterpene emissions in the PSI-model led to higher SOA (Fig. 5.2, right panel). Comparison of model results with measurements in Europe indicated that the bias for summer afternoon ozone mixing ratios higher than 50 ppb was lower when BVOC emissions were calculated with MEGAN, especially in southern Europe. For mixing ratios lower than 50 ppb however, the PSI-model showed a better performance. Mean bias between measured and modelled total organic aerosol was 8–90% lower by PSI model compared to MEGAN. Differences between the results of two simulations with different BVOC models were not only in ozone and SOA but also in particulate nitrate and sulfate, suggesting that the oxidant concentrations available for the formation of secondary inorganic aerosols might be affected significantly by the BVOC emissions, as also shown in Aksoyoglu et al. [2].

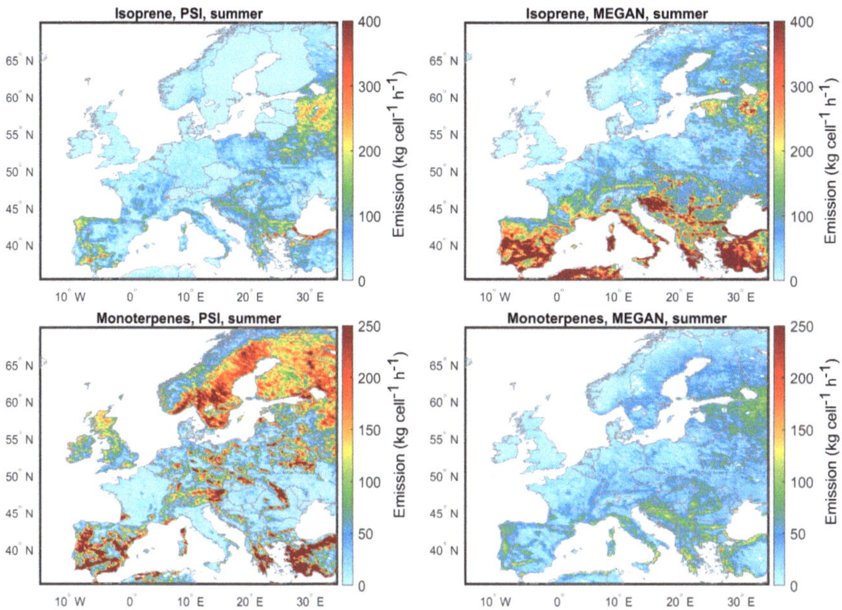

Fig. 5.1 Isoprene (upper panels) and monoterpene (lower panels) emissions (kg cell^{-1} h^{-1}) in July 2011 estimated by PSI-Model (left) and MEGAN (right)

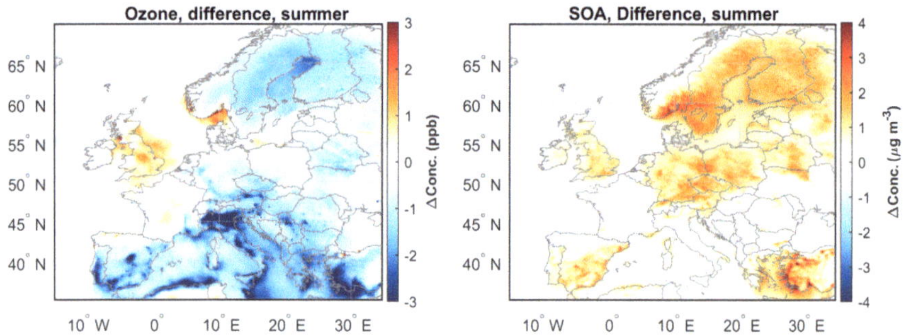

Fig. 5.2 Difference in ozone (ppb) (left) and in SOA (μg m^{-3}) (right) in July 2011 between PSI-Model and MEGAN (PSI-model—MEGAN)

5.4 Conclusion

In this study, the European air quality in 2011 was simulated by CAMx using two biogenic volatile organic compound (BVOC) emission models: MEGAN and PSI-model. The results showed that MEGAN generates more isoprene, but much less monoterpene emissions than the PSI-model in Europe probably due to their different land use and vegetation cover. In spite of much higher isoprene emissions generated by MEGAN compared to PSI-model, difference in ozone was relatively small (<10%) while three times higher monoterpene emissions in the PSI-model generated significantly more SOA (~110%) in summer. Comparison with measurements suggested a better performance with the PSI-model for organic aerosols while MEGAN showed a better agreement with measurements for high ozone levels. Particulate nitrate and sulfate concentrations were also affected by the BVOC emission model used. The results of this study emphasize the importance of BVOC emissions in air quality simulations and model inter-comparison studies.

Questions and Answers

Questioner: Rostislav Kouznetsov

Question: Why do the state borders appear quite clearly in the BVOC emission maps of your model? It is especially well seen in isoprene emission maps from Scandinavia and Baltic states.

Answer: This is the case only for isoprene and it comes from using country-specific forest fractions, historically based on data from Simpson et al. (1999). In PSI-model the main isoprene emitting tree species is oak (a small fraction also emitted by Norway spruce). According to the country-specific data used, the isoprene-relevant forest fractions are zero in Sweden and Baltic countries, leading to clear borders in isoprene emission maps. This can be avoided in future by using other types of forest data.

Questioner: Vanisa Surapipith

Question: Is there any plan to improve PSI model? In particular, it is our intention to query for a possibility to do any necessary field campaign to improve the input for the biogenic VOC emission estimation. There is an interest to explore further this issue in our research institute in Thailand, there are very few studies so far in southeast Asia.

Answer: No, there is no such plan at the moment. PSI-model was developed originally only for Switzerland on a high-resolution domain using Swiss tree inventories and later it was extended to cover Europe.

Acknowledgements We are grateful to the following people: H. Denier van der Gon at TNO for providing the anthropogenic emissions, F. Canonaco, C. O'Dowd, J. J. Ovadnevaite, Y. Schmale, N. Marchand, O. Favez, S. Gilardoni, MC. Minguillón and K. Florou for sharing the ACSM/AMS measurement data. Our thanks extend to ECMWF, CSCS and Ramboll.

References

1. S. Andreani-Aksoyoglu, J. Keller, Estimates of monoterpene and isoprene emissions from the forests in Switzerland. J. Atmos. Chem. **20**, 71–87 (1995). https://doi.org/10.1007/BF01099919
2. S. Aksoyoglu, G. Ciarelli, I. El-Haddad, U. Baltensperger, A.S.H. Prévôt, Secondary inorganic aerosols in Europe: sources and the significant influence of biogenic VOC emissions, especially on ammonium nitrate. Atmos. Chem. Phys. **17**, 7757–7773 (2017). https://doi.org/10.5194/acp-17-7757-2017
3. B. Bessagnet, G. Pirovano, M. Mircea, C. Cuvelier, A. Aulinger, G. Calori, G. Ciarelli, A. Manders, R. Stern, S. Tsyro, M. García Vivanco, P. Thunis, M.T. Pay, A. Colette, F. Couvidat, F. Meleux, L. Rouïl, A. Ung, S. Aksoyoglu, J.M. Baldasano, J. Bieser, G. Briganti, A. Cappelletti, M. D'Isidoro, S. Finardi, R. Kranenburg, C. Silibello, C. Carnevale, W. Aas, J.C. Dupont, H. Fagerli, L. Gonzalez, L. Menut, A.S.H. Prévôt, P. Roberts, L. White, Presentation of the EURODELTA III intercomparison exercise—evaluation of the chemistry transport models' performance on criteria pollutants and joint analysis with meteorology. Atmos. Chem. Phys. **16**, 12667–12701 (2016). https://doi.org/10.5194/acp-16-12667-2016
4. A.B. Guenther, X. Jiang, C.L. Heald, T. Sakulyanontvittaya, T. Duhl, L.K. Emmons, X. Wang, The model of emissions of gases and aerosols from nature version 2.1 (MEGAN2.1): an extended and updated framework for modeling biogenic emissions. Geosci. Model. Dev. **5**, 1471–1492 (2012). https://doi.org/10.5194/gmd5-1471-2012
5. L.H. Hildebrandt Ruiz, G. Yarwood, Interactions between organic aerosol and NOy: influence on oxidant production, Final Report for AQRP project 12-012 (2013)
6. L.W. Horowitz, S. Walters, D.L. Mauzerall, L.K. Emmons, P.J. Rasch, C. Granier, X. Tie, J.F. Lamarque, M.G. Schultz, G.S. Tyndall, A global simulation of tropospheric ozone and related tracers: description and evaluation of MOZART, version 2. J. Geophys. Res. **108**, 4784 (2003)
7. J.J.P. Kuenen, A.J.H. Visschedijk, M. Jozwicka, H.A.C. Denier van der Gon, TNO-MACC_II emission inventory a multi-year (2003–2009) consistent high-resolution European emission inventory for air quality modeling. Atmos. Chem. Phys. **14**, 10963–10976 (2014). https://doi.org/10.5194/acp-14-10963-2014
8. D.C. Oderbolz, S. Aksoyoglu, J. Keller, I. Barmpadimos, R. Steinbrecher, C.A. Skjøth, C. Plaß-Dülmer, A.S.H. Prévôt, A comprehensive emission inventory of biogenic volatile organic compounds in Europe: improved seasonality and land-cover. Atmos. Chem. Phys. **13**, 1689–1712 (2013). https://doi.org/10.5194/acp-13-1689-2013

9. D. Simpson, A. Benedictow, H. Berge, R. Bergström, L.D. Emberson, H. Fagerli, C.R. Flechard, G.D. Hayman, M. Gauss, J.E. Jonson, M.E. Jenkin, A. Nyíri, C. Richter, V.S. Semeena, S. Tsyro, J.P. Tuovinen, Á. Valdebenito, P. Wind, The EMEP MSC-W chemical transport model-technical description. Atmos. Chem. Phys. **12**, 7825–7865 (2012). https://doi.org/10.5194/acp-12-7825-2012
10. W.C. Skamarock, J.B. Klemp, J. Dudhia, D.O. Gill, D.M. Barker, M.G. Duda, X.-Y. Huang, W. Wang, J.G. Powers, A description of the advanced research WRF Version 3, Mesoscale and Microscale Meteorology Division, National Center for Atmospheric Research, Boulder, Colorado, USA (2008)

Chapter 6
A Proof-of-Concept for Linking the Global Meteorological Model, MPAS-a with the Air Quality Model, CMAQ

David Wong, Hosein Foroutan, Jonathan E. Pleim, O. Russell Bullock Jr., Robert C. Gilliam, Jerold A. Herwehe, Christian Hogrefe and George Pouliot

Abstract Researchers who perform air quality modeling studies usually do so on a regional scale. Typically, the boundary conditions are generated by another model which might have a different chemical mechanism, spatial resolution, and/or map projection. Hence, a necessary conversion/interpolation takes place which introduces additional error. In a broader sense, air pollution is a global issue, thus, limited area modeling on the regional scale is not well suited to represent long-range transport from key source regions in other parts of the world. We have developed a prototype system to link the Model for Prediction Across Scales—Atmosphere (MPAS-A) with the Community Multiscale Air Quality (CMAQ) model to address these short-comings. Pollutant transport is conducted within MPAS-A, rather than in CMAQ,

D. Wong (✉) · J. E. Pleim · O. Russell Bullock Jr. · R. C. Gilliam · J. A. Herwehe · C. Hogrefe · G. Pouliot
Computational Exposure Division, National Exposure Laboratory, US EPA, Ann Arbor, MI, USA
e-mail: wong.david-c@epa.gov

J. E. Pleim
e-mail: pleim.jon@epa.gov

O. Russell Bullock Jr.
e-mail: bullock.russell@epa.gov

R. C. Gilliam
e-mail: gilliam.robert@epa.gov

J. A. Herwehe
e-mail: herwehe.jerry@epa.gov

C. Hogrefe
e-mail: hogrefe.christian@epa.gov

G. Pouliot
e-mail: pouliot.george@epa.gov

H. Foroutan
Department of Biomedical Engineering and Mechanics, Virginia Tech, Blacksburg, VA, USA
e-mail: hosein@vt.edu

C. Mensink et al. (eds.), *Air Pollution Modeling and its Application XXVI*,
Springer Proceedings in Complexity, https://doi.org/10.1007/978-3-030-22055-6_6

to provide consistency with the meteorological processes. A coupler has been constructed to facilitate data exchange between the two models. Initial test simulations show reasonable results when compared with observational data.

6.1 Introduction

Air quality modeling studies using the Community Multiscale Air Quality modelling system (CMAQ) performed at the US EPA, are typically done on a regional scale. The air quality model is driven by a meteorological model since it requires various fields such as wind components, pressure, temperature, humidity, and many others as input. The offline system (i.e. sequential simulation of meteorology followed by chemistry and transport) carries a list of shortcomings [4] which drove the development of the WRF-CMAQ two-way coupled model with direct aerosol radiative effects.

Studies have shown that pollutants are frequently transported over hundreds or thousands of kilometers to affect specific receptors. Satellite images have shown a smoke plume from the 2002 Alaskan wild fires was transported down to southern Texas and then advected toward the eastern seaboard. Researchers employ domain nesting (i.e. the main study domain is enclosed by a larger domain of coarser resolution), which is usually repeated two to four times, to bring in influence from long distance sources. However, nesting for air quality modeling is usually applied in one direction only. Recently, CMAQ has been extended to the hemispheric scale [2] to help address long range transport issues.

Boundary conditions are a major challenge for regional scale modelling. In practice, boundary conditions are usually created from a global chemistry model such as GEOS-Chem. To make this global model output suitable to serve as boundary conditions for a regional model is not a straightforward matter. It involves a number of steps such downscaling, map projection conversion, and chemical species mapping. These steps introduce various degrees of error.

Currently in the offline version of CMAQ or in the WRF-CMAQ coupled model, chemical species transport processes take place within CMAQ and the advection and diffusion algorithms are not the same as in the meteorological model. Thus, this inconsistency is an additional concern in CMAQ modeling. This new development, coupling a global meteorological model, MPAS-A, with an air quality model, CMAQ, focuses on addressing the above issues: boundary conditions and transport inconsistency.

6.2 Implementation

6.2.1 Transport

In the MPAS-AQ prototype, all horizontal advection processes are performed in the meteorological core, MPAS-A, which uses a centroidal Voronoi (nominally hexagonal) mesh. It uses a C-grid staggering, where scalar prognostic variables such as potential temperature, density, water vapor mixing ratios, and any species concentrations are defined at the cell centers, while wind speed is defined by values perpendicular to cell faces. In our coupled online meteorology-air quality framework, species concentrations are updated at cell centers after all CMAQ processes (excluding advection). MPAS-A then calculates species flux divergence by summing fluxes over all faces of a cell, which gives an updated (i.e. next dynamical time-step) value of concentration due to advection. We emphasize that this procedure is consistent for all types of scalars in MPAS-A resulting in a completely consistent transport treatment in our MPAS-AQ model.

6.2.2 Emissions

Emissions for the MPAS-AQ simulations were processed from several underlying databases and following a similar approach as described in Xing et al. [5] for H-CMAQ simulations. Anthropogenic emissions from nine sectors were obtained from the HTAP_v2 database [1]. Biomass burning emissions were obtained from the EDGARv4.2 database. Finally, climatological biogenic volatile organic compounds (VOCs) and lightning-produced nitrogen oxides (NO_x) emissions were obtained from the GEIA (Global Emission Inventory Activity) database. Vertical allocation is also considered based on different emission categories. Temporal allocation (weekly and diurnal in the case of monthly emission totals, and monthly, weekly, and diurnal in the case of annual emission totals) followed the approach described in Xing et al. [5]. Chemical speciation was based on North American speciation profiles as applied to the 2010 emission inventories described in Pouliot et al. [3] and Janssens-Maenhout et al. [1]. To transfer the emissions from their native $0.1° \times 0.1°$ grid to the MPAS-AQ unstructured mesh, each of the 3,600 $\times$ 1,800 grid cells of the gridded emissions fields were assigned to the closest MPAS-AQ cell based on the latitude and longitude of the MPAS-AQ cell center. If multiple emission grid cells were assigned to the same MPAS-AQ cell, their emissions were summed. Note that the biogenic and lightning emissions were first interpolated from $1.0° \times 1.0°$ to $0.1° \times 0.1°$ and then assigned to the closest MPAS-AQ cells.

6.2.3 Data Exchange and I/O

A coupler was constructed to facilitate data exchange between the MPAS-A and AQ models. This coupler also provides other functions: converts perturbation pressure to pressure and potential temperature to temperature that the AQ model needs, computes other AQ model specific data, such as level height, rearranges data dimension order (MPAS-A is k i and AQ is i k), and injects O_3 analysis data from the Global Forecast System (GFS) into the highest model layers according to a predefined lowest layer threshold and nudging algorithm.

In addition, a new input/output (I/O) system is implemented in this coupled model to allow the AQ model to output any information independent of the MPAS-A I/O system. This extra layer of flexibility reduces user burden to modify the model registry file prior to code compilation and allows the user to control which files to output via the run script.

6.3 Preliminary Model Results

A month of simulation from July 1, 2013 to July 31, 2013 was completed based on the emissions constructed using the methods described above. For simplicity, the same emission file was replicated for each day of the month. The global non-uniform 90–25 km mesh (finer resolution over North America) was used. The simulation was initialized with clean initial conditions and was run using 128 processor cores, requiring an average of 144 min to simulate one day.

The first 14 days were treated as a spinup period. Figure 6.1 shows the mean bias of 8 h max ozone (O_3) in the second half of the simulation period. The eastern US

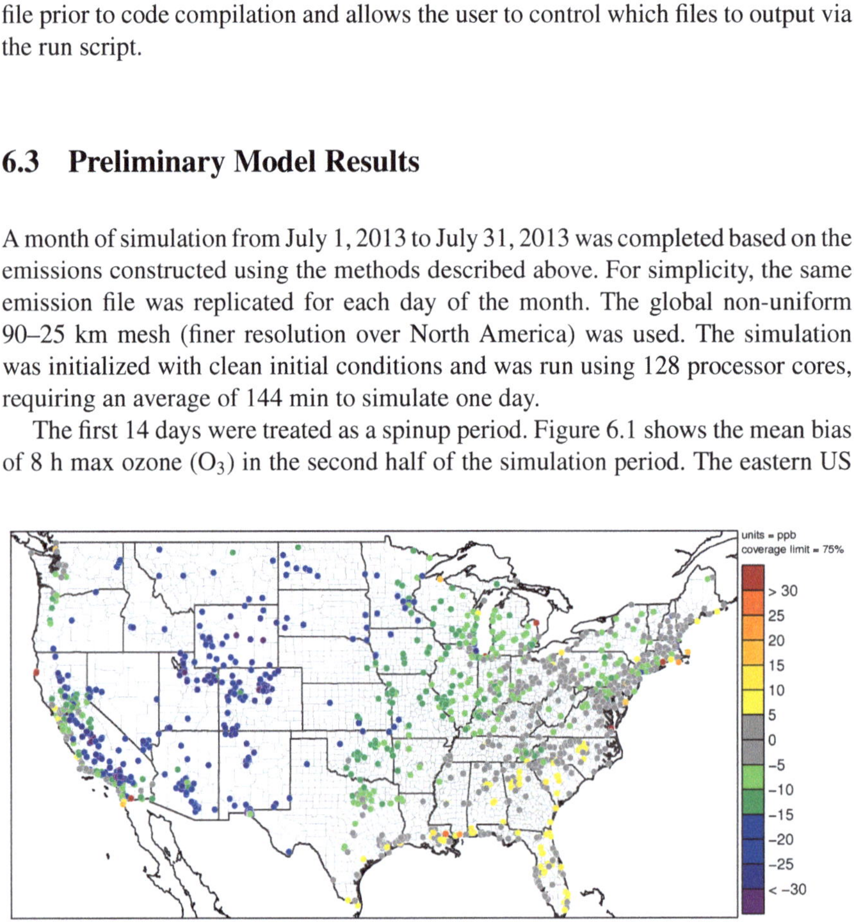

Fig. 6.1 Mean bias of 8 h max O_3 for 15–29 July 2013

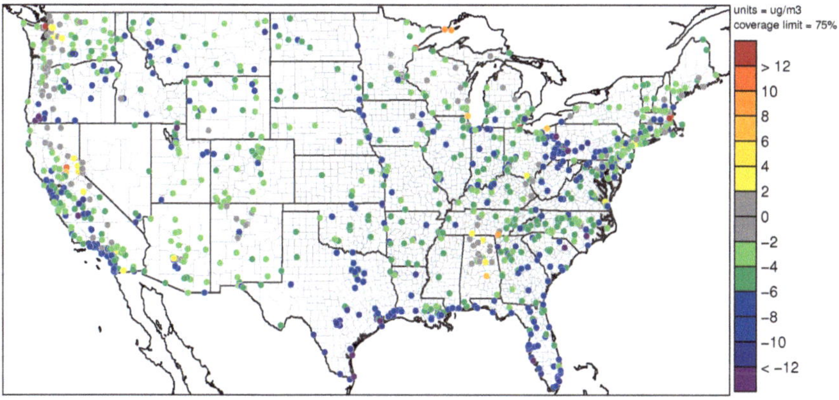

Fig. 6.2 Mean bias of $PM_{2.5}$ for 15–29 July 2013

shows quite acceptable results, while on the west coast and the Rocky Mountains region, higher negative bias was produced.

Model-predicted fine particulate matter ($PM_{2.5}$) (Fig. 6.2) shows a general under-prediction but still reasonable results considering the short study period and inadequate spinup.

6.4 Conclusions

This paper discussed results from the newly developed global meteorological model, MPAS-A, coupled with the CMAQ air quality model. With a one-month simulation, results show reasonable model performance despite clean initial conditions, a short spinup period, and lack of temporal variation in emission input. Next steps will include a full year simulation (2016) with seasonally varying emissions.

QUESTIONER: Anthony Dore
QUESTION: You started your talk by demonstrating downscaling (i.e. CMAQ 81 km -> 9 km -> 3 km). How flexible is the MPAS hexagonal grid in incorporating regional variation in grid resolution?
ANSWER: MPAS is a global atmospheric model that uses unstructured spherical centroidal Voronoi meshes with uniform or non-uniform resolution. In the non-uniform case, MPAS offers global coverage with seamless scale refinement to regional and local scales. Currently, NCAR has not released such mesh generator software. If a user interested in a particular mesh, he/she can send a request to NCAR to obtain one.

Disclaimer The views expressed in this presentation are those of the authors and do not necessarily reflect the views or policies of the U.S. EPA.

References

1. G. Janssens-Maenhout, M. Crippa, D. Guizzardi, F. Dentener, M. Muntean, G. Pouliot, T. Keating, Q. Zhang, J. Kurokawa, R. Wankmüller, H. Denier van der Gon, J.J.P. Kuenen, Z. Klimont, G. Frost, S. Darras, B. Koffi, M. Li, HTAP_v2.2: a mosaic of regional and global emission grid maps for 2008 and 2010 to study hemispheric transport of air pollution. Atmos. Chem. Phys., **15**, 11411–11432. https://doi.org/10.5194/acp-15-11411-2015 (2015)
2. R. Mathur, J. Xing, R. Gilliam, G. Sarwar, C. Hogrefe, J. Pleim, G. Pleim, S. Roselle, T.L. Spero, D.C. Wong, J. Young, Extending the Community Multiscale Air Quality (CMAQ) modeling system to hemispheric scales: overview of process considerations and initial applications. Atmos. Chem. Phys., May 2017, https://doi.org/10.5194/acp-2017-329
3. G. Pouliot, H.A.C.D. van der Gon, J. Kuenen, J. Zhang, M.D. Moran, P.A. Makar, Analysis of the emission inventories and model-ready emission datasets of Europe and North America for phase 2 of the AQMEII project. Atm. Environ. **115**, 345–360, https://doi.org/10.1016/j.atmosenv.2014.10.061 (2015)
4. D.C. Wong, J. Pleim, R. Mathur, F. Binkowski, T. Otte, R. Gilliam, G. Pouliot, A. Xiu, D. Kang, WRF-CMAQ two-way coupled system with aerosol feedback: software development and preliminary results. Geosci. Model Dev. **5**, 299–312 (2012)
5. J. Xing, R. Mathur, J. Pleim, C. Hogrefe, C.-M. Gan, D.C. Wong, C. Wei, R. Gilliam, G. Pouliot, Observations and modeling of air quality trends over 1990–2010 across the Northern Hemisphere: China, the United States and Europe. Atmos. Chem. Phys. **15**, 2723–2747 (2015). https://doi.org/10.5194/acp-15-2723-2015

Chapter 7
Long-Term Trends in Sulfur and Reactive Nitrogen Deposition Across the Northern Hemisphere and United States

Rohit Mathur, Yuqiang Zhang, Christian Hogrefe and Jia Xing

Abstract We model the changes in wet and dry deposition amounts of reactive nitrogen and sulfur over the 1990–2010 period using the WRF-CMAQ modeling system. WRF-CMAQ simulations for this 21-year period were conducted over a domain covering the northern hemisphere using a horizontal resolution of 108 km and a nested domain over the contiguous U.S. using a grid of 36 km resolution. The impacts of contrasting changes in emission patterns across the Northern Hemisphere, i.e., reductions in North America and Europe vs. increases across regions in Asia, on changing deposition amounts over terrestrial and aquatic ecosystems in these regions is analyzed.

7.1 Introduction

A detailed understanding of the distribution and fate of atmospheric sulfur (SO_x) and reactive nitrogen compounds (NO_y and NH_x) is desirable given their role in determining tropospheric acidic substances and particulate matter budgets and potential nutrient loading effects in sensitive ecosystems resulting from their atmospheric deposition. The ultimate fate of airborne SO_x, NO_y and NH_x is removal by wet scavenging and dry deposition, which in turn lead to a variety of environmental effects including altering net primary production, acidification, eutrophication and other nutrient loading effects. The gas-particle partitioning of airborne sulfur and reactive nitrogen regulates their transport distances since dry deposition velocity for fine particles is relatively low, and consequently their primary atmospheric sink is wet scavenging. Changing emissions patterns of NO_x, SO_2, and NH_3 have likely altered both their atmospheric transport distances as well as deposition patterns and amounts.

R. Mathur (✉) · Y. Zhang · C. Hogrefe
Computational Exposure Division, National Exposure Research Laboratory, U.S. Environmental Protection Agency, RTP, NC, Washington, D.C., USA
e-mail: mathur.rohit@epa.gov

J. Xing
School of Environment, Tsinghua University, Beijing, China

C. Mensink et al. (eds.), *Air Pollution Modeling and its Application XXVI*,
Springer Proceedings in Complexity, https://doi.org/10.1007/978-3-030-22055-6_7

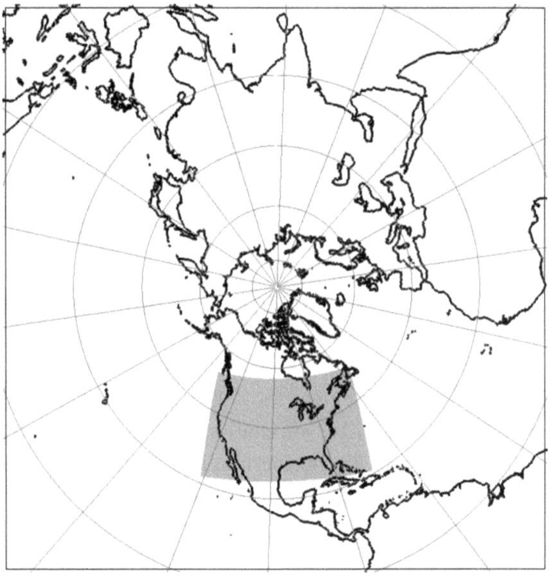

Fig. 7.1 The Northern Hemisphere (outer) and nested Continental U.S. modeling domain (shaded)

7.2 Model Setup

The WRF-CMAQ modeling system was used to simulate the tropospheric composition and wet and dry deposition of various atmospheric species over a 21-year period spanning 1990–2010. Model simulations were performed over a domain encompassing the entire Northern Hemisphere (Fig. 7.1), set on a polar stereographic projection and discretized with a horizontal grid spacing of 108 km [1]. Space and time varying lateral boundary conditions from these large scale simulations were then provided to a 36 km resolution simulation for the same period over the Continental U.S. (CONUS) domain (shaded region in Fig. 7.1). Year specific emissions for the Northern Hemispheric domain were derived from the EDGARv4.2 global emission inventory [2] while those for the CONUS domain were based on the inventory of Xing et al. [3].

Hourly dry and wet deposition amounts from the WRF-CMAQ simulations were aggregated to seasonal and annual totals for each of the 21 years and their spatial distributions and temporal trends were analyzed.

7.3 Results and Discussion

Model estimated 1990–2010 trends in annual total deposition (wet + dry) of sulfur and inorganic nitrogen (IN) across the Northern Hemisphere are shown in Fig. 7.2. The directionality, magnitude and spatial variability of the model estimated wet deposition trends for sulfur and IN were found to be in reasonable agreement with

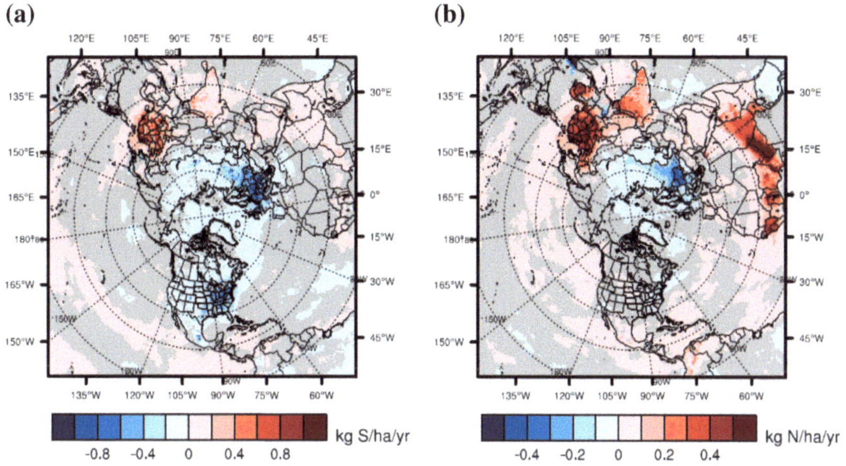

Fig. 7.2 1990–2010 trends in annual total wet + dry deposition across the Northern Hemisphere for **a** total sulfur (SO_2+SO_4^{2-}) expressed in kg S/ha/yr and **b** total inorganic nitrogen (NO_y+NH_x) expressed in kg N/ha/yr. Grey shaded areas represent locations with p-value >0.05 for the standard two-tailed Student T-test, i.e., areas where the modeled trend estimates were not significant at the 95% confidence level

those inferred from NADP measurements in the U.S. and EMEP measurements in Europe (not shown). Strong decreasing trends in atmospheric sulfur deposition are noted across the continental U.S., western Europe, and regions of the north Atlantic Ocean influenced by continental outflow, in response to the significant reductions in SO_2 emissions over the past two decades. In contrast, rising SO_2 emissions during this period resulted in increasing sulfur deposition trends across large parts of Asia as well as large portions of the Pacific influenced by Asian outflow. The spatial distributions of the trends in IN deposition are influenced by trends in both NO_x and NH_3 emissions. In Europe where emissions of both species have seen reductions during this period, we note reductions in IN deposition with contrasting increases in regional deposition across Asia in response to growth in emissions of both NO_x and NH_3.

The IN deposition across the U.S. however depicts decreasing trends in the northeastern U.S. and outflow over the Atlantic but increasing trends in the mid-western states. This is further explained by trends in atmospheric deposition of oxidized and reduced nitrogen which are illustrated in Fig. 7.3. Reductions in NO_x emissions resulted in systematic decreases in oxidized nitrogen with larger decreasing rates in the eastern than the western U.S. The increasing IN deposition trends over the east and central states were caused by NH_X deposition increases which in turn arise from increasing NH_3 emissions in these regions over the past two decades.

As illustrated in Fig. 7.4, strong increasing trends in dry deposition amounts of NH_x across the U.S. occurred during the 1990–2010 period and likely arise both from increasing NH_3 emissions but also perhaps from reduced transport distances. Reductions in SO_2 and NO_x emissions and their oxidation products have likely

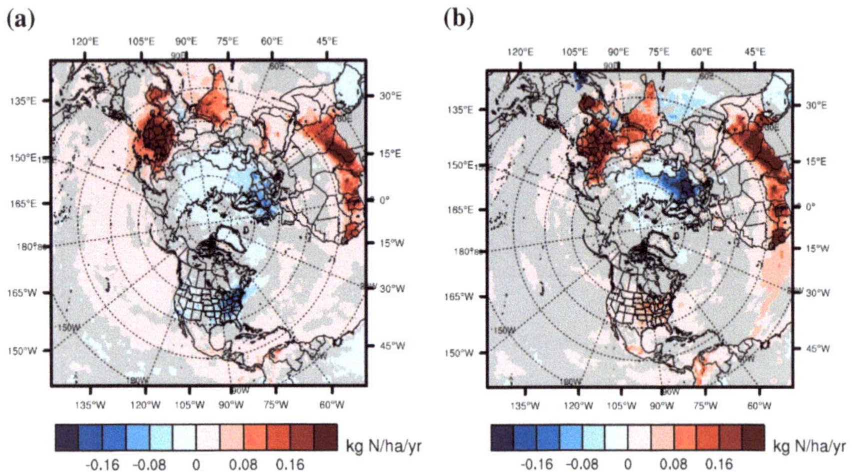

Fig. 7.3 1990–2010 trends in annual total wet + dry deposition (in kg N/ha/yr) across the Northern Hemisphere for **a** oxidized nitrogen (NO_y), and **b** reduced nitrogen (NH_x)

Fig. 7.4 1990–2010 trends in annual reduced nitrogen (NH_x) dry deposition (in kg N/ha/yr)

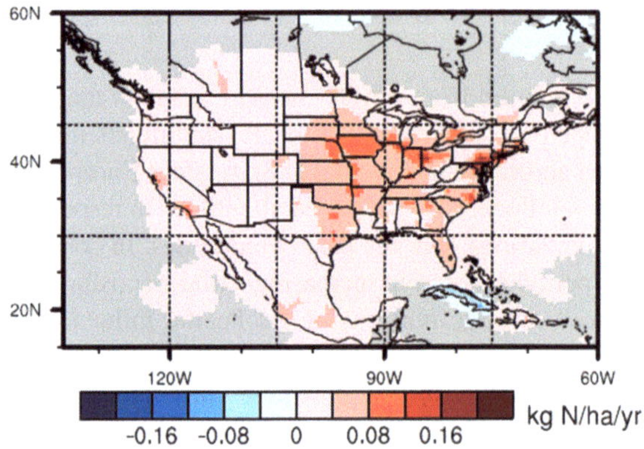

decreased the amounts of NH_x partitioning to the aerosol phase where scavenging by rain is the primary sink. Consequently, more NH_x remains in the gas-phase and dry deposits closer to the source regions as suggested in Fig. 7.4.

QUESTIONER: Talat Odman
QUESTION: The N-critical load in the northeast has decreased while it remained the same (or increased) in eastern North Carolina. What is this telling us in terms of sources of N?
ANSWER: The trends in total-N deposition are influenced by individual trends in reduced and oxidized-N deposition. In the northeast U.S., the 1990–2010 trends in N-deposition are dominated by trends in oxidized-N deposition while those in eastern North Carolina are dominated by trends in reduced-N deposition. Consequently, reductions in NO_x emissions across the northeastern U.S. have resulted in decreasing trends in N-deposition in the region. In contrast, increasing NH_3 emissions in eastern North Carolina resulted in increase in reduced-N and total-N deposition in that region.

QUESTIONER: Camilla Geels
QUESTION: Did you consider changes in land-use—could be important for dry deposition. If not, do you have any idea how significant they are?
ANSWER: To our knowledge no single dataset exists that describes in a consistent manner the changes in land-use for the multi-decadal period analyzed here. To minimize uncertainties associated with using different land-use data sets for different periods during 1990–2010, we decided to use the same land-use characterization for the entire period. While we acknowledge the likely impact of changing land use on simulated atmospheric dynamics and dry deposition amounts, we do not have a detailed assessment of the magnitude of the impact for this period. However, one could speculate that since both NH_3 and HNO_3 dry deposit efficiently to most surfaces, and since changes in land-use have been gradual, it is likely that the assumption of time invariant land-use does not have a significant impact on our estimated trends and results. Nevertheless, sensitivity simulations would need to be conducted to confirm this in a more quantitative manner.

Disclaimer The views expressed in this paper are those of the authors and do not necessarily reflect the views and policies of U.S. Environmental Protection Agency.

References

1. R. Mathur, J. Xing, R. Gilliam, G. Sarwar, C. Hogrefe, J. Pleim, G. Pouliot, S. Roselle, T.L. Spero, D.C. Wong, J. Young, Extending the Community Multiscale Air Quality (CMAQ) modeling system to hemispheric scales: overview of process considerations and initial applications. Atmos. Chem. Phys. **17**, 12449–12474 (2017). https://doi.org/10.5194/acp-17-12449-2017
2. J. Xing, R. Mathur, J. Pleim, C. Hogrefe, C.-M. Gan, D.C. Wong, C. Wei, R. Gilliam, G. Pouliot, Observations and modeling of air quality trends over 1990–2010 across the Northern Hemisphere: China, the United States and Europe. Atmos. Chem. Phys. **15**, 2723–2747 (2015). https://doi.org/10.5194/acp-15-2723-2015
3. J. Xing, J. Pleim, R. Mathur, G. Pouliot, C. Hogrefe, C.-M. Gan, C. Wei, Historical gaseous and primary aerosol emissions in the United States from 1990 to 2010. Atmos. Chem. Phys. **13**, 7531–7549 (2013). https://doi.org/10.5194/acp-13-7531-2013

Chapter 8
Trend Analysis of Air Pollution and Nitrogen Deposition Over the Netherlands Using the EMEP4NL and OPS Model

Eric van der Swaluw, Wilco de Vries, Roy Wichink Kruit, Jan Aben, Massimo Vieno, Hilde Fagerli, Peter Wind and Addo van Pul

Abstract A trend analysis is performed over the period 2006–2015 for concentration and deposition of nitrogen compounds over the Netherlands. The analysis is performed with high resolution (~1 km) model simulations with the Gaussian plume model OPS and the grid model EMEP4NL. Both models use the same MACC III emission distribution for countries outside of the Netherlands, and spatially more detailed emissions for the Netherlands itself. Emission totals per SNAP sector per country are used over the period 2006–2015, according to the latest CEIP expert estimates. The OPS model is driven with yearly specific meteorological fields provided by the Royal Netherlands Meteorological Institute (KNMI), while EMEP4NL is driven by meteorological output data from the open source WRF meteorological model. Results from the model calculations are first compared with measurements. Next, the focus of the analysis will be on the effect of atmospheric chemistry on trend analysis of nitrogen components, like ammonia and ammonium in the atmosphere, and the dry and wet nitrogen deposition to the surface: OPS strongly parameterizes the chemistry, whereas EMEP4NL uses a state-of-the-art chemistry scheme. The influence of atmospheric chemistry on modeled trends in concentration and deposition is determined by comparing the trend calculations of both models, and their connection with trends in emissions of precursor gases over the period 2006–2015.

E. van der Swaluw (✉) · W. de Vries · R. W. Kruit · J. Aben · A. van Pul
RIVM, PO Box 1, 3720 BA Bilthoven, The Netherlands
e-mail: Eric.van.der.Swaluw@rivm.nl

M. Vieno
Centre for Ecology and Hydrology, Penicuik, UK

H. Fagerli · P. Wind
EMEP MSC-W, Norwegian Meteorological Institute, Oslo, Norway

C. Mensink et al. (eds.), *Air Pollution Modeling and its Application XXVI*,
Springer Proceedings in Complexity,
https://doi.org/10.1007/978-3-030-22055-6_8

8.1 Introduction

The Netherlands ranks second in value of agricultural export [2]; hence, the impact of nitrogen deposition on ecosystems is a relevant issue in the Netherlands. Recently there have been a few papers published [3, 5], which focus on the question whether the reported decline in ammonia emissions in the Netherlands is reflected in atmospheric concentration time series. The reason to focus on concentrations instead of deposition is that measurements of deposition are very scarce, whereas measurements of ammonia concentration are available on a much denser scale in the Netherlands.

There is no direct *linear* relation between ammonia emissions and concentrations, because the ammonia concentrations in the atmosphere are also influenced by meteorology, atmospheric chemistry and the dependence of deposition on the acidity of the surface on which the deposition takes place. Both the atmospheric chemistry and the acidity of the surface also change over the periods considered, because they both depend on the concentration of both sulphur and nitrogen components in respectively the atmosphere and the surface layer.

Wichink Kruit et al. [5] use the OPS model in order to quantify the effects of meteorology and physicochemical processes on atmospheric ammonia concentrations. The OPS model is a source-receptor model, which combines a Gaussian plume model approach for local applications with a Lagrangian trajectory model for long-range transport. Chemistry in the OPS model is parametrized and therefore less robust than, for example, a chemical module in a Eulerian chemistry transport model.

Here we adopt the EMEP4NL model in order to calculate ammonia concentrations and related components over the Netherlands over a period of 10 years, i.e. from 2006 to 2015. The calculations are performed on an embedded grid with the highest resolution of 2.0×1.3 km over the Netherlands.

8.2 The Models Used Over the Period 2006–2015: EMEP4NL and OPS

The EMEP model [1] is employed in this paper (as EMEP4NL), in order to make calculations over the Netherlands at high resolution for the purpose of a trend analysis. We follow the same approach as Vieno et al. [4], who couples the output of the WRF model for the meteorological data in order to drive the EMEP model in off-line mode over the UK (as EMEP4UK). An embedded grid is used with four layers; the coarsest one is over the whole of Europe at a resolution of a half degree, the highest resolution is over the Netherlands, with a resolution of 2.0×1.3 km.

Emissions are taken from MACC III for countries outside of the Netherlands, and spatially more detailed emissions for the Netherlands itself. Emission totals per SNAP sector per country are used over the period 2006–2015, according to the latest CEIP expert estimates.

A comparison of the EMEP4NL model output is made with results from the OPS-model over the same period, for which the same emissions are used. The OPS model is driven with yearly specific meteorological fields provided by the Royal Netherlands Meteorological Institute (KNMI).

8.3 Some First Preliminary Results of Trend Run with EMEP4NL and OPS

The calculations as performed with the EMEP4NL and OPS model over 2006–2015 give a plethora of output data like concentrations of ammonia and other precursors; secondary inorganic aerosols; particulate matter; wet and dry deposition of sulphur and nitrogen. The focus of this trend analysis will initially be on the different nitrogen components like ammonia, ammonium, nitrate, and dry and wet nitrogen deposition and, in particular, the relation of these components to the decrease in ammonia emissions in the Netherlands over the period 2006–2015. An extensive discussion of this whole topic is beyond the scope of this abstract, so we only show some preliminary results here.

Figure 8.1 (top panel) shows the wet deposition of NH_4 over the whole period as measured (LML/MAN) and calculated (EMEP4NL). Measurements over this period have been taken at 8–11 monitoring stations distributed more or less homogeneously over the Netherland. One can see there is an underestimation of 1.35 with the EMEP4NL model, which is partly due to the fact that precipitation is underestimated by the WRF model. The spatial representation however is quite good. Figure 8.1 (bottom panel) shows the trend of (normalized) ammonia concentrations over the period 2006–2015 for respectively measurements (LML/MAN), the EMEP4NL model and the OPS model.

8.4 Conclusion

The comparison between measurements and calculations of ammonia concentrations shows good agreement for both models, EMEP4NL and OPS. Also, the models agree well in this respect. The spatial distribution of the wet deposition of ammonia is well represented by EMEP4NL, but the deposition is underestimated by the model. A detailed analysis of the trends of all the different nitrogen components and their relation to the decreasing ammonia emissions in the Netherlands, is currently being performed. A presentation of more detailed results of this analysis is beyond the scope of this short paper, and will hence be presented elsewhere.

Acknowledgements This research was financed by the Dutch Ministry of Agriculture, Nature and Food Quality.

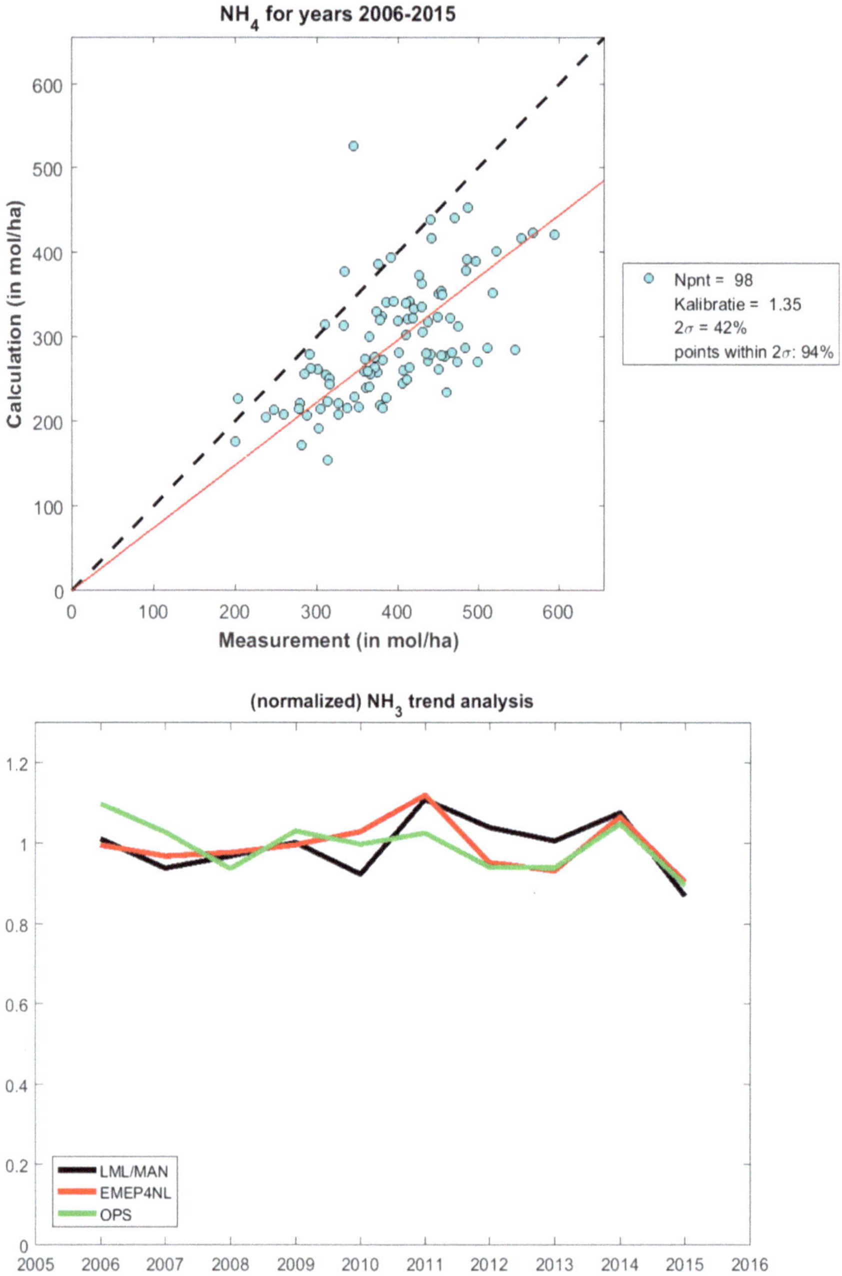

Fig. 8.1 Top panel: scatter plot of measured and calculated values of the wet deposition of NH_4 at all LML stations over the period 2006–2015 for the EMEP model. Bottom panel: the trend of NH_3 over the period 2006–2015 according to measurements (LML/MAN), EMEP4NL and OPS. The concentrations have been normalized in the same way as in van Zanten et al. [3]

References

1. D. Simpson, A. Benedictow, H. Berge, R. Bergström, L.D. Emberson, H. Fagerli, C.R. Flechard, G.D. Hayman, M. Gauss, J.E. Jonson, M.E. Jenkin, A. Nyíri, C. Richter, V.S. Semeena, S. Tsyro, J.-P. Tuovinen, Á. Valdebenito, P. Wind, The EMEP MSC-W chemical transport model—technical description. Atmos. Chem. Phys. **12**, 7825–7865 (2012)
2. The Economist, August 2014. Farming in the Netherlands: Polder and Wiser, vol. 23. The Economist (Sevenum: The Economist Group). Retrieved 29 August 2014
3. M.C. van Zanten, R.J. Wichink Kruit, R. Hoogerbrugge, E. van der Swaluw, W.A.J. van Pul, Trends in ammonia measurements in the Netherlands over the period 1993–2014, Atmospheric Environment, 148 (2017)
4. M. Vieno, M.R. Heal, S. Hallsworth, D. Famulari, R. Doherty, M., A.J. Dore, Y.S. Tang, C.F. Braban, D. Leaver, M.A. Sutton, S. Reis, The role of long-range transport and domestic emissions in determining atmospheric secondary inorganic particle concentrations across the UK, Atmos. Chem. Phys. **14**, 8435–8447. https://doi.org/10.5194/acp-14-8435-2014 (2014)
5. R.J. Wichink Kruit, J. Aben, W. de Vries, F. Sauter, E. van der Swaluw, M.C. van Zanten, W.A.J. van Pul, Modelling trends in ammonia in the Netherlands over the period 1990–2014, Atmospheric Environment, **154** (2017)

Chapter 9
Atmospheric Contribution to Eutrophication of the Baltic Sea

Jerzy Bartnicki

Abstract Nitrogen and phosphorus are two main nutrients responsible for eutrophication of the Baltic Sea. Almost all phosphorus is entering the sea via rivers, whereas 20–30% of nitrogen is deposited from the air. Therefore, there is a need for monitoring atmospheric nitrogen deposition to the Baltic Sea. Time series of annual nitrogen depositions to the Baltic Sea have been calculated for the period 1995–2015 with the same version of the EMEP MSC-W model. They show significant inter annual fluctuations due to changes in meteorological conditions from one year to another. To reduce the influence of meteorological conditions on the results the so called "normalized" depositions have been also calculated. They indicate a clear decline of annual depositions of oxidized nitrogen to the Baltic Sea and only minor decline of reduced nitrogen depositions in the considered period. Emissions from Germany and Poland are the main sources contributing to deposition of oxidized nitrogen to the Baltic Sea basin followed by the ship traffic on the Baltic Sea and on the North Sea. Transportation and combustion are the main emission sectors contributing to oxidised nitro-gen deposition, whereas, agriculture is the dominating emission sector contributing to reduced nitrogen deposition.

9.1 Introduction

Eutrophication is one of the major environmental problems for the European Seas. It is especially a threat to the shallow seas, such as the Baltic Sea with the average depth of 54 m. Excessive amounts of nutrients entering the Baltic Sea water can cause the growth of algae and other aquatic plants, leading to negative effects such as oxygen depletion in the sea. Two nutrients are mainly responsible for eutrophication of the Baltic Sea: phosphorus and nitrogen. Almost all phosphorus is entering the sea via rivers, whereas 20–30% of nitrogen is deposited to the Baltic Sea from the air [1]. Therefore, there is a need for monitoring of nitrogen deposition to the Baltic

J. Bartnicki (✉)
Norwegian Meteorological Institute, Henrik Mohns Plass 1, 0313 Oslo, Norway
e-mail: jerzy.bartnicki@met.no

C. Mensink et al. (eds.), *Air Pollution Modeling and its Application XXVI*, Springer Proceedings in Complexity,
https://doi.org/10.1007/978-3-030-22055-6_9

Sea based on measurements and model calculations. In the frame of a long-term cooperation between Helsinki Commission and EMEP, nitrogen depositions to the Baltic Sea Basin and its sub-basins are calculated every year in the Meteorological Synthesizing Centre-West (MSC-W) of EMEP in Oslo [2]. Time series of annual nitrogen depositions to the Baltic Sea have been calculated for the period 1995–2015. They show significant inter annual fluctuations due to changes in meteorological conditions from one year to another. The so called "normalized" depositions have been also calculated to reduce the influence of meteorological variability. Additional important task for MSC-W is the identification of main emission sources contributing to nitrogen deposition into the entire basin and sub-basins of the Baltic Sea. Here we present and discuss the main emission sources contributing to annual nitrogen deposition into the Baltic Sea basin in the year 2014.

The EMEP/MSC-W model, a multi-pollutant 3D Eulerian Chemical Transport Model, has been used for all nitrogen computations presented here. It has been documented in detail in [4] and in the annual chapters on model updates in subsequent EMEP status reports with the last one in 2017 [5].

The model is regularly evaluated against measurements from the EMEP network under the LRTAP convention (e.g. [3]). The performance of the EMEP/MSC-W model can be considered as state-of-the-art over a large range of both gaseous species and particulate matter.

9.2 Atmospheric Nitrogen Deposition to the Baltic Sea

Time series of annual and normalized depositions of oxidised and reduced nitrogen for the period 1995–2015 are shown in Fig. 9.1. In annual depositions of oxidized and reduced nitrogen there is a large inter-annual variability caused by changing meteorological conditions which is to large extent limited in calculated normalized depositions. In addition to actual and normalized depositions, the minimum and

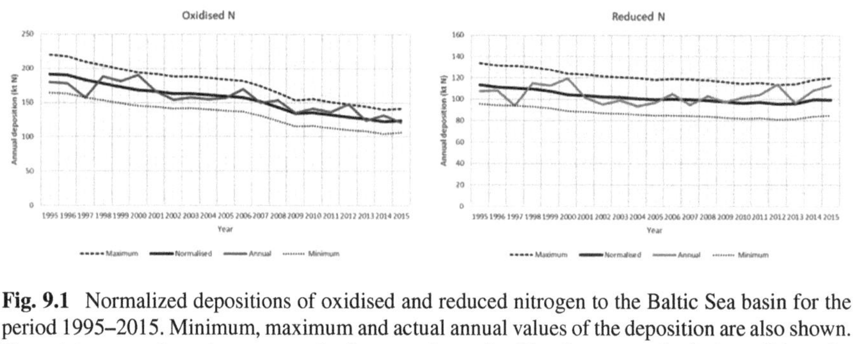

Fig. 9.1 Normalized depositions of oxidised and reduced nitrogen to the Baltic Sea basin for the period 1995–2015. Minimum, maximum and actual annual values of the deposition are also shown. The minimum and maximum annual values are determined by the meteorological conditions for each year

maximum values, corresponding to the best and worst meteorological conditions, have been also calculated, indicating meteorological uncertainty.

Compared to 1995, annual deposition of oxidised nitrogen in 2015 is 33% lower. The normalized deposition of oxidised nitrogen is even more declining in the same period −35%. So, there is clear decline in the deposition of oxidised nitrogen to the Baltic Sea basin in the period 1995–2014. Uncertainty indicated by a relative distance between minimum and maximum deposition for each year is 29%.

In case of reduced nitrogen deposition to the Baltic Sea basin, annual deposition in 2015 is 5% higher in 2015 than at the beginning of the period in 1995. The normalised deposition of reduced nitrogen is declining (12%) in 2015 compared to 1995 deposition. However, there is a slight increase in the reduced nitrogen deposition at the end of the period. Uncertainty indicated by a relative distance between minimum and maximum deposition for each year is slightly higher in case of reduced nitrogen compared to oxidized nitrogen −35%.

9.3 Main Sources of Nitrogen Deposition to the Baltic Sea

Altogether there are more than 50 emissions sources in the EMEP domain which contribute to nitrogen deposition to the Baltic Sea basin. The source allocation budget is presented for the latest available year 2014, but the results are similar for the previous years [1]. The top ten sources contributing to annual oxidized and reduced nitrogen deposition to the Baltic Sea basin in 2014 are shown in Fig. 9.2.

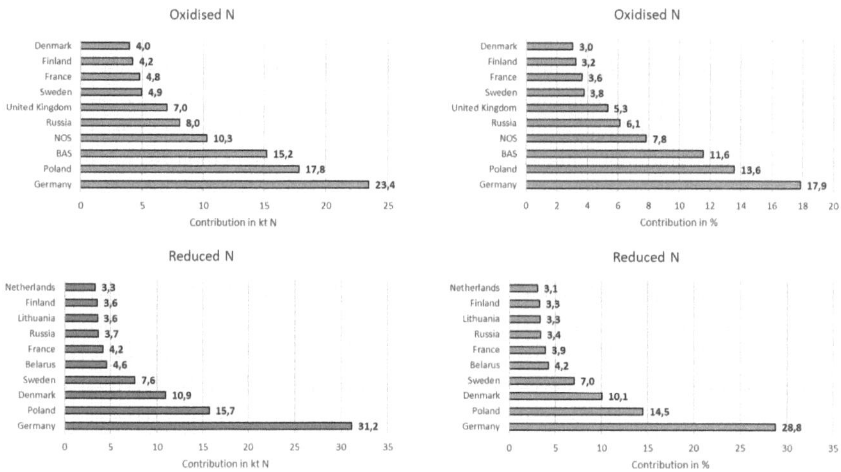

Fig. 9.2 Top ten sources with highest contributions to annual depositions of oxidised and reduced nitrogen into the Baltic Sea basin in the year 2014

Germany and Poland are the main sources contributing to oxidized nitrogen deposition to the Baltic Sea basin with 23.4 kt N and 17.8 kt N, respectively. This correspond to 17.9 and 13.6% of the total annual oxidized nitrogen deposition to the Baltic Sea basin in 2014 which is 131 kt N. There is also a significant contribution from the ship traffic on the Baltic Sea (15.2 kt N) and ship traffic on the North Sea (10.3 kt N), corresponding to 11.6% and 7.8%, respectively. In addition to Germany and Poland, Russia, Sweden and Denmark are the HELCOM Contracting Parties among ten major contributors. Nitrogen emissions in all HELCOM countries together contribute 51% to oxidised nitrogen deposition. Two distant emission sources (United Kingdom and France) can be found among top ten contributors.

In case of reduced nitrogen deposition to the Baltic Sea basin, Germany dominates the list with 31.2 kt N corresponding to 28.8% contribution to annual deposition in 2014. Three other HELCOM countries are next on the list: Poland, Denmark and Sweden with 15.7 kt N, 10.9 kt N and 7.6 kt N, respectively corresponding to 14.5%. 10.1% and 7.0% contribution. Altogether HELCOM countries contribute 74% to reduced nitrogen deposition. In general contribution of the sources located close to the Baltic Sea is more significant in case of reduced nitrogen deposition than in case of oxidized nitrogen deposition. However, three distant sources: Belarus, France and Netherlands are also present on the Top Ten list for reduced nitrogen deposition.

Transportation and combustion are the main emission sectors contributing to oxidised nitrogen deposition, whereas, agriculture is the dominating emission sector contributing to reduced nitrogen deposition.

9.4 Conclusions

Atmospheric nitrogen deposition to the Baltic Sea basin has been calculated with the EMEP/MSC-W model for each year of the period 1995–2015. Compared to 1995 there is clear reduction of oxidized nitrogen deposition in the year 2015, both in annual (−33%) and normalized deposition (−35%). The situation is different in case of reduced nitrogen deposition. However, the normalised deposition is declining (12%) in 2014 compared to 1995 deposition, the annual deposition is 5% higher in 2014 than in 1995. Also, uncertainty related to variable meteorology is higher for reduced nitrogen deposition than for oxidized nitrogen deposition.

Germany and Poland are the main sources contributing to oxidized nitrogen deposition to the Baltic Sea basin with 23.4 kt N and 17.8 kt N, respectively. This correspond to 17.9% and 13.6% contribution, respectively. There is also a significant contribution from the ship traffic on the Baltic Sea (15.2 kt N) and North Sea (10.3 kt N), corresponding to 11.6% and 7.8%, respectively.

In case of reduced nitrogen deposition to the Baltic Sea basin, Germany dominates the list with 31.2 kt N corresponding to 28.8% contribution to annual deposition in 2014. Three other HELCOM countries follow Germany on the list: Poland, Denmark and Sweden with 15.7 kt N, 10.9 kt N and 7.6 kt N, respectively corresponding to 14.5%. 10.1% and 7.0% contribution.

Transportation and combustion are the main emission sectors contributing to oxidized nitrogen deposition, whereas, agriculture is the dominating emission sector contributing to reduced nitrogen deposition.

Questions and Answers

Questioner name: Valerie Garcia

Q: Were you able to discern sources of the P inputs into the Baltic Sea from rivers?

A: Only atmospheric input is discussed in this study, but the information about P inputs into the Baltic Sea from rivers is available in the HELCOM reports.

Questioner name: Eric van der Swaluw

Q: The normalized reduced nitrogen deposition is going down, whereas the reduced nitrogen deposition is going up. Did you do a statistical test of the significance of the two trends over the period considered?

A: Yes. The statistical tests indicate that there is no trend in annual deposition of reduced nitrogen and that there is a declining trend in normalized deposition of reduced nitrogen with significance level 0.001.

Acknowledgements The HELCOM Convention has financed a part of the work presented here. The author is indebted to the scientific team at MSC-W for their help and inspiring discussions.

References

1. J. Bartnicki, *Atmospheric deposition to the Baltic Sea*. Monographs of the Institute of Meteorology and Water Management. Warsaw, Poland, 126 p. (2014)
2. J. Bartnicki, A. Gusev, W. Aas, M. Gauss, J.E. Jonson, Atmospheric supply of nitrogen, cadmium, mercury, lead, and PCDD/Fs to the Baltic Sea in 2015. EMEP Centres Joint Report for HELCOM. EMEP/MSC-W Technical Report 2/2017. Norwegian Meteorological Institute. Oslo, Norway. Available in the web: http://www.emep.int/publ/helcom/2017/index.html (2017)
3. M. Gauss, S. Tsyro, H. Fagerli, A.-G. Hjellbrekke, W. Aas, Acidifying and eutrophying components, Supplementary material to EMEP Status Report 1/2017, available online at www.emep.int, The Norwegian Meteorological Institute, Oslo, Norway (2017)
4. D. Simpson, A. Benedictow, H. Berge, R. Bergström, L.D. Emberson, H. Fagerli, C.R. Flechard, G.D. Hayman, M. Gauss, J.E. Jonson, M.E. Jenkin, A. Nyíri, C. Richter, V.S. Semeena, S. Tsyro, J.-P. Tuovinen, Á. Valdebenito, P. Wind, The EMEP MSC-W chemical transport model—technical description. Atmos. Chem. Phys. **12**, 7825–7865 (2012). https://doi.org/10.5194/acp-12-7825-2012
5. D. Simpson, R. Bergström, H. Imhof, P. Wind, Updates to the EMEP/MSC-W model, 2016–2017. *In Transboundary particulate matter, photo-oxidants, acidifying and eutrophying components*. EMEP Status Report 1/2017. The Norwegian Meteorological Institute, Oslo, Norway (2017)

Chapter 10
Modelling the Concentration of Ammonia and Exceedance of the Critical Level in the UK

Anthony Dore, Jane Hall, Ed Rowe, Oliver Pescott, Edward Carnell, Samuel Tomlinson, Ulrike Dragosits, Sim Tang, Janet Simkin, Amy Stephens, Christine Braban, William Bealey and Mark Sutton

Abstract The FRAME atmospheric chemistry transport model was applied to calculate the concentration of NH_3 at a 1 km resolution over the UK. The results showed that the 1 μg m^{-3} critical level for NH_3 was exceeded for 60% of the UK land area and the 3 μg m^{-3} critical level was exceeded for 3% of the land area. Model simulations using historical emissions suggested that average NH_3 concentrations in the UK have increased by a factor of 2.5 between 1970 and 2010 due to both an increase in NH_3 emissions and large reductions in SO_2 emissions causing a reduction in the availability of H_2SO_4 to react with NH_3.

10.1 Introduction

Atmospheric nitrogen deposition is known to pose a global threat to biodiversity. Whilst levels of NO_x emissions have declined significantly in recent decades in many European countries, efforts to reduce atmospheric emissions of NH_3 have been less effective. The critical level is defined as a threshold value above which significant harmful effects may occur. For the impact of NH_3 on natural ecosystems, the critical level is set by the United Nations Economic Comission for EuropE at 1 μg m^{-3} for lichens and bryophytes and 3 μg m^{-3} for all other plants [1].

A. Dore (✉) · E. Carnell · S. Tomlinson · U. Dragosits · S. Tang · A. Stephens · C. Braban · W. Bealey · M. Sutton
Centre for Ecology & Hydrology, Edinburgh, Scotland, UK
e-mail: todo@ceh.ac.uk

J. Hall · E. Rowe
Centre for Ecology & Hydrology, Bangor, Wales, UK

O. Pescott
Centre for Ecology & Hydrology, Wallingford, UK

J. Simkin
Newcastle University, Newcastle upon Tyne, UK

C. Mensink et al. (eds.), *Air Pollution Modeling and its Application XXVI*, Springer Proceedings in Complexity,
https://doi.org/10.1007/978-3-030-22055-6_10

10.2 Methodology

The Fine Resolution Atmospheric Multi-pollutant Exchange (FRAME) model is a Lagrangian atmospheric chemistry transport model which uses annually averaged meteorology (wind direction frequency rose and annual precipitation map) to calculate annual average deposition of nitrogen and sulphur as well as gas and particulate concentrations. The model uses spatially distributed emissions of NH_3, NO_x and SO_2 and simulates gas to particle conversion through both dry phase and aqueous phase chemistry. Dry and wet deposition to the surface are simulated using scavenging coefficients and a canopy resistance parameterisation respectively. The model can be run at either a 1 km or a 5 km resolution over the British Isles.

The model was run at a 5 km resolution using historical estimates of emissions of NH_3, NO_x and SO_2 for the UK for the years 1970, 1990, 2010 and future projections of emissions for the year 2030. During more recent years high resolution (1 km) emissions data have become available and the model was run at 1 km resolution for the years 2009, 2010, 2011, 2012, 2013 and 2014. The results were used to calculate long-term trends in NH_3 concentrations in the UK and the area of the country with exceedance of the critical level for NH_3. Evaluation of model performance was undertaken by comparison with measurements from the United Kingdom Eutrophying and Acidifying Pollutants monitoring network (https://uk-air.defra.gov.uk/networks/network-info?view=ukeap).

10.3 Results

The spatial distribution of NH_3 concentrations modelled at a 1 km resolution for the UK is illustrated in Fig. 10.1 with the correlation with measurements of annually averaged NH_3 concentration from 87 sites. NH_3 concentrations are highest in the intensive agricultural regions of England and Northern Ireland and lowest in the remote upland regions of northern Scotland. The correlation with measurements gave a Pearson correlation coefficient of $r = 0.71$ and a normalised mean bias of 0.17. Although the model was run at a relatively high spatial resolution of 1×1 km, there is still considerable sub-grid variability in the NH_3 concentration in the rural environment due to the uneven spatial distribution of agricultural land and natural ecosystems. However Hallsworth et al [3] demonstrated that the model achieved a better correlation with measurements of NH_3 in semi-natural areas with 1×1 km resolution modelled data than with 5×5 km resolution data. Although the model employs relatively simple dynamic, chemical and dry deposition schemes compared to more complex Eulerian models, it was able to obtain a good correlation with NH_3 concentrations compared with other models as part of a model inter-comparison exercise [2]. This good performance could be attributed to the high vertical resolution in the model of 1 m near the surface.

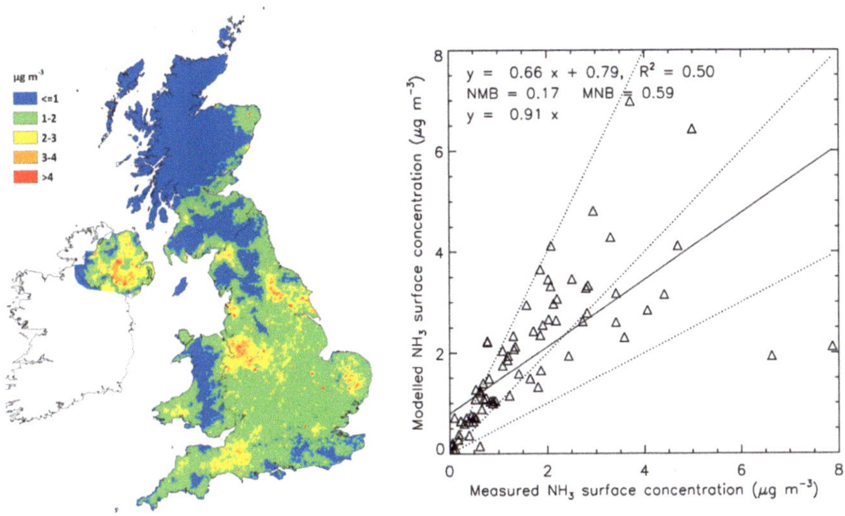

Fig. 10.1 Modelled annually averaged concentration of NH_3 in the UK for the years 2012–14. (left); Model correlation with measurements of NH_3 concentration for the year 2013 (right)

The 1 × 1 km resolution data were used to calculate the exceedance of the critical level as a three-year rolling average during the period 2009–2014. Summary statistics are shown in Table 10.1 for 2012–14. This demonstrates that large areas of the UK (60%), as well as nature reserves, have NH_3 concentrations which exceed the 1 $\mu g\ m^{-3}$ critical level for sensitive species, with a significantly lower area (2.7%) exceeding the higher 3 $\mu g\ m^{-3}$ critical level for all other species.

The risk posed by NH_3 to natural ecosystems is dependent on the spatial distribution of habitat types that are of conservation value in relation to agricultural areas with high NH_3 emissions. As illustrated in Fig. 10.2 conservation habitats associated with lowland regions (calcareous grassland, broadleaf, beech, oak and other wood) have large areas with exceedance of the 1 $\mu g\ m^{-3}$ critical level whereas those associated with upland ecosystems (acid grassland, heath, bog and montane) have much smaller areas with exceedance.

The historical simulations of NH_3 concentration in the UK showed that whilst there was a 49% increase in NH_3 emissions between 1970 and 1990, average

Table 10.1 % area of the UK land areas with exceedance of the 3 $\mu g\ m^{-3}$ and 1 $\mu g\ m^{-3}$ critical levels for 2012–2014 (SAC: Special Area of Conservation, SPA: Special Protection Area, SSSI: Site of Special Scientific Interest)

Metric	1 $\mu g\ m^{-3}$	3 $\mu g\ m^{-3}$
% UK land area	60	2.7
% N-sensitive habitats	22	0.5
% SACs	58	5.4
% SPAs	50	6.1
% SSSIs	68	2.7

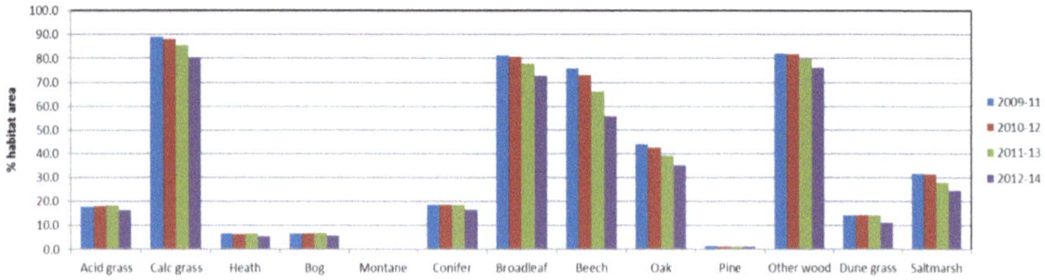

Fig. 10.2 % area of broad habitats in the UK with exceedance of the 3 μg m^{-3} critical level for NH_3 during 2009–2014

NH_3 concentrations in the UK increased by 120%. Emissions of NH_3 fell by 17% between 1990 and 2010 during which period average NH_3 concentrations remained unchanged. The model results were in overall agreement with analysis of measured NH_3 concentrations from the national monitoring network (87 sites) which also showed no overall statistically significant change in NH_3 concentration between 1998 and 2014 [4]. This non-linear response of NH_3 concentration to emission change was attributed to the major decrease in SO_2 emissions (by 94%) between 1970 and 2010 resulting in a decreased availability of H_2SO_4 to react with NH_3. Future projections for the year 2030 suggest that with little change in NH_3 emissions, but SO_2 emissions continuing to decline, national NH_3 concentrations will increase by 10% relative to 2010.

10.4 Conclusion

The results from an atmospheric chemistry transport model showed that the NH_3 concentration exceeds 1 μg^{-3} over 60% of the UK land area, posing a threat to sensitive species such as lichens and bryophytes, particularly in the lowland intensive agricultural regions of England and northern Ireland. Historical simulations indicate that average NH_3 concentrations in the UK have increased by a factor of 2.2 since 1970, driven by both increased NH_3 emissions and decreased SO_2 emissions. Preliminary analysis of the spatial distribution of bryophytes however shows that species richness is highest in the more humid upland and west coastal regions of the UK which tend to coincide with areas of low agricultural activity where NH_3 concentrations are below the 1 μg^{-3} critical level.

Questions and Answers

Questioner Name: **Clemens Mensink**

Q: Were there any abatement measures taken in the UK over the last twenty years and if so did you see any impact?

A: Reductions in ammonia emissions during the last 20 years have been caused mostly by economic factors leading to a reduction in livestock numbers. At a national scale monitoring data reveals no significant trends in NH_3 concentrations. However in areas dominated by pig and poultry emissions, where emissions reductions were highest, measurements showed a decrease in NH_3 concentrations.

Questioner Name: **Stefano Alessandrini**
Q: Do you simulate ammonia emissions one by one? How do you deal with background substances reacting with ammonia in the model? Why don't you perform the bias correction site by site? The bias correction might be site specific?
A: Ammonia emissions are distributed spatially by livestock category as well as fertiliser and non-agricultural emissions. See below regarding the chemical reaction scheme. There could be a justification to use a spatially variable correction to the modelled ammonia concentrations, based on geography and climatology or indeed dominant livestock category, though this hasn't been checked However the bias for any two model-measurement points is strongly influenced by the precise location of the measurement site and the highly variable local ammonia concentration at a model sub-grid resolution. The bias for all sets of paired model-measurement points was therefore used collectively to obtain an overall statistically significant bias.

Questioner Name: **Ted Russell**
Q: In the U.S. ammonia emissions have increased along with a decrease in acidic precursors, and ammonia levels have stayed relatively constant, similar to your code. If one looks at the time scale analysis, the deposition of ammonia is rather fast, faster than the lifetime of ammonia sulphate. This suggests that the ammonia gas levels would be in a pseudo state between emissions and deposition, buffering any major change in ammonia gas concentration changes. Does your model have a thermodynamic aerosol module and how does it show ammonia gas changes when sulphate and nitrate is reduced?
A: In the model H_2SO_4 and HNO_3 are generated by dry and aqueous phase oxidation reactions. NH_3 reacts instantaneously to form ammonium sulphate in a one-way reaction whilst the reaction between HNO_3 and NH_3 to form ammonium nitrate is represented as a two way equilibrium reaction. The sensitivity of ammonia concentration and deposition to changes in NO_x and SO_2 emissions has not been calculated specifically. However on a national scale both the model and measurements broadly suggest that a reduction of 20% in NH_3 emissions corresponds to no change in NH_3 concentrations whilst SO_2 and NO_x emissions have decreased significantly during this period.

Acknowledgements This work was funded by the UK Natural Environment Research Council and the UK Department for Environment, Food and Rural Affairs.

References

1. J.N. Cape, L.J. van der Eerden, L.J. Sheppard, I.D. Leith, M.A. Sutton, Evidence for changing the Critical Level for ammonia. Environ. Pollut. **157**(3), 1033–1037 (2009)
2. A.J. Dore, D. Carslaw, C. Braban, C. Cain, C. Chemel, C. Conolly, R.G. Derwent, S.J. Griffiths, Hayman G. Hall, S. Lawrence, S.E. Metcalfe, A. Redington, D. Simpson, M.A. Sutton, P. Sutton, Y.S. Tang, M. Vieno, M. Werner, J.D. Whyatt, Evaluation of the performance of different atmospheric chemical transport models and inter-comparison of nitrogen and sulphur deposition estimates for the UK. Atmos. Environ. **119**, 131–143 (2015)
3. S. Hallsworth, A.J. Dore, W.J. Bealey, U. Dragosits, M. Vieno, S. Hellsten, Y.S. Tang, M.A. Sutton, The role of indicator choice in quantifying the threat of atmospheric ammonia to the 'Natura 2000' network. Environ. Sci. Policy **13**, 671–687 (2010)
4. Y.S. Tang, C.F. Braban, U. Dragosits, A.J. Dore, I. Simmons, N. van Dijk, R.I. Smith, J. Poskitt, M.G. Pereira, P.O. Keenan, H. Carter, C. Conolly, K. Vincent, R.I. Smith, M.R. Heal, M.A. Sutton, Drivers for spatial, temporal and long-term trends in atmospheric ammonia and ammonium in the UK. Atmos. Chem. Phys. **18**, 705–733 (2018)

Chapter 11
Stratospheric Age-Of-Air and SF6 Simulations with Silam

Rostislav Kouznetsov, Mikhail Sofiev, Julius Vira and Gabriele Stiller

Abstract The spatial distribution of Age of Air (AoA) in the stratosphere has been extensively used to evaluate and compare general circulation and chemical transport models. We performed multi-decade simulations of SF6 and several AoA tracers in the atmosphere with Silam chemistry-transport model driven with ERA-Interim reanalysis. The resulting distributions of AoA agree well with each other and with simulations made earlier, however disagree with AoA derived from observed SF6 concentrations in polar areas. The simulations of SF6 were made in two variants: fully passive passive, and with account for mesospheric depletion and for gravitational separation. The results indicate a good agreement with observations by MIPAS satellite instrument for non-passive SF6, and have substantial discrepancies in polar areas for passive SF6. Thus we conclude that the discrepancy between SF6 and modelled AoA originates from the violation of the assumption of the SF6 passiveness.

11.1 Introduction

Age of air (AoA) at some point in stratosphere is defined as the average time elapsed since molecules of air there had last been in troposphere. The spatial distribution of AoA in the stratosphere is a good indicator of global atmospheric circulation and had been extensively used to evaluate and compare general circulation and chemical transport models in the stratosphere [10]. In particular, numerous studies use AoA as a diagnostics of stratospheric circulation.

R. Kouznetsov (✉) · M. Sofiev
Finnish Meteorological Institute, Helsinki, Finland
e-mail: Rostislav.Kouznetsov@fmi.fi

R. Kouznetsov
Obukhov Inst. of Atmospheric Physics, Moscow, Russia

J. Vira
Cornell University, Ithaca, NY, USA

G. Stiller
Karlsruhe Institute of Technology, Karlsruhe, Germany

C. Mensink et al. (eds.), *Air Pollution Modeling and its Application XXVI*,
Springer Proceedings in Complexity,
https://doi.org/10.1007/978-3-030-22055-6_11

AoA can be inferred form the observations of some tracer that changes in the troposphere and then remains unchanged along the transport of an air parcel in the stratosphere. One of such tracers is sulphur hexafluoride SF_6 that is practically passive in the troposphere and in stratosphere. SF_6 accumulates in the atmosphere and have increased by more than an order of magnitude since 1960s [7], and thus the delay between SF_6 mixing ratio in troposphere and in arbitrary point in stratosphere gives AoA at that point. The AoA derived from observations of an atmospheric tracer in assumption of its passivity is called "apparent AoA", which is subject for corrections to get "true" AoA [3].

There exist substantial discrepancies between the AoA derived from observations and AoA derived from various modelling approaches: the experimental studies [3] report ages derived from SF_6 observations that exceed 10 years in polar regions, whereas the model studies [2, 6] report them below 6 years. In the current study we used the SILAM Chemistry-Transport Model (CTM) driven by ERA-interim reanalysis to simulate both AoA and the procedure of its evaluation from observations of SF_6 distribution in the stratosphere. The main goal was to identify the causes of disagreement between simulations and observations of AoA and quantify relative importance of mesospheric depletion of SF_6, gravitational separation of SF_6 in upper stratosphere for accurate inferring of AoA from observations.

11.2 Materials and Methods

SILAM (System for Integrated modeLling of Atmospheric coMposition) is an off-line chemistry-transport model [8]. Global Eulerian simulations were performed for the period of 1980–2015 with 1.44° resolution. The model was driven with **ERA-interim reanalysis** [1], that has T255 spectral resolution ($\sim$0.72°) and 60 hybrid sigma-pressure levels [1], having nominal pressure at the uppermost level of 10 Pa. Same levels were used in SILAM, except for the uppermost layer bounded with pressure levels of 20 and 10 Pa.

In this study two SF_6 tracers are used: fully passive "sf6pass" and "sf6" that is subject for gravitational separation and destruction in the mesosphere, that was implemented via effective lifetime of 30 days in the uppermost model layer. Both tracers had the same emissions [7]. The inventory covers 1970–2008, and has been extrapolated with linearly growing trend until 2015. The reference AoA was derived from "ideal age" [10], a tracer whose mixing ratio linearly increases with time everywhere and continuously forced to zero at the surface.

The simulation results has been evaluated with the results of SF_6 retrieval from the limb-viewing **MIPAS** instrument [9] operated on-board of the Envisat satellite in 2002-2012. The conversion of retrieved SF_6 to AoA was performed by referring to a global mean of in situ SF_6 measurements at the surface as provided by NOAA/ESRL [4].

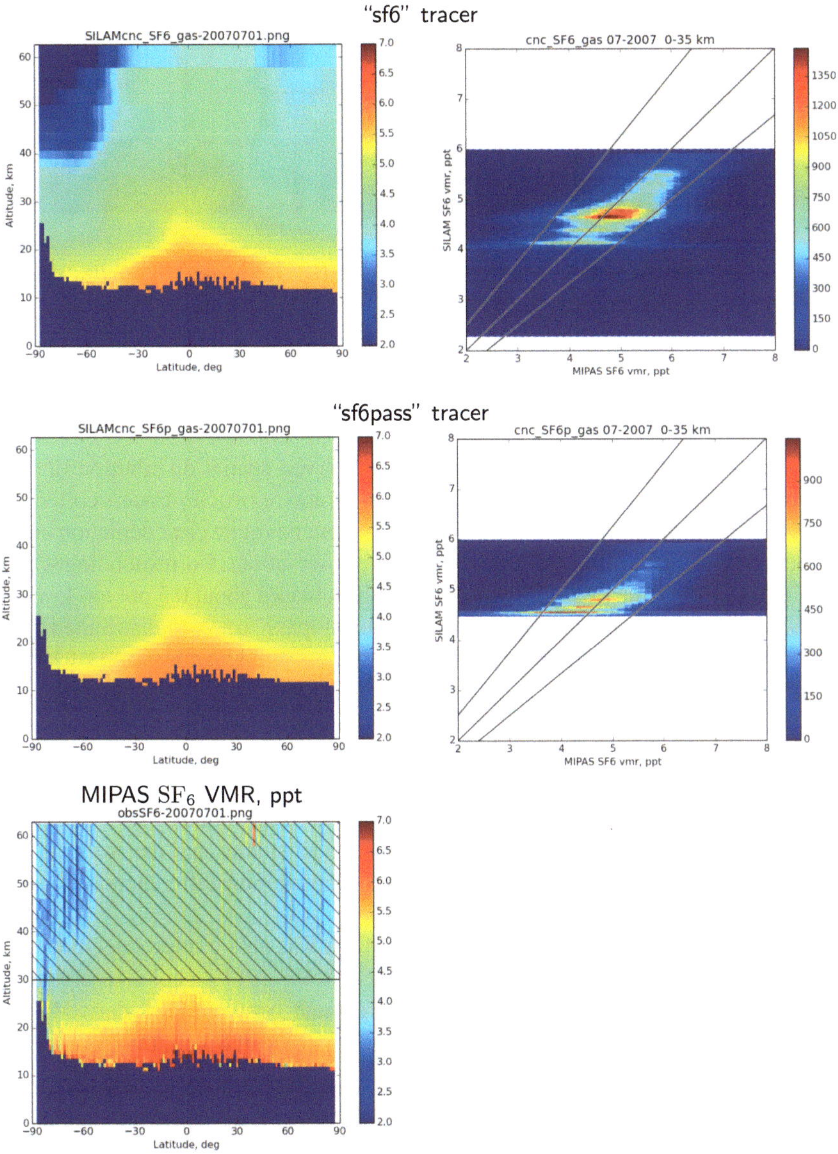

Fig. 11.1 SF_6 mixing ratios (in ppt) simulated and observed with MIPAS in June 2007, and corresponding scatter plots. The color-bars at scatter plots give number of the collocated profile points for each cluster. Lines indicate the MIPAS instrument noise error: 20% range of the retrieved mixing ratios [3]. The values from above 30 km, not recommended for use by KIT-IMK team, are shaded

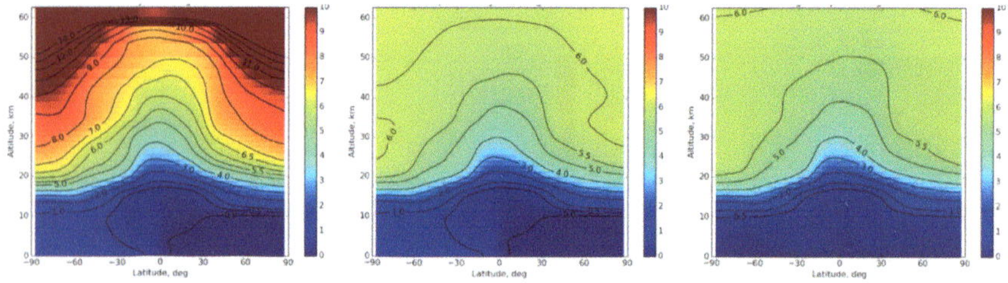

Fig. 11.2 Zonal mean distributions of the apparent AoA (in years) for 2007 "sf6" (left), "sf6pass" tracers and AoA from ideal-age tracer (right)

11.3 Results

The simulations were able to reproduce the observed spatial distribution of SF_6. Fig. 11.1 shows the zonal-mean of volume mixing ratio of two SF_6 tracers collocated with MIPAS retrievals for June 2007. The "sf6" tracer has quite clear depletion inside the polar vortex that qualitatively agrees with the observations. Corresponding scatter plot indicates a good agreement with model values biased about 0.5 ppt too low, but still well within the errors of observations. The "sf6pass" tracer is distributed quite uniformly in the upper stratosphere. I has much smaller dynamic range of VMRs than "sf6" and than observed SF_6.

The apparent Age-Of-Air derived from the "sf6pass" tracer (Fig. 11.2, left) agrees well with AoA derived from MIPAS observations [3]. It indicates older than 10 years air in polar stratosphere, which is much higher than values reported by all model studies we are aware of. The apparent Age-Of-Air derived from the "sf6" (Fig. 11.2, center) is much smaller, and agrees with other model studies and the AoA derived from "ideal-age" tracer (Fig. 11.2, right). The small remaining differences are caused by non-linear growth of VMR of SF6 with time.

The following **conclusions** can be made. Silam was able to reproduce both SF_6 distribution in stratosphere and the AoA. The effects of SF_6 depletion are much stronger than it was assumed in [3]. The agreement between AoA derived from "sf6pass" and from "ideal-age" indicates the consistency of Silam transport.

QUESTION: You have calculated mean age with several methods using forward simulations. One could, however, obtain the age spectra with inverse simulations from Green functions, in a similar manner as it was done by Holzer (1999) [5]. Is that method applicable to your study?

ANSWER: Indeed from Green functions age spectra can be calculated. It would be, however, too expensive computationally to calculate temporal evolution of 3D fields of age spectra.

References

1. D.P. Dee, S.M. Uppala, A.J. Simmons, P. Berrisford, P. Poli, S. Kobayashi, U. Andrae, M.A. Balmaseda, G. Balsamo, P. Bauer et al., The ERA-Interim reanalysis: configuration and performance of the data assimilation system. Q. J. Roy. Meteorol. Soc. **137**(656), 553–597 (2011). https://doi.org/10.1002/qj.828
2. M. Diallo, B. Legras, A. Chdin, Age of stratospheric air in the era-interim. Atmos. Chem. Phys. **12**(24), 12133–12154 (2012). https://doi.org/10.5194/acp-12-12133-2012
3. F.J. Haenel, G.P. Stiller, T. von Clarmann, B. Funke, E. Eckert, N. Glatthor, U. Grabowski, S. Kellmann, M. Kiefer, A. Linden, T. Reddmann, Reassessment of mipas age of air trends and variability. Atmos. Chem. Phys. Discuss. **15**(10), 14685–14732 (2015). https://doi.org/10.5194/acpd-15-14685-2015
4. B.D. Hall, G.S. Dutton, D.J. Mondeel, J.D. Nance, M. Rigby, J.H. Butler, F.L. Moore, D.F. Hurst, J.W. Elkins, Improving measurements of SF_6 for the study of atmospheric transport and emissions. Atmos. Measurement Techn. **4**(11), 2441–2451 (2011). https://doi.org/10.5194/amt-4-2441-2011
5. M. Holzer, Analysis of passive tracer transport as modeled by an atmospheric general circulation model. J. Climate **12**(6), 1659–1684 (1999). https://doi.org/10.1175/1520-0442(1999)012<C1659:AOPTTA>2.0.CO;2
6. M. Krol, M. de Bruine, L. Killaars, H. Ouwersloot, A. Pozzer, Y. Yin, F. Chevallier, P. Bousquet, P. Patra, D. Belikov, S. Maksyutov, S. Dhomse, W. Feng, M.P. Chipperfield, Age of air as a diagnostic for transport time-scales in global models. Geoosci. Model Devel. Discuss. **2017**, 1–33 (2017). https://doi.org/10.5194/gmd-2017-262
7. M. Rigby, J. Mhle, B.R. Miller, R.G. Prinn, P.B. Krummel, L.P. Steele, P.J. Fraser, P.K. Salameh, C.M. Harth, R.F. Weiss et al., History of atmospheric SF6 from 1973 to 2008. Atmos. Chem. Phys. **10**(21), 10305–10320 (2010). https://doi.org/10.5194/acp-10-10305-2010
8. M. Sofiev, J. Vira, R. Kouznetsov, M. Prank, J. Soares, E. Genikhovich, Construction of the silam Eulerian atmospheric dispersion model based on the advection algorithm of michael galperin. Geosci. Model Dev. **8**(11), 3497–3522 (2015). https://doi.org/10.5194/gmd-8-3497-2015
9. G.P. Stiller, T. von Clarmann, M. Höpfner, N. Glatthor, U. Grabowski, S. Kellmann, A. Kleinert, A. Linden, M. Milz, T. Reddmann, T. Steck, H. Fischer, B. Funke, M. López-Puertas, A. Engel, Global distribution of mean age of stratospheric air from MIPAS SF_6 measurements. Atmos. Chem. Phys. **8**(3), 677–695 (2008). https://doi.org/10.5194/acp-8-677-2008
10. D. Waugh, Age of stratospheric air: Theory, observations, and models. Reviews of Geophysics **40**(4) (2002). https://doi.org/10.1029/2000rg000101

Chapter 12
Spatio-Temporal Monitoring and Modelling of Birch Pollen in Belgium

Andy Delcloo, Willem W. Verstraeten, Sebastien Dujardin, Nicolas Bruffaerts, Marijke Hendrickx, Rafiq Hamdi and Mikhail Sofiev

Abstract Air quality is primordially affected by anthropogenic emissions and has a tremendous impact on human health with more than 6 million premature deaths worldwide in 2015 (Landrigan et al. in The Lancet Commission on pollution and health, 2017) [3]. Biogenic emissions of aerosols such as pollen also impact the human wellbeing. The industrialized world suffers from a global increase in the burden of allergic respiratory diseases. Air pollution can influence both allergens and allergic subjects by increasing the immune reaction, and/or by an intensified biogenic emissions. In Europe, a quarter of the population suffers from pollinosis, whereas in some countries the prevalence is over 40%. To date, pollen of various trees and grasses in Belgium are monitored by the Belgian Scientific Institute for Public Health (Sciensano) at five stations on a daily basis. This sparse sampling cannot cover the spatial representativeness of the airborne pollen. Chemistry Transport Models (CTM's) are therefore an interesting tool to both quantify and forecast its spatial and temporal distribution. Here we show the results of the spatio-temporal modelled birch pollen over Belgium using the CTM SILAM. This model is driven by 2008 ECMWF meteorological data and a MACC-III birch tree fraction map showing the spatial distribution of potential pollen sources. Pollen modelling is based on the temperature degree days approach.

A. Delcloo (✉) · W. W. Verstraeten · R. Hamdi
Royal Meteorological Institute of Belgium, Brussels, Belgium
e-mail: Andy.Delcloo@meteo.be

A. Delcloo · R. Hamdi
Department of Physics and Astronomy, Ghent University, Ghent, Belgium

S. Dujardin
Department of Geography, University of Namur, Namur, Belgium

N. Bruffaerts · M. Hendrickx
Mycology and Aerobiology Unit, Sciensano, Brussels, Belgium

M. Sofiev
Finnish Meteorological Institute (FMI), Helsinki, Finland

C. Mensink et al. (eds.), *Air Pollution Modeling and its Application XXVI*,
Springer Proceedings in Complexity,
https://doi.org/10.1007/978-3-030-22055-6_12

12.1 Introduction

12.1.1 Methods and Data

Pollen are biogenic aerosols with a diameter of typically 5–50 times larger than conventional atmospheric aerosols depending on vegetation type. Birch pollen can be seen as a PM22 aerosol. Chemistry transport models such as SILAM (System for Integrated modeLling of Atmospheric coMposition) are able to simulate the dispersion of pollen based on processes such as wind advection (transport with air masses), mixing due to turbulence, gravitational settling (dry deposition), and scavenging with precipitation (wet deposition) [6].

The simulation of birch pollen levels in the air requires the quantification of the spatio-temporal emission sources of birch pollen at the surface in the model domain. Stated otherwise, a map with the areal fraction of birch trees is highly necessary as an underlying data set or input. At the European scale such a map was first compiled by Sofiev et al. [5]. Another issue in modelling pollen levels is the evaluation of pollen emission timing and intensity. Pollen modelling with SILAM is based on the temperature degree days approach or the thermal time flowering model. The parameterization of flowering follows a principle of two thresholds (start and end of the flowering season) for the temperature sum [4], which assumes that the timing of birch flowering is mostly driven by accumulated ambient temperature during a certain time period. The cumulative fraction of pollen released from the beginning of a year until a certain time is piecewise linear and proportional to the temperature sum during the main flowering season. Short-term meteorological conditions such as wind speed, relative humidity and precipitation rate will affect the amount of pollen in the air. Precipitation and humidity suppress the pollen release and threshold values are used to compute reduction factors. Typically, the lower and upper thresholds from relative humidity are 50 and 80%. For precipitation the lower and upper thresholds are 0 and 0.5 mm h^{-1} (the grid cell average rate). At the saturation wind speed of 5 m s^{-1} the pollen release rate is maximal. At a wind speed around 1 m s^{-1}, no pollen emissions occur.

12.1.2 Observational Data

ECMWF ERA-INTERIM meteorological data is used to drive the transport model. Pollen levels simulated by SILAM are evaluated using observations of pollen data taken from the Belgian Scientific Institute for Public Health (Sciensano). In 2008 SIPH monitored pollen concentrations at four stations (De Haan: 50.824523 N, 4.382457 E; Antwerp: 51.212683 N, 4.397888 E; Brussels: 50.824523 N, 4.382457 E; Charleroi: 50.408 N, 4.444 E) on a daily basis.

From Flemish and Walloon forest inventory data the relative diameter at breast height (DBH) for birch trees are derived at each sampling plot (11.080 plots in the Walloon region, 2.147 plots in the Flemish region) [1]. Based on the statistical relationship between DBH and birch tree canopy coverage [2], the areal fraction at each inventory site was assessed.

12.2 Results and Discussions

Since pollen levels in SILAM are determined by the abundance of birch trees in Belgium, having an up-to-date map of birch tree fractions is essential. Figure 12.1 illustrates the spatial distribution of potential birch pollen sources using the reference map and the updated version using forest inventory data.

Using a first preliminary version of updated birch tree fraction map for Belgium and by updating the starting and ending dates of the birch pollen season into the SILAM transport model, the correlation (R^2) of the observed versus the modelled pollen levels increased with 23% on average with respect to the MACC birch map. Figure 12.2 shows the time series of observed (Sciensano) and modelled (SILAM RMI) pollen counts for the birch pollen season of 2008. Large overestimations are observed at the location of De Haan which is situated at the coastline. Large pollen underestimations occur over Antwerp. The very high observed peaks in pollen concentrations are not well captured by the model. Backward trajectory analysis shows that the surface air masses that reach Antwerp originates from regions in Middle Europe where high birch fractions are more likely.

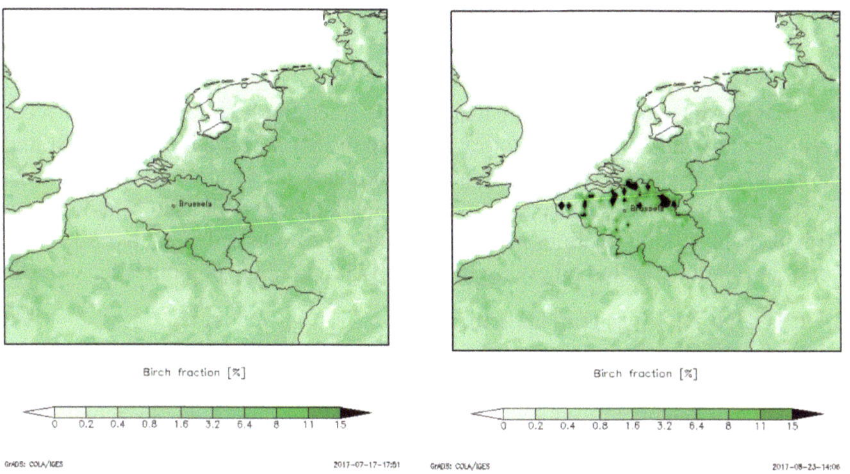

Fig. 12.1 Left panel, the birch fraction map accordingly to the MACC project; Right panel, a preliminary updated version for the Belgian territory using Flemish and Walloon forest inventory data

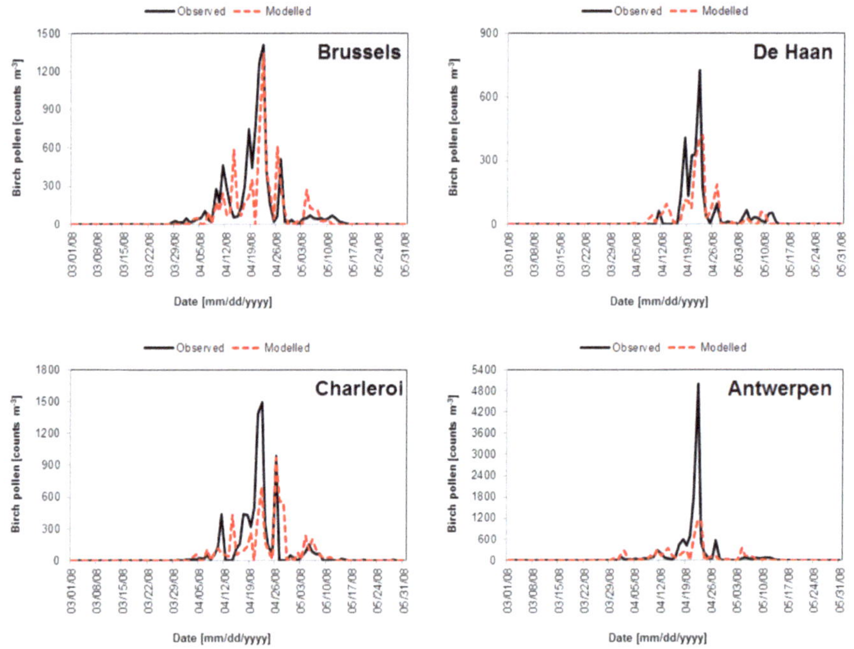

Fig. 12.2 Time series of observed (SIPH) (black line) and modelled (SILAM RMI) (dotted red line) pollen counts for the pollen season of 2008 (March–May) at four locations in Belgium using the updated birch tree fraction map

12.3 Conclusions

We provide an update of the areal birch tree fractions based on the most recent Flemish and Walloon forest inventory data into the SILAM model to better estimate the pollen concentrations in Belgium. A preliminary backward trajectory analysis, however, suggests that surface air masses originating from regions in Middle Europe with likely high birch fractions might contribute to unexplained high observed pollen peaks in Belgium. Stated otherwise, not all observed pollen are domestic.

QUESTIONER: Silvia Trini Castelli
QUESTION: With reference to the plume arriving in Belgium and traced back as transported from Easter European birch forests, given the long distance one could expect that pollen may be intercepted, blocked and even deposited. So, actually, how much of the released pollen may effectively reach Belgium?
ANSWER: This can be estimated by applying a source-receptor sensitivity study. This requires a lot of additional computations and is beyond the scope of this study. The SILAM model takes already into account dry- and wet deposition processes and also considers both local as well as transported emissions from far/short range pollen emissions. The back trajectory analysis only gives an indication on the origin of the air masses, which may or may not contain birch pollen.

QUESTIONER: Heinke Schlünzen
QUESTION: How do you determine the total number of pollen emitted per catkin (tree) ahead of the year, since the total number is quite different between different years?
ANSWER: In this study we only simulated 2008. We are working on a new methodology to estimate these inter-seasonal variations among different successive years. For now, a fixed potential flux value is set as default.

Acknowledgements This research was partly funded by the Belgian Science Policy Office (BELSPO) in the frame of the Belgian Research Action through Interdisciplinary Networks Brain (BRAIN.be) programme—project RespirIT (BR/154/A1/RespirIT) and partly funded by the Royal Meteorological Institute of Belgium. We acknowledge the discussions with the other members of the RespirIT.

References

1. S. Dujardin, C. Linard, N. Dendoncker, Research note on mapping allergenic tree species in Belgium. BR/154/A1/RespirIT, Belgian research action through interdisciplinary networks (2017)
2. G.E. Hemery, P.S. Savill, S.N. Pryor, Applications of the crown diameter–stem diameter relationship for different species of broadleaved trees. For. Ecol. Manag. **215**, 285–294 (2005)
3. P.J. Landrigan et al., The Lancet Commission on pollution and health (2017). http://dx.doi.org/10.1016/S0140-6736(17)32345-0. Accessed 19 Oct 2017
4. T. Linkosalo, H. Ranta, A. Oksanen, P. Siljamo, A. Luomajoki, J. Kukkonen, M. Sofiev, A double-threshold temperature sum model for predicting the flowering duration and relative intensity of Betula pendula and B. pubescens. Agric. For. Meteorol. **150**(12), 6–11 (2010). https://doi.org/10.1016/j.agrformet.2010.08.007
5. M. Sofiev, P. Siljamo, H. Ranta, A. Rantio-Lehtimäki, Towards numerical forecasting of long-range air transport of birch pollen: theoretical considerations and a feasibility study. Int. J. Biometeorol. **50**, 392–402 (2006). https://doi.org/10.1007/s00484-006-0027-x
6. M. Sofiev, P. Siljamo, H. Ranta, T. Linkosalo, S. Jaeger, A. Rasmussen, A. Rantio-Lehtimaki, E. Severova, J. Kukkonen, A numerical model of birch pollen emission and dispersion in the atmosphere. Description of the emission module. Int. J. Biometeorol. **2013**(57), 45–58 (2013). https://doi.org/10.1007/s00484-012-0532-z

Chapter 13
Development and Verification of a New Meteo-Dispersive Modelling System for Accidental Releases in the Italian Territory: SMART

Andrea Bisignano, Silvia Trini Castelli and Piero Malguzzi

Abstract A new modelling system, SMART, is under development for the simulation of accidental releases dispersion. The interfacing code ARAMIS was originally created to interface the non-hydrostatic atmospheric model MOLOCH and the Lagrangian stochastic dispersion model SPRAY. Here, a comparison between simulations with the new modelling suite SMART and the RMS modelling system, applied in several previous assessment studies, is presented for a case of a release in complex terrain in southern Italy. The new suite is planned to be adopted as a tool for emergency response purposes in any part of the Italian territory at any time.

13.1 Introduction

At the CNR-ISAC daily numerical forecasts are issued, by applying the meteorological models developed by the MODAT modelling group. The forecasts are produced in support of research on the atmospheric circulation and composition and in order to test and improve the meteorological models operating from the global to the local scales. At the regional scale and for the full Italian territory, MOLOCH model is producing forecasts with high spatial detail, grid size of 1.25 km, integrating the non-hydrostatic, fully compressible equations for the atmosphere (http://www.isac.cnr.it/dinamica/projects/forecasts/moloch/). Given the availability of these high-resolution forecasts, we developed a new interface, ARAMIS, between MOLOCH and the Lagrangian particle dispersion model SPRAY, and tested it in some case studies in different sites of Italy. The final goal is to build a modelling suite that can forecast the dispersion of possible accidental releases in any part of the Italian territory at any time.

A. Bisignano · S. Trini Castelli (✉)
CNR- Institute of Atmospheric Sciences and Climate, Turin, Italy
e-mail: s.trinicastelli@isac.cnr.it

P. Malguzzi
CNR- Institute of Atmospheric Sciences and Climate, Bologna, Italy

C. Mensink et al. (eds.), *Air Pollution Modeling and its Application XXVI*,
Springer Proceedings in Complexity,
https://doi.org/10.1007/978-3-030-22055-6_13

Here, preliminary simulations with the new SMART (Spray—Moloch Atmospheric Regional Tool) suite are compared to the results obtained with the RMS (RAMS-MIRS-SPRAY) modelling system for a case of one-day emissions in the complex terrain of the Agri valley in southern Italy. The investigation focuses on assessing the effect of the different meteorology on the pollutant dispersion. Comparisons between the meteorological variables, the plume dynamics and the concentration fields are presented and discussed.

13.2 The Case Study and the Simulations

In order to proceed with preliminary tests of the SMART suite, we considered a one-day case, November 30 in 2013, referring to a recent impact assessment study [2], where the RMS modelling system was applied to simulate the dispersion of the pollutant emitted by the COVA oil refinery plant in the Agri valley. This last is located in the south-western sector of Basilicata Region in southern Italy, it is orientated NW-SE and is bordered on both sides by the Apennine Mountains, thus it is characterized by complex and heterogeneous terrain.

Four nested 3D grids were used in RAMS atmospheric model, respectively with 48, 12, 4 and 1 km horizontal grid size, all with 35 levels on a stretched vertical grid (first level at 24 m, top of domain at 22 km). A one-year simulation was run for 2013 and to reduce the time needed to perform the yearly simulation, RAMS analysis fields were acquired from previous runs over Italy for the two coarse domains of 48 and 12 km resolution. They were then used as input and nudging on hourly basis for the two nested domains at 4 and 1 km resolution. This implies that the two-way nesting was not active from the two finest to the two coarsest grids. From RAMS fields on the finest 1-km resolution domain (45×30 km^2), MIRS boundary-layer code calculated the surface layer and turbulent variables needed by SPRAY Lagrangian particle dispersion model.

A selection of the MOLOCH outputs at the grid resolution of 1.25 km, stored for year 2013, was gathered for a domain 152×176 km^2 with 47 stretched vertical levels (first level at 33 m, top of domain at 11 km). In addition, a devoted MOLOCH run was performed using a finer resolution of 0.5 km on a domain 172×180 km^2, with the same vertical levels. In both cases, the new interfacing code ARAMIS was run to elaborate MOLOCH fields in the format useful for SPRAY model and to calculate the turbulent variables needed by it.

In SPRAY dispersion simulation, the Lagrangian particles were emitted every 30 s considering the emission from the different stacks as point sources. The number of the released particles is established to generate a minimum concentration associated at the single particle of 0.005 μg m^{-3} for NO_x, CO and SO_2. Such minimum value is appropriate for reproducing the concentrations with a good detail. The heights of the stacks vary between 12 and 33 m and high temperatures and exit velocities characterize the emissions, leading to strong plume rises in SPRAY model.

13.3 Results and Discussion

In Fig. 13.1 a comparisons between RAMS and MOLOCH outputs for temperature and wind speed is presented for two locations. The differences between the two models are not negligible, MOLOCH showing for this single day a better performance on average. RAMS produces more smoothed trend, probably connected to the approach used in this configuration, where the large-scale forcing from the coarse grids are not integrated by the small scale feedback from the finest grids. Differences are found also between MOLOCH outputs at the two resolutions, revealing a more enhanced variability when using 0.5 km resolution, as it can be expected.

In Fig. 13.2 the maps of the daily mean of the ground level concentration are plotted in the simulation domains for NO_x. Comparing the concentration distributions, we

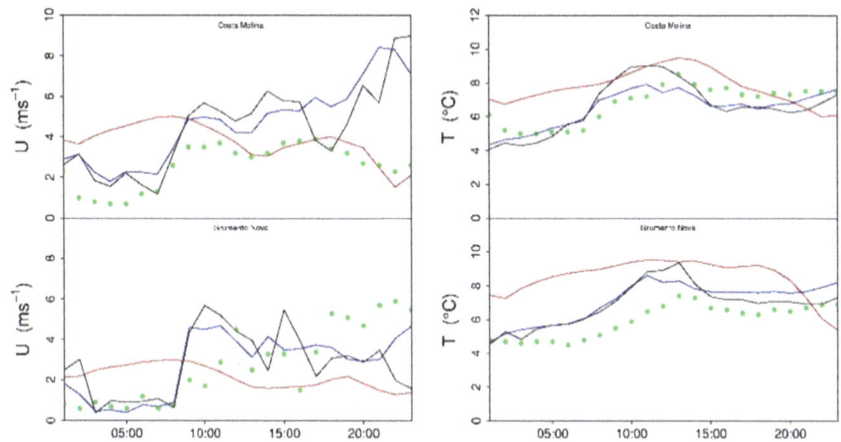

Fig. 13.1 Wind speed (left) and temperature (right) at Costa Molina and Grumento Nova. Green dots: observations; red line: RAMS; blue and black lines: MOLOCH at 1.25 and 0.5 km grid size

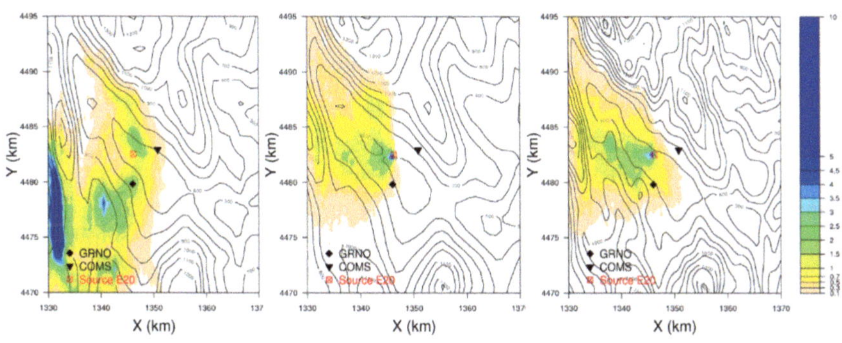

Fig. 13.2 NO_X ground level concentrations for RMS (left), SMART at 1.25 km (centre), SMART at 0.5 km (right). Grumento Nova (diamond) and Costa Molina (triangle) are also indicated

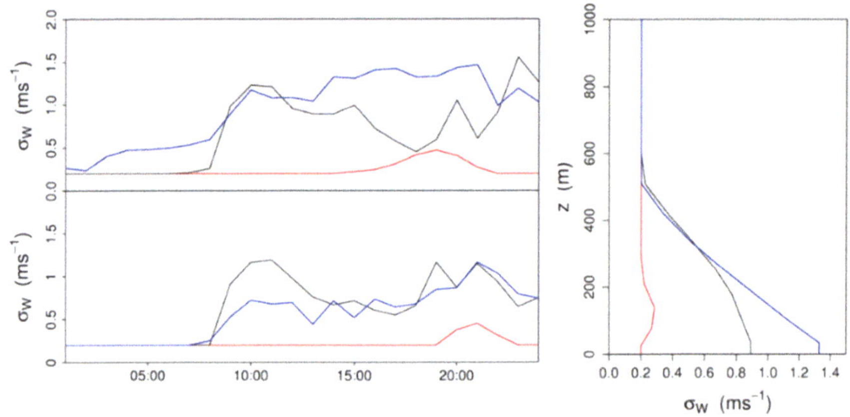

Fig. 13.3 Vertical wind velocity standard deviation at about 500 m from the source at two heights H (left) and as profiles at 14 a.m. (right). Red line: RMS ($H = 24$ m top and 291 m bottom), Blue and black lines: SMART ($H = 33$ m top and 254 m bottom) at 1.25 and 0.5 km grid size

notice that maximum values are found mostly far from the source for RMS and close to the source for SMART. RMS plumes are transported at high vertical layers, due to the plume rise; they are not much diffused and generate higher ground level concentration at larger distances from the plant sites, where they hit the orography. SMART plumes appear instead to be more diffused already close to the sources, in particular in the vertical. This leads to an efficient dispersion and lower concentration values are found on the slope of the mountains. This aspect is confirmed looking at the standard deviations of the vertical component of the wind velocity in Fig. 13.3, calculated respectively by MIRS and ARAMIS from the turbulent kinetic energy output by RAMS and MOLOCH, using the same formulations. RAMS diffusion is low, always close to a minimum value and with a peak in the late afternoon, while MOLOCH produces much larger values and a marked daily cycle. The low values of the turbulent kinetic energy in RAMS simulations, and their effects on the dispersion, were discussed in previous works [1, 3]. Since the turbulence parameterization is a key aspect for dispersion modelling, further investigations are ongoing to address the differences between RMS and SMART, evaluating their performances against observed data.

QUESTIONER: Denis Dionne.
QUESTION: Have you considered using your predictive model to allow industry to manage operations and demonstrate compliance or absence of nuisance.
ANSWER: The dispersion model SPRAY has been already adopted by environmental protection agencies and even industries as a modelling tool to monitor and assess the impact of pollutant releases connected to operational activities. The SMART modelling system is conceived to work on the full Italian territory as a tool for response and protection measures. As such, it will be run and handled by our Institute, which is a national public research institution. The objective is to provide the results

of the impact prediction and assessment to the civil protection officers and to the implicated industries, as support to the emergency response activities.

QUESTIONER: Shuzan Ren.
QUESTION: In your sensitivity test to different ABL parameterization schemes, do you use the same meteorological fields (wind and temperature)?
ANSWER: For the SMART suite, in the ARAMIS interfacing module we used the meteorological and turbulence fields provided by the MOLOCH model for applying different parameterizations of the wind velocity standard deviations. This allows addressing the differences between the schemes. Then, we have compared the same parameterizations in the RMS system but using the meteorological and turbulence fields output by the RAMS atmospheric model. This provides an intercomparison between the two modelling systems, SMART and RMS.

References

1. T. Hara, S. Trini Castelli, R. Ohba, C.J. Tremback, Validation studies of turbulence closure schemes for high resolutions in mesoscale meteorological models—a case of gas dispersion at the local scale. Atmos. Environ. **43**, 3745–3753 (2009)
2. C. Mangia, A. Bisignano, M. Cervino, L. Mortarini, S. Trini Castelli. Modeling air quality impact of pollutants emitted by an oil/gas plant in complex terrain in view of a health impact assessment. Air Qual. Atmos. Health **12**(4), 491–502 (2019)
3. S. Trini Castelli, T. Hara, R. Ohba, C.J. Tremback, Validation studies of turbulence closure schemes for high resolutions in mesoscale meteorological models. Atmos. Environ. **40**, 2510–2523 (2006)

Chapter 14
Comprehensive Modelling and Visualization of Particulate Matter in Support of Air Quality Management in Prince George, British Columbia, Canada

Peter L. Jackson, Dennis Fudge, Bruce Ainslie, John Spagnol, Christophe Corbel, Andreas Veira, Volker Schunicht and Brayden Nilson

Abstract Prince George British Columbia has among the highest levels of ambient particulate matter (PM) in western Canada. In order to effectively lower ambient PM levels, management agencies need to be able to attribute ambient levels to specific sources, which can then be targeted for reduction. Dispersion modelling can be used to attribute sources to ambient levels, but one issue is that emissions from many PM sources are poorly known and must be estimated. The Calpuff dispersion modelling

P. L. Jackson (✉) · J. Spagnol · B. Nilson
Environmental Science and Engineering Programs, University of Northern British Columbia, Prince George, BC V2N 4Z9, Canada
e-mail: peter.jackson@unbc.ca

J. Spagnol
e-mail: jspagnol@shaw.ca

B. Nilson
e-mail: nilson@unbc.ca

D. Fudge
Saskatchewan Ministry of Environment, Regina, Canada
e-mail: dennis.fudge@gov.sk.ca

B. Ainslie
Environment and Climate Change Canada, Ottawa, Canada
e-mail: bruce.ainslie@canada.ca

C. Corbel
Deepki, Paris, France
e-mail: christophe@ccorbel.com

A. Veira
German Meteorological Service, Offenbach, Germany
e-mail: andreas.veira@dwd.de

V. Schunicht
Anjoda Technical Solutions, Prince George, BC, Canada
e-mail: volker@anjoda.com

C. Mensink et al. (eds.), *Air Pollution Modeling and its Application XXVI*, Springer Proceedings in Complexity,
https://doi.org/10.1007/978-3-030-22055-6_14

system was applied to all known sources of PM_{10}, $PM_{2.5}$, NO_x and SO_2 affecting the Prince George airshed over the three-year period 2003–2005. The model results were evaluated by comparison with ambient levels as well as with results from a speciation study using Positive Matrix Factorization and Chemical Mass Balance techniques. This resulted in constraints on the emissions of some of the more poorly characterized sources. In addition, a web-based visualization and scenario tool was developed for air quality managers to enable them to make science-based decisions to improve air quality. The results of the modelling and source attribution will be discussed, and the web-based scenario tool demonstrated.

14.1 Introduction

Prince George, British Columbia (BC) Canada is a city of 74,000 [1] located at the junction of the Fraser and Nechako Rivers in a valley about 150 m below the BC Central Interior Plateau. Prince George has among the highest levels of $PM_{2.5}$ in Western Canada, although it generally meets the BC ambient air quality objective of 8 $\mu g\ m^{-3}$, it occasionally exceeds the 24 h average standard of 25 $\mu g\ m^{-3}$. Three kraft pulp mills (east and northeast of the downtown central business district), and other industrial sources in Prince George contribute significant amounts of $PM_{2.5}$ to the airshed, mainly through combustion of biomass [2]. Other important anthropogenic sources of $PM_{2.5}$ in the Prince George area include open burning of wood debris, wood burning for residential heating during the winter, cooking from commercial and household stoves, locomotive emissions, and diesel and gasoline vehicle combustion engines [2], as well as anthropogenic dust sources such as road dust and fugitive dust. The wind-sheltered valley area of Prince George experiences frequent temperature inversions associated with stagnant air conditions that often result in elevated pollution levels [4]. Winter meteorological conditions can result in more frequent temperature inversions in the valley, less atmospheric mixing from daytime solar radiation heating of the surface, and consequently higher levels of $PM_{2.5}$ pollution [2, 4]. Between 2005 and 2016, deseasonalized monthly mean Prince George $PM_{2.5}$ levels have been decreasing at 0.16 $\mu g\ m^{-3}\ year^{-1}$, while 98th percentile levels have been decreasing at 0.66 $\mu g\ m^{-3}\ year^{-1}$ [3]. Based on a wind-sector analysis, these improvements were largely driven by environmental upgrades at major industrial sources to the northeast of the Prince George downtown.

In order to better understand the contribution of specific sources to ambient $PM_{2.5}$ levels in Prince George and guide air quality management, a community-wide modelling study was conducted to simulate the transport and dispersion of all identifiable $PM_{2.5}$ sources. In order to help air quality managers understand and visualize the model results, a web-based decision support tool, called AirQuest was developed. This paper describes the modelling study and the visualization tool.

14.2 Methods

The Calpuff modelling system was used to simulate the dispersion and transport of particulate matter (PM_{10} and $PM_{2.5}$), as well as SO_2 and NO_x, with a goal of accurately simulating both primary and secondary $PM_{2.5}$ in the Prince George airshed. In total, the transport and dispersion of about 1500 individual sources were simulated at hourly resolution from 2003–2005. Of these, there were 33 industrial sources with a total of about 350 emitting units. Calmet was used to specify the meteorological conditions, with surface observations from six stations and vertical sounding data from one station used as input. The computational domain was 24 grid points (west–east) by 36 grid points (south–north) set at 1 km resolution with a 500 m resolution receptor grid plus ten discrete receptors (at monitoring locations) nested within it, for a total of 1883 receptors.

A major challenge in conducting a community-wide dispersion modelling study of PM is accurately specifying all sources in the airshed. While the major industrial sources are relatively well quantified based on permit levels and stack monitoring, most sources are unmonitored and must be estimated using emission factors and activity levels. To evaluate the modelling system, including the emission inventory, we compare the model outputs with meteorological observations, ambient air pollution levels at several monitoring stations, and with data and information from a speciation study and receptor modelling that was conducted in 2005. Because of uncertainty in PM emissions from many sources, an iterative approach was used in refining the emission inventory. First the "best guess" emissions were modelled and the resulting ambient concentrations compared with speciation data and ambient concentrations. In addition, scaling factors for each source were calculated by optimizing the fit between modelled concentration and ambient concentrations at monitoring locations.

Dispersion model output quantifies the link between ambient concentrations at all receptors to the specific emission sources that contribute to those ambient concentrations. Consequently model output can be used to identify the most important sources to target for reduction: an analysis of dispersion model output is the ideal tool to effectively manage air quality. Because modelled ambient concentrations scale linearly with the emission rate, changes in emissions can be assessed by linearly scaling the modelled ambient levels by the same amount, without the need to re-run the dispersion model. This was exploited by developing the web-based AirQuest decision support tool for air quality mangers to use to run source reduction scenarios, without having to re-run Calpuff.

14.3 Results and Discussion

The evaluation of model output from the 2010 iteration of the model suggested that emissions from road dust, locomotives, and restaurants were over-estimated, while fugitive dust emissions from commercial operations, gravel pits and industrial storage were under-estimated. Consequently, more care was taken in estimating the emissions from these sources, resulting in emissions shown in Fig. 14.1. As a result of these emission changes, the model does a better job of predicting the distribution of ambient concentrations at monitoring locations (Fig. 14.2). The AirQuest decision support tool was designed to allow users to run emission change scenarios and visualize the resulting ambient concentrations spatially, and statistically at monitoring locations (Fig. 14.3).

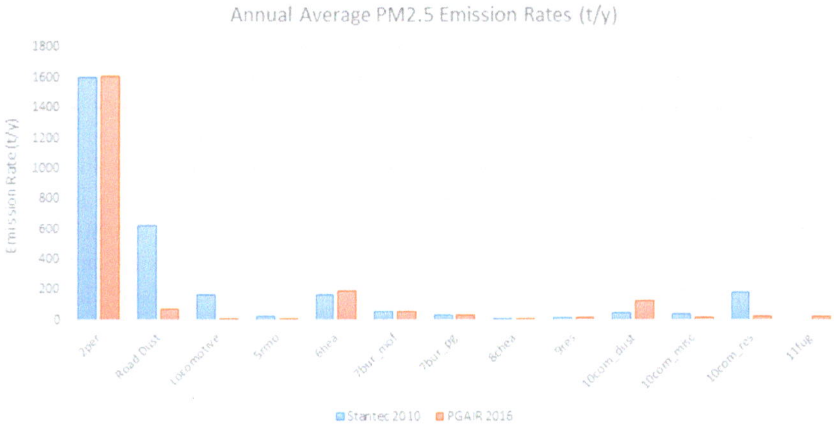

Fig. 14.1 Total emissions before (Stantec 2010 in blue on the left), and after (PGAIR 2016 in red on the right) refining the emission inventory. The source category codes are as follows. "2 per" are industrial emissions, "5rmo" are mobile emissions from vehicles, "6hea" and "8chea" are residential and commercial heating emissions, "7bur_mof" and "7bur_pg" are burning of wood waste in and around Prince George. "9res" are miscellaneous residential sources, "10com_dust" and "10com_misc" are commercial dust and miscellaneous sources, "10com_res" are restaurant emissions, and "11fug" are fugitive dust emissions

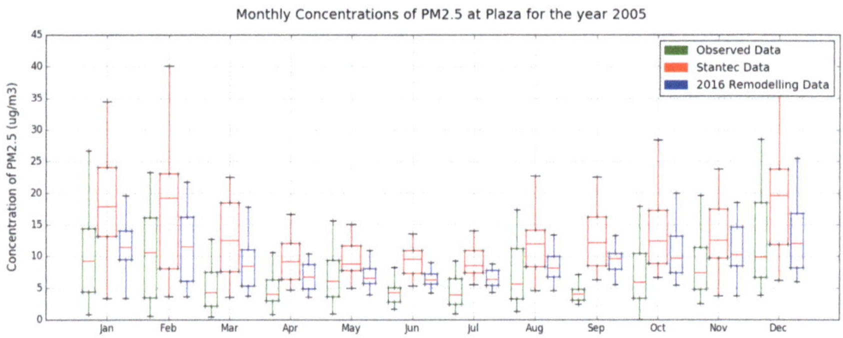

Fig. 14.2 Comparison of monthly PM2.5 levels in 2005 at the Plaza 400 monitor location in downtown Prince George: observed (green on the left), modelled in 2010 (labelled Stantec, in red in the middle), and remodelled in 2016 with revised emissions (blue on the right)

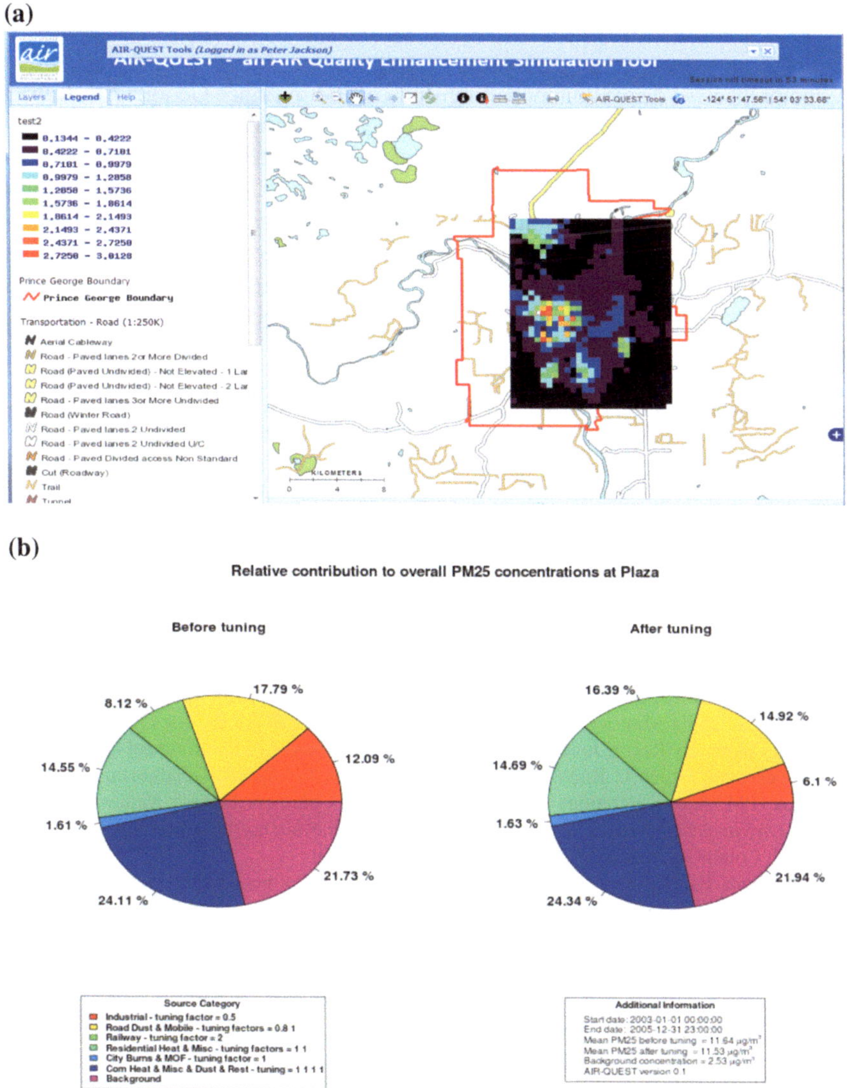

Fig. 14.3 Examples of AirQuest **a** spatial scenario output and **b** statistical scenarios at a monitoring location. The "before tuning" panel indicates the Calpuff model output, while the "After tuning" panel indicates the source reduction scenario in which industrial emissions were reduced to 50% and road dust to 80%

14.4 Conclusion

A comprehensive dispersion modelling study of PM in Prince George, BC, Canada was designed to help manage air quality. The greatest challenge was in accurately specifying all the sources of PM in the airshed. Using an iterative approach in which model results are evaluated by comparison with various observations, the emissions of more poorly characterized sources were revised, resulting in a better simulation of PM. In order to help air quality managers make use of the dispersion model results, a web-based visualization tool (AirQuest) was developed in which users can run source reduction scenarios and visualize the resulting concentrations spatially and statistically.

Acknowledgements Funding was from: BC Ministry of Environment, City of Prince George, Environment Canada, Northern Health, Canfor, and the Natural Science and Engineering Research Council of Canada.

References

1. BCStats, 2016 Census of Population and Housing (2016). http://www.bcstats.gov.bc.ca/StatisticsBySubject/Census/2016Census/PopulationHousing.aspx. Accessed 17 Mar 2017
2. British Columbia Ministry of Environment Environmental (BCMoE), Air Quality in Prince George Summary Report June 2016. British Columbia Ministry of Environment (2016), 23 p
3. P.L Jackson, J. Albino, C. Birch, B. Nilson, J. Pawluk and T. Tereshchak, Trends in fine particulate matter (PM2.5) concentrations in Prince George, British Columbia, Canada. *Natural Resources and Environmental Studies Institute. Research Extension Note*, no. 11 (University of Northern British Columbia, Prince George, BC, Canada, 2017), 21 p
4. M. Noullett, P.L. Jackson, M. Brauer, Winter measurements of personal exposure and ambient fine particle mass, sulphate and light absorbing components in a northern community. Atmos. Environ. **40**, 1971–1990 (2006)

Chapter 15
Source Localization of Ruthenium-106 Detections in Autumn 2017 Using Inverse Modelling

Pieter De Meutter, Johan Camps, Andy Delcloo and Piet Termonia

15.1 Introduction

In late September and October 2017, Ru-103 and Ru-106 have been detected throughout the northern hemisphere by national environmental radioactivity monitoring networks and by the International Monitoring System that is being established to verify compliance with the Comprehensive Nuclear-Test-Ban Treaty. Ru-103 (half-life: 39.26 d) and Ru-106 (half-life: 373.6 d) are radioactive particulates that have no natural sources and for which there is no measurable global background. Based on the fact that only Ru-106 (and, in much lower concentrations and at fewer places, Ru-103) but no other fission products such as iodine and cesium have been measured, a nuclear accident can be excluded.

The *Institut de Radioprotection et de Sûreté Nucléaire* (IRSN) has published a study on the possible source regions based on measurements exchanged via the *Ring of Five* network, an informal network of experts. They have employed forward atmospheric transport modelling over a limited domain to test which single grid box source can best explain the observations. They found that the detected Ru-106 is likely coming from an area between the Volga and the Urals [1].

P. De Meutter (✉) · J. Camps
Belgian Nuclear Research Institute, Boeretang 200, 2400 Mol, Belgium
e-mail: pieter.de.meutter@sckcen.be

J. Camps
e-mail: johan.camps@sckcen.be

P. De Meutter · A. Delcloo · P. Termonia
Royal Meteorological Institute of Belgium, Ringlaan 3, 1180 Brussels, Belgium
e-mail: andy.delcloo@meteo.be

P. De Meutter · P. Termonia
Department of Physics and Astronomy, Ghent University, Krijgslaan 281, 9000 Ghent, Belgium
e-mail: piet.termonia@meteo.be

C. Mensink et al. (eds.), *Air Pollution Modeling and its Application XXVI*,
Springer Proceedings in Complexity,
https://doi.org/10.1007/978-3-030-22055-6_15

In this paper, we assess the possible source areas for the Ru-106 based on measurements from the International Monitoring System (IMS) for the verification of the Comprehensive Nuclear-Test-Ban-Treaty that detected Ru-106 throughout the northern hemisphere. We use an adjoint approach and quantify meteorological uncertainty using the ensemble approach.

15.2 Data and Methods

Meteorological data from the Integrated Forecasting System operational at ECMWF have been used. The extracted data had horizontal grid spacings of 1° and 137 non-uniform vertical levels up to 0.01 hPa. The inverse modelling was performed using 282 Ru-106 detections and non-detections from the International Monitoring System.

Inverse modelling involves finding a source term $\mathbf{x}(x, y, z, t)$ based on a vector of observations $\mathbf{y}$:

$$\mathbf{y} = \mathbf{M}\,\mathbf{x} \tag{15.1}$$

here, $\mathbf{M}$ is the source-receptor-sensitivity matrix which is obtained by running the atmospheric transport and dispersion model Flexpart [2] in backward mode [3]. Wet deposition is taken into account. In practice, no perfect match between the observed and simulated activity concentrations is possible, since both the source-receptor-sensitivity matrix and the observations contain uncertainties. Instead, the disagreement is minimised using an optimisation procedure. A cost function is defined to quantify the disagreement. It is assumed that the true source lies in one of the grid boxes having the lowest cost function value. There is no need to rerun the atmospheric transport model during the optimisation: only the source term $\mathbf{x}(x, y, z, t)$ needs to be varied until a sufficiently good match is found with $\mathbf{y}$.

We assume that the Ru-106 detections originated from a single point source in the lowest model level. We perform the optimisation for each grid box separately in the lowest model level (each grid box is thus assumed to be a source; since this does not involve rerunning the atmospheric transport and dispersion model, this procedure is fast). The result is a cost function value for each grid box, and an associated optimal source term. Grid boxes with a low cost function are assumed to be possible source locations.

The cost function that has been used here, is the geometric variance (15.2; index i goes over all n observations). Reference [4] found that this cost function performed well for inverse modelling over a large domain using noisy measurements. A parameter α has been added which allows dealing with non-detections (α has been given a value of 0.005 mBq/m^3, which corresponds roughly to the detection limit of the Ru-106 observations).

$$costfunction(\mathbf{x}) = \exp\left(\frac{1}{n}\sum_{i=1}^{n}\left(\log(y_i + \alpha) - \log(M_{ij}x_j + \alpha)\right)^2\right) \tag{15.2}$$

15.3 Results

Figure 15.1 shows the cost function values resulting from the optimisation. A distinct region of possible source locations can be seen between the Volga and the Ural mountains. According to the inverse modelling, the Ru-106 release should have been between 10^{14} and 10^{15} Bq.

The Ensemble Data Assimilation (EDA) system of ECMWF has been used to estimate the effect of meteorological uncertainty on the source localisation. The EDA system consists of 26 independent lower-resolution 4D-Var assimilations, of which 25 use perturbed observations, sea-surface temperatures and model physics [5]. By adding and subtracting the perturbations from the ensemble mean, we obtained 50 perturbed and 1 unperturbed members. For each ensemble member, Flexpart has been run and the inverse modelling is applied on each of the resulting 51 **M** matrices. Only a subset of observations is used to limit the computational cost. To visualize uncertainty, a threshold has been applied to the resulting 51 cost function maps to discriminate possible source regions from other regions. Since each member of the EDA system is by construction equally likely, the 51 cost function maps can be readily used to construct grid point-wise probability maps, shown in Fig. 15.2.

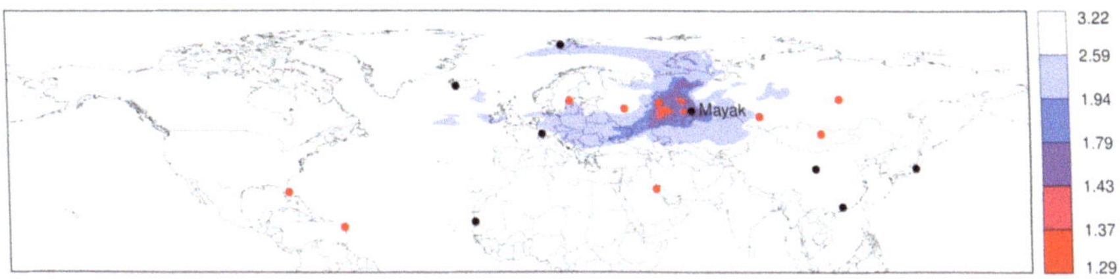

Fig. 15.1 Grid box cost function values resulting from the optimisation. The black dots show IMS stations where no Ru-106 was measured; the red dots show stations where Ru-106 was measured. The location of the nuclear facility Mayak is also shown. The legend values correspond to the cost function quantiles of 0, 0.05, 0.1, 0.5, 1, 10 and 100%

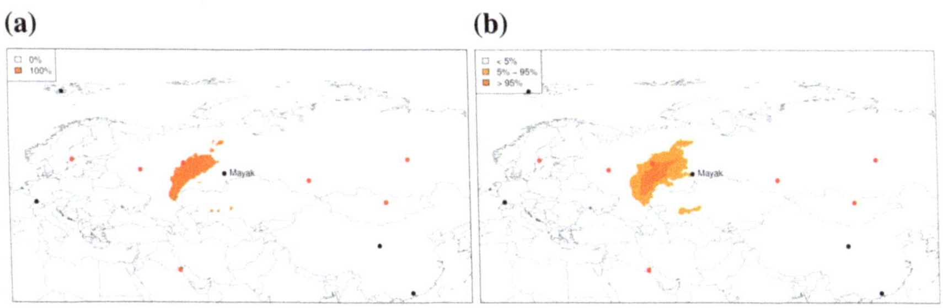

Fig. 15.2 Probability of being a likely source for each grid box separately using **a** the unperturbed member only and **b** the full ensemble. The black dots show IMS stations where no Ru-106 has been measured; the red dots show IMS stations where Ru-106 was measured. The location of the nuclear facility Mayak is also shown

15.4 Discussion

The Ru-106 detections made in autumn 2017 provide a very interesting case for inverse atmospheric transport modelling since (i) the release must have been strong, resulting in detections throughout the northern hemisphere and (ii) the absence of a Ru-106 background that could contaminate the signature of the event of interest. The detections span more than three orders of magnitude, so that a cost function is needed that treats high and low detections with similar importance. The IMS observations of Ru-106 are found to be compatible with a single grid box source located in the area between the Volga and the Ural mountains. According to the inverse modelling, the Ru-106 release could have been up to 1 PBq. The results of this study are in agreement with the study published by IRSN, although a different methodology and different observations have been used.

Acknowledgements One of the authors (P De Meutter) acknowledges funding from Engie under contract number JUR2015-28-00.

References

1. Detection of Ruthenium 106 in France and in Europe: Results of IRSNs Investigations. http://www.irsn.fr. Accessed 29 Jan 2018
2. A. Stohl, C. Forster, A. Frank, P. Seibert, G. Wotawa, Technical note: the Lagrangian particle dispersion model FLEXPART version 6.2. Atmos. Chem. Phys. (2005). https://doi.org/10.5194/acp-5-2461-2005
3. P. Seibert, A. Frank, Source-receptor matrix calculation with a Lagrangian particle dispersion model in backward mode. Atmos. Chem. Phys. (2000). https://doi.org/10.5194/acp-4-51-2004
4. G. Cervone, P. Franzese, Monte Carlo source detection of atmospheric emissions and error functions analysis. Comput Geosci. (2010). https://doi.org/10.1016/j.cageo.2010.01.007
5. M. Bonavita, E. Hlm, L. Isaksen, M. Fisher, The evolution of the ECMWF hybrid data assimilation system. Q. J. Roy. Meteor. Soc. (2016). https://doi.org/10.1002/qj.2652

Chapter 16
Comparing the ISORROPIA and EQSAM Aerosol Thermodynamic Options in CAMx

Bonyoung Koo, Swen Metzger, Pradeepa Vennam, Chris Emery, Gary Wilson and Greg Yarwood

Abstract The Comprehensive Air quality Model with extensions (CAMx) has been employing ISORROPIA for inorganic aerosol thermodynamic calculations. Recently, the Equilibrium Simplified Aerosol Model (EQSAM) was added to CAMx as an alternative to ISORROPIA for computing concentrations of the inorganic ions: sulfate, nitrate, ammonium, sodium and chloride. By design, EQSAM has more streamlined algorithms than ISORROPIA, which may lead to different model concentrations and faster run times. We apply CAMx using both thermodynamic equilibrium modules for simulations of the continental US with 12 km grid resolution for winter and summer months. Model predictions of inorganic ions by both algorithms are compared and model performance is evaluated against ambient data collected by speciated PM monitoring networks. The overall model run times are also compared. Our purpose is to introduce EQSAM as a newly available option in CAMx and provide information relevant to choosing between the two available aerosol thermodynamics modules in CAMx taking into consideration the scheme's performance attributes as well as the requirements of each model application.

B. Koo · P. Vennam · C. Emery · G. Wilson · G. Yarwood (✉)
Bay Area Air Quality Management District, 375 Beale Street, Suite 600, San Francisco, CA 94105, USA
e-mail: gyarwood@ramboll.com

B. Koo
e-mail: bkoo@ramboll.com

P. Vennam
e-mail: pvennam@ramboll.com

C. Emery
e-mail: cemery@ramboll.com

G. Wilson
e-mail: gwilson@ramboll.com

S. Metzger
ResearchConcepts io GmbH, Freiburg im Breisgau, Germany
e-mail: swen.metzger@eco-serve.de

C. Mensink et al. (eds.), *Air Pollution Modeling and its Application XXVI*, Springer Proceedings in Complexity,
https://doi.org/10.1007/978-3-030-22055-6_16

16.1 Introduction

Atmospheric particulate matter (PM) or aerosol has been recognized to have adverse effects on human health [1] and visibility [6]. Atmospheric PM is a complex mixture of inorganic ions, carbonaceous material, crustal elements, trace metals and water. The inorganic components, which are mainly comprised of sulfate (SO_4^{2-}), nitrate (NO_3^-), ammonium (NH_4^+), sodium (Na^+) and chloride (Cl^-), are important contributors to $PM_{2.5}$ (PM with aerodynamic diameter less than or equal to 2.5 μm) mass globally [11].

Atmospheric PM models typically assume thermodynamic equilibrium to determine partitioning of volatile inorganic components such as NO_3^- and NH_4^+ between the gas and aerosol phases. A variety of thermodynamic equilibrium models have been developed ([8] and references therein). Among them, ISORROPIA [4, 10] is widely used by regional and global chemical transport models because of its numerical efficiency. ISORROPIA reduces computational costs by dividing the relative humidity (RH) and composition space into subdomains that minimize the number of equations to be solved. However, ISORROPIA computes activity coefficients via an iterative procedure that adds to computational cost. Solving thermodynamic equilibrium for inorganic ions remains one of the most computationally demanding processes in large scale air quality modeling.

EQSAM4clim [8], originally developed for climate simulations, is based on a single solute coefficient approach [7] that efficiently parameterizes single solution hygroscopic growth accounting for aerosol water uptake from the deliquescence RH up to supersaturation. EQSAM4clim extends the single solute coefficient approach to treat water uptake for multi-component mixtures. The advantage of this approach is that the gas-aerosol partitioning and the mixed solution water uptake can be solved analytically eliminating the need for iterations, which brings potentially significant speed-up. EQSAM4clim has been implemented and evaluated in a global chemistry climate model [9].

The Comprehensive Air quality Model with extensions (CAMx; www.camx.com) has provided alternative schemes for several model processes, but ISORROPIA has been the only approach for determining thermodynamic equilibrium partitioning of inorganic PM components. In this study, we implemented EQSAM4clim in CAMx as an alternative to ISORROPIA, and evaluated model predictions of inorganic $PM_{2.5}$ components by EQSAM4clim and ISORROPIA over a continental US modeling domain.

16.2 Modeling Platform

The latest version of CAMx (version 6.40) was applied to simulate two month-long episodes (January and July 2011) from a US Environmental Protection Agency (EPA) 2011 modeling platform [3]. The modeling grid covers the continental US with

12 km horizontal grid resolution and extends vertically up to approximately 50 mbar (20 km). The Carbon Bond Version 6 Revision 4 (CB6r4) chemistry mechanism [2] was selected for gas-phase chemistry. Inorganic thermodynamic equilibrium partitioning is modeled by ISORROPIA (version 1.7) and EQSAM4clim (version 10). Both ISORROPIA and EQSAM4clim assumed a metastable aerosol liquid phase following Guo et al. [5].

16.3 Results and Discussion

Figure 16.1 compares monthly average $PM_{2.5}$ NO_3^- and NH_4^+ concentrations predicted by ISORROPIA and EQSAM4clim for January. Both models agree fairly well in predicting spatial distributions and peak magnitudes of each species. In the summer month when less nitric acid partitions into the aerosol phase, EQSAM4clim tends to predict lower NO_3^- than ISORROPIA (not shown). Both models assume a negligible vapor pressure of sulfuric acid, essentially driving all sulfuric acid into the particle phase, resulting in little difference in SO_4^{2-}.

The model predictions of NO_3^-, NH_4^+ and Cl^- were evaluated against ambient measurements at the Interagency Monitoring of Protected Visual Environments (IMPROVE) sites and EPA's Chemical Speciation Network (CSN) sites. IMPROVE

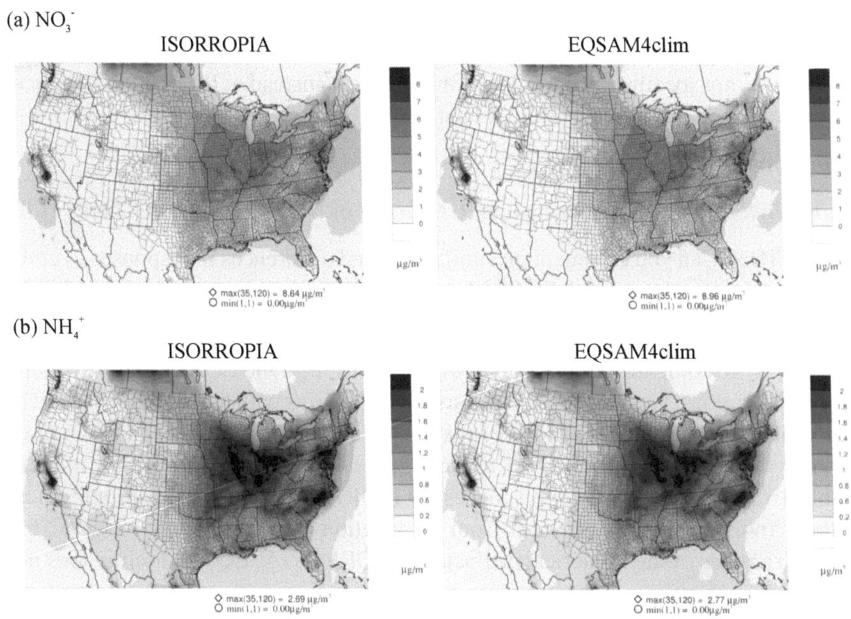

Fig. 16.1 January average concentrations of $PM_{2.5}$ NO_3^- and NH_4^+ by ISORROPIA and EQSAM4clim

Table 16.1 Normalized mean bias and error (NMB and NME) and correlation coefficient (r) statistics for 24-h average $PM_{2.5}$ NO_3^-, NH_4^+ and Cl^- by ISORROPIA (ISO) and EQSAM4clim (EQS) in January

Network	Species	NMB (%)		NME (%)		r	
		ISO	EQS	ISO	EQS	ISO	EQS
CSN	NO_3^-	3.1	1.4	51.4	50.0	0.50	0.51
	NH_4^+	−7.5	−9.7	40.6	39.4	0.63	0.65
	Cl^-	79.0	−12.9	169	115	0.14	0.15
IMPROVE	NO_3^-	43.6	34.3	86.9	82.2	0.69	0.70
	NH_4^+	7.6	3.7	45.0	43.7	0.82	0.82
	Cl^-	60.9	−29.3	192	124	0.29	0.28

sites are located in rural areas whereas CSN sites are mostly in urban/suburban areas. Table 16.1 summarizes performance statistics of the two models for January. Both models show relatively good NH_4^+ performance while somewhat over-predicting NO_3^- in rural sites. Cl^- is overestimated by ISORROPIA, but underestimated by EQSAM4clim. However, Cl^- is a minor component of $PM_{2.5}$ in most inland areas of the modeling domain.

This study, though limited, shows that the two thermodynamic schemes' results are sufficiently similar that either scheme could reasonably be selected. However, further analyses (e.g., model responses to emission changes) are desired to develop a larger evidence base for choosing which scheme to employ for a particular application. On our test simulations, using EQSAM4clim (although the current implementation in CAMx was not specifically optimized for efficiency) instead of ISORROPIA reduced the overall model runtime by 4% (January) to 7% (July).

Questions and Answers

QUESTIONER: Sarav Arunachalam, University of North Carolina at Chapel Hill

QUESTION: Can you comment on similarities and differences in responses to emission perturbations found with EQSAM and ISORROPIA that may help a CAMx user select which scheme to choose?

ANSWER: To compare model responses by ISORROPIA and EQSAM4clim to emission perturbations, we have conducted model simulations with 5 hypothetical sources added in Nevada, Idaho, Missouri, Pennsylvania and South Carolina, and compared impacts of the new sources. Both models generally agree for maximum impacts (Table 16.2). However, model responses by ISORROPIA are subject to numerical artifacts (noisy responses often far from the location of perturbations). Figure 16.2 shows examples of the numerical artifacts by ISORROPIA; EQSAM4clim does not show such artifacts.

Table 16.2 Maximum 24-h average responses to adding 5 hypothetical sources in January

Source	$PM_{2.5}$ NO_3^- ($\mu g/m^3$)		$PM_{2.5}$ NH_4^+ ($\mu g/m^3$)	
	ISORROPIA	EQSAM4clim	ISORROPIA	EQSAM4clim
Nevada	0.05	0.04	0.03	0.03
Idaho	0.02	0.02	0.01	0.01
Missouri	0.04	0.09	0.03	0.03
Pennsylvania	0.05	0.06	0.02	0.02
S. Carolina	0.05	0.05	0.02	0.02

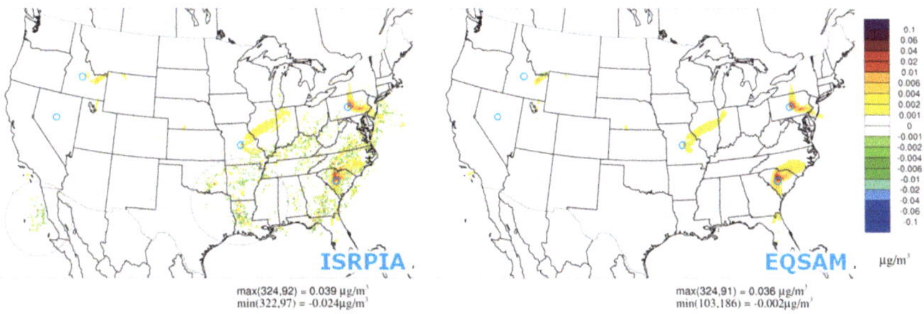

Fig. 16.2 24-h average $PM_{2.5}$ NO_3^- responses to adding 5 hypothetical sources (blue circles) using ISORROPIA and EQSAM4clim on January 15; responses are similar downwind of the blue circles; dotted circles indicate noisy responses using ISORROPIA that are not seen using EQSAM4clim

References

1. C.I. Davidson, R.F. Phalen, P.A. Solomon, Airborne particulate matter and human health: a review. Aerosol Sci. Technol. **39**, 737–749 (2005). https://doi.org/10.1080/02786820500191348
2. C. Emery, B. Koo, W.C. Hsieh, A. Wentland, G. Wilson, G. Yarwood, EPA Contract EPD12044; WA 4-07: Meteorological, Photochemical, and Dispersion Modeling Support. Task 7: Update Carbon Bond Chemical Mechanism. Technical Memorandum prepared for US Environmental Protection Agency (2016). http://www.camx.com/files/emaq4-07_task7_techmemo_r1_1aug16.pdf
3. EPA, Air Quality Modeling Technical Support Document for the 2015 Ozone NAAQS Preliminary Interstate Transport Assessment. Office of Air Quality Planning and Standards, US Environmental Protection Agency, Research Triangle Park, NC, USA (2016). https://www.epa.gov/airmarkets/air-quality-modeling-technical-support-document-2015-ozone-naaqs-preliminary-interstate
4. C. Fountoukis, A. Nenes, ISORROPIA II: a computationally efficient thermodynamic equilibrium model for K^+–Ca^{2+}–Mg^{2+}–$NH_4{}^+$–Na^+–SO_4^2–NO_3–$Cl^-$$H_2O$ aerosols. Atmos. Chem. Phys. **7**, 4639–4659 (2007). https://doi.org/10.5194/acp-7-4639-2007
5. H. Guo, A.P. Sullivan, P. Campuzano-Jost, J.C. Schroder, F.D. Lopez-Hilfiker, J.E. Dibb, J.L. Jimenez, J.A. Thornton, S.S. Brown, A. Nenes, R.J. Weber, Fine particle pH and the partitioning of nitric acid during winter in the northeastern United States. J. Geophys. Res. Atmos. **121**, 10355–10376 (2016). https://doi.org/10.1002/2016JD025311

6. W.C. Malm, "Introduction to Visibility", Air Resources Division, National Park Service. Cooperative Institute for Research in the Atmosphere, NPS Visibility Program, Colorado State University, Fort Collins, CO, USA (1999). https://www.epa.gov/sites/production/files/2016-07/documents/introvis.pdf
7. S. Metzger, B. Steil, L. Xu, J.E. Penner, J. Lelieveld, New representation of water activity based on a single solute specific constant to parameterize the hygroscopic growth of aerosols in atmospheric models. Atmos. Chem. Phys. **12**, 5429–5446 (2012). https://doi.org/10.5194/acp-12-5429-2012
8. S. Metzger, B. Steil, M. Abdelkader, K. Klingmüller, L. Xu, J.E. Penner, C. Fountoukis, A. Nenes, J. Lelieveld, Aerosol water parameterisation: a single parameter framework. Atmos. Chem. Phys. **16**, 7213–7237 (2016). https://doi.org/10.5194/acp-16-7213-2016
9. S. Metzger, M. Abdelkader, K. Klingmüller, B. Steil, J. Lelieveld, Comparison of Metop PMAp Version 2 AOD Products using Model Data. Final Report for EUMETSAT ITT 15/210839. Department of Atmospheric Chemistry, Max Planck Institute for Chemistry, Mainz, Germany (2016). http://bit.ly/2Epxf9b
10. A. Nenes, S.N. Pandis, C. Pilinis, ISORROPIA: a new thermodynamic equilibrium model for multiphase multicomponent inorganic aerosols. Aquat. Geochem. **4**, 123–152 (1998). https://doi.org/10.1023/A:1009604003981
11. G. Snider, C.L. Weagle, K.K. Murdymootoo, A. Ring, Y. Ritchie, E. Stone, A. Walsh, C. Akoshile, N.X. Anh, R. Balasubramanian, J. Brook, F.D. Qonitan, J. Dong, D. Griffith, K. He, B.N. Holben, R. Kahn, N. Lagrosas, P. Lestari, Z. Ma, A. Misra, L.K. Norford, E.J. Quel, A. Salam, B. Schichtel, L. Segev, S. Tripathi, C. Wang, C. Yu, Q. Zhang, Y. Zhang, M. Brauer, A. Cohen, M.D. Gibson, Y. Liu, J.V. Martins, Y. Rudich, R.V. Martin, Variation in global chemical composition of $PM_{2.5}$: emerging results from SPARTAN. Atmos. Chem. Phys. **16**, 9629–9653 (2016). https://doi.org/10.5194/acp-16-9629-2016

Chapter 17
Using Higher Order Sensitivity Approaches to Assess Aircraft Emissions Impacts on O_3 and $PM_{2.5}$

Calvin Arter and Sarav Arunachalam

Abstract This study utilized an advanced sensitivity analysis, the higher order Decoupled Direct Method (HDDM-3D) as implemented in the Community Multiscale Air Quality Model (CMAQ) to quantify the impacts of aviation emissions during the landing and takeoff (LTO) cycle at nine individual airports; five located in regions of attainment of O_3 and $PM_{2.5}$ NAAQS: Boston Logan (BOS), Kansas City (MCI), Raleigh Durham (RDU), Seattle-Tacoma (SEA), and Tucson (TUS); and four located in regions of nonattainment: Chicago O'Hare (ORD), Hartsfield-Jackson Atlanta (ATL), John F. Kennedy (JFK), and Los Angeles (LAX). Fuel burn changes needed at the four nonattainment airports ranged from −14.9 (357,185 less tons) to −3.3 (55,715 less tons) times less fuel burned and from 1.6 (29,826 more tons) to 3.1 (79,584 more tons) times more fuel burned to reduce ambient $PM_{2.5}$ by 0.1 $\mu g/m^3$ and O_3 by 1 ppb, respectively. Fuel burn changes needed at the five attainment airports ranged from 20.4 (39,516 more tons) to 48.0 (397,180 more tons) times more fuel burned and from −449.0 (−477,734 less tons) to −24.0 (46,648 less tons) times less fuel burned to increase ambient $PM_{2.5}$ by 0.1 $\mu g/m^3$ and O_3 by 1 ppb, respectively. Using these estimates for a range of airports, we demonstrate an illustration of how HDDM-based sensitivity calculations can be used to develop source specific impacts on potential attainment designations for a region.

17.1 Introduction

The aviation sector has seen substantial growth in the past decade with air carriers flying 631 million passenger miles in the year 2015, resulting in an increase of around

---

C. Arter (✉)
Department of Environmental Sciences and Engineering, University of North Carolina at Chapel Hill, Chapel Hill, NC, USA
e-mail: arterca@email.unc.edu

C. Arter · S. Arunachalam
Institute for the Environment, University of North Carolina at Chapel Hill, Chapel Hill, NC, USA
e-mail: sarav@email.unc.edu

C. Mensink et al. (eds.), *Air Pollution Modeling and its Application XXVI*, Springer Proceedings in Complexity,
https://doi.org/10.1007/978-3-030-22055-6_17

11% over the past ten years [1]. This growth is expected to continue with the Federal Aviation Administration forecasting a 2.1% increase in U.S. carrier passenger growth each year for the next 20 years, and a 2.1% and 3.5% growth in system traffic in revenue passenger miles over the next 20 years for domestic travel and international travel, respectively [2].

This projected growth can place a burden on atmospheric air quality as aircraft attributable emissions become a larger component of all anthropogenic emissions. Aircraft emissions are composed primarily of nitrogen oxides (NO_X), sulfur oxides (SO_X), volatile organic compounds (VOC), primary elemental carbon (PEC), primary organic carbon (POC), and primary sulfate (PSO_4). Both LTO and cruise-mode emissions can lead to the formation of air pollutants such as O_3 and $PM_{2.5}$ affecting populations near airports. One study on 99 U.S. airports estimates an increase in premature deaths due to aviation emissions from 75 deaths in 2005 to 460 deaths in 2025 [3].

17.1.1 Methods

In this study we utilize an Eulerian atmospheric chemical transport model (CTM), CMAQv5.0.2 [4], to quantify the concentration and transport of PM, O_3, and other pollutants in a 36 × 36 km grid cell resolution domain.

Various sensitivity analyses in the atmospheric CTM framework are used for guiding policy and environmental scenarios. We make use of a sensitivity analysis approach, (DDM-3D) [5–7], that excels in describing sensitivities to multiple input parameters as well as sensitivities to small variations. Both of these conditions are important for modeling aviation emissions since we aim to describe impacts due to varying aircraft emission species and the aviation emission sector is much smaller than other anthropogenic emission sectors. Within DDM-3D, derivatives are taken at each time step, which calculate the change in concentration of a chemical species with respect to some change in an input parameter. Output from DDM-3D is in the form of sensitivity coefficients, which express the derivatives taken at each time step.

First order sensitivity coefficients describe linear changes in concentrations with respect to changing emissions. In our case of air pollutants attributable to aviation emissions, we need to be concerned with air pollutant species that may not be linearly dependent on aviation emissions. The chemistry surrounding tropospheric $PM_{2.5}$ and O_3 formation is far more complicated than what can be expressed with only first order changes. Hence, we extend our DDM-3D analysis framework to higher order decoupled direct method in three dimensions (HDDM-3D) in order to calculate second order sensitivity coefficients. In doing so, we hope to capture more of the non-linearities in the chemistry related to $PM_{2.5}$ and O_3 formation from aviation emissions.

For our study, six precursor emissions that are responsible for the formation of $PM_{2.5}$ and two precursor emissions that are responsible for the formation of O_3 are chosen as sensitivity parameters for our HDDM-3D analyses. Three gas phase species; NO_X, SO_2, VOC, and three particle phase species; PEC, POC, and PSO_4 are directly emitted from aircrafts and they all can lead to the formation of $PM_{2.5}$ in the atmosphere. For O_3 formation, we consider NO_X and VOC emissions to be the only precursors.

17.1.2 Airport Selection

Our domain consists of the continental United States (CONUS) with a 148 × 112 grid of 36 km × 36 km resolution. We utilize the FAA's Aviation Emission Design Tool (AEDT) [8]; Wilkinson et al. [9] for constructing aircraft flight segments, and LTO segments are processed into gridded emission rate files using AEDTProc [10]. The five airports in attainment regions were chosen based on having no attainment status (maintenance, marginal, nonattainment) for O_3, $PM_{2.5}$, and PM_{10}; had at least 0.05% of annual passenger boardings, designated as being small hubs according to the FAA's VALE [11]; and that represented the greatest geographic and climatic diversity while servicing major metropolitan areas (MSA population >1,000,000 people). We selected four of the top five airports in total number of enplanements for our group of airports in nonattainment regions. Simulations are performed for the months of January and July in 2005 with a 10 day spin up for each month. January and July are chosen to approximately represent winter and summer characteristics seen during each half of the year. O_3 sensitivities are considered only for July simulations.

17.2 Results

17.2.1 Sensitivities

We calculated first and second order sensitivities at the location of each airport's grid cell within our domain. For brevity we show only the first order sensitivities of O_3 and $PM_{2.5}$ to their respective precursor emissions at each airport's grid cell. Figure 17.1 shows first order O_3 sensitivities with respect to aircraft VOC and NO_X emissions for the month of July. O_3 sensitivities are dominated by NO_X emissions in the airport's grid cell and the net negative sum of both VOC and NO_X sensitivities indicate the titration of O_3 in the airport's grid cell from NO_X emissions.

Figure 17.2 shows both first order $PM_{2.5}$ sensitivities to VOC, NO_X, SO_2, PSO_4, PEC, and POC aircraft emissions (bottom row) and the breakdown of these $PM_{2.5}$ sensitivities by the constituents that make up the total $PM_{2.5}$. $PM_{2.5}$ sensitivities can vary depending on the airport with the airports located in nonattainment regions having a much larger impact on the formation of $PM_{2.5}$.

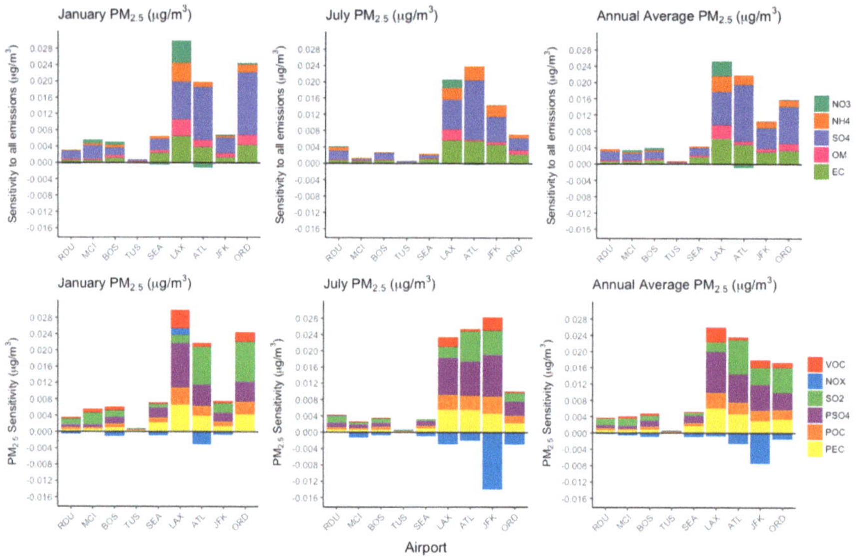

Fig. 17.1 First order $PM_{2.5}$ sensitivities disaggregated by output species (top) and precursor species (bottom) at grid cell containing airport

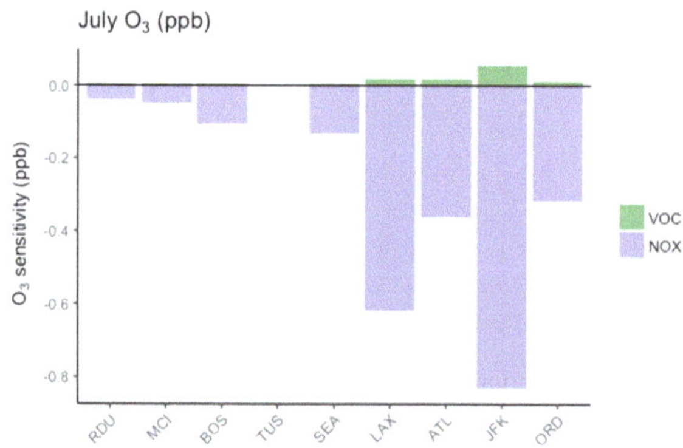

Fig. 17.2 First order O_3 sensitivities disaggregated by precursor at grid cell containing airport for the month of July

17.2.2 Fuel Burn Reductions or Increases

Utilizing Taylor series expansions with the sensitivities calculated at each airport's grid cell, we are able to calculate the emission reductions or increase needed to decrease or increase O_3 by 1 ppb or $PM_{2.5}$ by 0.1 $\mu g/m^3$. These reductions in emissions have been expanded to reductions in fuel burn by relating the total amount of SO_2 emitted in each airport's grid cell to the amount of fuel burned in each grid cell

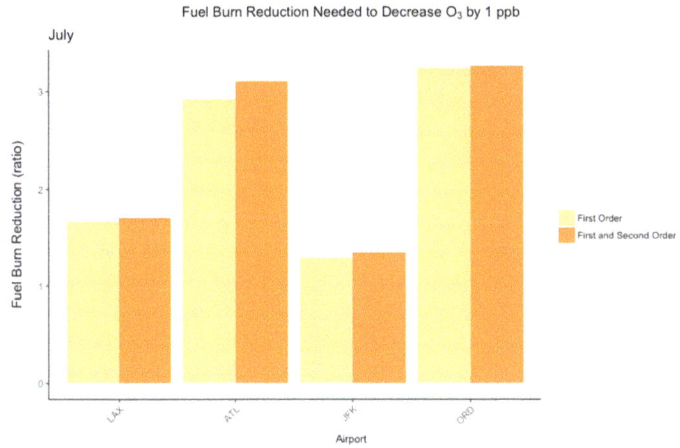

Fig. 17.3 Fuel burn reductions to decrease O_3

through:

$$SO_2 = \frac{FSC}{1000} \times \frac{100 - E}{100} \times \text{FUELBURN} \times \frac{64}{32} \quad (17.1)$$

as described in the Guidance on AEDT Aircraft Emissions for use in Climate and Air Quality Analyses.

Figure 17.3 shows the fuel burn reduction needed to decrease O_3 by 1 ppb at our airports located in regions of nonattainment. Positive values indicate that an increase in fuel burn is needed to decrease O_3 at the grid cells containing the airports. Figure 17.4 shows what is needed for a decrease of $PM_{2.5}$ by 0.1 $\mu g/m^3$ at the same airports. Figure 17.5 shows fuel burn increase needed to increase O_3 by 1 ppb at our airports located in regions of attainment. Negative values indicate a reduction in fuel burn is needed to increase O_3 in the airports' grid cells. Figure 17.6 shows what is needed for an increase of $PM_{2.5}$ by 0.1 $\mu g/m^3$ at the same airports.

17.3 Conclusion

We have utilized HDDM-3D as implemented in CMAQv5.0.2 to quantify the impacts of airport-specific aviation emissions on the formation of $PM_{2.5}$ and O_3. The application of sensitivity coefficients to individual airports and precursor species allows for a more tailored approach in assessing air quality and health impacts as the aviation sector continues to grow. Fuel burn reduction or increased amounts can vary up to an order of magnitude depending on the airport; and for the case of O_3 reduction or increase, the opposite trend is needed in each case to achieve the reduced or increased amount due to titration effects from NO_X emissions. Our approach presented here

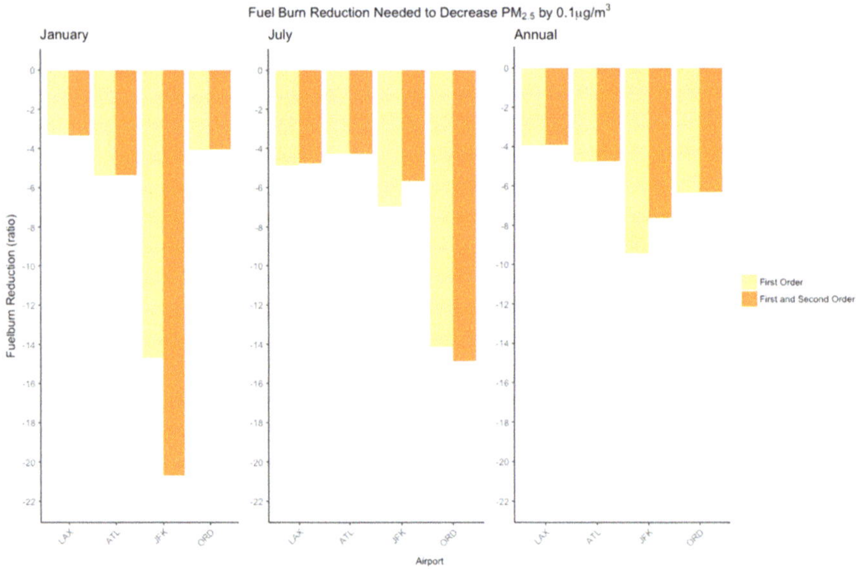

Fig. 17.4 Fuel burn reductions to decrease $PM_{2.5}$

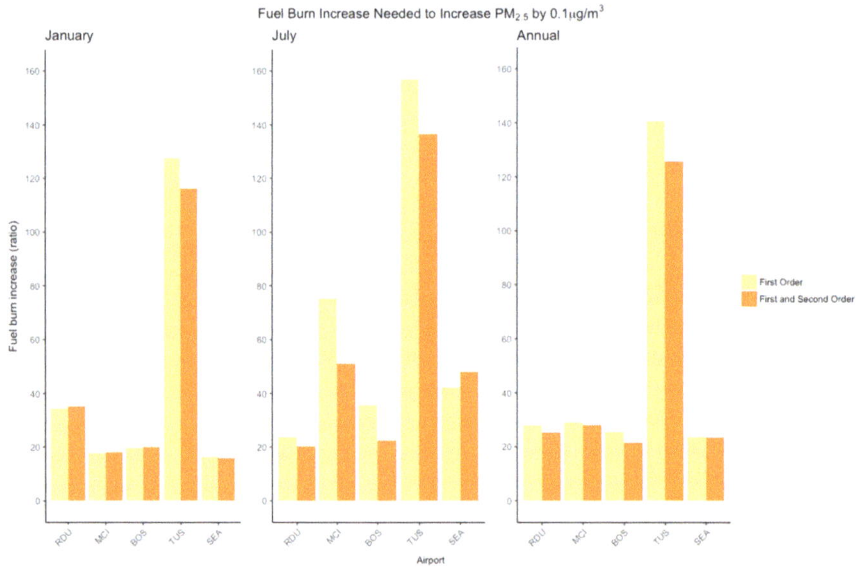

Fig. 17.5 Fuel burn increases to increase $PM_{2.5}$

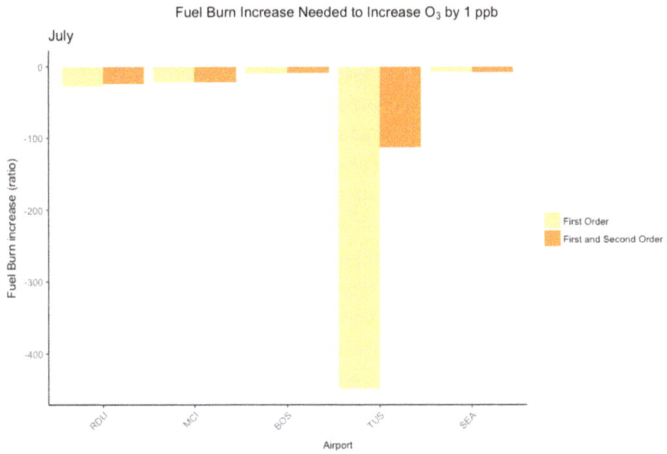

Fig. 17.6 Fuel burn increases to increase O_3

can easily be expanded to look at additional source sectors to specifically assess the impacts of a single source on the region's attainment designations.

This work was funded by the US Federal Aviation Administration (FAA) Office of Environment and Energy as a part of ASCENT Project 19 under grants to UNC. Any Opinions, findings, and conclusions or recommendations expressed in this material are those of the authors and do not necessarily reflect the views of the FAA or other ASCENT Sponsors.

References

1. U.S. Department of Transportation, *T-100 segment data*, Bureau of Transportation (2016), https://www.transtats.bts.gov/Fields.asp?TableID=293
2. U.S. Federal Aviation Administration (2016), *FAA Aerospace Forecast 2016-2036*,https://www.faa.gov/data_research/aviation/aerospace_forecasts/media/FY2016-36_FAA_Aerospace_Forecast.pdf
3. J.I. Levy, M. Woody, B.H. Baek, U. Shankar, S. Arunachalam, Current and future particulate-matter-related mortality risks in the United States from aviation emissions during landing and takeoff. Risk Anal. **32**, 237–249 (2012)
4. D. Byun, J. Ching, *Science Algorithms of the EPA Models-3 Community Multiscale Air Quality (CMAQ) Modeling System* (1999), EPA/600/R-99/030
5. A.M. Dunker, The decoupled direct method for calculating sensitivity coefficients in chemical kinetics. J. Chem. Phys. **81**, 2385 (1984)
6. S.L. Napelenok, D.S. Cohan, M.T. Odman, S. Tonse, Extension and evaluation of sensitivity analysis capabilities in a photochemical model. Environ. Model. Softw. **23**, 994 (2008)
7. S.L. Napelenok, D.S. Cohan, Y. Hu, A.G. Russell, Decoupled direct 3D sensitivity analysis for particulate matter (DDM-3D/PM). Atmos. Environ. **40**, 6112 (2006)
8. C. Roof, G.G. Fleming, Aviation environmental design tool (AEDT), in *22nd Annual UC Symposium on Aviation Noise and Air Quality*, pp. 1–30 (2007)

9. J.T. Wilkerson, M.Z. Jacobson, A. Malwitz, S. Balasubramanian, R. Wayson, G. Fleming, A.D. Naiman, S.K. Lele, Analysis of emission data from global commercial aviation: 2004 and 2006. Atmos. Chem. Phys. **10**, 6391–6408 (2010)
10. B.H. Baek, S. Arunachalam, M. Woody, P.L. Vennam, M. Omary, F. Binkowski, G. Fleming, A New Interface to Model Global Commercial Aircraft Emissions from the FAA Aviation Environmental Design Tool (AEDT) in Air Quality Models in 11th Annual CMAS Conference, Chapel Hill, NC, 15–17 October 2012
11. U.S. Federal Aviation Administration (2016), *List of Commercial Service Airports in the United States and their Nonattainment and Maintenance Status,* https://www.faa.gov/airports/environmental/vale/media/valeeligibleairports.xlsx

Chapter 18
Development and Current Status of the GEM-MACH-Global Modelling System at the Environment and Climate Change Canada

Jack Chen, Diane Pendlebury, Sylvie Gravel, Craig Stroud, Irena Ivanova, Jean DeGranpré and David Plummer

Abstract The GEM-MACH-Global model is a global online meteorology-chemistry system currently being developed at the Department of Environment and Climate Change Canada (ECCC). The model is an extension of the Department's operational, regional GEM-MACH numerical weather and air quality prediction system. The objectives for its development are to improve our understanding of the long range transport and fate of criteria air contaminants, and to improve our forecasting system by providing chemical boundary conditions for the regional air quality forecast system, and background fields for global chemical data assimilation (O_3 and NOy species). For this purpose, GEM-MACH-Global was recently updated with a comprehensive photolysis module (JVAL14-MESSy) and a detailed gas-phase chemistry module based on the SAPRC07C mechanism. Compared to its original ADOM2 chemistry mechanism, the revised gas-phase chemistry is more explicit, with new species and ~15 additional reactions important in the upper troposphere and lower stratosphere (UTLS) region. Furthermore, a lightning emission module

J. Chen (✉) · D. Pendlebury · S. Gravel · C. Stroud · I. Ivanova · J. DeGranpré
Air Quality Research Division, Environment and Climate Change Canada, Ontario, Canada
e-mail: jack.chen@canada.ca

D. Pendlebury
e-mail: diane.pendlebury@canada.ca

S. Gravel
e-mail: sylvie.gravel@canada.ca

C. Stroud
e-mail: craig.stroud@canada.ca

I. Ivanova
e-mail: irena.ivanova@canada.ca

J. DeGranpré
e-mail: jean.degrandpre@canada.ca

D. Plummer
Canadian Centre for Climate Modelling and Analysis, Environment and Climate Change Canada, Ontario, Canada
e-mail: david.plummer@canada.ca

C. Mensink et al. (eds.), *Air Pollution Modeling and its Application XXVI*,
Springer Proceedings in Complexity, https://doi.org/10.1007/978-3-030-22055-6_18

was incorporated to represent NOx emissions aloft. These changes were evaluated with a 2010 annual simulation on a 400 × 200 global-grid. The simulation included inputs from 2010 HTAP global anthropogenic emissions, GFEDv3 biomass burning emissions and ECCC's operational meteorological analyses. The presentation will describe the current state-of-science development of GEM-MACH-Global and show comparisons results of the annual simulation.

18.1 GEM-MACH-Global Model

GEM-MACH is the air quality model of Environment and Climate Change Canada (ECCC). It is a chemistry extension to the current Department's operational numerical weather prediction system [1, 2]. The model is an online, one-way coupled meteorology-chemistry model with a latitude-longitude (Arakawa C) grid and a staggered-hybrid vertical coordinate system. The model has 80 vertical layers with model top at 0.1 hPa. Approximately 9 layers are below the first 1 km. GEM-MACH's atmospheric chemistry components include full process representation of oxidant chemistry with gas-, aqueous-, heterogeneous-chemistry, and dry and wet deposition of gases. Aerosol chemistry and physics are from the Canadian Aerosol Module [3]. The gas-phase chemistry is based on the ADOM2 mechanism which was shown to offer an adequate representation of the chemistry of a polluted continental troposphere [4] but was inadequate for global scale representation as it lacks several key reactions in the UTLS region that are necessary for using the model forecasts as trial fields towards operational NOy and O_3 chemical data assimilation.

The following sections describe the work in progress to improve the formulation of GEM-MACH for its use at the global scale (GEM-MACH-Global). Innovations include a revised gas-phase mechanism, the adoption of a flexible photolysis module and the introduction of lightening NOx emissions. Preliminary results of an annual evaluation of the model are presented.

18.1.1 New Gas-Phase Chemistry

A new chemical mechanism based on the condensed version of SAPRC-07, CS07A [5] was incorporated into GEM-MACH-Global utilizing the Kinetic PreProcessor (KPPv2.2.3) with the rodas3 Rosenbrock solver [6]. To CS07A, the following reactions and their rate equations were added specifically for the UTLS region:

$O_2 + hv \rightarrow O_3P$	: jval_O_2	(1)
$NO + hv \rightarrow O1d + N$	: jval_NO	(2)
$N_2O + hv \rightarrow O1d + N_2$	: jval_N_2O	(3)
$N_2O_5 + hv \rightarrow NO_2 + NO_3$	: jval_N_2O_5	(4)
$N_2O_5 + hv \rightarrow NO + O_3P + NO_3$	: jval_NO_2NOO	(5)
$CH_4 + hv \rightarrow 2HO_2 + CO + H_2\ \{+ 2H\}$	: jval_CH_4	(6)
$N + NO \rightarrow N2 + O_3P$	: a = 2.10E-11, b = 0.0, c = 100.0	(7)*
$N + O2 \rightarrow NO + O_3P$	: a = 1.50E-11, b = 0.0, c = −3600.0	(8)*
$N + NO_2 \rightarrow N_2O + O3P$	: a = 5.80E-12, b = 0.0, c = 220.0	(9)*
$O1D + N_2O \rightarrow 2.0\ NO$	: a = 7.25E-11, b = 0.0, c = 20.0	(10)*
$O1D + N_2O \rightarrow N2 + O_2$	: a = 4.63E-11, b = 0.0, c = 20.0	(11)*
$O1D + O_3 \rightarrow O3P + 1.5\ O_2$	: 1.2 E-10	(12)

* Arrhenius equation for reaction rate: ARR = a * EXP(-b/Temp) * (Temp/300.0)^c

Additionally, isoprene oxidation reactions based on Pye et al. [7], not considered in the original CS07A, were added in anticipation of future updates to secondary organic aerosol module:

$yIOH + NO \rightarrow NO$	: a = 2.60e-12, b = −380.0, c = 0.0	(13)*
$yIOH + HO_2 \rightarrow HO_2 + IOOH$	: a = 3.80e-13, b = −900.0, c = 0.0	(14)*
$IOOH + OH \rightarrow OH + IEPX$	: a = 1.90E-11, b = −390.0, c = 0.0	(15)*
$IOOH + OH \rightarrow 0.387\ yIOH + OH$	: a = 4.75E-12, b = −200.0, c = 0.0	(16)*
$IEPX + OH \rightarrow OH$	: a = 5.78E-11, b = 400.0, c = 0.0	(17)*

* Arrhenius equation for reaction rate: ARR = a * EXP(-b/Temp) * (Temp/300.0)^c

Since reactive gases formed from halogen sources are not represented in the mechanism, upper level stratospheric ozone production is derived using the LINOZ pre-determined linearized flux-based ozone boundary condition parameterization [8].

18.1.2 New Photolysis Module

The photolysis module based on the Modular Earth Submodel System (JVAL14-MESSy) of Sander et al. [9] was incorporated to provide photolysis rates (Jval) for the new reactions with consideration of upper atmosphere influences of actinic flux and variable ozone column. Photolysis rates were calculated using the actinic flux ratio approach of Landgraf and Crutzen [10]. The wavelength range covers 178 nm to 682 nm with 8 wavelength bins. Input ozone column, cloud radiative properties

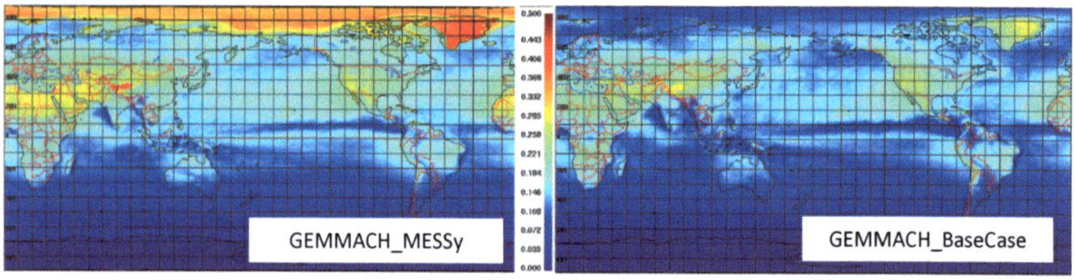

Fig. 18.1 Mean summer (JJA) surface level JNO2 photolysis rate [sec^{-1}] from GEM-MACH MESSy module (left) and the GEM-MACH base-case (right)

and cloud fraction were calculated online at runtime. Comparisons with ADOM2 Jval showed spatial differences both near the surface and aloft (Fig. 18.1), resulting in general increases in domain surface O_3 in the summer, and reductions in winter months.

18.1.3 Lightning NOx Emission

Lightning is calculated using a parameterization based on the upward flux of ice at 440 hPa scaled by fractional cloud cover [11]. Other parameterizations were also tested, but this parameterization gave the best overall geographic distribution and the best seasonal variation compared with the LIS/ODT climatology of lightning flash rate for the GEM-MACH Global configuration. NO_x emissions due to lightning are assigned a value of 500 mol $flash^{-1}$ for mid latitude lightning strikes [12], and 260 mol $flash^{-1}$ for tropical lightning strikes to account for the differences in strike length between tropical and mid latitude storms [13]. The cloud-to-ground (CG) versus intra-cloud (IC) lightning ratio is calculated based on Price and Rind [14] using cold-cloud thickness, but here we assume that the same efficiency for both IC and CG lightning flashes. The vertical distribution of the NO_x emissions is based on profiles from Ott et al. [15] that differentiate between ocean lightning, and tropical, subtropical and mid latitude continental lightning. The profiles are scaled by cloud top height so that NO_x emissions are not located above the cloud top. Figure 18.2 shows the annual mean lightning NOx emissions for a simulation of 2010 for both geographical location and vertical distribution. Under these conditions, the model is tuned to produce an annual average lightning flash rate of 46 flashes day^{-1} globally, and produces a total of 7.4 Tg N per year.

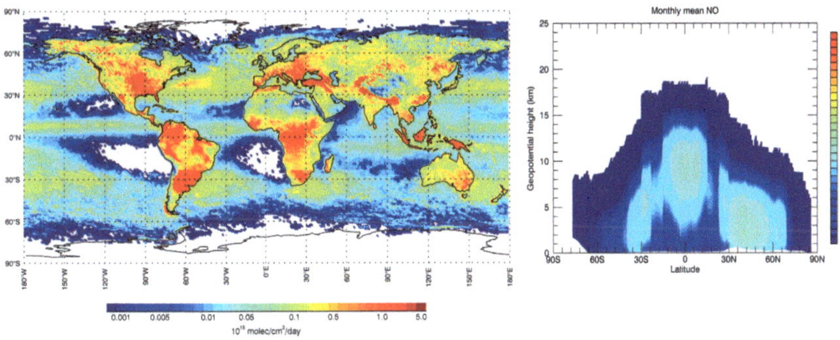

Fig. 18.2 Annual mean lightning NOx emissions for 2010 in total column amount in molecules cm^{-2} day^{-1} (left) and monthly zonal mean NOx in ppbv (right)

18.2 Annual Simulation

The updated GEM-MACH-global system is applied for evaluation to an annual simulation over 2010. The anthropogenic emissions were derived from the Hemispheric Transport of Air Pollution (HTAPv2) dataset. Biogenic emissions are modelled online with the BEIS3 algorithm with updates to provide a global vegetation land cover. Biomass burning emissions are included using the inventory from the Global Fire Emissions Database (GFEDv3). Additional chemical conditions for long-lived species such as CH4, and N2O are prescribed using monthly varying climatologies. Figure 18.3 show summer surface NO and O_3 concentration differences as result of MESSy photolysis module updates taking the MESSy case output minus the base case. Figure 18.4 show JJA zonal average HNO3 with ADOM2 and revised conditions. Additional results are forth coming with comparisons with analysis fields.

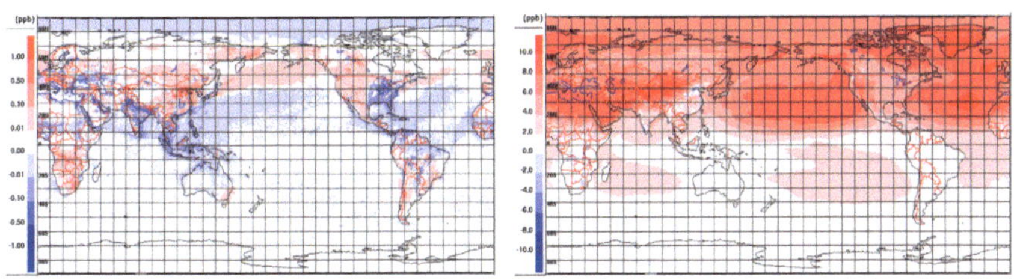

Fig. 18.3 Average summer (JJA) surface NO (left) and O_3 (right) concentration changes for GEM-MACH-Global base case with the new MESSy photolysis module

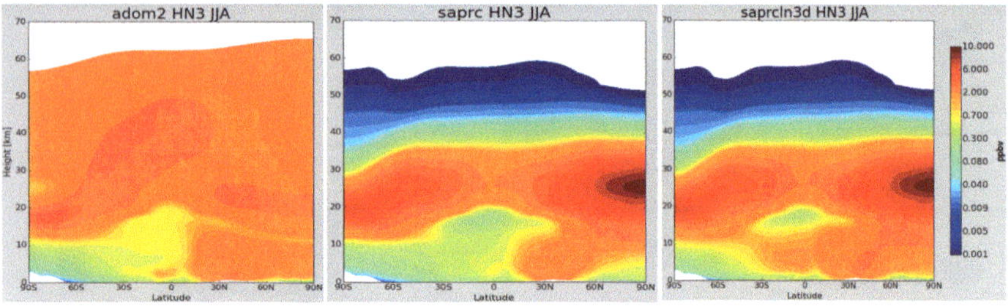

Fig. 18.4 Zonal average summer (JJA) HNO_3 concentrations for GEM-MACH-Global base case chemistry (left), SAPRC chemistry (middle), and SAPRC + lightning NOx emissions (right)

References

1. M. Charron et al., The stratospheric extension of the canadian global deterministic medium-range weather forecasting system and its impact on tropospheric forecasts. Mon. Weather Rev. **140**, 1924–1944 (2012)
2. J. Côté et al., The operational CMC–MRB global environmental multiscale (GEM) model. Part I: design considerations and formulation. Mon. Weather Rev. **126**, 1373–1395 (1998)
3. S.-L. Gong et al., Canadian Aerosol Module: a size segregated simulation of atmospheric aerosol processes for climate and air quality models: 1. Module Dev. J. Geophys. Res. **108**(D1), 4007 (2003)
4. W.R. Stockwell, F.W. Lurmann, Intercomparison of the ADOM and RADM gas-phase chemical mechanisms. Electrical Power Research Institute Topical Report, EPRI, 3412 Hillview Avenue, Palo Alto, CA., 254 p (1989)
5. W.P.L. Carter, Development of a condensed SAPRC-07 chemical mechanism. Atmos. Environ. **44**, 5336–5345 (2010)
6. A. Sandu, R. Sander, Technical note: simulating chemical systems in Fortran90 and Matlab with the Kinetic PreProcessor KPP-2.1. Atmos. Chem. Phys. **6**, 187–195 (2006). https://doi.org/10.5194/acp-6-187-2006
7. Havala O.T. Pye et al., Epoxide pathways improve model predictions of isoprene markers and reveal key role of acidity in aerosol formation. Environ. Sci. Technol. **47**(19), 11056–11064 (2013)
8. C.A. McLinden et al., Stratospheric ozone in 3-D models: a simple chemistry and the cross-tropopause flux. J. Geophys. Res. Atmos. **105**, 14653–14665 (2000)
9. R. Sander et al., The photolysis module JVAL-14, compatible with the MESSy standard, and the JVal PreProcessor (JVPP). Geosci. Model Dev. **7**, 2653–2662 (2014)
10. J. Landgraf, P.J. Crutzen, An efficient method for online calculations of photolysis and heating rates. J. Atmos. Sci. **55**, 863–878 (1998)
11. D.L. Finney et al., Using cloud ice flux to parametrise large-scale lightning. Atmos. Chem. Phys. **14**, 12665–12682 (2014). https://doi.org/10.5194/acp-14-12665-2014
12. R.C. Hudman, et al. Surface and lightning sources of nitrogen oxides over the United States: magnitudes, chemical evolution, and outflow. J. Geophys. Res. **112**, D12S05 (2007). https://doi.org/10.1029/2006jd007912
13. H. Huntrieser et al., Lightning-produced NOx in tropical, subtropical and mid latitude thunderstorms: new insights from airborne and lightning observations. Geophys. Res. Abstr. SRef-ID:1607-7962/gra/EGU06-A-03286 (2006)
14. C. Price, D. Rind, What determines the cloud-to-ground lightning fraction in thunderstorms? Geophys. Res. Lett. **20**, 463–466 (1993). https://doi.org/10.1029/93GL00226
15. L.E. Ott et al., Production of lightning NOx and its vertical distribution calculated from three-dimensional cloud-scale chemical transport model simulations. J. Geophys. Res. **115**, D04301 (2010). https://doi.org/10.1029/2009JD011880

Chapter 19
Hamilton Airshed Modelling System

Anthony Ciccone, Janya Kelly and James Wilkinson

Abstract The objective of this study is to develop the Hamilton Airshed Modelling System (HAMS) providing a platform to better understand the processes and contributions to Hamilton's air quality, informing future policy and human health impact decisions. Air quality in an urban airshed is influenced by local, regional and transboundary sources, geography and meteorology. HAMS must handle different emission sources, and the transportation and dispersion of emissions, to achieve realistic simulations of local impacts on air quality. HAMS relies on the development of two key data sets. The meteorology dataset impacts the transportation, transformation, dispersion, and deposition of pollutants over the challenging terrain in the Hamilton area. The emissions dataset represents the local and regional sources and contaminants influencing the air quality with contributions from industrial, commercial, residential, biogenic, and transportation sources. The Community Multi-scale Air Quality (CMAQ) model combines these datasets in a nested a one-way grid formation from regional (36 km) to local (1.33 km) scales and validates modeled output against observations. Recent local studies indicate that mobile and industrial sources are the primary emission sources. This study further examines the source contributions of mobile, industrial and background sources to local impacts on air quality.

A. Ciccone · J. Kelly (✉)
Golder Associates Ltd, Mississauga, ON L5N 7K2, Canada
e-mail: jkelly@golder.com

A. Ciccone
e-mail: aciccone@golder.com

J. Wilkinson
Terra-Technologies, Eugene, OR 97401, USA
e-mail: terrakit.jw@gmail.com

C. Mensink et al. (eds.), *Air Pollution Modeling and its Application XXVI*,
Springer Proceedings in Complexity,
https://doi.org/10.1007/978-3-030-22055-6_19

19.1 Introduction

The objective of this study is to develop and implement the Hamilton Airshed Modelling System (HAMS), which will provide a platform to better understand the processes and contributions to Hamilton's air quality and aid to inform future policy and human health impact decisions. Air quality in an urban airshed is influenced by local, regional and transboundary sources, as well as by the geographical features and prevailing meteorological conditions. An airshed modelling system must be capable of handling the different emission sources, the complex meteorology, and the transportation and dispersion of emissions to achieve realistic simulations of local impacts on air quality.

The air quality in the Hamilton Region was modelled by tracking the emission, dispersion and chemical transformation of twenty selected contaminants in the airshed, with a focus on NO_X, SO_2, O_3 and particulate matter ($PM_{2.5}$ and PM_{10}). Emphasis was also placed on benzene and benzo(a)pyrene (B[a]P) for health and regulatory reasons. The modelling is completed using four levels of nested spatial grids, providing both regional and local contributions to the air quality. The Hamilton airshed is part of the larger southern Ontario airshed, with influences from both south-western Ontario and the United States. The influences each have different emission profiles with respect to contaminants, quantity of emissions, and types of sources (e.g., stacks, roadways, area), as well as the specific location of the sources.

19.2 Modelling System Description

The HAMS relies on the development of two key data sets: meteorology and emissions. To represent the transport, and aid in the representation of the chemical transformation, dispersion and deposition of pollutants, a meteorological data set is needed, which will include the unique and challenging influences of the terrain in the Hamilton area. To represent the sources and compounds influencing air quality, an emissions inventory is required which includes both local and regional sources and accounts for contributions from human-made and natural sources.

The Weather Research and Forecasting (WRF) Nonhydrostatic Mesoscale Model (NMM) model version 3.6.1 [3] was selected to generate 3-dimenisional hourly meteorological data for the dispersion and chemical modelling. Detailed analysis demonstrated that WRF did well at replicating observed atmospheric values of temperature, mixing ratio, wind speed and wind direction. Monthly and annual analyses demonstrated a high level of accuracy in reproducing observations, even in the region surrounded by the Great Lakes (Tier IV).

Gridded, hourly emissions estimates of speciated compounds were prepared primarily with the Sparse Matrix Operator Kernel Emissions version 3.6 (SMOKE v3.6) [2] for emission and source data including: human-made emissions data covering industrial, commercial, residential, mobile, and non-road (commercial marine shipping, locomotives and aircraft). Biogenic sources were prepared using the Model of Emissions of Gases and Aerosols from Nature (MEGAN; [4]) biogenic emissions model. For the on-road mobile sources, the emissions rates were based on input mobile source activity data, emission factors estimated using the Motor Vehicle Emission Simulator (MOVES2014 [5]), and outputs from transportation travel-demand models. Emissions data sets were derived from multiple agencies (e.g., United States Environmental Protection Agency (USEPA), Environment and Climate Change Canada and Ministry of Environment and Climate Change) as well as other municipal and regional agencies and industries in Hamilton.

The modelling system was based on the Community Multi-scale Air Quality v5.0.2 (CMAQ v5.0.2; [1] model. CMAQ simulates the various chemical and physical processes influencing the local air quality, including the emission, dispersion and chemical transformation of pollutants. The CMAQ modelling system consists of four pre-processors to the CTM, which provide information on the meteorology, meteorology-dependant chemical reactions and the concentration of species at the boundary and in the background (i.e. non-local sources). The modelling system was developed for four one-way nested modelling tiers with resolutions of 36, 12, 4 and 1.33 km. The highest resolution tier (Tier IV) was selected to allow for the focus on the Hamilton region. HAMS was run for 2012 with meteorology and emissions developed to be consistent with the air quality in 2012.

19.3 Modelling Performance Results

Model Performance results were measured against publicly available monitoring data in Tier IV in both paired and unpaired statistical analyses. The modelled values are taken from the grid cell that contains the monitoring site. Spatial displays across the Tier IV domain and time series at monitoring sites were also considered.

The modelling system provides generally reasonable results given the complexity of the model and inputs. CMAQ tends to over-predict concentrations with the exception PM_{10} where the model under-predicts based on the mean bias. The under-prediction of PM_{10} is likely due to unaccounted local sources generating fugitive dust. Model results are within a factor of two for all compounds (i.e., a fractional bias $\leq 67\%$).

Time series analysis shows the modelled concentrations following the observed concentrations, with a tendency to over-predict during the winter season. The over-prediction of the NO_2 is evident throughout the year and is consistent at approximately 10 ppb, which suggest that transboundary emissions are influencing the results.

The spatial displays show a large difference between the maximum daily value (over the entire year) and the annual average of the daily values. This indicates that the region experiences high magnitude events but the average conditions are much lower, by at least an order of magnitude. Generally, maximum concentrations occurred along

Burlington Ave E and at the intersection of Highway 403 and Highway 8, depending on the compound of interest. Ozone and NO_2 spatial plots show concentrations correlating with the major roadways.

19.4 Source Apportionment Results

To complete the source apportionment for HAMS a "zero-out" approach was used where emissions from selected source groups are removed from the inventory. The importance of that source is assessed by evaluating the change in ambient air quality as the difference (absolute and percentage) in compound concentration between the base case (with the source group) minus the scenario without the source group. Simulations were carried out evaluate the contribution to the air quality compounds of interest from: on-road mobile sources; industrial sources; non-road mobile (rail, airport and marine) sources; and transboundary only (all sources within Tier IV were removed).

The source apportionment simulations were carried out for a selected month of each season for 2012 as well as for an ensemble average of all four months to represent a year for the Tier IV domain. The Tier III domain results were used for initial conditions and a minimum of a five day spin up time was used for each simulation. Results indicate a strong transboundary influence in the Hamilton region, with transboundary influences contributing upwards of 90% for particulate matter ($PM_{2.5}$ and PM_{10}). Local industrial emissions contribute less than 20% by compound to air quality in the Hamilton region, except for B(a)P where it is the main source (47%). Local on-road sources are a major contributor (34%) to NO_2 levels in the domain, after transboundary (44%). Transportation related emissions are the major contributor to transboundary emissions for all compounds except SO_2 which is dominated by industrial sources.

19.5 Conclusion

The results of the HAMS demonstrate that the model is conservative in the simulation of air quality levels within the Hamilton region. While most compounds are over-predicted, they are within a factor of two of the observations. The source apportionment demonstrates a strong transboundary influence across the domain, heavily influenced by transportation (on-road and non-road) emissions. Local industrial emissions have a comparatively small influence, apart from B(a)P. Further studies are needed to strengthen the source apportionment and provide more detailed information to support policy development to improve the air quality in the Hamilton region.

Acknowledgements Golder would gratefully like to acknowledge the following contributions which made the Hamilton Airshed Modelling System possible. The initial and boundary conditions from GEOS-CHEM were provided by Dr. Barron Henderson and his research team at the University in Florida. Environment and Climate Change Canada provided the SMOKE ready national emissions inventory for Canada, as well as Paul Makar and Junhua Zhang providing technical expertise on processing emissions in SMOKE and interpreting the impact on modelled concentrations. Finally, Golder would like to thank the Project Steering Committee and the Hamilton Industrial Environmental Association for providing continued direction during the project.

References

1. CMAS, Operational Guidance for the Community Multiscale Air Quality (CMAQ) Modeling System (2015). Retrieved from https://www.airqualitymodeling.org/index.php/CMAQ_version_5.0_(February_2010_release)_OGD
2. CMAS, Community Modeling & Analysis System. SMOKE v.3.6 (2014). Retrieved from https://www.cmascenter.org/smoke/
3. NCAR, Weather Research and Forecasting Model (2018). Retrieved from https://www.mmm.ucar.edu/weather-research-and-forecasting-model
4. UCAR, Model of Emissions of Gases and Aerosols from Nature (MEGAN) (2015). Retrieved from http://lar.wsu.edu/megan/index.html
5. USEPA, Motor Vehicle Emission Simulator (2014). Retrieved from http://www.epa.gov/otaq/models/moves/

Chapter 20
On the Urban Canopy Effects in Regional Climate Simulations—An Inter-Model Comparison and Potential for Prediction

Tomas Halenka, Peter Huszar, Michal Belda, Jan Karlicky and Tereza Novakova

Abstract To assess the impact of cities and urban surfaces on climate, the modeling approach is often used with inclusion of urban parameterization in land-surface interactions. This is especially important when going to higher resolution, which is common trend both in operational weather prediction and regional climate modelling. Inclusion of urban related effects can differ significantly across the land-surface models and urban canopy parameterizations. For adaptation and mitigation measures applied in big cities, especially in connection to climate change perspective, it is important to assess this uncertainty as well as to analyse the effects, which can affect air quality situation within the cities. These are main tasks of the new project URBI PRAGENSI. We performed experiments to assess urban effects on climate over central Europe for the decade 2001–2010, using two regional climate models (RegCM4 and WRF) in 10 km resolution driven by ERA-Interim reanalyses, three surface schemes (BATS and CLM4.5 for RegCM4 and Noah for WRF) and five urban canopy parameterizations available: one bulk urban scheme, three single layer and a multilayer urban scheme.

20.1 Introduction

The role of cities is increasing and will continue to increase in future, as the population within urban areas is growing faster. More than 50% of world population has been living in cities, the number of big cities is steadily growing and the outlook for 2030 is more than 60% population in cities above 300,000 [21]. In Europe, it is now about 75%, and about 84% of population shall be living in urban areas in the middle of 21st century. It means that weather and climate change effects are affecting significant and increasing part of population via urban environment.

T. Halenka (✉) · P. Huszar · M. Belda · J. Karlicky · T. Novakova
Department of Atmospheric Physics, Faculty of Mathematics and Physics, Charles University, 182 00 Prague, Czech Republic
e-mail: tomas.halenka@mff.cuni.cz

C. Mensink et al. (eds.), *Air Pollution Modeling and its Application XXVI*,
Springer Proceedings in Complexity,
https://doi.org/10.1007/978-3-030-22055-6_20

Under increasing resolution of numerical models, which is common trend both in regional climate modelling and numerical weather prediction, the role of proper parameterization of urban processes is rising as well. Model grid-boxes at scales of 10 km and less can be fully filled with urban land use and for bigger cities even different parts of the city with different parameters can be resolved. The inclusion of the specific urban processes affecting energy balance and transport, i.e. heat, humidity, momentum fluxes, is vital to get the urban environment effects on urban canopy layer correctly, which is necessary for further proper planning of adaptation and/or mitigation options by city authorities, as well as to inform citizens on the real conditions in the city, especially under some specific events like heat waves, smog conditions etc.

The artificial urban surfaces clearly differ from natural surfaces and processes involved represent additional sinks and sources of momentum, heat and moisture affecting the properties of local atmosphere and having specific impact on the meteorological conditions, which is a well-known phenomenon since the 1980's [5, 12, 13]. One of the most comprehensively studied aspects of urban surfaces effect is the Urban Heat Island (UHI) phenomenon. UHI forms as a result of modified energy budget due to the canyon-like geometry of the canopy layer and the specific thermal parameters of the artificial surfaces [12]. Due to their decreased albedo, urban surfaces store more heat compared to rural areas and after sunset this heat is released with a reduced intensity because of the decreased sky-view factor [6], which makes UHI typical especially for late afternoon and nighttime. Moreover, there is missing cooling due to lack of evaporation.

There are further impacts of urban surfaces on other meteorological parameters. Richards and Oke [17] studied the changes of surface humidity, while e.g. Roth [18] focused on the impact on roughness and turbulence. Many studies dealt with the structure of the urban boundary layer including the impact on the height of the planetary boundary layer [15] In addition, urban-meteorology [10] and wind speeds [7]. Urbanization-triggered changes in precipitation and hydrological processes got as well into the attention of research [19, 20]. In addition, urban-meteorology interaction may significantly influence air-quality ([8], with further references). These influences are connected within the phenomenon of UHI having a direct impact on the human health [16] and, in general, on the comfort of living in the cities, especially in context of vulnerable minorities or groups of population.

To provide a reliable, numerical modeling based perspective of the UHI phenomenon, and of other related impacts (e.g. on wind speed/direction, precipitation, mixing layer height, etc.), the complex nature of the mechanical, thermo-dynamical and radiative processes has to be represented in models. The inclusion of urban-canopy-models (UCM) is necessary in this regard [1, 2, 9, 14] which are specially designed to parameterize the processes specific to the urban environment that are not resolvable at the model's scale. Most of the modelling studies focused on a particular city with minor interest in the impact on regional scale further from the urban area itself and mostly just for case studies of several days. For more references see Huszar et al. [9], where we attempted long term view on the urban effects in regional climate simulation for central Europe. Building upon this study we further analyse and compare the effects of different urban paramterizations in two atmospheric models, regional climate model RegCM and WRF.

20.2 Experiments Setup

The regional climate model RegCM4.2 [4] was used with its BATS land surface scheme [3] extended by Huszar et al. [9], following Chen et al. [2], with the Single-layer Urban Canopy Model (SLUCM) that accounts for the most relevant processes specific to the urban environment including the anthropogenic heat release. In newer version RegCM4.5 used with CLM4.5 scheme, internal option of urban parameterization is included in terms of SLUCM by Oleson et al. [14]. In addition, model WRF in the adaptation for regional climate simulations was tested with all the options of urban parameterizations available, i.e. bulk version, where actually there is are no urban proceses included, except the land use parameters and properties are set with proper values adequate the artificial surfaces in the urban environment. Similar implementation of SLUCM [2] is another option in WRF, as well as the more complex multi-layer scheme BEP-BEM including energetics of buildings as well, following Martilli et al. [11]. The simulations for the central European region have been performed for the years of 2001–2010 in 10 km resolution, with all the settings. The comparison of urban effects in the individual simulations is presented against the "control" simulations when urban land use was replaced by typical land use of the urban vicinity.

20.3 Results and Conclusions

In Fig. 20.1 we present the comparison of simulations with selected urban parameterizations against the experiment with no-urban"control" setting. It is clearly seen that the effects of UHI differ significantly depending on the parameterization choice and model implementation, however, common features of UHI behavior are reproduced reasonably well. In summer night, as expected, under specific conditions of heat waves, the intensity of UHI can achieve for mid-size cities like Prague quite high values of about 5 °C, which is in accordance with observations. For larger cities, or with higher urbanization due to more heavy population, like e.g. Budapest, the simulations can reach up to 10 °C intensity which is value reffered to in literature.

The models performance in terms of daily course of temperature for all the simulations and selected urban area of Prague has been improved and urban effects due to urban parameterization are clearly identified. Even in WRF cases the bulk parameterization can reasonably reproduce the temperature effect, except for winter, where anthropogenic heat release dominates, which is best reproduced by BEP-BEM parameterization. However, the results for surface wind and mixing height clearly

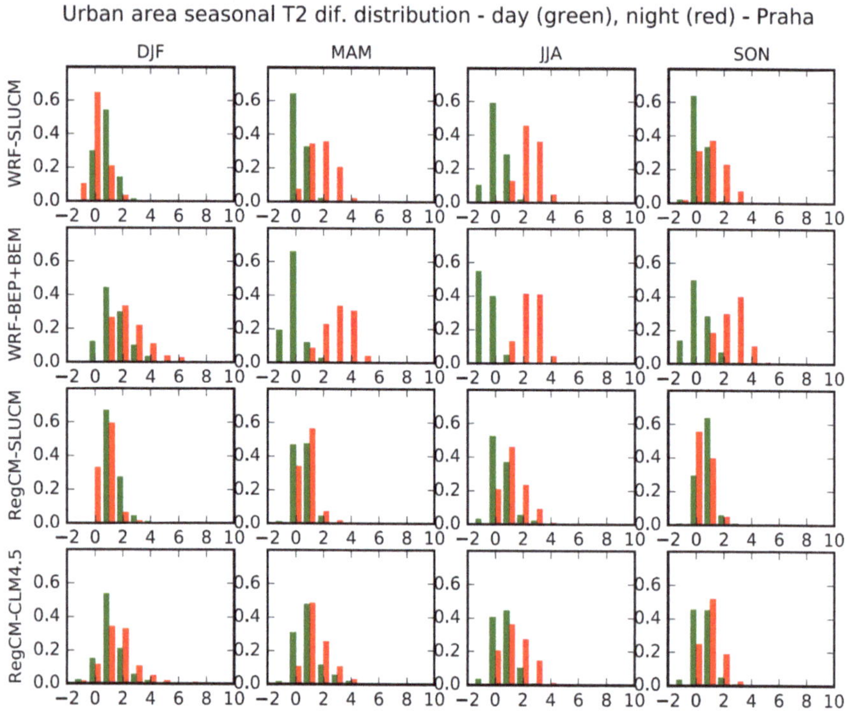

Fig. 20.1 The seasonal distribution of UHI effect for individual urban parameterizations in Prague during 2001–2010

show that simplistic treatment of the urban environment is not adequate in any season, with clear effect of the process based urban parameterizations decreasing the values of surface wind by the urban infrastructures and increasing the mixing height. This is clearly revealing the significance of use the proper parameterization of urban processes, especially in connection to air quality applications.

Acknowledgements The authors wish to thank for support under Programme OP PPR with the project URBI PRAGENSI, CZ.07.1.02/0.0/0.0/16_040/0000383, as well as local support of Charles University in Programme Progress, Q16-Environmental Research

References

1. A. Baklanov, U. Korsholm, A. Mahura, C. Petersen, A. Gross, ENVIRO-HIRLAM: on-line coupled modelling of urban meteorology and air pollution. Adv. Sci. Res. **2**, 41–46 (2008)
2. F. Chen, H. Kusaka, R. Bornstein, J. Ching, C.S.B. Grimmond, S. Grossman-Clarke, T. Loridan, K.W. Manning, A. Martilli, S. Miao, D. Sailor, F.P. Salamanca, H. Taha, M. Tewari, X. Wang, A.A. Wyszogrodzki, C. Zhang, The integrated WRF/urban modelling system: development,

evaluation, and applications to urban environmental problems. Int. J. Clim. **31**(273–288), 2011 (2011)
3. R.E. Dickinson, A. Henderson-Sellers, P. Kennedy, Biosphere–atmosphere transfer scheme (BATS) version 1 as coupled to the NCAR community climate model. *Tech Rep*, National Center for Atmospheric Research Tech Note NCAR.TN-387 + STR, NCAR, Boulder, CO (1993)
4. F. Giorgi, E. Coppola, F. Solmon, F., L. Mariotti, M. Sylla, X. Bi, N. Elguindi, G. T. Diro, V. Nair, G. Giuliani, S. Cozzini, I. Guettler, T.A. O'Brien, A. Tawfi k, A. Shalaby, A. Zakey, A. Steiner, F. Stordal, L. Sloan, C. Brankovic, RegCM4: Model description and preliminary tests over multiple CORDEX domains, Clim. Res. **52**, 7–29 (2012)
5. J.M. Godowitch, J.K.S. Ching, J.F. Clarke, Evolution of the nocturnal inversion layer at an urban and nonurban site. J. Appl. Meteorol. and Clim. **24**, 791–804 (1985)
6. C.S.B. Grimmond, T.R. Oke, Comparison of heat fluxes from summertime observations in the suburbs of four North American cities. J. Appl. Meteorol. **34**, 873–889 (1995)
7. A. Hou, G. Ni, H. Yang, Z. Lei, Numerical analysis on the contribution of urbanization to wind stilling: an example over the greater beijing metropolitan area. J. Appl. Meteor. Climatol. **52**, 1105–1115 (2013)
8. P. Huszar, M. Belda, T. Halenka, On the long-term impact of emissions from central European cities on regional air quality. Atmos. Chem. Phys. **16**, 1331–1352 (2016)
9. P. Huszar, T. Halenka, M. Belda, M. Zak, K. Sindelarova, J. Miksovsky, Regional climate model assessment of the urban land-surface forcing over central Europe. Atmos. Chem. Phys. **14**, 12393–12413 (2014)
10. A. Martilli, Numerical study of urban impact on boundary layer structure: sensitivity to wind speed, urban morphology, and rural soil moisture. J. Appl. Meteor. **41**, 1247–1266 (2002)
11. A. Martilli, A. Clappier, M.W. Rotach, An urban surface exchange parameterization for mesoscale models. Bound. Layer Meteorol. **104**, 261–304 (2002)
12. T.R. Oke, The energetic basis of the urban heat island. Q. J. R. Meteorol. Soc. **108**, 1–24 (1982)
13. T.R. Oke, Boundary Layer Climates, 2nd edn., Routledge (1987)
14. K.W. Oleson, J.B. Bonan, J. Feddema, M. Vertenstein, C.S.B. Grimmond, An urban parameterization for a global climate model, Part I: Formulation and evaluation of two cities. J. Appl. Meteorol. Clim. **47**, 1038–1060 (2008)
15. M. Piringer, Exploring the urban boundary layer by sodar and tethersonde. Phys. Chem. Earth Pt. B. **26**, 881–885 (2001)
16. C.E. Reid, M.S. O'Neill, C.J. Gronlund, S.J. Brines, D.G. Brown, A.V. Diez-Roux, J. Schwartz, Mapping community determinants of heat vulnerability. Environ. Health Perspect. **117**, 1730–1736 (2009)
17. K. Richards, T.R. Oke, Validation and results of a scale model of dew deposition in urban environments. Int. J. Climatol. **22**, 1915–1933 (2002)
18. M. Roth, Review of atmospheric turbulence over cities. Q. J. Roy. Meteor. Soc. **126**, 941–990 (2000)
19. C.M. Rozoff, W.R. Cotton, J.O. Adegoke, Simulation of St. Louis, Missouri, land use impacts on thunderstorms. J. Appl. Meteorol. **42**, 716–738 (2003)
20. J.M. Shepherd, H. Pierce, A.J. Negri, Rainfall modification by major urban areas: observations from spaceborne rain radar on the TRMM satellite. J. Appl. Meteor. **41**, 689–701 (2002)
21. United Nations, Department of Economic and Social Affairs, Population Division (2014). World Urbanization Prospects: The 2014 Revision

Part II
Local and Urban Scale Modeling

Chapter 21
Atmospheric Pollution: Experience from Mexico City and Santiago de Chile

Luisa T. Molina, Wenfang Lei, Miguel Zavala, Victor Almanza, Agustin Garcia, Pablo Saide and Marcelo Mena-Carrasco

Abstract Megacities are now a common phenomenon in many regions around the world and present major challenges for the global environment. The concentrations of people and their activities have resulted in higher demand for energy and consumption of fossil fuels, leading to air pollution that affects public health and visibility, causes regional haze and acid deposition, and alters the earth's climate. Recent advances in real-time pollutant measurement technologies and improved air quality models are allowing scientists to better understand the emission sources of pollutants and the complex atmospheric processes leading to severe air pollution, and providing policy makers the tools for designing cost effective mitigation strategies. This study addresses the effects of megacities and urban complexes on the Earth's atmosphere, using Mexico City and Santiago as examples of cities that have been actively managing their air quality. Both cities demonstrate the types of environmental problems experienced by many urban centers and confront similar challenges to solving them.

L. T. Molina (✉) · W. Lei · M. Zavala · V. Almanza
Molina Center for Energy and the Environment, La Jolla, CA, USA
e-mail: ltmolina@mce2.org; ltmolina@mit.edu

W. Lei
e-mail: wflei@mce2.org

M. Zavala
e-mail: miguelz@mce2.org

V. Almanza
e-mail: valmanza@mce2.org

A. Garcia
Center for Atmospheric Sciences, National University of Mexico, Mexico City, Mexico
e-mail: agustin@atmosfera.unam.mx

P. Saide
Department of Atmospheric and Oceanic Sciences and Institute of the Environment and Sustainability, University of California, Los Angeles, CA, USA
e-mail: saide@atmos.ucla.edu

M. Mena-Carrasco
Climate Change Group, World Bank, Washington, D.C., USA
e-mail: mmenacarrasco@worldbank.org

C. Mensink et al. (eds.), *Air Pollution Modeling and its Application XXVI*,
Springer Proceedings in Complexity,
https://doi.org/10.1007/978-3-030-22055-6_21

With appropriate planning, dedicated scientific research, robust emissions control policies, and effective access to advanced technologies and financial support, these urban centers also have the opportunity to manage the growing population sustainably while reducing atmospheric pollution and its impacts.

21.1 Introduction

Megacities, typically defined as urban agglomerations with more than 10 million inhabitants, are now a common phenomenon in many regions around the world. As of April 2018, 55% (4.3 billion) of the global population (7.8 billion) reside in urban areas. The world's urban population is projected to increase 2.5 billion by 2050 due to population growth and high urbanization rate, with almost 90% of this increase occurring in Asia and Africa. Currently there are 33 megacities in the world; this is projected to increase to 43 by 2030, with most of them in the developing regions [16]. While most of the world's urban population still lives in cities of fewer than 10 million inhabitants, many of these cities are growing faster than the megacities. The concentrations of people and their activities have resulted in higher demand for energy and consumption of fossil fuels, leading to atmospheric pollution that affects public health, causes regional haze and acid deposition, and alters the earth's climate. On the other hand, as centers of economic growth, higher education and technology innovation, these urban centers also offer opportunities to manage the growing population sustainably and reduce the atmospheric pollution and its impacts.

Air pollution in megacities and large urban centers is strongly influenced by topography, meteorology, and a wide array of natural and anthropogenic sources including industrial activities, residential combustion, solvents use and transport, among others. The sources of emissions vary substantially among megacities, notably by geographical region. For example, the transport sector accounts for most of the emissions in megacities and large urban centers in Europe and the Americas, whereas emissions from Asia and Africa are largely from residential biofuel use [1]. Mexico City, Mexico and Santiago, Chile are presented as examples of urban centers that have been active in managing their air quality. Both cities have demonstrated the types of environmental problems experienced by large urban centers, encountered similar challenges, and have been successful in implementing emissions reduction strategies.

21.2 Mexico City Metropolitan Area

The Mexico City Metropolitan Area (MCMA) extends over 7600 km^2 and is home to more than 21 million inhabitants, making it the largest metropolitan area in North America and one of the largest in the world. With an altitude of 2240 m a.s.l., an opening to the Mexican Plateau to the north, and a mountain gap to the southeast, the MCMA is surrounded by mountain ridges and volcanoes that separate it from adjacent urban areas (see Fig. 21.1). The MCMA currently includes Mexico City, 59 municipalities of the State of Mexico and one municipality of the State of Hidalgo; it has undergone huge transformations in urbanization and demographics over the past decades. The expansion of the MCMA is not unique in the region; the neighboring metropolitan areas of Puebla, Tlaxcala, Toluca, Cuernavaca, Queretaro and Tula are also extending their territories, producing a contiguous urban complex known as the "Megalopolis" in the central region of Mexico with an estimated population of about 31 million (see Fig. 21.1).

There is a strong connection between the topography of the MCMA and the regional meteorology that ultimately impacts air pollution levels throughout the year [8, 9]. The "ozone season" occurs during the warm-dry season from March to June, with the highest ozone peaks occurring mostly during April and May. This period is characterized by high pressure systems that induce days with clear skies and weak wind. Intense solar radiation is also common in this period due to the MCMA's high altitude and low latitude, further inducing fast photochemical oxidant formation. Intense dust and biomass burning emissions during the warm-dry season can

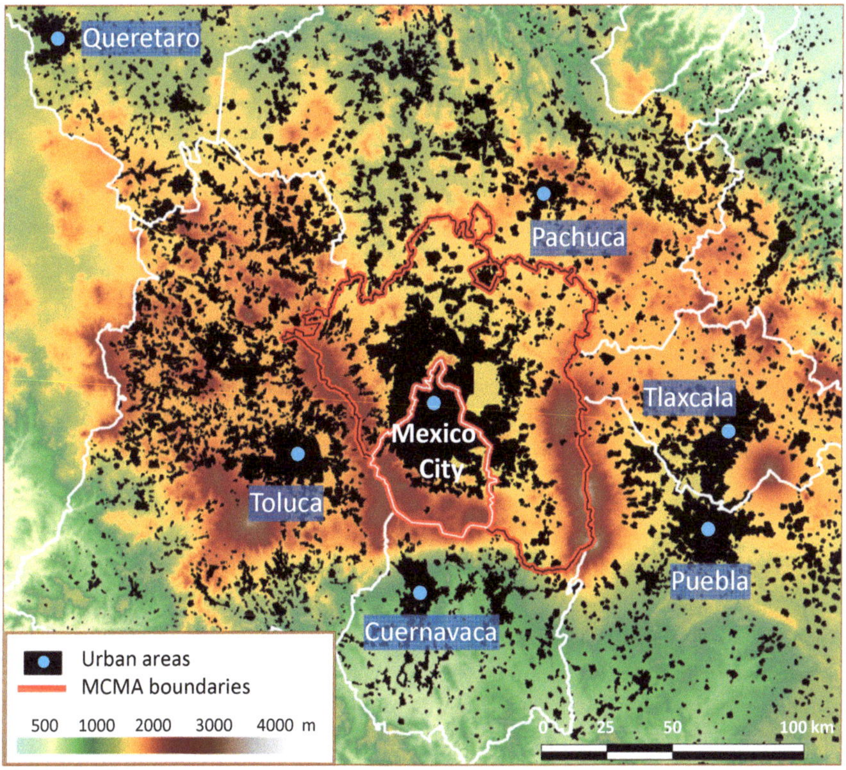

Fig. 21.1 Map of the Megalopolis and topography in the central region of Mexico. Urban areas are shown in black. Red lines show the MCMA boundaries

also increase particulate matter (PM) concentrations. The rainy season from June to October has lower PM concentrations; however, because of intense photochemistry occurring just before the afternoon showers, the ozone levels continues to be relatively high. The cool-dry season from November to February is characterized by high levels of PM due to strong surface inversions and weak wind that also induce higher concentrations of primary pollutants during the morning hours. As a result, air pollution is a concern for MCMA residents throughout the year.

In addition to the topographic and meteorological drivers of air pollution in the MCMA, the rapid population growth, uncontrolled urban development, high rates of motorization, industrialization and consumption of fossil fuels combined to severely impact the air quality in the region. With ozone peaked above 300 ppb 40–50 days a year and high concentrations of all criteria pollutants in the late 1980s, Mexico City was ranked as the most polluted megacity in the world [17]. The Mexico City government addressed the air pollution problem by establishing an air quality monitoring network, and developing and implementing air quality management programs that combined modifications to the regulatory framework with technological changes. In collaboration with the Integrated Program on Urban, Regional and Global Air Pollution at the Massachusetts Institute of Technology, the project "Integrated strategy for the air quality management in the Valley of Mexico 2001–2010" was also pivotal for understanding the drivers of air pollution in the MCMA and providing scientific basis for the design of air quality control strategies. Major measures implemented by the government in the transport sector included removal of lead from gasoline and its reformulation, mandatory use of catalytic converters in automobiles, reduction of sulfur content in diesel fuel, reinforcement of the vehicle inspection and maintenance program, and implementation of the air pollution contingency program and "no driving day" programs; other implemented measures in the industrial and residential sectors included substitution of fuel oil in industry and power plants with natural gas, and reformulation of liquefied petroleum gas for cooking and water heating [7]. These and other strategies led to significant reduction of ambient concentrations of criteria air pollutants in the MCMA (see Fig. 21.2). However, there are still many days with poor air quality with respect to O_3, PM_{10} and $PM_{2.5}$. It is necessary to implement new and effective strategies to continue improving air quality, especially in the residential and transport sectors.

21.2.1 Field Measurement Studies in the MCMA

The high levels of aerosols and intense photochemical activities in the MCMA's atmosphere and the unique combination of topographical and meteorological drivers have attracted several large international collaborative studies on the emissions, transformations, and transport of air pollutants since the 1990s. The Mexico City Air Quality Research Initiative (MARI) project obtained surface and vertical profile observations of air pollutants and meteorology during 1990–1994 to support air quality modeling studies [15]. The Aerosol and Visibility Evaluation (IMADA-AVER) campaign in

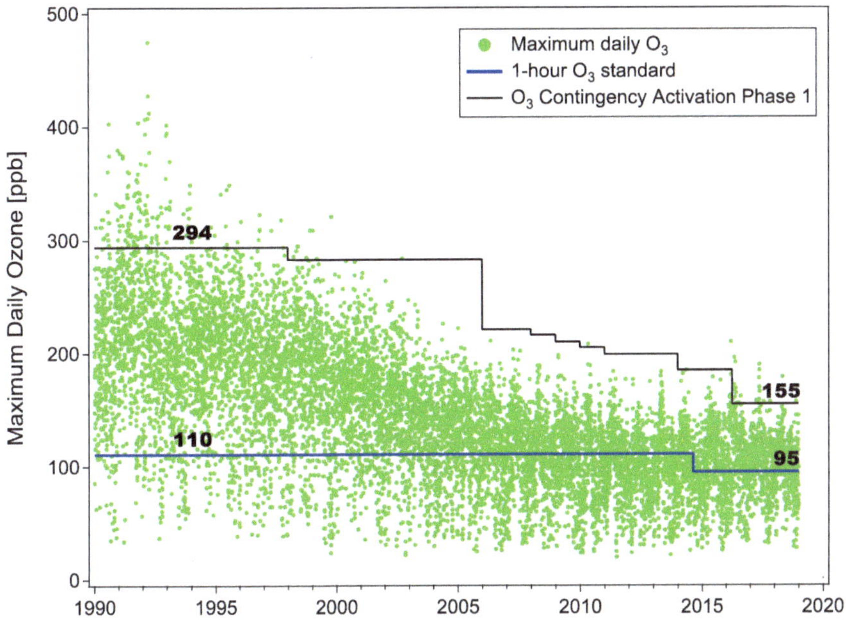

Fig. 21.2 Daily maximum O_3 concentrations (1990–2018) in the MCMA showing changes in the historical value for Phase 1 Contingency activation (black line) and 1-hour O_3 standard (blue line). The graph was generated using data from Mexico City Secretariat of Environment Air Quality Monitoring Network. https://www.sedema.cdmx.gob.mx/

February–March 1997 provided comprehensive measurements of particulate chemical composition and meteorological parameters [2, 3]. These campaigns highlighted the need for more extensive experimental data to update and improve the MCMA emissions inventory and to better understand the chemistry, dispersion and transport processes. The MCMA-2002/2003 campaigns in February 2002 and April 2003 addressed these issues by deploying research-grade instrumentation and using innovative exploratory mobile and fixed site measurement techniques to obtain extensive data of many oxidant precursors and photochemical intermediates, as well as meteorological parameters and emissions [8]. The MILAGRO (Megacity Initiative: Local and Global Research Observations) Campaign in March 2006 deployed a wide range of instruments at ground sites and on aircraft, together with satellite observations and provided extensive measurements of gas and aerosol chemistry, aerosol microphysics and optics, radiation and meteorology. The overall campaign was supported by meteorological forecasting and numerical simulations, the MCMA ambient air quality monitoring network, and meteorological data from Mexico National Meteorological Service [10, 13]. Over 200 peer-reviewed publications have resulted from MCMA-2003 and MILGRO-2006 field studies, contributing significantly to our understanding of the emissions and the atmospheric processes leading to the formation of ozone, secondary aerosols and other pollutants.

Earlier modeling studies suggested that O_3 production within the MCMA is generally VOC-limited. However, potential changes in VOC/NO_x ratios resulting from variations in VOC composition and contributions from key emission sources driven by recent technological and regulatory changes in the MCMA highlight the need to update our understanding of O_3 formation. Under a VOC-limited regime, reduction of VOCs could effectively reduce O_3 formation in the MCMA; reductions of NO_x emissions would not be as effective locally, but may reduce O_3 production at the regional scale. Long-range export of reactive nitrogen from Mexico City occurred primarily through the formation of peroxyacetyl nitrates (PANs), which can thermally decompose, regenerate NO_x and contribute to regional O_3 formation. The production of O_3 (a potent greenhouse gas) continued in the outflow for several days with potential climate impact.

The deployment of multiple instrumentation and sampling techniques in different sites during the MCMA-2003 and MILAGRO field campaigns provided a wealth of knowledge on the chemical composition and physical properties of different types of aerosols, including dust, biomass burning, and anthropogenic emissions and their chemical transformation. One of the main results was the observation during MCMA-2003 that secondary organic aerosol (SOA) production was higher than previously predicted with traditional SOA models, triggering active research for updating SOA formation mechanisms. SOA accounts for a large fraction of the organic aerosol burden; the formation, evolution, and climate impacts of SOA (and organic aerosols) remain one of the least understood aspects in atmospheric science and are important current research topics.

The results from the field studies and policy implications were presented to the Mexican authorities and were included in the current Air Quality Management Program (PROARE 2011–2020). Some of the key policy actions that the government has reinforced include: strengthening the controls of vehicular emissions with advanced technologies and fuel reformulation, diesel vehicles self-regulation program, enhancement of the mandatory vehicle inspection and maintenance program and the atmospheric environmental contingency program, improved public transportation and bike-sharing program. Recently, with the collaboration of Barcelona Supercomputing Center, the Mexico City government has developed an air quality forecasting system to alert the public of high pollution event 24 h in advance, as well as in evaluating emission reduction policies for air quality improvement and other co-benefits (climate, food security, etc.).

21.2.2 Central Mexico Megalopolis

MILAGRO was planned to investigate the evolution of emission plumes from megacities. All large cities emit large amounts of pollutants from multiple local sources that combine with emissions from neighboring cities. Mexico City was the dominant city in central Mexico and MILAGRO provided an opportunity to follow the urban emission plumes from the city center and to study how they change with time and space. However, this situation has changed recently due to the expansion of the urban areas beyond the MCMA forming the Megalopolis (see Fig. 21.1). The Megalopolis concentrates about 25% of the Mexican population, and a substantial fraction of the national gross domestic product is derived from the intensive commercial and industrial activities taking place in this region. However, the combination of poorly planned urban development, inefficient transport systems for a large number of people and goods, and the continuously growing unsustainable economic activities have exerted substantial land-use changes in the region, ultimately impacting the air quality and the environment. The different administrative and regulatory frameworks among the states of the Megalopolis further contribute to the lack of an integrated sustainable development programs. In 2013, the Megalopolis Environment Commission (CAMe) was created to coordinate regional policies and programs.

In recent years there have been several episodes of high O_3 concentrations in Mexico City and other entities of the Megalopolis. During a notable episode in March 2016, warm temperatures and still air left pollutants trapped in the basin led to high O_3 concentrations, forcing the authorities to declare a Phase 1 contingency, the first since 2005. CAMe developed an Air Quality Management Program (PROAIRE 2017–2030) for the Megalopolis to address the situation [12]. Subsequently a modeling study was performed by the Molina Center and collaborators to evaluate the impacts of the implementation of an integrated comprehensive emission control strategies identified in the PROAIRE 2017–2030 on ozone concentrations [6].

In this study three meteorological episodes were selected during the period of high O_3 season (dry-hot season) in the Megalopolis, reflecting the main pollutant transport conditions in the region and used as the baseline for the modeling and analysis of regional O_3 formation. Baseline emissions for the region from the Mexican National Emissions Inventory for the year 2013–2014 were provided by the Mexican authority. An integrated emission reduction scenario was designed that includes all the control strategies listed in the PROAIRE, targeting vehicle, residential, commercial, consumer products, solvents, automobile coatings, painting, wood-burning and regulations (reduction of 32% NO_x and 46% VOCs). The model-ready emission files were used to simulate O_3 concentrations in the Megalopolis for both the baseline and the integrated emissions reduction scenario for the three selected meteorological episodes using a version of the WRF-Chem air quality model [4, 14] developed by Li et al. [5].

Preliminary modeling results indicate substantial benefits (ozone reduction) in the northern and northeastern regions of Mexico City and in the cities of Toluca, Cuernavaca, Pachuca and Querétaro, but negative impacts in the central and southern regions of Mexico City and also in the Puebla and Tula regions (see Fig. 21.3). These results were consistent for the three meteorological episodes selected. However, it is important to note that the evaluation of the impacts on air pollutant concentrations strongly depends on the accuracy of the emission estimates used in air quality models. While the MCMA has been the subject of several intensive field studies and has an extensive ambient air quality monitoring network, air pollution studies are scarce in the other five states and the air quality monitor stations are not sufficient for analyzing trends. Therefore, it is important to perform field measurement-based studies of ozone

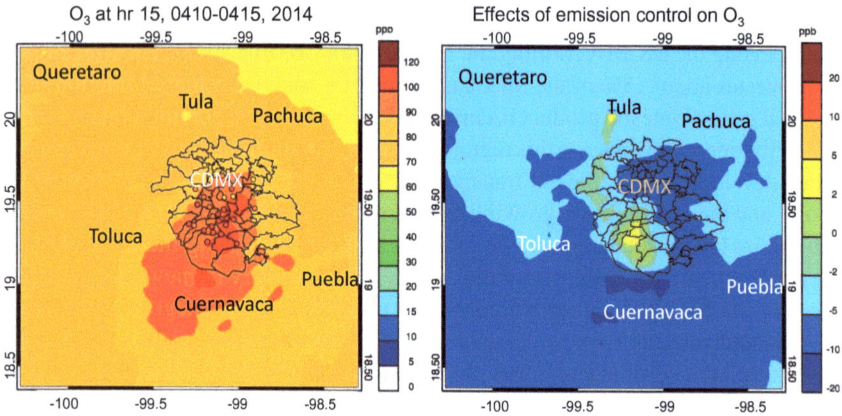

Fig. 21.3 Ozone simulation in the Megalopolis. Left panel: Comparison of measured (color-filled circles) and simulated (colored contours) O_3 concentrations during 10–15 April, 2014. Right panel: Simulated effects of the PROAIRE 2017–2030 emission reduction strategy on the O_3 concentrations during the same period

precursors as well as modeling studies to evaluate the uncertainties in the emission inventories in the region.

21.3 Santiago de Chile

Santiago is the capital and the largest city of Chile with about 6.5 million inhabitants, representing about 36% of the Chilean population. Most of the city lies between 500 m and 650 m a.s.l. in the middle of a bowl-shaped valley surrounded by two mountain ranges: the Andes Mountains and the Cordillera de la Costa. Santiago has a Mediterranean climate with hot and dry summers from November to March and cool and humid winters from June to August. The unique geographical location and meteorological patterns impede the ventilation and dispersion of air pollutants within the valley, making Santiago particularly susceptible to high levels of air pollution. Uncontrolled urban expansion, a growing economy, increasing motorization and industrialization have contributed to poor air quality in Santiago, especially during the winter months. Until recently, Santiago was ranked as one of the most polluted cities in the world and frequently confronted air-quality alerts and pollution emergencies. Since the early 1990s, the Chilean government has taken numerous actions to reduce emissions from key sources and has implemented an air pollution alert system based on the maximum atmospheric PM concentration. According to the Chilean Ministry of the Environment, as a result of the implementation of the long-term mitigation measures on industry, transportation and heating, the annual $PM_{2.5}$ concentration in Santiago has decreased by 72% between 1989 and 2016, as shown in Fig. 21.4.

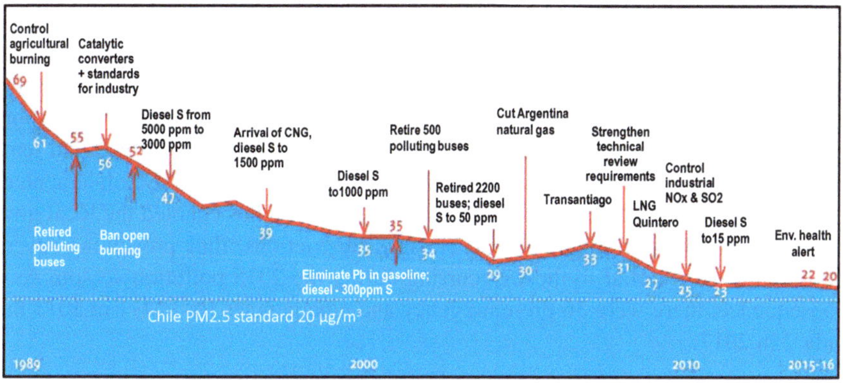

Fig. 21.4 Historical reduction of annual $PM_{2.5}$ in Santiago (*Credit* M. Mena 2017)

Nevertheless, the challenge of reducing air pollution remains. In June of 2015, a severe pollution episode forced the government to declare the first environmental emergency since 1999. In November 2017, a new air pollution control program came into effect establishing new $PM_{2.5}$ annual and daily standards, which was intended to reduce premature mortality of 2200 cases per year. The program includes stringent restrictions on diesel vehicle emissions (Euro 6 standard for new buses and light-duty vehicles, retrofit requirements for non-road vehicles, and ban on non-Euro 5 diesel light-duty vehicles), ban of wood burning in metropolitan Santiago, stricter emission standards on industry, and for the first time a ban on agricultural and open burning, and control of ammonia emissions from pork, chicken and egg producers. The Chilean government has further committed to improve air quality in Santiago with the introduction of 200 new electric buses.

21.3.1 Air Quality Forecasting System in Chile

Air quality forecasts are actively used in Chile for managing air pollution episodes. The current Chilean national standard for $PM_{2.5}$ is 50 $\mu g/m^3$ for 24 h average and episode categories correspond to Alert (80–110 $\mu g/m^3$), Pre-emergency (110–170 $\mu g/m^3$), and Emergency (>170 $\mu g/m^3$). Air quality forecasting tools are included in the decision-making process and are required for predicting pollution episodes for Santiago to protect the public from exposure to harmful levels of pollutants. This has stimulated active research to improve the accuracy of air quality forecasting.

Recently, a system capable of forecasting $PM_{2.5}$ episodes up to three days in advance for multiple cities in Central and Southern Chile was developed by Saide et al. [11]. The system uses the WRF-Chem model configured with a nested 2-km grid spacing domain to predict weather and inert tracers. Because of high correlation between observed CO and PM during episodes, two inert tracers (traffic and wood-burning stoves) were used to represent primary pollutants from anthropogenic sources. The tracers are converted to hourly $PM_{2.5}$ concentrations using an observationally based calibration, which reduces substantially the computation time relative to a full chemistry model. The conversion considers processes occurring in some of these cities, including contribution of stoves by city, weekend effect, and temperature dependence. The forecasting system has been successfully implemented; the advanced information about pollutant concentrations has enabled the local government to make contingency-based emission restrictions, thus preventing severe pollution events similar to those occurring during the 2015 pollution season. As a consequence, there were 36 pre-emergency and emergencies for $PM_{2.5}$ in 2013 but only 2 in 2017.

21.4 Conclusion

The experiences of Mexico City and Santiago demonstrate that air quality measurements and modeling studies are essential for improving our understanding of air pollution and its impacts, as well as providing sound scientific basis for policy makers to design effective emission reduction strategies for public health and environmental protection. Many of the emission reduction measures implemented in both cities also provide climate co-benefits, e.g., stricter emission standards for diesel vehicles and banning of wood-burning stoves will reduce black carbon, one of the key short-lived climate forcers; and new electric bus fleet will cut greenhouse gases substantially.

Recently, many Chinese cities, including the two large megacities Beijing and Shanghai, have been suffering serious air pollution. Ambient air particulate matter and household air pollution from solid fuel used are two of the leading risk factors of mortality in China. Through scientific research for understanding the factors that influence air pollution, and regional coordinated air pollution control actions implemented by the Chinese government authority, the concentration of atmospheric pollutants in several major cities have decreased substantially. Nevertheless, air pollution levels still exceed China's own air quality standards. Chinese government has continued to develop and implement emissions reduction strategies and has allocated significant resources to fight air pollution.

Currently, many urban centers, especially in the developing world, are facing similar challenges of population growth and economic development. The examples presented in this study illustrate that with appropriate planning, dedicated scientific research, robust emissions control policies, and effective access to advanced technologies and financial support, these urban centers also have the opportunities to manage their growing population and development sustainably while protecting the environment.

Questions and Answers

Questioner: Arlene Fiore
Question: Have there been specific policy actions addressing long-lived greenhouse gases (i.e. CO_2) in addition to synergistic efforts that also address air pollution (e.g., BC)?
Answer: In addition to implementing air quality management programs, the Government of Mexico City has also taken actions to mitigate greenhouse gases since 2000. Some of the largest emission reduction measures include biogas capture projects and waste-managements. In the transportation sector, the Metrobus (Bus Rapid Transit) initiated in 2005 was one of the most highly used lines in the world; it was the first project to sell emission reductions (CO_2eq) as part of its financing. These are complemented by fleet renewal projects for taxis and buses (including introduction of hybrid and electric vehicles). The current Climate Change Action Plans (2014–2020) aimed to reduce 10 million tons of CO_2eq of emissions by implementing strategic programs including renewable energy (e.g., solar panel), energy efficiency programs for public and private buildings (lighting, heating, etc.). As has been demonstrated in many cities around the world, many of the climate action plans also have substantial air quality co-benefits and vice versa. Therefore. it is important to integrate air quality and climate-stabilization goals in the design of environmental policy to realize potential synergistic benefits.

Questioner: Valerie Garcia
Question: Do you think the availability of low-cost air quality sensors will help with filling the gap in air quality monitoring?
Answer: Yes, low-cost air quality sensors have become increasingly important, not only for professional researchers who are using them for air quality monitoring, but also for citizen scientists who are looking to make measurements in the communities. They are affordable (low-cost) and easy to use. However, there is a need to ensure that the data generated from these sensors are reliable, accurate and reproducible. There are many commercially available low-cost sensors and several institutions have conducted performance evaluations for some of the sensors, for example, World Meteorological Organization has prepared an assessment on "Low-cost sensors for the measurement of atmospheric composition: overview of topic and future applications."

Acknowledgements The Molina Center is grateful to the Mexican Ministry of Environment and Natural Resources, the National Institute of Ecology and Climate Change, and the Mexico City Environment Secretariat for providing the emissions inventories and the air quality data used in the modeling study.

References

1. A. Baklanov, L.T. Molina, M. Gauss, Megacities, air quality and climate. Atmos. Environ. **126**, 235–249 (2016). https://doi.org/10.1016/j.atmosenv.2015.11.059
2. J.C. Doran, X. Bian, S.F.J. de Wekker et al (1998) The IMADA-AVER boundary layer experiment in the Mexico City area. Bull. Amer. Meteorol. Soc **79**, 2497–2508
3. S.A. Edgerton, X. Bian, J.C. Doran et al., Particulate air pollution in Mexico city: a collaborative research project. J. Air Waste. Manag. Assoc. **49**, 1221–1229 (1999)
4. G. Grell, S.E. Peckham, R. Schmitz, S.A. McKeen, G. Frost, W.C. Skamarock, E. Eder, Fully coupled 'online' chemistry within the WRF model. Atmos. Environ. **39**, 6957–6975 (2005)
5. G. Li, M. Zavala, W. Lei, A.P. Tsimpidi, V.A. Karydis, S.N. Pandis, M.R. Canagaratna, L.T. Molina, Simulations of organic aerosol concentrations in Mexico City using the WRF-CHEM model during the MCMA-2006/MILAGRO campaign. Atmos. Chem. Phys. **11**, 3789–3809 (2011). https://doi.org/10.5194/acp-11-3789-2011
6. LTMCE2 (LTM Center for Energy and the Environment) (2017) Evaluación de los impactos en la concentración de ozono por la aplicación de estrategias integradas de control de emisiones en la Megalópolis. Final Report INECC/LPN-009/2017, https://www.gob.mx/cms/uploads/attachment/file/328623/CAMe_Ozono_Modelacion_Reporte_Final.pdf. Accessed 14 Sept 2018
7. L.T. Molina, M.J. Molina (eds.), *Air Quality in the Mexico Megacity: An Integrated Assessment* (Kluwer Academic Publishers, Dordrecht, The Netherlands, 2002), p. 384
8. L.T. Molina, C.E. Kolb, B. de Foy, B.K. Lamb, W.H. Brune, J.L. Jimenez, R. Ramos-Villegas, J. Sarmiento, V.H. Paramo-Figueroa, B. Cardenas, V. Gutierrez-Avedoy, M.J. Molina, Air quality in North America's most populous city—OVerview of MCMA-2003 campaign. Atmos. Chem. Phys. **7**, 2447–2473 (2007). https://doi.org/10.5194/acp-7-2447-2007
9. L.T. Molina, B. de Foy, O. Vazquez-Martinez, V.H. Paramo-Figueroa (2009) Air quality, weather and climate in Mexico City. WMO Bull. **58**, 48–53. https://library.wmo.int/pmb_ged/bulletin58-1_en.pdf
10. L.T. Molina, S. Madronich, J.S. Gaffney, E. Apel, B. de Foy, J. Fast, R. Ferrare, S. Herndon, J.L. Jimenez, B. Lamb, A.R. Osornio-Vargas, P. Russell, J.J. Schauer, P.S. Stevens, R. Volkamer, M. Zavala, An overview of the MILAGRO 2006 campaign: Mexico City emissions and their transport and transformation. Atmos. Chem. Phys. **10**, 8697–8760 (2010). https://doi.org/10.5194/acp-10-8697-2010
11. P.E. Saide, M. Mena-Carrasco, S. Tolvett, P. Hernandez, G.R. Carmichael (2016) Air quality forecasting for winter-time $PM_{2.5}$ episodes occurring in multiple cities in central and southern Chile J. Geophys. Res. Atmos. 121, 558–575
12. SEMARNAT (2017) El Programa de gestión federal para mejorar la calidad del aire de la Megalópolis, 2017–2030 (PROAIRE de la Megalopolis 2017–2030), https://framework-gb.cdn.gob.mx/data/institutos/semarnat/Programa_de_Gestión_Federal_2017-2030_final.pdf. Accessed 14 Sept 2018
13. H.B. Singh, W.H. Brune, J.H. Crawford, F. Flocke, D.J. Jacob (2009) Chemistry and transport of pollution over the Gulf of Mexico and the Pacific: spring 2006 INTEX-B campaign overview and first results. Atmos. Chem. Phys. **9**, 2301–2318. https://doi.org/10.5194/acp-9-2301-2009
14. W.C. Skamarock, J.B. Klemp, J. Dudhia, D.O. Gill, D.M. Barker, M.G. Duda, X.-Y. Huang, W. Wang, J.G. Powers (2008) A description of the advanced research WRF version 3. NCAR Technical Note NCAR/TN-475 + STR
15. G.E. Streit, F. Guzman, Mexico City air quality: progress of an international collaborative project to define air quality management options. Atmos. Environ. **30**, 723–733 (1996)
16. UN Population Division (2018) World Urbanization Prospects 2018, https://population.un.org/wup/. Accessed 14 Sept 2018
17. UNEP/WHO (United Nations Environment Program/World Health Organization) (1992) Urban Air Pollution in Megacities of the World. Blackwell Publisher, Oxford

Chapter 22
Overview of the Change in NO_2 Assessment Maps During the Last 15 Years in Flanders: Problems Encountered and Solutions

Wouter Lefebvre, Bino Maiheu, Hans Hooyberghs, Stijn Vranckx, Felix Deutsch, Stijn Janssen, Karen van de Vel, Guido Cosemans, Peter Viaene, Jean Vankerkom, Marlies Vanhulsel, Filip Lefebre, Wim Peelaerts, Bart Degraeuwe, Clemens Mensink, Stijn Van Looy, Guy Driesen and Nele Smeets

Abstract During the last 15 years, the official assessment maps in Flanders have improved from a map consisting of only measured data points to a high resolution assessment which covers the complete area and takes into account several types of sources and street canyons. In order to improve this level of detail, multiple steps

---

W. Lefebvre (✉) · B. Maiheu · H. Hooyberghs · S. Vranckx · F. Deutsch · S. Janssen · K. van de Vel · G. Cosemans · P. Viaene · J. Vankerkom · M. Vanhulsel · F. Lefebre · W. Peelaerts · B. Degraeuwe · C. Mensink · S. Van Looy · G. Driesen · N. Smeets
VITO, Boeretang 200, 2400 Mol, Belgium
e-mail: wouter.lefebvre@vito.be

B. Maiheu
e-mail: bino.maiheu@vito.be

H. Hooyberghs
e-mail: hans.hooyberghs@vito.be

S. Vranckx
e-mail: stijn.vranckx@vito.be

F. Deutsch
e-mail: felix.deutsch@vito.be

S. Janssen
e-mail: stijn.janssen@vito.be

P. Viaene
e-mail: peter.viaene@vito.be

M. Vanhulsel
e-mail: marlies.vanhulsel@vito.be

F. Lefebre
e-mail: filip.lefebre@vito.be

W. Peelaerts
e-mail: wim.peelaerts@vito.be

C. Mensink et al. (eds.), *Air Pollution Modeling and its Application XXVI*, Springer Proceedings in Complexity,
https://doi.org/10.1007/978-3-030-22055-6_22

were taken. First of all, a land use regression model was introduced at an hourly scale at 4×4 km^2 resolution. Secondly, a Gaussian model was added for both point and line sources, correcting for emission double counting. Finally, a street canyon model was added to the chain, leading to improved resolution in these street canyons. In this work, we will discuss the problems encountered in these years such as how to account for double counting of emissions, how to correct the locations of the simplified road network and how to determine when street canyon calculations must be performed and how we solved them.

22.1 Introduction

Assessing the air quality for a whole region is not an easy task. One way of doing this is using measurements. This results in a list of locations where the concentrations are known. However, it is difficult to determine for which areas these measurements are representative [4]. Furthermore, a sufficiently dense network covering the territory with air quality sensors will need a lot of resources. Therefore, model data can be used to complement the information provided by the measurements. However, it is well-known that model results are biased and uncertain. As a result, a hybrid approach combining measurements and model data is often applied. This paper denotes the different steps that have been taken in the region of Flanders (northern Belgium) in order to arrive at the current high-resolution air quality maps. It shows the problems that were encountered and the solutions that were found to deal with them.

22.2 The Initial State

In the middle of the first decade of the 21st century, two tools were available for assessment in Flanders. The first were the measurements of the official measurement

C. Mensink
e-mail: clemens.mensink@vito.be

S. Van Looy
e-mail: stijn.vanlooy@vito.be

G. Driesen
e-mail: guy.driesen@vito.be

N. Smeets
e-mail: nele.smeets@vito.be

B. Degraeuwe
Air and Climate Unit, Directorate C—Energy, Transport and Climate, European Commission—Joint Research Centre, Brussels, Belgium
e-mail: bart.degraeuwe@ec.europa.eu

network of the Flanders Environment Agency (VMM). The other were low-resolution model data (15 × 15 km^2) provided by a chemical transport model (CTM) BelEUROS. The raw CTM model data was (still is) quite poor. This was partly due to the fact that the spatial scale of the CTM model is not adequate to represent the high degree of spatial variability in a densely populated area such as Flanders. The model also exhibited time- and spatial-dependent biases. Therefore, the first course of action was to improve the quality of the background concentration map.

22.3 Introducing a Better Background Map

In order to improve the quality of the low-resolution background map, and in order to improve the resolution, an hourly land-use regression model was introduced [2, 3]. This model combines, in an intelligent way, the measured data with the land use, resulting in hourly 4 × 4 km^2 maps of the concentration field for ozone, PM_{10}, $PM_{2.5}$, SO_2 and NO_2. More recently, NH_3 and BC were added to the model. The core of the interpolation model consists of a land use indicator, the β-parameter. This parameter is based on the local land use around the measurement locations and is then used to detrend spatially the measurements. This detrending eliminates the local character of the measurements (at a scale of 4 × 4 km^2). The resulting field can then be interpolated over the area (using kriging interpolation). Afterwards, for every location, the results are retrended again, using the β-parameter to add the local characteristics into the results.

The direct use of measurements results in low biases and it has been extensively shown by leaving-one-out validation and independent validation sets that the resulting RIO-model delivers high quality results, with low biases, low RMSE values and high correlations (Fig. 22.1).

22.4 Increasing the Resolution

In general, it is known that land-use regression models (LUR) such as RIO offer a cost-effective methodology for air quality assessment and, if the available proxy data is at high resolution, can be used at high resolution. However, they are prone to overfitting [9] which can be reduced by increasing the number of measurement data points in the region. However, in order to obtain good results, an extensive monitoring network is needed covering the resolutions for which the resulting map is made. In other words, for the 4 × 4 km^2 map that was constructed above, the current telemetric network is sufficient. However, for high resolution maps for the region of Flanders, the number of stations would need to increase quite strongly. This is feasible if low-cost measurement systems, such as passive samplers, can be used, but if hourly data is needed the cost would be very high. In addition the quality of

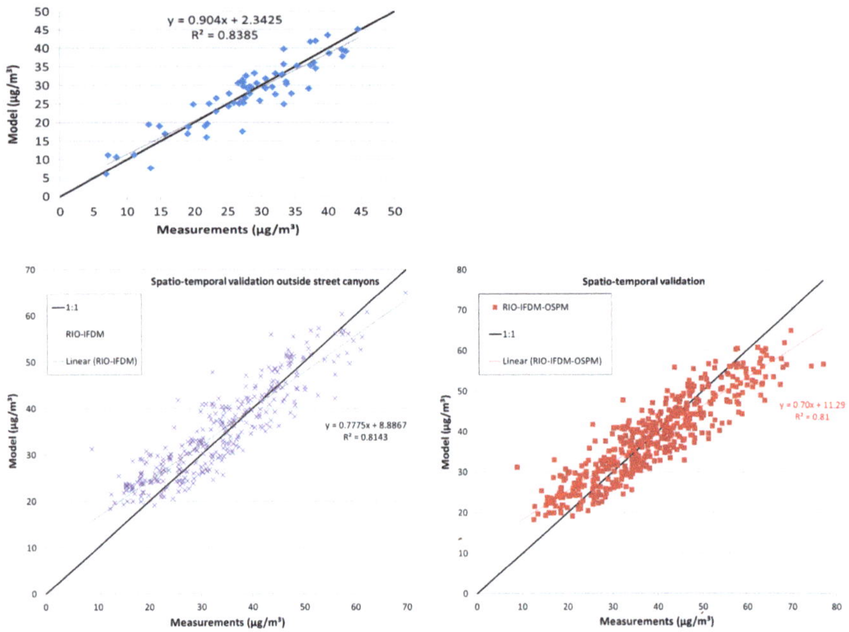

Fig. 22.1 Validation of the RIO model (upper figure), the RIO-IFDM model (lower left) and the RIO-IFDM-OSPM model (lower right). The RIO-validation is done over all non-traffic measurement stations in Belgium while the other validations are done over an independent NO_2-validation dataset (period July 2011–June 2012, two-weekly measurements) using all non-street canyon stations (lower left) and all stations (lower right)

low-cost sensors is still not sufficient to use them as replacements for the telemetric network (e.g. [8]).

Therefore, it was decided that instead of increasing the resolution of the land use regression model, it would be more interesting to couple the land-use regression model to a Gaussian plume model (IFDM, Immission Frequency Distribution Model, e.g. [6]). This model is applied for the well-known point sources and line sources. The majority of the emissions for NO_x, but only a minority for $PM_{2.5}$ is represented by those sources.

It is important when making this combination to correct for double counting. Indeed, the emissions are both present in the background concentration model (as this is based on interpolation of measurements in which all emissions are included) and in the higher resolution Gaussian model. Lefebvre et al. [5] therefore introduced a scheme in order to eliminate this double counting. The Gaussian model is calculated on a regular grid over a certain domain cell. In this cell, the average of the Gaussian model result is calculated and subtracted from the background concentrations. As such, the concentrations in the background cell are calculated without the emissions already present in the forefront model. Thereafter the corrected background concentrations are added to the detailed concentrations calculated by the Gaussian model. This is done hour by hour, accounting for chemistry [6]. The emissions for the Gaussian model are a combination of the known point sources and traffic emissions on the major roads. These are calculated using the COPERT emission factors and traffic flows and speeds provided by the Flanders Traffic Control Center (VVC). In order to validate the model, independent validation datasets using passive samplers were collected by the Flemish Environmental Agency (VMM). It was shown that the combined RIO-IFDM model provides good validation results for the non-street canyon locations in these datasets (Fig. 22.1). Furthermore, validation showed the importance of good traffic data for this type of exercise.

22.5 Including Street Canyons

Measurements have shown that the concentrations were still strongly underestimated, mainly for NO_2 and PM_{10}, due to the presence of street canyons. Therefore, we have added the street canyon model OSPM (Operational Street Pollution Model, [1]) to the chain, resulting in the RIO-IFDM-OSPM model. However, it was found that the location of traffic data that was available was sometimes slightly off (several meters to sometimes about 20 m) the true location. For the open-road concentration fields this was not so important, but when combining the road segments with the building locations to determine the location of the street canyons, this was problematic.

This problem was solved using an in-house developed algorithm combining on the one hand the VVC-dataset on traffic intensities having slightly incorrect road locations with on the other hand a dataset with correct road locations but without traffic intensities. This algorithm solves the large majority of the location problems in the dataset. At some specific complex locations a one-time manual intervention was needed.

Once the roads were shifted to their correct locations an in-house developed preprocessor was applied, providing it with the traffic data and the building heights. This preprocessor crawls over all line segments and marks at regular distances (typically 20 m) a dot. At this location, the preprocessor checks if the location is a street canyon location, and if so, determines the street canyon characteristics (such as width and height of the buildings) which are then provided to the OSPM-model.

The same problem for possible double counting of emissions is present, as emissions are both present in the Gaussian IFDM-model as in the OSPM-model. Therefore, a similar method avoiding double counting as between RIO and IFDM is applied.

The RIO-IFDM-OSPM model results in two sets of concentration values: one at non-street canyon-locations from the IFDM-model and one at street-canyon locations for the OSPM-model. In order to make one map, and avoid concentrations spilling out of the street canyons due to interpolation, a method for postprocessing was devised which is similar to a jigsaw-puzzle. Indeed, the concentrations in and out of the street canyons are treated separately and the result is then added together. The complete description of this postprocessing can be found in [7].

The resulting concentrations have been compared to the independent measurement campaign described above, integrating now also the street canyon locations of that campaign into the validation. It shows a high correlation between measurements and model, a low bias and a low RMSE (Fig. 22.1).

22.6 Conclusions

It was shown that using a combination of tools high-resolution and high quality air pollution maps can be made for assessment of air quality in a region. Several problems were encountered during the creation of these maps: avoidance of double counting, deficiencies in the basic data (locations of the roads) and difficulties in pre- and postprocessing street canyon data. The solutions applied for all of these problems in order to reach the current high resolution assessment maps were discussed.

Q & A

QUESTIONER: Sebnem Aksoyoglu

QUESTION: NO_2 concentrations are very high. Probably O_3 is not a problem there, but what about nitrate concentrations? Are they high?

ANSWER: O_3 is indeed not often a problem in Flanders. Although we have some days exceeding the 180 $\mu g/m^3$ value, the problem is limited to severe heath waves. Nitrate concentrations are often high, which leads to creation of secondary particulate matter in combination with the high ammonia emissions in the region.

References

1. R. Berkowicz, OSPM—A parameterised street pollution model. Environ. Monit. Assess. **65**(1/2), 323–331 (2000)
2. J. Hooyberghs, C. Mensink, G. Dumont, F. Fierens, Spatial interpolation of ambient ozone concentrations from sparse monitoring points in Belgium. J. Environ. Monit. **8**, 1129–1135 (2006)
3. J. Janssen, G. Dumont, F. Fierens, C. Mensink, Spatial interpolation of air pollution measurements using CORINE land cover data. Atmos. Environ. **42**, 4884–4903 (2008)
4. O. Kracht et al., Spatial representativeness of air quality monitoring sites—Outcomes of the FAIRMODE intercomparison exercise, JRC Technical report, in press (2018)
5. W. Lefebvre et al., Validation of the MIMOSA-AURORA-IFDM model chain for policy support: modeling concentrations of elemental carbon in Flanders. Atmos. Environ. **45**, 6705–6713 (2011)
6. W. Lefebvre, B. Degraeuwe, C. Beckx, M. Vanhulsel, B. Kochan, T. Bellemans, D. Janssens, G. Wets, S. Janssen, I. de Vlieger, Panis L. Int, S. Dhondt, Presentation and evaluation of an integrated model chain to respond to traffic- and health-related policy questions. Environ. Modell. Softw. **40**, 160–170 (2013)
7. W. Lefebvre, M. Van Poppel, B. Maiheu, S. Janssen, E. Dons, Evaluation of the RIO-IFDM-street canyon model chain. Atmos. Environ. **77**, 325–337 (2013)

8. L. Spinelle, M. Gerboles, G. Kok, S. Persijn, T. Sauerwald, Review of portable and low-cost sensors for ambient air monitoring of benzene and other volatile organic compounds. Sensors **17**(7), 1520 (2017)
9. TRANSPHORM, Transport related air pollution and health impacts—Integrated methodologies for assessing particulate matter—Deliverable D2.2.4—Spatially distributed source contributions for health studies: comparison of dispersion and land use regression models (2010), http://www.transphorm.eu/Portals/51/Documents/Deliverables/New%20Deliverables/D2.2.4_v6.pdf

Chapter 23
Modelling the Potential of Green Infrastructures to Reduce the Impact of Climate Change on Air Quality at Microscale

Vera Rodrigues, Sandra Sorte, Sílvia Coelho, Sandra Rafael, Ana Ascenso, Myriam Lopes, Ana Isabel Miranda and Carlos Borrego

Abstract This work focus on the assessment of green infrastructures benefits on air quality in Porto urban area, applying the CFD model VADIS to a particular area within the city. Three scenarios have been considered: (i) the baseline refers to the current morphological characteristics of the area; (ii) a green scenario comprises the replacement of built-up areas by green areas; and (iii) a green scenario corresponding to the implementation of green roofs. The results of baseline simulations shows a good agreement with local measurements with a NMSE of 0.4, 0.6 and 2.1 for CO, NO_2 and PM10 concentrations, respectively. The benefits of green infrastructures on air quality are assessed for future medium-term climate scenarios (2041–2070), applying a cascade of numerical models, from global to urban scale, the WRF-VADIS modelling system. Future climate data point out a decrease in the number of days with moderate to strong wind speed, and an increase in the number of days recording low wind speed conditions. The assessment of green infrastructures effects on air quality under future climate focus on low wind speed conditions. The results clearly show the disturbances exerted by green infrastructures, which are positively or negatively affecting mainly the adapted areas and their close surroundings.

23.1 Introduction

Computational Fluid Dynamics (CFD) models are a powerful tool to assess the local microclimate, considering in detail the morphological characteristics of the urban environment. In particular, CFD have been applied to assess the role of vegetation

---

V. Rodrigues (✉) · S. Sorte · S. Coelho · S. Rafael · A. Ascenso · M. Lopes ·
A. I. Miranda · C. Borrego
CESAM and Department of Environment and Planning, University of Aveiro,
3810-193 Aveiro, Portugal
e-mail: vera.rodrigues@ua.pt

C. Mensink et al. (eds.), *Air Pollution Modeling and its Application XXVI*,
Springer Proceedings in Complexity,
https://doi.org/10.1007/978-3-030-22055-6_23

in mitigating air pollution in urban areas, mainly considering the effects of trees induced by mechanical drag and/ or including the removal capacity of trees by deposition and filtration mechanisms [1, 2]. Although the effects of green infrastructures (GI) on the atmospheric dynamics have been widely studied, a deeper knowledge is still required regarding the overall perturbations induced by trees on turbulent flows and, consequently, on pollutants dispersion. Furthermore, the study of climate change (CC) ranges from global scale down to the regional and urban scale [3, 4]. Nevertheless, the full understanding of the impact of CC on urban microclimate and air quality at local scale remains a challenge with a lack of accurate predictions. This study aims to improve our understanding of how the changes on global and regional circulations will affect local scale urban microclimate, assessing the impact of GI on air quality under future climate scenarios for Porto's urban area.

23.2 Methodology

A cascade of numerical models, from global to local scale, was applied to Porto urban area. The numerical model Weather Research and Forecasting (WRF) [5] was applied for 4 nested domains, with 27, 9, 3 and 1 km resolution. Two simulations were performed, a representative simulation of the recent past climate, and a representative simulation of the future medium-term climate assuming the Representative Concentration Pathway Scenario RCP8.5. The CFD model VADIS was then applied to Porto urban area to assess local scale flow dynamics and air quality [6]. VADIS is applied to an area defined in the city centre of Porto turned into a computational domain of 1300 m × 1300 m × 150 m (Fig. 23.1), which includes the air quality station Francisco Sá Carneiro-Campanhã, mainly influenced by road-traffic emissions. The CFD simulations were performed with a grid resolution of 3 m × 3 m × 3 m. Figure 23.1a shows the computational domain for the baseline simulations. Figure 23.1b presents the implementation area for the green scenarios, identified by the red rectangles, which corresponds to 35% of the current built-up area located in the Southeast part of the domain. The green parks scenario corresponds to the replacement of the built-up area by green parks, while the green roofs scenario includes the implementation of green roofs within the buildings of the selected area.

Hourly averaged CO, NO_x and PM10 emission rates from on-road transport vehicles were calculated using the *Transport Emission Model for Line Sources* (TREM). The hourly emission rates are estimated for each road segment, considering the local information on traffic counting data. Vehicle counting data was acquired using automatic devices installed in seven distinct locations. Empirical rates expressing the relation with the traffic in the surrounding roads were applied in case of roads without available data [6]. The high emission levels are registered during the morning until evening, while the low emission levels are estimated during the night. Several baseline CFD simulations were performed for 24th September 2010, due to the availability of traffic counting data. The meteorological inflow data for these CFD simulations were obtained from mesoscale simulations using WRF model. The future

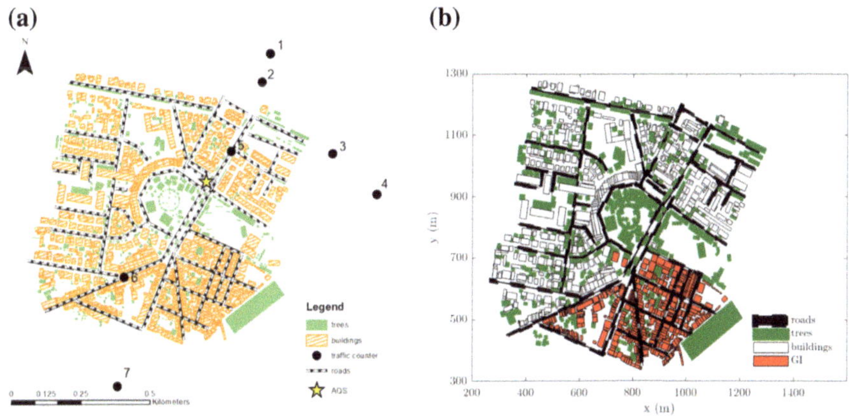

Fig. 23.1 Computational domain for the baseline simulations (**a**), including the location of the traffic sensor devices (1–7), as well as the air quality station, and the implementation area for the green scenarios (**b**)

climate meteorological projections presented by Rodrigues et al. [6] show an important reduction of wind speed in the first autumn months, compared to the recent past climate. The simulation results show a slight decrease in the average number of days with moderate to strong wind speed and a slight increase in the average number of days recording low wind speed conditions.

23.3 Numerical Results

The CFD performance is evaluated using the NMSE. The NMSE is equal to 0.4, 0.6 and 2.1 for CO, NO_2 and PM10 concentrations. Therefore, the simulation results denote the capability of the CFD model to accurately simulate CO and NO_2 concentrations, while results point out a weakness of the CFD to simulate PM10 concentrations, when only considering the emissions from road traffic.

Additionally, several CFD simulations were performed to assess the effects of GI on flow dynamics and, consequently, on CO, NO_2 and PM10 dispersion under future climate scenarios, focusing on distinct wind directions and low wind speed conditions. Figure 23.2 presents the differences of wind speed and PM10 concentrations (as example) between the green and the baseline scenarios, for a typical prevailing wind from North with a low wind speed of 2.5 m s^{-1}. Figure 23.2a shows the wind speed absolute differences at 1.5 m high between the green parks scenario and the baseline. Figure 23.2b shows the wind speed absolute differences between the green roofs and the baseline, at 10.5 m. The implementation of green parks leads to a maximum increase of wind speed of 2.8 m s^{-1} and a maximum decrease of 7.4 m s^{-1}. The implementation of green roofs promotes a maximum reduction of 9.5 m s^{-1} in the

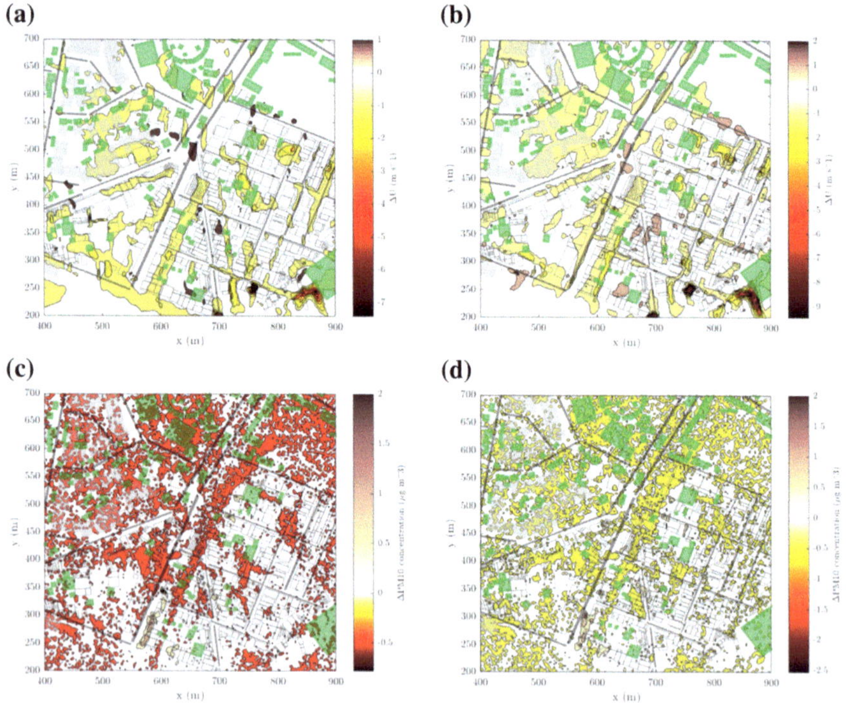

Fig. 23.2 Horizontal iso-contour plots of the differences between the green and the baseline scenarios. Differences of wind speed **a** for the green parks scenario at 1.5 m high and **b** for the green roofs scenario at 10 m high (the average height of buildings within the domain). Differences of PM10 concentrations **c** for the green parks scenario at 1.5 m high and **d** for the green roofs scenario at 10.5 m high

wind speed and a maximum increase of 2.4 m s^{-1}. Figure 23.2c and d show the absolute differences of PM10 at 1.5 m high between the green parks and the baseline, and at 10.5 m for the green roofs scenario. The implementation of green parks and green roofs promotes a maximum reduction of 0.8 and 0.4 μg m^{-3} in PM10 concentrations, respectively. On contrary, PM10 concentrations maximum increase is equal to 2.4 μg m^{-3} in green parks scenario and 2.2 μg m^{-3} in green roofs scenario. Therefore, the implementation of green infrastructures promotes slight increases and decreases of wind speed and, consequently, on pollutants concentration. The extreme variations of wind speed and pollutants concentrations are recorded within the implemented green areas and their surroundings. The results point out similar impact of green parks and green roofs for the same implementation area and considering the height at which the GI are implemented. However, at pedestrian level the impact of green parks will be more relevant.

23.4 Conclusion

This study presents a CFD model as a valid tool to assess climate change effects at local scale, and the role of green infrastructure as adaptation measures to increase resilience of Porto's urban area. The CFD model presents a good performance for the simulation of CO and NO_2 concentrations, while the results point out some weakness to simulate particulate matter dispersion. Therefore, the understanding of the influence of turbulent flow dynamics and the distinct emission sources contributing to fine particulate matter pollution still presenting a set of uncertainties. The results are an important contribution regarding the impact of green parks and green roofs at street level, both presenting similar effects on flow dynamics and pollutants dispersion.

Questions and Answers

Questioner: David Niemi

Question: *Was the impact of green areas on the turbulence and wind tunneling between large buildings also considered?*

Answer: The impact of green areas on the turbulence and wind velocity components was also assessed for the same case study, using as inflow wind conditions data from a recent meteorological year. These results were already published in the chapter book [6], and some other results are also being included in the articles entitled "Numerical and physical modeling of the impact of GI on flow dynamics in urban street canyons" and "CFD modeling of the impact of GI on flow dynamics and air quality in urban areas", both authored by Rodrigues et al., to be published at the Building and Environment and the Atmospheric Environment journals, respectively.

Questioner: Shuzhan Ren

Question: *CFD is such a high resolution model; when being used for forecast the results can be very sensitive to the initial conditions from downscaling regional model results. Have you test such sensitivity?*

Answer: The sensitivity of the initial meteorological conditions from regional model results was assessed in terms of wind velocity components and turbulence. These sensitivity tests were performed by comparison of the regional model results with meteorological measurements for recent meteorological years. Furthermore, the performance and accuracy of the CFD model was previously evaluated in terms of flow dynamics by Rafael et al. (2018) in the paper entitled "Impacts of green infrastructures on aerodynamic flow and air quality in Porto's urban area", published in the Atmospheric and Environment journal. In this paper, the authors initialize a set of CFD simulations with meteorological data from WRF results, using a grid cell from WRF simulations nearest the locations of the CFD domain boundaries.

Questioner: Fabian Lewartz

Question: *Have you tried to evaluate the impact of vertical green infrastructures? Do you think they can improve air quality in a 50-m radius, as advertised by a company selling green walls?*

Answer: In this work, it was only considered the impact of green infrastructures, such as green parks and green roofs. Although, the impact of vertical green infrastructures may also be assessed following the same procedure applied in this work. Nevertheless, it is required to evaluate a particular case study in order to determine a quantitative radius for the impact of these infrastructures, taking into account that the impact of these infrastructures on air quality strongly depends on the specific meteorological conditions and on the morphological characteristics of each case study.

Acknowledgements Thanks are due for the financial support to CESAM (UID/AMB/50017 – POCI-01-0145-FEDER-007638), to the FCT/MCTES through national funds (PIDDAC), and the co-funding by the FEDER, within the PT2020 Partnership Agreement and Compete 2020. The authors also acknowledge the Portuguese "Ministério da Ciência, Tecnologia e Ensino Superior" for the PhD Grant of Sandra Sorte (SFRH/BD/117164/2016).

References

1. J.H. Amorim, V. Rodrigues, R. Tavares, J. Valente, C. Borrego, CFD modelling of the aerodynamic effect of trees on urban air pollution dispersion. Sci. Tot. Environ. **461–462**, 541–551 (2013)
2. S. Janhäll, Review on urban vegetation and particle air pollution—Deposition and dispersion. Atmos. Environ. **105**, 130–137 (2015)
3. K. Markakis, M. Valari, A. Colette, O. Sanchez, O. Perrussel, C. Honore, R. Vautard, Z. Klimont, S. Rao, Air quality in the mid-21st century for the city of Paris under two climate scenarios: from the regional to local scale. Atmos. Chem. Phys. **14**, 7323–7340 (2014)
4. E. Sá, H. Martins, J. Ferreira, M. Marta-Almeida, A. Rocha, A. Carvalho, S. Freitas, C. Borrego, Climate change and pollutant emissions impacts on air quality in 2050 over Portugal. Atmos. Environ. **131**, 209–224 (2016)
5. W.C. Skamarock, J.B. Klemp, J. Dudhia, D.O. Gill, D.M. Barker, X.Y. Huang, W. Wang, J.G. Powers, A description of the advanced research WRF version 3. NCAR/TN-475+STR (Colorado, USA, 2008), p. 113
6. V. Rodrigues, S. Rafael, S. Sorte, S. Coelho, H. Relvas, B. Vicente, J. Leitão, M. Lopes, A.I. Miranda, C. Borrego, Adaptation to climate change at local scale: a CFD study in Porto urban area, in *Computational Fluid Dynamics—Basic Instruments and Applications in Science*, ed. by A. Ionescu (InTechOpen, 2018), pp. 163–186

Chapter 24
Impact of Urban Land Use and Anthropogenic Heat on Air Quality in Urban Environments

Shuzhan Ren, Craig Stroud, Stephane Belair, Sylvie Leroyer, Michael Moran, Junhua Zhang, Ayodeji Akingunola and Paul Makar

Abstract In the GEM-MACH URBAN project, Environment and Climate Change Canada (ECCC) high-resolution (2.5-km) Global Environment Multiscale-Modelling Air-quality and Chemistry (GEM-MACH) model and the Town Energy Balance Model (TEB) are being employed to examine the impact of the urban surface exchange scheme on the transport and diffusion of air pollutants in large cities such as Toronto, New York and Detroit. Simulation results show that while the TEB scheme causes O3 mixing ratios to increase, it leads to a decrease of CO and NOx mixing ratios and air quality health index (AQHI) values in the urban centers in both summer (July) and winter (January) months. The TEB scheme also has a big impact on the vertical diffusion coefficient, atmospheric boundary layer (ABL) height and air temperature. Comparisons against ECCC's meteorological and air quality (AQ)

S. Ren (✉) · C. Stroud · M. Moran · J. Zhang · A. Akingunola · P. Makar
Air Quality Research Division, Science and Technology Branch, Environment and Climate Change Canada, Toronto, ON, Canada
e-mail: shuzhan.ren@canada.ca

C. Stroud
e-mail: craig.stroud@canada.ca

M. Moran
e-mail: mike.moran@canada.ca

J. Zhang
e-mail: junhua.zhang@canada.ca

A. Akingunola
e-mail: ayodeji.akingunola@canada.ca

P. Makar
e-mail: paul.makar@canada.ca

S. Belair · S. Leroyer
Meteorological Research Division, Science and Technology Branch, Environment and Climate Change Canada, Dorval, QC, Canada
e-mail: stephane.belair@canada.ca

S. Leroyer
e-mail: sylvie.leroyer@canada.ca

C. Mensink et al. (eds.), *Air Pollution Modeling and its Application XXVI*, Springer Proceedings in Complexity,
https://doi.org/10.1007/978-3-030-22055-6_24

observation networks suggest that the inclusion of TEB scheme improves the forecasts of both surface temperature and pollutant mixing ratios.

24.1 Introduction

With rapid urbanization, the population in cities has been increasing dramatically around the globe. The urbanization leads to the significant change in urban meteorological conditions such as meso- and microscale urban heat island (UHI) effects, urban flooding, precipitation, humidity, fog, visibility, street canyon winds and surface energy fluxes. In addition, the change of urban meteorology and climate can cause elevated concentration levels for gaseous pollutants and aerosols, and consequently change the urban air quality. These changes are the direct consequence of the change of land coverage [1] and large amount of energy consumption from heating or cooling buildings and from local transportation within big cities.

As the majority of exceedances of air quality (AQ) standards occur in urban areas, accurate predictions of the major pollutants are extremely important for the health of the urban population. In the GEM-MACH URBAN project, the Canadian air quality model (GEM-MACH) and the town energy balance model (TEB) [2] are employed to investigate the impact of the urban canopy on the production and transport of pollutants within urban areas. Some results are shown in the following sections.

24.2 Urbanizing GEM-MACH with TEB

The TEB model can describe the complex urban fabric and the heat and momentum exchange mechanism between the urban surface and the atmosphere. Figure 24.1 shows some TEB parameters describing the urban fabric in the Great Toronto Area (GTA). These parameters are important for computing the heat storage within the urban canopy. It can be seen from Fig. 24.1 that downtown Toronto has a very high density of high buildings, large aspect ratio and roughness height. A climatological sensible heat flux from traffic is used. The magnitude of the traffic heat flux in urban centers is about 20 W/m^2.

24.3 Impacts of TEB on Temperature and Air Quality Health Index

The energy balance above the urban canopy can be changed significantly by the trapped short and long wave radiations by buildings within the urban canopy, and by anthropogenic activities (e.g., [2, 3]). This change leads to higher temperature in

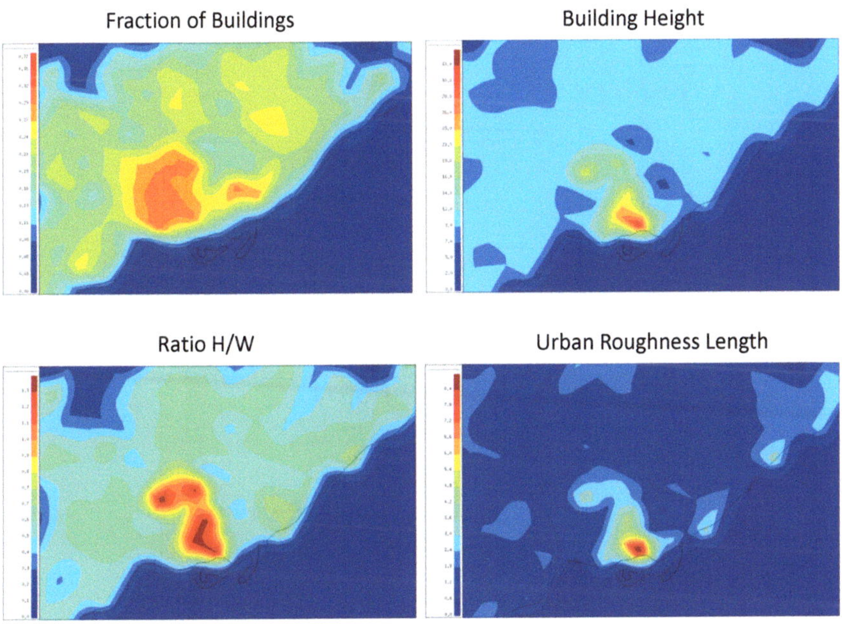

Fig. 24.1 Parameters characterizing the urban fabric in GTA including fraction of building, building height, aspect (height to width) ratio and urban roughness length

the urban area than the surrounding rural area, the so called urban heat island (UHI) effect [4].

The impact of TEB on urban temperature can be identified by comparing the simulation results with TEB and without TEB against temperature observed during the Pan Am games period (July, 2015). Figure 24.2 shows the diurnal variation of the monthly mean temperature bias from the two simulations at different locations in GTA. It is clear from the figure that at nighttime TEB generates UHI effect, and it significantly reduces the model temperature bias (observation minus model predication (O-P)) in both downtown and uptown areas. Noticeable improvements can also be identified in suburban and waterfront areas.

In addition to the impact on urban temperature, urban canopy has impact on the transport of chemical pollutants within the ABL by changing the turbulent mixing in vertical and the ABL height, as well as the production of chemical pollutants by changing the temperature and possibly altering solar radiation.

Figure 24.3 shows the diurnal variations of the monthly mean difference in air quality health index (AQHI) between the TEB simulations and non-TEB simulations in four big cities (within the model domain). It can be seen from the figure that while TEB can lead to reduction of AQHI, the patterns of reduction are different in each city and different season.

On weekdays in January, large reductions of the AQHI by TEB occur in the early morning and around noon in Toronto, New York and Detroit. Another big reduction

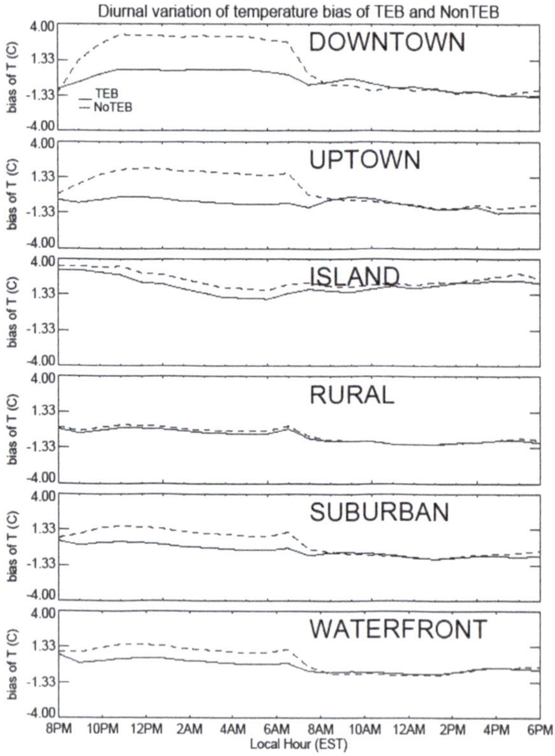

Fig. 24.2 Diurnal variation of the monthly mean temperature bias of the simulations with TEB (solid lines) and without TEB (dashed lines) over six land cover categories

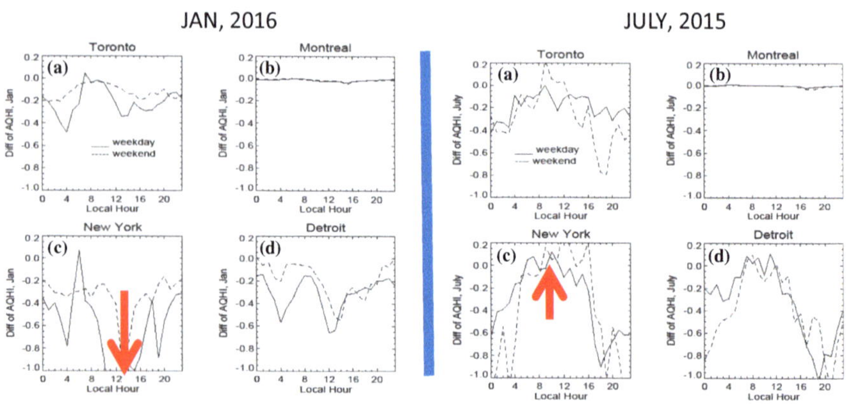

Fig. 24.3 Diurnal variation of the differences of monthly mean AQHI between the simulations with and without TEB in the four urban centers during weekday (solid lines) and weekend (dashed lines) in January 2016 (left) and July 2015 (right)

occurs at around 8 pm in New York. The reduction of AQHI in New York is much larger than the reduction in Toronto and Detroit. In July however, the reductions of AQHI occur in the early morning in the three big cities and large reductions occur not only in New York, but also in Detroit. Contrary to the big reduction at noon in January, AQHI has virtually no change at noon in New York and Detroit.

Figure 24.3 also shows that TEB has very different impact on AQHI during weekday and weekend. In January, the reduction of AQHI in early morning in weekday is much larger than the reduction in weekend in the three big cities. However, in July the reduction in early morning in weekend is much larger than the reduction in early morning in weekday in New York and Detroit.

24.4 Conclusions and Future Plans

Our simulations show that the inclusion of TEB in the GEM-MACH simulations can significantly reduce the cold temperature bias in the urban areas, and has big impacts on turbulent mixing coefficient and the ABL height (not shown). The simulation results also show that the change of meteorological condition leads the reduction of the AQHI in urban centers in both winter and summer.

In the current simulations, the climatologic traffic heat flux is used. In the future work, the diurnal variation of the traffic heat flux will be used to better describe the impact of the traffic patterns in urban areas on the UHI effect. In addition, the surface meteorological analysis data will also be used in future simulations to provide more accurate information of surface energy balance.

Question and Answer

Questioner: Alexander Baklanov

Questions:

1. Did you consider urban soil moisture processes and their effects?
2. In polluted cities, effects of urban aerosols on UBL and in particular UBL height can be comparable with UHI effects. Did you consider aerosol feedbacks in your studies?
3. You use the TEB model in the simulations of Masson [2], which was done for NWP and meteorological applications. For NWP applications, building effects on mechanical turbulence and on wind velocity structure in urban canopy is also important. In NEM/BEP parameterizations by Martilli et al. (2002) and latest version of TEB such effects are included more accurately. Did you analyzed/included these effects and other parameterizations?

Answer to (1): The air quality model (GEM-MACH) includes the land surface scheme which takes into account of urban soil moisture processes and their effects.

Answer to (2): The feedback of urban aerosols on simulations is not considered in this study. However, Wanmin Gong is going to discuss this effect in her talk. The same model is employed in her work.

Answer to (3): The TEB model used in this study is a single model with all the exchanges (of heat, momentum and chemical species) between the urban canopy and atmosphere occurring at the last model level. The scheme proposed by Martilli et al. can handle exchanges at several model levels. It may provide a better way describing the spatial distribution of air pollutants in urban areas particularly in the downtown core with high density of high-rise buildings.

References

1. S. Blair, L.P. Crevier, J. Mailhot, B. Bilodeau, Y. Delage, Operational implementation of the ISBA land surface scheme in the Canadian regional weather forecast model. Part I: warm season results. J. Hydrometeor. **4**, 352–370 (2003)
2. V. Masson, A physical based scheme for urban energy budget in atmospheric models. Bound. Layer Meteor. **94**, 357–397 (2000)
3. S. Leroyer, S. Blair, J. Mailhot, I.B. Strachan, Micro-scale numerical prediction over Montreal with the Canadian external urban modeling system. J. Appl. Meteor. Climate **50**, 2410–2428 (2011)
4. T.R. Oke, *Boundary Layer Climates* (Methuen, 1987), p. 435

Chapter 25 The Impact of Port Operations on Air Quality in Piraeus and the Surrounding Urban Areas

Nicolas Moussiopoulos, George Tsegas and Eleftherios Chourdakis

Abstract The dominant influence of shipping emissions on air quality of port cities and their surrounding urban areas has been demonstrated in a number of case studies, based on both assessment of monitoring data and the application of modelling tools. In the present work, chemical dispersion calculations are used for assessing the impact of marine traffic on air quality of Piraeus, Greece, and the neighbouring areas of the Athens urban agglomeration. A comprehensive marine emissions inventory is compiled using a bottom-up methodology on the basis of AIS traffic data, as well as activity and source parameter data obtained from other national databases, covering both merchant and passenger traffic in the area. A set of dispersion calculations are performed using the MEMO/MARS-aero chemical dispersion model for the Attica region, revealing the dominating contribution of the port area on the pollutant levels over the southern part of the city of Piraeus. Assessment of a hypothetical scenario, involving the implementation of cold ironing for ships at berth, indicates that emissions in the hoteling phase are significantly reduced, resulting to a notable reduction of concentrations in the port and the surrounding areas.

25.1 Introduction

This work aims at investigating the air quality impact of shipping activities in the Piraeus Port. In order to achieve this objective, marine traffic emissions near and in the port of Piraeus were calculated for the year 2015. The MEMO/MARS-aero dispersion model was then applied, in order to simulate the dispersion and chemical transformation of pollutants over the city of Piraeus and the Greater Athens Area (GAA). Two emissions scenarios were studied, corresponding to the present situation and a future application of mitigation technologies for ships at berth.

N. Moussiopoulos (✉) · G. Tsegas · E. Chourdakis
Laboratory of Heat Transfer and Environmental Engineering, Department of Mechanical Engineering, Aristotle University, 54124 Thessaloniki, Greece
e-mail: moussio@eng.auth.gr

C. Mensink et al. (eds.), *Air Pollution Modeling and its Application XXVI*,
Springer Proceedings in Complexity,
https://doi.org/10.1007/978-3-030-22055-6_25

25.2 Methodology and Case Setup

In the first part of the Piraeus study, emissions from marine traffic were calculated for two distinct areas, corresponding to the main port area of Piraeus and the designated anchorage area in the Saronic gulf (Fig. 25.1b). Exhaust emissions from passenger and cargo marine traffic were calculated for each individual vessel inside these areas during the reference year 2015. For the calculation of emissions and emission factors, a bottom-up approach was used, following the guidelines of the EMEP/EEA 2016 guidebook [1]. Emissions were aggregated separately for the cruise, manoeuvring, hoteling and loading/unloading phases, for both passenger and cargo traffic entering the two areas.

Activity data were obtained by collecting traffic information in the two designated areas through the Automatic Identification System (AIS) database for the entire year 2015.

Following the Tier-3 approach as specified in the guidebook, separate emission factors were calculated for the exhaust emissions of the main and auxiliary engines of each vessel according to the engine and fuel type and the operating phase (cruising, maneuvering, hoteling, loading/unloading). On the other hand, detailed information about engine type and fuel were not available for each individual vessel, therefore this information was apportioned to the individual traffic categories (see first column in Table 25.1) on the basis of statistical data contained in the European Maritime Safety Agency (EMSA) Central Ship Database (URL1).

In order to estimate the time spent by each vessel in every operating phase of the route, the total travelling time inside the two designated area polygons was apportioned according to the typical cruising speed, maneuvering time and time at berth for each of the above 9 vessel categories.

For calculating the air quality impact of the marine traffic, dispersion calculations were performed in a nested grid configuration, covering the entire area of the Attica

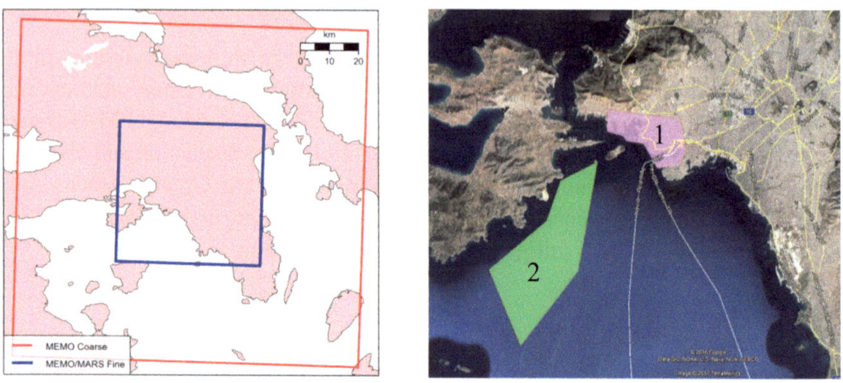

Fig. 25.1 Coarse and fine computational grid used in the dispersion simulations (left) and locations of the designated port (1) and anchorage (2) areas (right) for the Piraeus study

Table 25.1 Calculated exhaust emission factors and specific fuel consumption of the main engine, aggregated per vessel category

Ship type	Cruise				Manoeuvring/hoteling			
	Emission factors (g/kWh)			Specific fuel consumption (g fuel/kWh)	Emission factors (g/kWh)			Specific fuel consumption (g fuel/kWh)
	NO_x	NMVOC	PM_{10}		NO_x	NMVOC	PM_{10}	
Liquid bulk	15.86	0.57	1.44	199.18	12.68	1.71	2.33	219.40
Dry bulk carrier	16.58	0.59	1.62	196.33	13.24	1.78	2.38	216.40
Container	16.66	0.59	1.63	195.91	13.31	1.78	2.38	215.95
General cargo	14.60	0.53	1.13	203.67	11.67	1.59	2.20	224.09
Ro ro cargo	13.42	0.49	0.88	209.56	10.70	1.47	2.12	230.39
Passenger	12.41	0.46	0.73	218.13	9.80	1.38	2.14	239.72
Fishing	12.20	0.46	0.32	203.38	9.80	1.39	0.96	223.42
Tugs	11.74	0.34	0.33	203.92	9.38	1.01	1.00	224.01
Others	13.60	0.47	0.83	203.30	10.88	1.41	1.69	223.57

Peninsula (120 km coarse grid) in a resolution of 2 km, and the Greater Areas of Athens and Piraeus, including the port and anchorage polygons (50 km fine grid) in a resolution of 500 m. The non-hydrostatic meteorological model MEMO [2] and the Eulerian chemical dispersion model MARS-aero [3] were used to calculate concentrations of O_3, NO_2, PM_{10} and $PM_{2.5}$, following a meteorological classification scheme that uses eight representative days to compose the annual average concentration fields. The impact of marine emissions was quantified by means of a zero-out scheme, whereby dispersion calculations are repeated for a baseline scenario that excludes marine traffic. Baseline emissions for the rest of the activities in the Attica region were based on a localized emissions inventory for 2008, previously compiled in the framework of the TRANSPHORM project [4] and updated for 2015 using appropriate scaling factors. An additional emissions scenario was examined, corresponding to a future implementation of "cold ironing" for all passenger and cargo shops in the port of Piraeus. Under this scenario, every ship spending more than two hours in the hoteling, loading and unloading phases was assumed to deactivate its auxiliary engines for the entire duration of this phase, minus a period of one hour needed for connecting and disconnecting to the power grid ashore. The implementation of this scenario results in a reduction of about 60% on the total NO_2 and PM_{10} emissions during the time spent in the port area, i.e. excluding emissions in the anchorage area.

25.3 Results

In Fig. 25.2, average NO_2 and PM_{10} concentrations calculated using the full emissions set are shown (a, d). It is evident that the Piraeus area has a dominating contribution

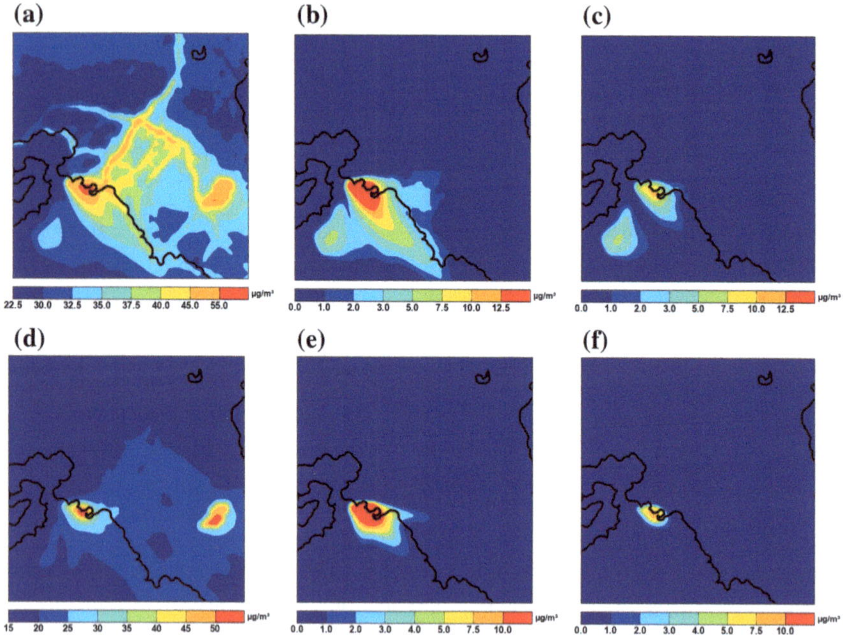

Fig. 25.2 Annual average NO_2 (**a**, **b**, **c**) and PM_{10} (**d**, **e**, **f**) concentrations calculated for the year 2015: absolute baseline concentrations (**a**, **d**), shipping contribution under the baseline scenario (**b**, **e**), and shipping contribution under the cold ironing scenario (**c**, **f**)

in the pollutant levels on the surrounding coastal areas, leading to concentrations of NO_2 and PM_{10} that exceed the annual limit values of 40 $\mu g/m^3$. The secondary dominant source visible in the eastern part of the Attica peninsula is the Athens International Airport with its associated road traffic emissions.

A more precise quantification of the contribution of the port and anchorage areas was obtained by assuming zero marine emissions and subtracting the resulting concentration fields from the "baseline" field. The result of this calculation is shown in Fig. 25.2b, e. It is again noteworthy that the port area has a contribution in excess of 18 $\mu g/m^3$ for NO_2 and of 22 $\mu g/m^3$ for PM_{10} as calculated for the downtown area of Piraeus. For both pollutants, the spatial extent of the port's impact includes at least part of the densely-populated coastal area, already burdened by local traffic sources. On the other hand, the annual average port plumes seem to have minor effect on the air quality status near the center of Athens, while the impact of traffic in the anchorage area appears to be contained within the boundaries of this offshore area. The implementation of "cold ironing" is found to have a significant effect in the port and surrounding areas, reducing NO_2 concentrations by 11 $\mu g/m^3$ (Fig. 25.2c), which corresponds to a 61% reduction of the shipping contribution in the port area. In the case of PM_{10} the calculated reduction amounts to 16 $\mu g/m^3$ (Fig. 25.2f), corresponding to a 73% reduction of the contribution of shipping traffic in the port area.

25.4 Conclusions

The impact of exhaust emissions from marine traffic on the air quality of Piraeus and the surrounding areas was investigated. A bottom-up scheme was used to establish shipping emissions based on emission factors calculated for different classes of vessels and spatial activity data obtained from the Automatic Identification System (AIS), while maps of pollutants concentrations were obtained by applying the MEMO/MARS-aero chemical dispersion system. The calculated concentrations indicate that the port has a dominant contribution to the air pollution in the nearby areas, while the corresponding PM_{10} and NO_2 concentrations have extended local maxima higher than the legislative limit values. Calculations under a scenario of cold ironing for ships in berth indicated a potential reduction by more than 60% of the shipping sector contributions to the observed concentrations. These results point to the need to prioritize emissions mitigation and other control measures related to the port activities of Piraeus.

Q & A

QUESTIONER: Jukka-Pekka Jalkanen (FMI)

QUESTION: What was assumed for the ship fuel in this study? The EU sulphur directive regulates the fuel used while at berth. Was that taken into account?

ANSWER: The fuel-dependent emission factors for the main and auxiliary engines were adapted from the inventories used in TRANSPHORM project (2014) and as such may not accurately reflect current emission trends. On the other hand, the selection of 2015 as reference year and corresponding activity data for our calculations indicates that the shipping emissions are consistently calculated. In any case, it is trivial to apply correction factors as part of the presented methodology in order to accurately represent fuel specifications consistent with present or planned legislative frameworks.

QUESTIONER: Talat Odman (GA TECH)

QUESTION: The title of your talk mentions "port operations" but it appears you only focused on ship emissions. What about cargo handling equipment (cranes, forklifts etc.)? Are they all electrified at Piraeus? Also, trucks and locomotives that come into the port to distribute the cargo (and passengers/vehicles) may be considered as part of the port operations.

ANSWER: Although cargo handling in Piraeus is indeed highly electrified, this is a valid remark with respect to our treatment of the road cargo- and passenger traffic around the port. Of course, the traffic emissions on the public road network (including the roads surrounding the port) are included in the calculation as part of the "city road" emission sector. But the contribution of these emissions near- and in the port area are not explicitly apportioned to the "port operation" sector in the dispersion results. This could be a point of improvement in a future extension of the study.

QUESTIONER: Sebnem Aksoyoglu (PSI)

QUESTION: What is the temporal variation of the contribution from ships to PM2.5 concentrations? Does it follow the same temporal variation as emissions?

ANSWER: The temporal variation of the shipping contribution indeed has a very high correlation with the emission temporal profiles, as provided by the AIS data. Effects of secondary production/transformation were present but we have not attempted to resolve this effect due to the complicated dependence of concentration fields on local meteorological flows.

Acknowledgements The authors acknowledge the contribution of Mr. Michael Drakoulas in the calculation of emissions for the Piraeus port.

References

1. EEA, *EMEP/EEA Air Pollutant Emission Inventory Guidebook* (European Environmental Agency, 2016)
2. N. Moussiopoulos, I. Douros, G. Tsegas, S. Kleanthous, E. Chourdakis, An air quality management system for policy support in Cyprus. Adv. Meteorol. (2012), https://doi.org/10.1155/2012/959280 (Hindawi Publishing Corporation)
3. N. Moussiopoulos, I. Douros, G. Tsegas, S. Kleanthous, E. Chourdakis, An air quality management system for Cyprus. Glob. Nest J. **12**, 92–98 (2010)
4. J. Kuenen, H. Denier van der Gon, A. Visschedijk, H. van der Brugh, High resolution European emission inventory for the years 2003–2007. TNO report TNO-060-UT-2011-00588 (Utrecht, 2011)
5. URL1, http://www.emsa.europa.eu/related-projects/central-ship-database.html

Chapter 26
Development and Implementation of an Online Chemistry Module to a Large Eddy Simulation Model for the Application in the Urban Canopy

Sabine Banzhaf, Basit Khan, Renate Forkel, Emmanuele Russo, Farah Kanani-Sühring, Klaus Ketelsen, Mona Kurppa, Matthias Mauder, Björn Maronga and Siegfried Raasch

Abstract Large-Eddy Simulation (LES) models are so far barely applied to dispersion and chemical transformation of pollutants in urban air quality studies. Within the joint project MOSAIK (Modellbasierte Stadtplanung und Anwendung im Klimawandel/Model-based city planning and application in climate change) a new LES based state-of-the-art microscale urban climate model PALM-4U, has been developed. The new model includes both gas phase and aerosol chemistry. For practical applications, our approach is to go beyond the simulation of single street canyons to chemical transformation, advection and deposition of air pollutants in the larger urban canopy. First LES results of a test case for an urban quarter of Berlin (Germany) are presented.

S. Banzhaf (✉) · E. Russo
Institute of Meteorology, Freie Universität Berlin, Berlin, Germany
e-mail: sabine.banzhaf@met.fu-berlin.de

B. Khan · R. Forkel · M. Mauder
Karlsruher Institut für Technologie, Institute of Meteorology and Climate Research - Atmospheric Environmental Research, Garmisch-Partenkirchen, Germany

F. Kanani-Sühring · B. Maronga · S. Raasch
Institute of Meteorology and Climatology, Leibniz University Hannover, Hannover, Germany

K. Ketelsen
Independent Software Consultant, Hannover/Berlin, Germany

M. Kurppa
University of Helsinki, Helsinki, Finland

B. Maronga
Geophysical Institute, University of Bergen, Bergen, Norway

C. Mensink et al. (eds.), *Air Pollution Modeling and its Application XXVI*,
Springer Proceedings in Complexity,
https://doi.org/10.1007/978-3-030-22055-6_26

26.1 Introduction

More than half of the world's population lives in urban settlements. The rapid urbanization exacerbates the negative consequences and vulnerability of the urban climate by replacement of the natural conditions, an excessive and concentrated consumption of resources and a high energy demand resulting in increased air pollution. Accurate representation of emission, dispersion, chemical transformation and removal of air pollutants in the urban canopy requires fine-scale turbulence-resolving simulations that can explicitly resolve building structures, surface heat fluxes at building facades, street canyons and terrain variations.

LES models explicitly resolve relevant scales of turbulent motion, so that these models can capture the inherent unsteadiness of atmospheric turbulence and advection. Nevertheless, LES models are so far rarely applied to urban air quality studies, in particular for the investigation of chemical transformation of pollutants.

Within the joint project MOSAIK a new urban micro-scale model is developed under the lead of the Institute of Meteorology and Climatology at the Leibniz Universität Hannover. The new urban micro-scale model PALM-4U is based on the state-of-the-art LES model PALM [5]. PALM-4U includes a fully coupled 'online' chemistry module. In the following, the chemistry module is described and results of a sample simulation are presented.

26.2 PALM-4U Chemistry Module

Chemistry is fully online coupled into PALM-4U. Automatic generation of the chemistry code with the Kinetic Pre-Processor (KPP, [1]) allows for high flexibility concerning the complexity of the applied chemical mechanism. A modified version of the KP4 post-processor [3] is used for optimizing the KPP-generated code and adapting it for PALM-4U.

Currently PALM-4U includes the following chemistry options:

- *CBM4* (Carbon Bond Mechanism, Gery et al. [2], 32 compounds, 81 reactions)
- *SMOG* (a simple photochemical smog mechanism, 12 compounds, 12 reactions)
- *SIMPLE* (further simplification of SMOG, 9 compounds, 7 reactions)
- *PHSTAT* (photo-stationary state only, 3 compounds, 2 reactions)
- *PASSIVE* (just 2 passive tracers, no chemical reactions)

A simple photolysis parameterization is implemented in PALM-4U. So far shading effects are not taken into account but will be implemented in near future.

Furthermore, a two-way LES-LES nesting is implemented in PALM-4U which makes it possible to zoom in smaller areas (e.g. urban extracts of $1 \times 1\ \mathrm{km}^2$) for high resolution simulation (of e.g. 1×1 m).

26.3 Case Study for an City Quarter of Berlin, Germany

A 9 h long test simulation starting on 21st of July at 5:00 UTC was carried out for a model domain with 96 × 96 grid points and 10 m grid width. In the vertical the domain extends up to 1500 m. The domain covered a small area of Berlin around the Ernst-Reuter-Platz, a junction with some high buildings and heavy car traffic. Only gas-phase chemistry was switched on for the simulation shown here. The chosen chemistry option *SIMPLE* includes the following reactions:

$$NO_2 + hv \rightarrow NO + O_3 \tag{26.1}$$

$$O_3 + hv \rightarrow 2OH + O_2 \tag{26.2}$$

$$NO + O_3 \rightarrow NO_2 \tag{26.3}$$

$$RH + OH \rightarrow RO_2 + H_2O \tag{26.4}$$

$$RO_2 + NO \rightarrow NO_2 + RCHO + HO_2 \tag{26.5}$$

$$HO_2 + NO \rightarrow NO_2 + OH \tag{26.6}$$

$$NO_2 + OH \rightarrow HNO_3 \tag{26.7}$$

Since detailed emission data are not available yet, only traffic emissions, which were parameterized depending on the street type classes from OpenStreetMap were considered (for main roads the following emission rates were assumed: NO = 65.93 $\mu g\ m^{-2}\ s^{-1}$; NO_2 = 28.26 $\mu g\ m^{-2}\ s^{-1}$; RH = 21.05 $\mu g\ m^{-2}\ s^{-1}$).

Selected results of the simulation are presented in Fig. 26.1. The figure shows surface concentrations of nitrogen dioxide (NO_2) and ozone (O_3) and matching vertical cross sections along the indicated line (in Fig. 26.1a) at 13:30 UTC. The vertical cross sections also include the vertical velocity as contours.

Both, NO_2 and O_3 vary in space and time due to the chemical transformation of the atmospheric trace gases and advection by turbulent motions. The near surface, concentrations of NO_2 are high due to the direct emission from traffic and conversion of nitrogen monoxide (NO) to NO_2. The daytime O_3, nitrogen oxides (NO_x) and VOCs (volatile organic compounds) relation is driven by a complex non-linear photochemistry. Figure 26.1 shows the typical VOC-sensitive regime where O_3 concentrations are depressed by NO_x titration.

The presented vertical profiles in Fig. 26.1c, f show that the chemical components are mixed upwards with the gradual increase of the boundary layer from simulation start towards the early afternoon.

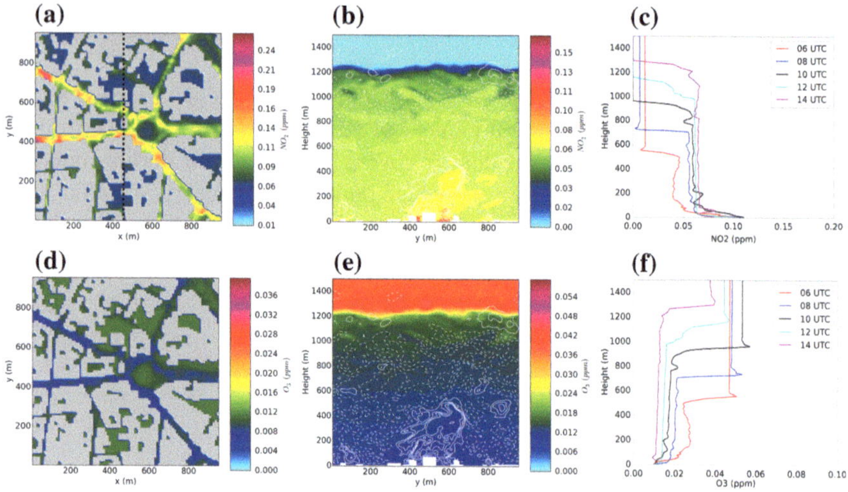

Fig. 26.1 Simulation area and concentrations of NO_2 (**a**) and O_3 (**d**) at 13:30 UTC, Z-Y-vertical cross section along the indicated line in (**a**) with concentrations (shaded color) of NO_2 (**b**) and O_3 (**e**) and vertical velocity w (−2.0 to 2.0 m s^{-1}, negative values as dashed lines) at 13:30 UTC and vertical profiles of NO_2 (**c**) and O_3 (**f**) at different times at location ★

26.4 Conclusion and Outlook

PALM-4U is currently under extensive development, which also holds for its chemistry module. Different mechanisms will be added and already implemented mechanisms undergo further testing. The model is able to successfully simulate chemical transformation and transport of the chemical species on the turbulence resolving scale. Results show the strong effect of turbulent structures on simulated pollutant concentration fields.

Further features are planned to be introduced in the near future: Realistic anthropogenic emissions as well as deposition will be implemented within the next months. External forcing from regional scale models is planned to be installed within the project run-time. Efforts will be made to speed up chemistry (vector version, larger time steps for chemistry). Accounting for shading effects within the photolysis parameterization is intended. The implementation of the SALSA aerosol module [4], which is coupled to the gas phase chemistry is currently being tested. SALSA comprises aerosol physics and aerosol chemistry. It uses 10 size bins in size space and includes up to 5 gas phase compounds and up to 7 PM compounds and water. Last named, a RANS-mode is also being developed to run the model at coarser resolution.

Acknowledgements MOSAIK is funded by the German Federal Ministry of Education and Research (BMBF) under grant 01LP1601A, with in the framework of Research for Sustainable Development (FONA).

References

1. V. Damian, A. Sandu, M. Damian, F. Potra, G.R. Carmichael, The kinetic preprocessor KPP—A software environment for solving chemical kinetics. Comput. Chem. Eng. **26**(11), 1567–1579 (2002)
2. M.W. Gery, G.Z. Whitten, J.P. Killus, M.C. Dodge, A photochemical kinetics mechanism for urban and regional scale computer modelling. J. Geophys. Res. **94**, 12925–12956 (1989). https://doi.org/10.1029/JD094iD10p12925
3. P. Jöckel, A. Kerkweg, A. Pozzer, R. Sander, H. Tost, H. Riede, A. Baumgaertner, S. Gromov, B. Kern, Development cycle 2 of the modular Earth submodel system (MESSy2). Geosci. Model Dev. **3**, 717–752 (2010). https://doi.org/10.5194/gmd-3-717-2010
4. H. Kokkola, H. Korhonen, K.E.J. Lehtinen, R. Makkonen, A. Asmi, S. Järvenoja, T. Anttila, A.-I. Partanen, M. Kulmala, H. Järvinen, A. Laaksonen, V.-M. Kerminen, SALSA—A sectional aerosol module for large scale applications. Atmos. Chem. Phys. **8**, 2469–2483 (2008). https://doi.org/10.5194/acp-8-2469-2008
5. B. Maronga, M. Gryschka, R. Heinze, F. Hoffmann, F. Kanani-Sühring, M. Keck, K. Ketelsen, M.O. Letzel, M. Sühring, S. Raasch, The parallelized large-eddy simulation model (PALM) version 4.0 for atmospheric and oceanic flows: model formulation, recent developments, and future perspectives. Geosci. Model Dev. **8** (2015), https://doi.org/10.5194/gmd-8-2515-2015

Chapter 27
Potential Impact of a Low Emission Zone on Street-Level Air Quality in Barcelona City Using CALIOPE-Urban Model

Jaime Benavides, Albert Soret, Marc Guevara, Carlos Pérez-García Pando, Michelle Snyder, Fulvio Amato, Xavier Querol and Oriol Jorba

Abstract Barcelona city (Spain) has a very high vehicle density (approx. 5500 vehicles km^{-2}) being the majority diesel (64%). Barcelona traffic stations report chronic exceedances of nitrogen dioxide (NO_2) European annual regulatory limits since the year 2000. In December 2017, a Low Emission Zone (LEZ) has been implemented in Barcelona to restrict access to the ring-road area to gasoline-powered passenger cars before 2000, diesel passenger cars before 2006 and vans before Euro 1 (registered before 1994) during air pollution episodes. This policy is planned to become permanent on December 2020. This work is an initial step towards evaluating the impact of Barcelona LEZ on air quality using CALIOPE-Urban. CALIOPE-Urban is a street-scale modelling system that couples CALIOPE air quality mesoscale modelling system, which provides air quality forecasts at 1 km horizontal resolution over Barcelona city, with R-LINE. Here we evaluate CALIOPE-Urban and assess its sensitivity to structural reductions of NO_x emissions. We evaluate the coupled modeling system using observations from an experimental campaign in April to May 2013 in Barcelona city.

27.1 Introduction

A 68% of Barcelona citizens were exposed in 2016 to NO_2 levels above the annual limit [1]. To decrease the negative impact of air pollution on citizens health, Barcelona municipality implemented last December 2017 a Low Emission Zone (LEZ) that focus on its first stage on reducing NO_2 levels during air pollution episodes that are expected to occur 2–3 times a year [4]. This policy is planned to become permanent

J. Benavides · A. Soret · M. Guevara · C. P.-G. Pando · O. Jorba (✉)
Earth Sciences Department, Barcelona Supercomputing Center (BSC), Barcelona, Spain
e-mail: oriol.jorba@bsc.es

M. Snyder
Institute for the Environment, University of North Carolina at Chapel Hill, Chapel Hill, NC, USA

F. Amato · X. Querol
Institute of Environmental Assessment and Water Research, IDAEA-CSIC, Barcelona, Spain

C. Mensink et al. (eds.), *Air Pollution Modeling and its Application XXVI*,
Springer Proceedings in Complexity,
https://doi.org/10.1007/978-3-030-22055-6_27

on December 2020. In the literature, little evidence of the positive impact of LEZs on decreasing NO_2 levels has been found in part due to the limited understanding of diesel emission factors under real-world driving conditions [6]. To evaluate episodic and structural impact on air quality derived from LEZ implementation, a modelling system coupling regional to street scales can provide the necessary information of background pollutant levels and meteorological data to estimate local pollutant concentration gradients using a street-scale dispersion model. In this work, we evaluate CALIOPE-Urban system that combines CALIOPE [7] with R-LINE [8] adapted to Barcelona street geometrical conditions and we explore the sensitivity of NO_2 street concentration levels to a reduction of NO_x emissions as a prior step to analyze the impact of Barcelona LEZ.

27.2 Methodology

27.2.1 CALIOPE Air Quality System

The CALIOPE system integrates the Weather Research and Forecasting meteorological model (WRF), the BSC-CNS in-house High-Elective Resolution Modelling Emission System (HERMESv2.0; [5]), the Community Multiscale Air Quality Modeling System (CMAQ) and the mineral Dust REgional Atmospheric Model (BSC-DREAM8b). The mesoscale system runs over Europe at a 12-km horizontal resolution, Iberian Peninsula at 4-km, and the Barcelona domain at 1-km. CMAQ vertical levels are collapsed from the 38 WRF levels to 15 layers up to 50 hPa with six layers falling within the PBL. CMAQ version 5.0.2 with CB05 chemical mechanism and AERO5 aerosol scheme is used.

27.2.2 CALIOPE-Urban

The combination between CALIOPE and R-LINE, CALIOPE-Urban, requires the use of CMAQ for background concentration data and WRF for meteorological inputs. WRF bottom layer over the street of interest is used as boundary conditions for R-LINE local meteorology module. With respect to background concentrations, the upwind urban background scheme is applied. The upwind background scheme chooses CMAQ grid cell values depending on wind speed and direction. R-LINE hourly emissions inputs for each road segment are provided by HERMES that produces specific hourly emissions for each street segment. R-LINE is run using the numerical integration approach to solve Gaussian dispersion equations and the Generic Reaction Set (GRS) for $NO–NO_2–O_3$ chemical reactions [3]. Receptors are located at 3 meters (m) above ground level forming a regular grid of 10 m horizontal resolution to predict spatial patterns and in specific coordinates to compare with observations.

To evaluate predicted hourly NO_2 concentrations we use observations from an experimental campaign conducted in April and May 2013 in Barcelona [2]. During the campaign, mobile laboratories placed at the parking lane of several street segments measured air quality parameters at 3 m height. For this study, data gathered every 30 min at Valencia Street No. 445 is shown. This street presents a marked canyon pattern where building height to street width ratio is approximately 1.

As an initial step towards developing a methodology to assess the potential impact of the LEZ on NO_2 concentration levels, in this work we analyze the sensitivity of CALIOPE-Urban to potential reductions of NO_x emissions that could be caused by a policy to improve air quality in the city without taking into account the LEZ at this stage. In the analysis presented here, we only use a mesoscale model run that incorporates the full emissions in the region and we force the street-scale model to use as input explicit emissions for each street segment with a reduction of 15%. This is a preliminary methodology to address the question of the effectiveness of such measure. In a future work, the reductions will be considered both in the mesoscale and urban scale models.

27.3 Preliminary Results

We start by assessing the ability of the model to represent spatial and temporal variability of NO_2 concentrations in Barcelona, then we analyze the sensitivity to a potential reduction of NO_x emissions. Figure 27.1 shows the spatial detail provided by street-scale and mesoscale combined solutions over Eixample district for 8 UTC 9 April 2013. The increased resolution of street-scale makes possible to visualize NO_2 street gradients.

To visually analyze hourly NO_2 temporal variation of the models and to evaluate the sensitivity of the system to changes in NO_x emissions we use observations from the highly trafficked experimental campaign site of Valencia street, which has an average daily traffic of 32,500 vehicles and represents the typical street canyon pattern with an average of 20 m building height and 20 m street width, during the period 4–8 April 2013. This period and site are chosen because they represent the main NO_2 concentration patterns found during the experimental campaign in the most extended geometry of the city (i.e. Eixample district). Figure 27.2 shows observed NO_2 concentration levels (black) compared to CALIOPE (blue), CALIOPE-Urban (red) and for the street-scale model using reduced NO_x emissions of 15% (green). From Fig. 27.2 we can see a high difference in performance between mesoscale (blue) and street-scale (red) models. Overall mesoscale is systematically under-predicting the concentration during the central hours of the day and is not able to capture the peaks. In contrast, street-scale represents more precisely the measured diurnal variation, giving estimates on weekdays in accordance with observations levels and a better reproduction of peak values. For example, on the 8th at 6 am (UTC) a

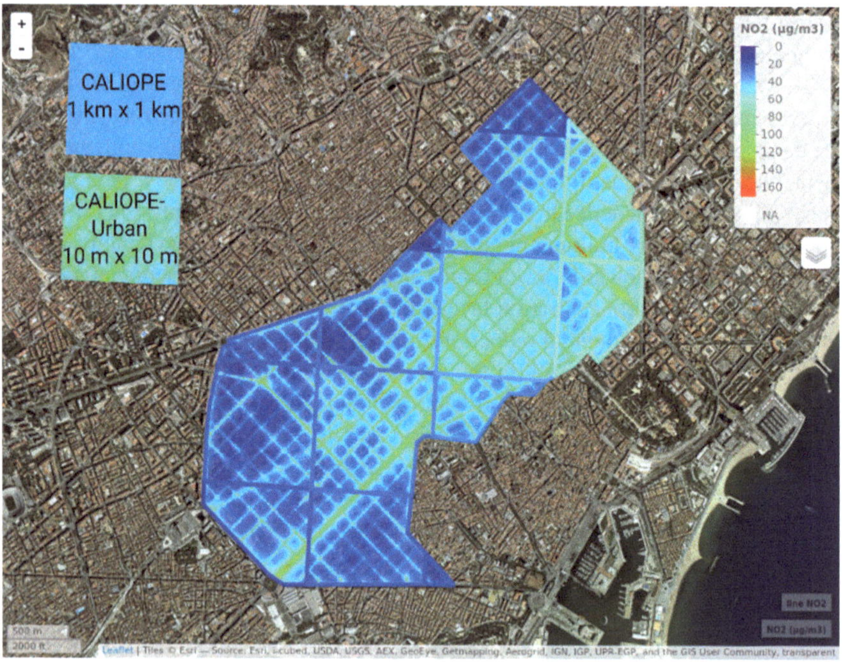

Fig. 27.1 NO_2 hourly level at 8 am (UTC) on 9th April 2013. Squares represent mesoscale 1-km grid cell values and the inside content street-scale concentrations

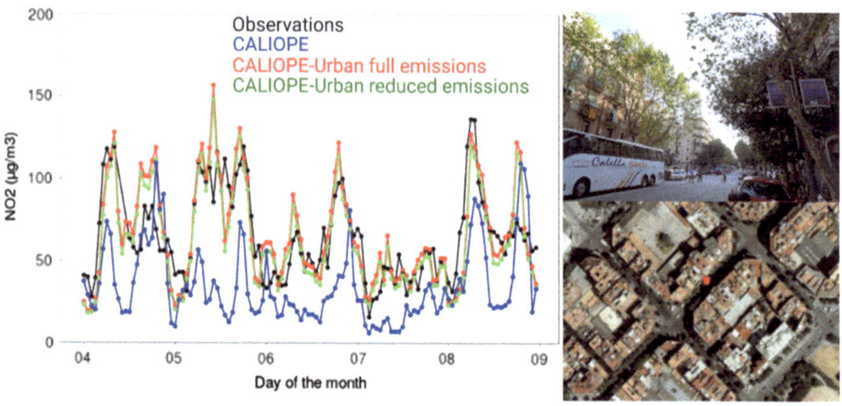

Fig. 27.2 NO_2 temporal variation for the period 4–8 April 2013 in Valencia Street No. 455. Left panel shows the comparison between model configurations. On the right side, the top image is the site street canyon and the bottom image is the site seen from above. (Observations shown in BLACK, street-scale model in RED, street-scale model with reduced NO_x emissions in GREEN and mesoscale in BLUE)

concentration peak found in observations (136.5 $\mu g\ m^{-3}$) is better reproduced by street-scale (128.7 $\mu g\ m^{-3}$) than by mesoscale (72.4 $\mu g\ m^{-3}$). It is also remarkable the improvement of CALIOPE-Urban during the central hours of the day.

With regard to NO_2 sensitivity to a 15% decrease in NO_x emissions, we see that NO_2 levels systematically decrease following the same dynamics as the street-scale model with full NO_x emissions. Furthermore, on a simulation executed for the entire period of April and May 2013 we find an average difference between the NO_2 concentration outputs of these two model configurations of 5 $\mu g\ m^{-3}$ producing lower concentration values with the reduced emission configuration, a significant reduction of NO_2 levels taking into account the limitation that CMAQ was executed with full emissions and used for background concentrations in both model runs.

27.4 On Going Work

Next steps are directed to decrease the emission uncertainties by integrating in the system new emission and traffic data that has been collected in Barcelona, permitting the inclusion of more realistic emission factors, traffic composition and intensity. Then, we will apply this new emission inventory to assess the potential impact on NO_2 concentration levels of the LEZ in Barcelona by simulating the effect of restricting the entrance to the ring-road to the banned vehicles. We will integrate the emissions in both the mesoscale and the local scale model runs.

Acknowledgements BSC researchers acknowledge the grants CGL2013-46736-R, CGL2016-75725-R and COMRDI15-1-0011-04 of the Spanish Government. Jaime Benavides PhD work is funded with a grant from the FPI Programme by the Spanish Ministry of the Economy and Competitiveness, call 2014, with reference EEBB-I-17-12296.

References

1. Agència de Salut Pública, Avaluació de la qualitat de l'aire a la ciutat de Barcelona. Technical report, Agència de Salut Pública, Barcelona (2016)
2. F. Amato, A. Karanasiou, P. Cordoba, A. Alastuey, T. Moreno, F. Lucarelli, S. Nava, G. Calzolai, X. Querol, Effects of road dust suppressants on PM levels in a mediterranean urban area. Environ. Sci. Tech. **48**(14), 8069–8077 (2014)
3. M. Azzi, G. Johnson, An introduction to the generic reaction set photochemical smog mechanism, in *Proceedings of the 11th International Clean Air Conference, August 1992* (1992), pp. 451–462
4. Barcelona City Council, Mesura de govern: Programa de mesures contra la contaminació atmosfèrica. Technical report (2016)
5. M. Guevara, F. Martínez, G. Arévalo, S. Gassó, J.M. Baldasano, An improved system for modelling Spanish emissions: HERMESv2.0. Atmos. Environ. **81**, 209–221 (2013)
6. C. Holman, R. Harrison, X. Querol, Review of the efficacy of low emission zones to improve urban air quality in European cities. Atmos. Environ. **111**, 161–169 (2015)

7. M.T. Pay, F. Martínez, M. Guevara, J.M. Baldasano, Air quality forecasts on a kilometer-scale grid over complex Spanish terrains. Geosci. Model Dev. **7**(5), 1979–1999 (2014)
8. M. Snyder, A. Venkatram, D. Heist, S. Perry, W. Petersen, V. Isakov, RLINE: a line source dispersion model for near-surface releases. Atmos. Environ. **77**, 748–756 (2013)

Chapter 28
Population Exposure to Emissions from Industry, Traffic, Shipping and Residential Heating in the Urban Area of Hamburg

Martin Otto Paul Ramacher, Matthias Karl, Armin Aulinger and Johannes Bieser

Abstract This study investigates the contributions of four major emission sources—industry, road traffic, shipping and residential heating—on air quality in the harbour city of Hamburg using a local-scale modelling system comprising meteorological, emissions and chemical transport models. Moreover, human exposure with regard to the overall air quality and the emissions sources under investigation was calculated. Based on detailed emission inventories and an evaluated CTM system, this study identifies road traffic as a major source of $PM_{2.5}$ pollution and exposure during the entire year and in almost all populated areas in Hamburg. Overall, the highest contributor to $PM_{2.5}$ concentrations is the industrial sector focussing on less populated areas.

28.1 Introduction

Urban air quality related issues are among the most critical societal concerns. Several studies prove a causal relation of emissions and ambient concentrations of pollutants in urban areas on short- and long-term health effects [9] and emphasize the need for emission reduction measures. Urban air quality management (UAQM) strongly depends on information about the spatial distribution of emissions and their sources, concentrations and exposure to health related pollutants. This information can be provided by city scale Chemical Transport Model (CTM) simulations, which are an indispensable tool to assess the current and future air quality. In addition, it is possible to identify the contribution of individual emission sources and sectors as well as the impact of regional background concentrations from surrounding regions, e.g. from long-range transports of $PM_{2.5}$ which are contributing largely to urban air quality [7].

M. O. P. Ramacher (✉) · M. Karl · A. Aulinger · J. Bieser
Department of Chemical Transport Modeling, Helmholtz-Zentrum Geesthacht, Institute of Coastal Research, Max-Planck-Strasse 1, 21502 Geesthacht, Germany
e-mail: Martin.Ramacher@hzg.de

C. Mensink et al. (eds.), *Air Pollution Modeling and its Application XXVI*,
Springer Proceedings in Complexity,
https://doi.org/10.1007/978-3-030-22055-6_28

This study investigates the contribution of the four major emission sources—industry (13%), road traffic (57%), shipping (10%) and residential heating (7%)—on $PM_{2.5}$ concentrations in the harbour city of Hamburg. Moreover, we will investigate the impact of regional $PM_{2.5}$ concentrations, due to missing source attribution of regional $PM_{2.5}$ in the existing Air Quality Plan of Hamburg. For this we use an urban modelling system comprising meteorological, emission and chemical transport model systems. Moreover, exposure of the population with regard to the overall air quality and the emissions sources under investigation was calculated. Based on this information, which is not available at this point, it is possible to identify sources and contributions of air pollution for the total urban area instead of evaluating the air quality situation with measurement stations.

28.2 Detailed Emission Inventories

Emission data are probably the most important input for chemistry transport model (CTM) systems [12] and have been identified as major source of improvement during previous model studies in local scale simulations for the city of Hamburg [14]. Therefore, detailed emission inventories for NO_x, O_3, CO, NMVOC, SO_2, PM_{10} and $PM_{2.5}$ have been gathered and created from various data sources to capture the major emission sources following SNAP (Selected Nomenclature for Air Pollution) of the European Environmental Agency (EEA).

Spatially gridded annual emission totals with a grid resolution of 1×1 km^2 were provided for this study by the German Federal Environmental Agency (Umweltbundesamt, UBA). The spatial distribution of the annual emission totals for the model domain has been done at UBA using the ArcGIS based software GRETA ("Gridding Emission Tool for ArcGIS"), which can generate regionalized emission data sets for all SNAP sectors for the complete area of the Federal Republic of Germany [16]. Hourly area emissions with 1-km horizontal resolution for SNAP categories 02 (domestic heating), 03 (commercial combustion), 06 (solvent and other product use), 08 (other mobile sources, not including shipping), and 10 (agriculture and farming) were derived from the UBA area emissions by temporal disaggregation using monthly, weekly and hourly profiles.

The gridded ship emission inventory (SNAP8) for the port area of Hamburg, is based on a bottom-up approach using activity data based on the Automatic Identification System (AIS) and activity based emission factors for NO_x, SO_2, CO, CO_2, hydrocarbons (VOC), and PM [1, 6, 10]. The calculated emission totals were spatially distributed on a 250 m $\times$ 250 m grid according to the ship routes inside the harbor area and ship-specific temporally distributed over a year on hourly time resolution according to the activity data.

The traffic emission inventory is based on 15,851 line source emissions of NO_x, NO_2, PM_{10} and $PM_{2.5}$ of the Hamburg road network provided by the city of Hamburg using a bottom-up approach with emission factors from HBEFA version 3.1 [17].

Industry emissions (SNAP 1, 4, 5, 9) are modeled as point sources based on emission reports (11. BImSchVG) combined with data set on European stacks by Pregger and Friedrich [13] to account for plume rise calculations within the CTM system.

28.3 Chemical Transport Model Set-Up and Evaluation of Reference Runs

The coupled meteorological and Chemistry Transport Model TAPM (The Air Pollution Model, Hurley [11] was used to calculate hourly three-dimensional concentration fields and sector contributions of multiple pollutants, including PM_{10} and $PM_{2.5}$. The hourly meteorological fields for the CTM study domain Hamburg (28 × 28 km^2) were obtained from the inner domain of a nested meteorological simulation with TAPM with a 1-km horizontal resolution (28 × 28 km^2 extent). The meteorological component of TAPM is an incompressible, non-hydrostatic, primitive equation model with a terrain following vertical sigma coordinate for 3-D simulations. The outer domain is driven by three-hourly ECMWF ERA5 synoptic scale reanalyses ensemble means on a longitude/latitude grid with 0.3 × 0.3°resolution. A vegetative canopy, soil scheme, and an urban scheme with 7 urban land use classes [11] are used at the surface, while radiative fluxes, both at the surface and at upper levels, are also included. The terrain elevation data was adopted from the German Digital Elevation Model [3] on 200-m horizontal resolution and the land use information was adopted from the CORINE land cover database [5] on 100-m horizontal resolution (Fig. 28.1).

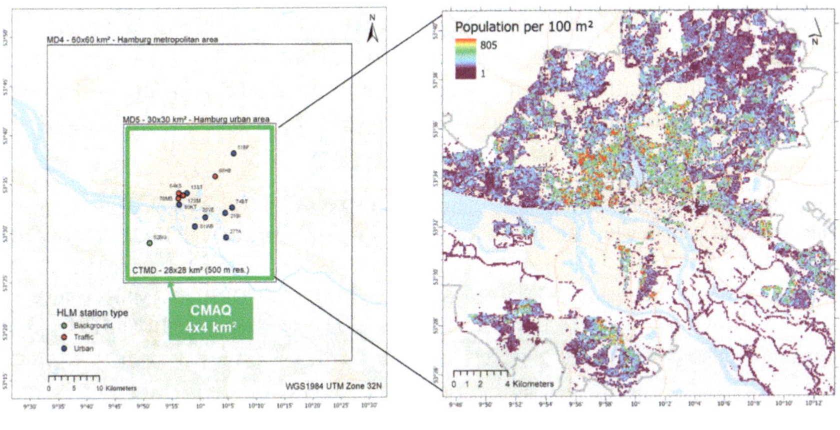

Fig. 28.1 Nested meteorological domains MD1–MD5, driven by ECMWF ERA5 synoptic fields (**a**). The CTM domain CTMD with 500 m resolution and 28 × 28 km^2 extent is nested within MD5 and uses by 4 × 4 km^2 CMAQ regional boundary conditions (**b**). Population density for Hamburg 2011 with 100 × 100 m^2 resolution

The chemical transport module of TAPM was applied for the Hamburg metropolitan area with a setup of 56 × 56 grid cells of 500 × 500 m^2 each. The chemistry in TAPM is based on the Generic Reaction Set (GRS) by [2]. To account for initial and boundary conditions of chemical compounds, TAPM was set-up as part of a one-way nested model chain, which coupled the model off-line to the CMAQv5.0.1 CTM (Byun und Schere [4] driven by COSMO-CLM mesoscale meteorological model version 5.0 [15] for the year 2012 using the ERA-Interim re-analysis as forcing data [8]. Thus, hourly concentrations are used at the boundaries to account for background concentrations in the CTM domain.

28.4 $PM_{2.5}$ Concentrations and Source Apportionment

Reference run outcomes for $PM_{2.5}$ including all emission sources in 2012 have been compared to air quality data from the Hamburg monitoring network (http://luft.hamburg.de/) at four available $PM_{2.5}$ measurement sites (Table 28.1). The statistical evaluation of annual $PM_{2.5}$ concentrations exhibits a good model performance with Pearson correlation coefficients of $\geq$0.5 and IOA between 0.63 and 0.70. The model tends to underestimate $PM_{2.5}$ concentrations at urban stations (13ST, 20VE, 61WB) with an NMB of −6 to −12%. At the only traffic station (64KS) the model highly underestimates the measured concentrations with an NMB of −38%.

Due to the few measurement stations for $PM_{2.5}$, comparisons for PM_{10} at nine measurements sites have also been performed which give similar performance results and trends. The modeled atmospheric concentrations of $PM_{2.5}$ show good statistical performances for annual, seasonal and daily averages, as well as the diurnal cycle (not shown here).

To identify the contribution of each emission sector to the overall air quality situation the per-turbation method has been used: Based on the evaluated reference run including all emission sources, four more simulations have been performed, each disregarding either traffic, shipping, industrial or residential heating emissions completely. By calculating the difference of the reference run with these simulations, the

Table 28.1 Annual model performance statistics of TAPM for $PM_{2.5}$ based on daily mean concentration at all stations with sufficient data availability in 2012 (derived with JRC Fairmode Delta Tool v5.6)

Station code	$Mean_O$ [$\mu g/m^3$]	$Mean_M$ [$\mu g/m^3$]	STD_O [$\mu g/m^3$]	STD_M [$\mu g/m^3$]	NMB [%]	CORR [−]	RMSE [$\mu g/m^3$]	IOA [−]
13ST	12.50	11.16	8.51	8.33	−10.65	0.51	8.41	0.70
20VE	12.02	11.20	8.59	8.63	−6.78	0.52	8.44	0.71
61WB	13.22	11.62	8.47	9.49	−12.01	0.50	9.16	0.69
64KS	18.41	11.47	9.22	8.48	−37.67	0.51	11.21	0.63

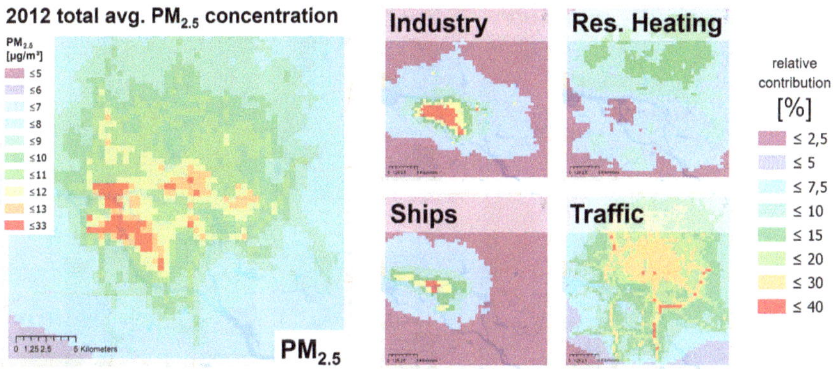

Fig. 28.2 Total averaged annual $PM_{2.5}$ concentration (left) and relative source contributions for industry, residential heating, shipping activities and traffic (right) in Hamburg 2012, calculated with the TAPM model

impact of each sector to the overall air quality is identified and the spatial distribution of each sectors contribution can be estimated.

The $PM_{2.5}$ concentration maxima (up to 33 $\mu g/m^3$) in the Hamburg area are mainly located south-west of the city-centre in the port and industrial area of Hamburg but also in nearby residential areas. These hotspot areas are mainly influenced by concentrations resulting from industrial emission sources (about 40%). In addition, with up to 30% shipping contributes significantly to the high concentrations in this area. Traffic related $PM_{2.5}$ concentrations are very low in these hotspots and are concentrated around highways and in the city centre, with contribution patterns of high impact in the city centre (30%) and decreasing impacts towards the outskirts (10%). The residential heating contributions to the annual average focus on the Northern residential areas of Hamburg with maximum 15%. Except from hotspots, the total averaged contribution of all local emission sources to $PM_{2.5}$ concentrations accounts for 70% of the total $PM_{2.5}$ concentrations. The remaining 30% are mainly due to long-range transport (Fig. 28.2).

28.5 $PM_{2.5}$ Exposure to Industry, Traffic, Shipping and Residential Heating

Based on the simulated $PM_{2.5}$ concentrations on $500 \times 500\ m^2$ grid, the exposure of population has been calculated by multiplying the concentration grids with gridded population density on a $100 \times 100\ m^2$ resolution. Highest individual exposure occurs in areas with high population density, which are in the city-centre as well as in the residential areas in the South of Hamburg. By analyzing the exposure, it becomes clear, that the pollution hot spots in the South-West of Hamburg are not populated.

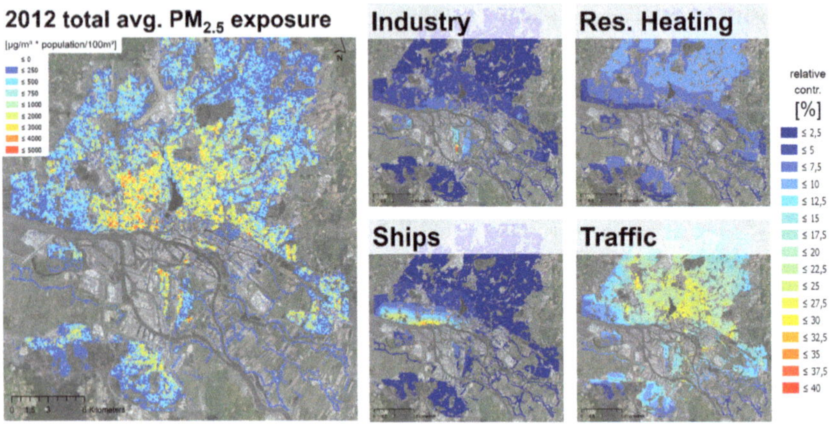

Fig. 28.3 Total averaged annual population exposure to $PM_{2.5}$ (left) and relative source contributions to population exposure for industry, residential heating, shipping activities and traffic (right) in Hamburg 2012, calculated with simulated TAPM concentrations (resampled to $100\ m^2$) and gridded population density

Nevertheless, the industrial sector has the highest contribution of about 40% to the densely populated district of Wilhelmsburg in the South of Hamburg. The traffic sector contributes with 15% in the outskirts and up to 30% in the center to the $PM_{2.5}$ exposure all over the city, while shipping has a big impact at the North West side of the river Elbe. When it comes to the impact of residential heating, there are higher concentrations and exposure in winter (Fig. 28.3).

28.6 Conclusions

Based on detailed emission inventories and an evaluated urban-scale CTM system, this study identifies road traffic as a major source of $PM_{2.5}$ pollution and exposure during the entire year and in almost all populated areas in Hamburg. Followed by industrial and shipping emissions, which have higher contribution in less populated areas. When it comes to the impact of residential heating on $PM_{2.5}$, there are higher concentrations and exposures (about +60%) in winter. The major contributor to $PM_{2.5}$ concentrations is the industrial sector focussing on less populated areas. While shipping emissions have their highest contribution to emissions in the harbor area, they still have a significant impact on close by residential areas. The spatial and temporal distribution of air pollutants from different sectors resulting from this study were not available so far and can support future urban air quality management in Hamburg.

References

1. A. Aulinger, V. Matthias, M. Zeretzke, J. Bieser, M. Quante, A. Backes, The impact of shipping emissions on air pollution in the Greater North Sea region—Part 1: current emissions and concentrations. Atmos. Chem. and Phys. Discuss. **15**(8), 11277–11323 (2015). https://doi.org/10.5194/acpd-15-11277-2015
2. M. Azzi, G.M. Johnson, M. Cope, An introduction to the generic reaction set photochemical smog mechanism, in Proceedings of the 8th International Clean Air Conference, 8th International Clean Air Conference, Melbourne, 6–11 May 1984, Melbourne.
3. BKG: Digitales Geländemodell Gitterweite 200 m, Bundesamt für Kartographie und Gedoäsie (2013). http://www.geodatenzentrum.de/docpdf/dgm200.pdf. Accessed 22 Nov 2017
4. D. Byun, K.L. Schere, Review of the Governing Equations, computational algorithms, and other components of the Models-3 Community Multiscale Air Quality (CMAQ) Modeling System. Appl. Mech. Rev. **59**(2), 51 (2006). https://doi.org/10.1115/1.2128636
5. CLC: Copernicus Land Monitoring Service (2012). http://land.copernicus.eu/pan-european/corine-land-cover/clc-2012/. Accessed 22 Nov 2017
6. Clean North Sea Shipping (CNSS), Final report—key findings and recommendations: Technical report (2014). http://cnss.no/wp-content/uploads/2014/03/CNSS_Final_Report_web.pdf. Accessed 7 Mar 2016
7. EEA 2016 – M. Gager; B. Ullrich, European Union Emission Inventory Report 1990–2014 Under the UNECE Convention on Long-Range Transboundary Air Pollution (LRTAP). Luxembourg: Publications Office (EEA report, No 16/2016) (2016)
8. B. Geyer, High-resolution atmospheric reconstruction for Europe 1948–2012: CoastDat2. Earth Syst. Sci. Data **6**(1), 147–164 (2014). https://doi.org/10.5194/essd-6-147-2014
9. Gerard Hoek, Ranjini M. Krishnan, R. Beelen, A. Peters, B. Ostro, B. Brunekreef, J.D. Kaufman, Long-term air pollution exposure and cardio- respiratory mortality: a review. Environ. Health Glob. Access Sci. Source **12**(1), 43 (2013). https://doi.org/10.1186/1476-069X-12-43
10. J.H.J. Hulskotte, H.A.C. Denier van der Goon, Fuel consumption and associated emissions from seagoing ships at berth derived from an on-board survey. Atmos. Environ. **44**(9), 1229–1236 (2010). https://doi.org/10.1016/j.atmosenv.2009.10.018
11. P.J. Hurley, *TAPM V4: Part 1: Technical description* (CSIRO, Aspendale, Vic, 2008)
12. V. Matthias, J.A. Arndt, A. Aulinger, J. Bieser, H. van der Denier Gon, R. Kranenburg, J. Kuenen, D. Neumann, G. Pouliot, M. Quante, Modeling emissions for three-dimensional atmospheric chemistry transport models. J. Air Waste Manag. Assoc. **68**(8), 763–800 (2018). https://doi.org/10.1080/10962247.2018.1424057
13. T. Pregger, R. Friedrich, Effective pollutant emission heights for atmospheric transport modelling based on real-world information. Environ. Pollut. (Barking, Essex: 1987) **157**(2), 552–560 (2009). https://doi.org/10.1016/j.envpol.2008.09.027
14. M.O.P. Ramacher, M. Karl, A. Aulinger, J. Bieser, V. Matthias, M. Quante, The impact of emissions from ships in ports on regional and urban scale air quality, in *Air Pollution Modeling and its Application XXV*, ed. by C. Mensink, G. Kallos (Springer, 2017)
15. B. Rockel, A. Will, A. Hense, The Regional Climate Model COSMO-CLM (CCLM), Meteorol. Z., 347–348 (2008)
16. C. Schneider, M. Pelzer, N. Toenges-Schuller, M. Nacken, A. Niederau, Gridding Emission Tool for ArcGIS (GRETA) (in German), ArcGIS basierte Lösung zur detaillierten, deutschlandweiten Verteilung (Gridding) nationaler Emissionsjahreswerte auf Basis des Inventars zur Emissionsberichterstattung. Umweltbundesamt, Texte 71/2016 (2016), pp. 1–223. ISSN 1862-4804
17. UBA: Handbook of Emission Factors for Road Transport (in German), HBEFA version 3.1, Umweltbundesamt Berlin (2010). http://www.hbefa.net/d/. Accessed 22 Nov 2017

Part III
Emission Modelling and Processing

Chapter 29
Model of Emissions of Gases and Aerosol from Nature Version 3 (MEGAN3) for Estimating Biogenic Emissions

Alex Guenther, Xiaoyan Jiang, Tejas Shah, Ling Huang, Sue Kemball-Cook and Greg Yarwood

Abstract Biogenic volatile organic compound (BVOC) emissions from terrestrial ecosystems drive distributions of several atmospheric constituents relevant to air quality and climate. BVOC emission rates can vary more than an order of magnitude over spatial scales of a few kilometers and time scales of less than a day which makes estimation of these emissions especially challenging. New improvements to the Model of Emission of Gases and Aerosols from Nature (MEGAN version 3) are described including (1) a transparent approach for assigning emission factors and other model parameters, (2) updated emission factors and algorithms based on recent measurements, and (3) treatments for previously unrepresented processes including stress induced emissions and canopy heterogeneity. The estimated emissions are compared to alternative model approaches and evaluated with aircraft measurements of concentrations and fluxes. Remaining gaps and priorities for future progress in biogenic organic emission modeling are also discussed.

29.1 Introduction

The overall goal of this study was to improve numerical model predictions of regional ozone and aerosol distributions by reducing uncertainties associated with quantita-

A. Guenther · X. Jiang
Department of Earth System Science, University of California, Irvine, CA 92697, USA
e-mail: alex.guenther@uci.edu

X. Jiang
e-mail: jiang.ut@gmail.com

T. Shah · L. Huang · S. Kemball-Cook · G. Yarwood (✉)
Ramboll, 7250 Redwood Boulevard Suite 105, Novato, CA 94945, USA
e-mail: GYarwood@ramboll.com

T. Shah
e-mail: TShah@ramboll.com

S. Kemball-Cook
e-mail: SKemballCook@ramboll.com

C. Mensink et al. (eds.), *Air Pollution Modeling and its Application XXVI*,
Springer Proceedings in Complexity,
https://doi.org/10.1007/978-3-030-22055-6_29

tive estimates of BVOC emissions. Although there have been advancements in the procedures used to simulate BVOC emissions, there are still major uncertainties that affect the reliability of air quality simulations. This includes significant gaps in our understanding of BVOC emissions and their implementation in numerical models including (1) isoprene emission factors, (2) missing compounds, and (3) and unrepresented processes including canopy heterogeneity and stress induced emissions. In this study, we developed new emission factors and incorporated missing BVOC compounds and unrepresented BVOC emission processes into the Model of Emissions of Gases and Aerosols from Nature (MEGAN) framework. To accomplish this, we developed a transparent and comprehensive approach to assigning isoprene and monoterpene emission factors and updated MEGAN to include additional BVOC and processes including stress induced emissions and canopy heterogeneity. We evaluated MEGAN BVOC emission inventories for Texas and surrounding regions using surface and aircraft observations and a photochemical model.

29.2 MEGAN Emission Factor Processor

One of the study objective was to develop a database system that provides a transparent approach for estimating BVOC emission factors. A transparent approach for determining emission factors and other variables would facilitate efforts to assess and improve biogenic emission models. An Emission Factor Processor for MEGAN3 (MEGAN-EFP) was developed as an open source Python code to ingest driving variables and calculate landscape scale biogenic emission factors and other model drivers. The approach is flexible so that users can use any available landcover and emissions data and can investigate the performance of various datasets. A data quality rating system was implemented so that the user can choose to omit low quality data. The framework was implemented for BVOC emissions, specific leaf area and emission light dependence fraction and can be extended to include other plant and soil traits that can vary spatially. The framework is illustrated in Fig. 29.1. The emission factors calculated with the MEGAN-EFP are highly transparent in that the landcover and emission data that they are based on can be traced back to specific publications.

29.3 Isoprene and Monoterpene Emission Factor Data

The MEGAN-EFP currently includes over 10,000 biogenic emissions measurements from more than 200 studies. There are many other measurements published in the literature and the database will be extended in the future and is expected to guide and motivate new measurements that can fill gaps in the existing database. About half of the isoprene and monoterpene emissions data in the MEGAN-EFP are from studies that were conducted during the decades between 1960 and 2000 and are referred to here as the Twentieth Century Emissions Database (TCED). The TCED is a synthe-

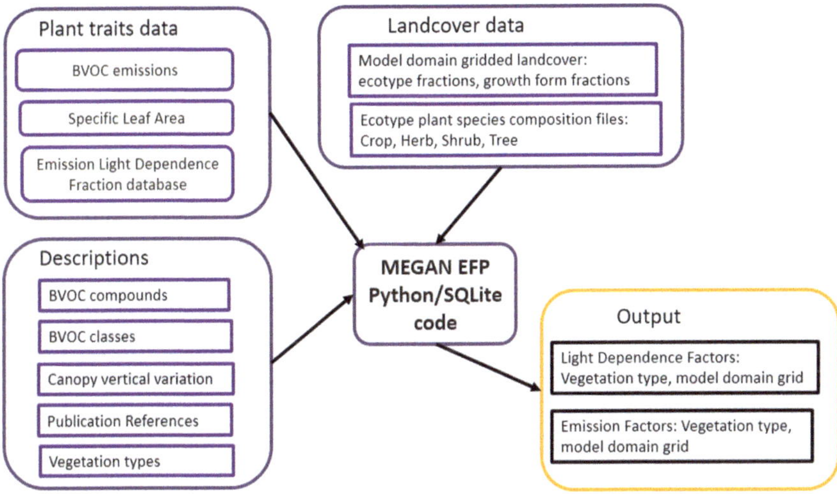

Fig. 29.1 Schematic of the MEGAN-EFP input and output data

sis of emissions measurements that include qualitative data, often only indicating whether a plant species emits or does not emit isoprene or total monoterpenes, or are highly uncertain and do not follow the protocols considered necessary to obtain high quality data as described by Niinemets et al. [4].

To demonstrate the MEGAN-EFP, we synthesized, reconciled and calculated isoprene and monoterpene (terpenoid) emission factors for Texas and the surrounding region. High quality enclosure measurements of isoprene emission were available for a large fraction of the tree species that comprise most of the total isoprene emissions in Texas and were integrated into the MEGAN-EFP. Utilizing the MEGAN-EFP, it was determined that almost all the Texas isoprene emitting trees are members of just three genera (oaks, sweetgum, tupelo/gum), all of which have had at least one species investigated with state-of-the-art techniques. We compared these emission factors with those used in the EPA's Biogenic Emission Inventory System [5] and MEGAN v2.1 [1]. The comparison demonstrated that the differences between BEIS and MEGAN isoprene emission estimates are primarily due to the specific leaf area (the leaf area to leaf mass ratio) estimates and reconciled discrepancies between leaf enclosure and aircraft estimates of isoprene emission factors. One tree genera, the oaks, is highly diverse with many species that have not been studied so the possibility exists that some of these oak species may emit isoprene at substantially different rates than those that have been studied. The assessment also revealed that there are no high-quality isoprene emission data for tree species other than the high isoprene emitters. This allows the possibility that some of the tree species thought to be non-emitters may have non-zero emission rates. The available monoterpene emissions data were incorporated into the MEGAN-EFP but there were no high-quality enclosure measurements of monoterpene emissions. Aircraft monoterpene flux measurements were incorporated into the database and currently are the best approach for constraining monoterpene emission factors.

29.4 MEGAN3 Sensitivity and Evaluation

A comparison of MEGAN2.1 and MEGAN3, with the emission factors generated by the initial runs of the MEGAN-EFP, indicated that the MEGAN3 isoprene emissions were considerably lower than MEGAN2.1 and in better agreement with aircraft flux and concentration measurements. Generally, there was a reduction in monoterpene emissions calculated in MEGAN3 relative to MEGAN2.1 that degraded agreement. Concentration comparisons using the Comprehensive Air Quality Model with extensions [6] indicated that estimates of monoterpenes aloft were often considerably lower using the MEGAN3 estimates. The cause of the reduction of monoterpene emissions in MEGAN3 is not known, but is under investigation. Assessment of the new stress algorithms (high temperature, low temperature, high winds, high ozone) incorporated into MEGAN3 indicated that there was relatively little impact on isoprene and monoterpenes using the initial parameterizations that are a conservative representation of these processes. These stresses are expected to have a greater impact on other BVOC.

29.5 Conclusions

MEGAN3 can be used to provide biogenic emissions estimates that are more accurate than MEGAN2.1 as demonstrated by comparison to aircraft flux measurements. MEGAN3 also facilitates assessing and improving individual model components including emission factors, canopy and soil environment conditions, and response functions which should lead to improvements over alternative models that are based on outdated emission factors and model algorithms.

The MEGAN3 framework can be used for assessing available landcover and emissions data and identifying gaps. The available landcover and emissions data that were incorporated for the initial database used for this project improved isoprene emissions for the regions investigated by the 2013 SAS aircraft study. Discrepancies between enclosure and aircraft measurements of isoprene emissions were reconciled.

Distributions of tree species, the major source of isoprene and monoterpene emissions, can be adequately estimated with existing landcover data in forested regions of southeast and southcentral US. Additional efforts are needed to improve tree species distributions in urban and savannah locations. Isoprene and other VOC emission studies have focused on closed canopy forests. The performance of models in open canopies in savannas and shrublands has not been well characterized and the existing landcover data, canopy environment and isoprene response functions may not be suitable for these landscapes.

29.6 Recommendations for Future Work

Forests are an important source of monoterpene emissions, but the rates remain highly uncertain. High quality measurements of monoterpene emission factors should be conducted to characterize the dominant vegetation. Emission factors for isoprene, including low or zero emissions, and other compounds, including sesquiterpenes and stress compounds, could be estimated by the same study.

The oaks are a diverse genus and include some European species that do not emit isoprene. The isoprene EF of more of the dominant oaks, including the savannah and shrub oaks that are relatively understudied, should be investigated with high quality measurements to quantify any within-genera variation.

Soil NO emissions are important in agricultural areas. Improved crop landcover and nitrogen fertilizer rate distributions should be incorporated into MEGAN along with advances in representation of other processes controlling soil NO emissions [2].

Further development and testing of MEGAN3 is recommended, including integration of additional MEGAN-EFP landcover and emissions data especially non-tree landcover, compounds other than isoprene and monoterpenes, and regions outside of southeastern and southcentral US.

Questions and Answers

Questioner: Clemens Mensink, Flemish Institute for Technological Research (VITO)

Question: How do you assign emission factor quality (J rating) – what are the criteria?

Answer: This is an important question that points directly to the need for community engagement in developing the MEGAN3 emission factor database. At this stage, only emission factors that were developed according to the Niinemets protocols [4] are assigned the highest quality rating (J = 4) with the majority of data assigned one of two low quality ratings: J = 0 for semi-quantitative screening studies conducted primarily in the 1960s–1980s using static branch enclosures to characterize understory and/or cut branches and J = 1 for more recent measurements that do not meet the Niinemets protocols. Having a quality rating provides motivation and opportunity for wider community engagement in assigning quality ratings as well as expanding the emission factor database. A primary motivation for developing the MEGAN3 emission factor preprocessor (EFP) is to encourage community-wide contributions that improve MEGAN3 for everyone.

Questioner: Paul Makar, Environment and Climate Change Canada

Question: Can the MEGAN canopy model provide information on light intensity at different heights within the forest canopy that could be used to drive air chemistry models?

Answer: The off-line version of MEGAN has a canopy environment model that explicitly simulates leaf-level light intensity and temperature on both sun and shade leaves at five canopy levels and could be adapted to drive photolysis rates for air quality models. [3] have demonstrated the importance of including in-canopy photolysis, chemistry, and turbulence in air quality model simulations and an effort to integrate the drivers for these processes with those for canopy BVOC emissions is needed.

Questioner: Xuesong Zhang, University of Toronto, Earth, Atmospheric and Planetary Physics
Question: It would be interesting to evaluate MEGAN3 VOC emissions over extensive natural forests such as Boreal, Amazon and Equatorial African regions.
Answer: We agree completely that more evaluation of MEGAN3 is essential and look forward to global model applications and additional observations that can provide such evaluation and, importantly, improve the MEGAN3 input databases and then share the improvements with the modeling community. A primary motivation for developing the MEGAN3 emission factor preprocessor (EFP) is to encourage community-wide contributions that improve MEGAN3 for everyone.

References

1. A.B. Guenther, X. Jiang, C.L. Heald, T. Sakulyanontvittaya, T. Duhl, L.K. Emmons, X. Wang, The Model of Emissions of Gases and Aerosols from Nature version 2.1 (MEGAN2.1): an extended and updated framework for modeling biogenic emissions. Geosci. Model Dev. **5**(6), 1471–1492 (2012)
2. R.C. Hudman, N.E. Moore, A.K. Mebust, R.V. Martin, A.R. Russell, L.C. Valin, R.C. Cohen, Steps towards a mechanistic model of global soil nitric oxide emissions: implementation and space based-constraints. Atmos. Chem. Phys. **12**, 7779–7795 (2012). https://doi.org/10.5194/acp-12-7779-2012
3. P.A. Makar, R.M. Staebler, A. Akingunola, J. Zhang, C. McLinden, S.K. Kharol, B. Pabla, P. Cheung, Q. Zheng, The effects of forest canopy shading and turbulence on boundary layer ozone **8**, 15243 (2017). https://doi.org/10.1038/ncomms15243
4. Ü. Niinemets, U. Kuhn, P.C. Harley, M. Staudt, A. Arneth, A. Cescatti, P. Ciccioli, L. Copolovici, C. Geron, A. Guenther, J. Kesselmeier, M.T. Lerdau, R.K. Monson, J. Peñuelas, Estimations of isoprenoid emission capacity from enclosure studies: measurements, data processing, quality and standardized measurement protocols. Biogeosciences **8**, 2209–2246 (2011). https://doi.org/10.5194/bg-8-2209-2011
5. G. Pouliot, J. Bash, Updates to Version 3.61 of the biogenic emission inventory system (BEIS), in *Presented at Air and Waste Management Association Conference*, Raleigh, NC (2015)
6. Ramboll, The Comprehensive Air quality Model with extensions (CAMx) homepage (2017). http://www.CAMx.com

Chapter 30
Modelling the Temporal and Spatial Allocation of Emission Data

Volker Matthias, Jan Arndt, Armin Aulinger, Johannes Bieser and Markus Quante

Abstract Atmospheric chemistry transport models (CTMs) need spatially and temporally resolved emission data as input. Atmospheric concentrations of pollutants as well as their deposition depend not only on the emitted amount but also on place and time of the emissions used for the model calculations. Available emission inventories, both regional and global ones, typically provide annual emissions of specific substances on a predefined grid. Often, this grid is of coarser resolution than the model grid and the temporal resolution is not higher than monthly. In addition, many species like volatile organic compounds (VOCs) or particulate matter (PM) are only given as lumped sums and not split into their chemical components. This requires further processing of the emissions in order to produce sufficiently resolved data sets for follow-up CTM runs. As a consequence, emission models were developed for the purpose of creating "model-ready" emissions. They use methods that depend on the emission sector and the additional data available for the disaggregation of the inventory data, e.g. land use and population density data. Recently, new methods for specific sectors like agriculture, residential heating and traffic have been developed. The most commonly used global and regional emission inventories are summarized and an overview of currently applied methods to spatially and temporally disaggregate emission inventory data is given. Particular emphasis is laid on the temporal disaggregation by presenting methods that allow the creation of individual time profiles for each model grid cell.

30.1 Introduction

Atmospheric chemistry transport models are applied for investigating the interactions between emissions from different sources and their influence on the spatial and temporal distribution of pollutant concentrations. These models need accurate

V. Matthias (✉) · J. Arndt · A. Aulinger · J. Bieser · M. Quante
Helmholtz-Zentrum Geesthacht, Institute of Coastal Research, Max-Planck-Strasse 1, 21502 Geesthacht, Germany
e-mail: volker.matthias@hzg.de

C. Mensink et al. (eds.), *Air Pollution Modeling and its Application XXVI*,
Springer Proceedings in Complexity,
https://doi.org/10.1007/978-3-030-22055-6_30

information on the chemical reactions, the meteorological situation and on the emissions of gases and particles into the atmosphere. Many studies have emphasized the importance of when and where substances are released into the atmosphere for the reliability of air quality predictions, for example [4, 5, 8].

Anthropogenic emission data is typically provided in emission inventories. They contain annual national totals for a number of emission sectors. Sometimes, the data is given on predefined grids, however, a temporal emission distribution is usually not given. Here, we present methods that are currently used to produce temporally and spatially resolved emission data for certain emission sectors. Most of them are top-down approaches that use surrogates like population density to spatially distribute emission data. Bottom-up methods require much more information about the properties of the emission source but they have a larger potential to produce detailed and realistic emission data.

30.2 Emission Inventories

Emission inventories contain data from many emission sources which are combined into aggregated emissions for a certain time span and region and for a number of reactive substances. Typically, they are given as annual sums for a specified country. A number of global and regional inventories with free data access exist. More information about their availability for chemistry transport model applications is given in a review paper by Matthias et al. [11].

Emissions are classified into sectors. In these sectors emissions from any type of units used for the same purpose and with similar properties are combined and their emissions are added up. Examples are, among many others, heating systems in households, power plants, small industrial facilities in a certain production sector or cars and trucks. Specific nomenclature systems exist for these emission sectors. They might be very broad like in the SNAP (Selected Nomenclature for Air Pollutants) system that contains 11 sectors or much more detailed like the European NACE system (Statistical Classification of Economic Activities in the European Community), the Nomenclature for reporting (NFR) system used in the United Nations Framework Convention on Climate Change (UNFCCC) or the NEI (National Emission Inventory) standards used in North America. They might contain several hundreds of sub-sectors and sub-sub-sectors. On the other hand in many regions of the world information about emissions to air are still very sparse.

The chemical components (or "species") included in emission inventories mainly consist of SO_2, NO_x, CO, NMVOC, NH_3 and PM. Some species are in fact mixtures of substances and need to be split into individual molecules (e.g. for $NO_x = NO + NO_2$) or groups of molecules (e.g. for NMVOCs) before they can be used in CTMs. For NMVOCs, it depends on the chemical mechanism implemented in the chemistry transport model, how a split needs to be done. For PM, sometimes components are defined like Black Carbon (BC) and Organic Carbon (OC), which are reported separately and treated individually in CTMs.

30.3 Emission Disaggregation

30.3.1 Horizontal and Vertical Disaggregation

Emission inventories that are given in gridded format like EDGAR [15] or the ECLIPSE emissions from the International Institute for Applied System Analysis (IIASA) [10] are provided on spatial resolutions between approximately 0.1 × 0.1 degrees and 0.5 × 0.5 degrees. Atmospheric chemistry transport models, on the other hand, often use other spatial resolutions and other map projections. Therefore, the emission data needs to be re-gridded.

This can be done with rather simple interpolation methods where no other information about the emission sources is considered. Other methods use so called proxy-data that contains spatial information related to the emissions for distributing them on a new grid. Taking for example the residential heating sector, the assumption here is that the emissions depend on the population density. The emissions from this sector would first be aggregated for a larger area, e.g. based on annual sales statistics for energy carriers in a specific country. Then, they would be distributed following population density data that is available on a very high resolution grid (down to 1 × 1 km^2). Afterwards, the data would be newly aggregated on the grid used for the chemistry transport model runs, if this is necessary.

Frequently, emissions depend on meteorological quantities, and here in the first place on ambient temperature. This is especially the case for residential heating emissions, but it also applies to emissions from agriculture or evaporative emissions from cars. Consequently, these emissions can be spatially—and temporally—disaggregated following the meteorological data that is necessary to run the chemistry transport model system. Aulinger et al. [1] presented a study where they disaggregated national emission totals for benzo(a)pyrene according to heating demand which is obviously temperature dependent.

Emission strengths do not vary only horizontally, there are also large differences in emission height. Large power plants, industrial facilities, and, e.g., ships have high stacks that release their emissions in high altitudes. In addition, the exit velocity and the temperature of the exhaust gas need to be considered when the emissions shall be vertically allocated. Bieser et al. [3] calculated more than 40,000 vertical emission profiles for six pollutants and various meteorological conditions using algorithms implemented in the SMOKE model [7]. These profiles were clustered into 73 groups using hierarchical cluster analysis. They contain effective emission heights for several SNAP sectors and can be used as representative emission heights in regional model applications for Europe. However, this approach could still be improved when more information about stack properties in a number of countries would be available.

30.3.2 Temporal Disaggregation

Most of the emission inventories provide only annual values, however, some contain seasonal variations with monthly emission maps (e.g. HTAP2.2 [9], ECLIPSE [10] or REAS 2.1 [14]). For model ready emission data, the temporal variation of specific source sectors is considered via the application of typical time profiles. They are typically divided into fixed monthly, weekly and daily variations with no dependence on season or location.

However, the spatial and temporal allocation of emissions can be improved, as some publications demonstrated for ammonia [2, 6, 17]. Skjoth et al. [17] and Backes et al. [2] proposed to split emissions from agriculture into animal husbandry, manure management and fertilizer use. Emissions from stables can be modelled according to ambient temperatures and wind velocity, manure management needs to follow legislative restrictions and fertilizer use depends on season and crop growing times. A similar approach was followed by Hendricks et al. [6] for agricultural emissions in the Netherlands.

Menut et al. [12] investigated diurnal cycles of traffic emissions in several European cities. They found significant differences between northern European cities and those in the South and considered this in their regional scale chemistry transport model calculations with the CHIMERE model. Mues et al. [13] followed a similar approach for traffic emissions in Germany. Both studies revealed similar results. While correlation coefficients for NO_2, O_3 and $PM_{2.5}$ were improved when the model results were compared to observations, the mean concentrations showed only small changes. Figure 30.1 shows the temporal variation of hydrocarbon (HC) emissions from traffic in Germany. The time profile was developed within the German project "Verkehrsentwicklung und Umwelt II (Traffic development and the environment II)" [16]. There, it was demonstrated that the largest part of the HC emissions stems from tank evaporation. These and the emissions due to cold starts are temperature dependent, which was represented in the emission data set.

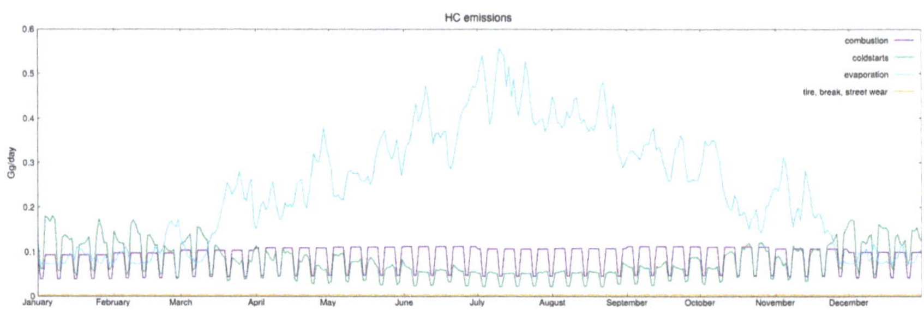

Fig. 30.1 Temporal distribution of daily emissions of volatile hydrocarbons (HC) from traffic in Germany

30.4 Summary and Outlook

Detailed emission data is crucial for 3D atmospheric chemistry transport model calculations. This concerns not only emission totals but also their temporal and spatial distribution. Because emissions cannot be constantly monitored for all relevant sources, well suited proxy data can help distributing bulk emissions from certain sectors. Heating emissions from households, emissions from animal husbandry as well as evaporative emissions from cars all depend on ambient temperatures and partly also on wind. Therefore, meteorological data often serves as good proxy for their spatio-temporal allocation. This way, the proxy data is calculated as in a bottom-up approach and then used to distribute the bulk emissions given on national level.

Large parts of the data that is collected for traffic monitoring could be used to improve the representation of these emissions in chemistry transport model systems in the future. This holds not only for cars and trucks but also for ship traffic and aviation. However, besides the availability of that data for research purposes, new data science technologies need to be developed for handling such large amounts of unstructured data.

Questions and Answers

QUESTIONER: Greg Yarwood

QUESTION: Are gridded inventories a limitation for industrial point source emissions in particular?

ANSWER: Yes, they are. The emissions will be taken as an average over the entire grid cell, which means that they are drastically diluted if the grid is rather coarse. This will have consequences for the dispersion of the emissions and the chemical transformations. In addition, if the location of a point source within the grid is not exactly given, it cannot be attributed to the correct grid cell on a finer grid. This might cause significant errors in the location of a source when the data is re-gridded.

References

1. A. Aulinger, V. Matthias, M. Quante, An approach to temporally disaggregate Benzo (a) pyrene emissions and their application to a 3D Eulerian atmospheric chemistry transport model. Water Air Soil Pollut. **216**(1–4), 643–655 (2011)
2. A. Backes, A. Aulinger, J. Bieser, V. Matthias, M. Quante, Ammonia emissions in Europe, Part I: development of a dynamical ammonia emission inventory. Atmos. Environ. **131**, 55–66 (2016)
3. J. Bieser, A. Aulinger, V. Matthias, M. Quante, H.A.C. Denier van der Gon, Vertical emission profiles for Europe based on plume rise calculations. Environ. Pollut. **159**(10), 2935–2946 (2011)
4. H.A.C. Denier van der Gon et al., Particulate emissions from residential wood combustion in Europe—revised estimates and an evaluation. Atmos. Chem. Phys. **15**, 6503–6519 (2015)
5. G. Frost et al., New Directions: GEIA's 2020 vision for better air emissions information. Atmos. Environ. **81**, 710–712 (2013)
6. C. Hendriks et al., Ammonia emission time profiles based on manure transport data improve ammonia modelling across north western Europe. Atmos. Environ. **131**, 83–96 (2016)

7. M.R. Houyoux et al., Emission inventory development and processing for the Seasonal Model for Regional Air Quality (SMRAQ) project. J. Geophys. Res. Atmos. **105**, 9079–9909 (2000)
8. U. Im et al., Evaluation of operational online-coupled regional air quality models over Europe and North America in the context of AQMEII phase 2. Part II: Particulate matter. Atmos. Environ. (2014). http://dx.doi.org/10.1016/j.atmosenv.2014.08.072
9. G. Janssens-Maenhout, M. Crippa, D. Guizzardi, F. Dentener, M. Muntean, G. Pouliot, T. Keating, Q. Zhang, J. Kurokawa, R. Wankmuller, H.D. van der Gon, J.J.P. Kuenen, Z. Klimont, G. Frost, S. Darras, B. Koffi, M. Li, HTAP_v2.2: a mosaic of regional and global emission grid maps for 2008 and 2010 to study hemispheric transport of air pollution. Atmos. Chem. Phys. **15**(19), 11411–11432 (2015)
10. Z. Klimont, K. Kupiainen, C. Heyes, P. Purohit, J. Cofala, P. Rafaj, J. Borken-Kleefeld, W. Schöpp, Global anthropogenic emissions of particulate matter including black carbon. Atmos. Chem. Phys. **17**(14), 8681–8723 (2017)
11. V. Matthias, J.A. Arndt, A. Aulinger, J. Bieser, H.A.C. Denier van der Gon, R. Kranenburg, J. Kuenen, D. Neumann, G. Pouliot, M. Quante, Modeling emissions for three dimensional atmospheric chemistry transport models. J. Air Waste Manag. Assoc. **68**(8), 763–800 (2018)
12. L. Menut, A. Goussebaile, B. Bessagnet, D. Khvorostiyanov, A. Ung, Impact of realistic hourly emissions profiles on air pollutants concentrations modelled with CHIMERE. Atmos. Environ. **49**, 233–244 (2012)
13. A. Mues et al., Sensitivity of air pollution simulations with LOTOS-EUROS to the temporal distribution of anthropogenic emissions. Atmos. Chem. Phys. **14**(2), 939–955 (2014)
14. T. Ohara, H. Akimoto, J. Kurokawa, N. Horii, K. Yamaji, X. Yan, T. Hayasaka, An Asian emission inventory of anthropogenic emission sources for the period 1980–2020. Atmos. Chem. Phys. **7**(16), 4419–4444 (2007)
15. J.G.J. Olivier, J.A.H.W. Peters, J. Bakker, J.J.M. Berdowski, A.J.H. Visschedijk, J.J. Bloos, EDGAR 3.2: Reference database with trend data of global greenhouse gas emissions for 1970–1995, in *Non-CO2 Greenhouse Gases: Scientific Understanding, Control Options and Policy Aspects* (2002), pp. 291–292
16. S. Seum, J. Bieser, S. Ehrenberger, Ul. Kugler, German and European Ground-Transport Emissions in three different scenarios until 2040. Transport Research Arena, TRA 2018, 16–19. April 2018, Wien, Austria (2018)
17. C.A. Skjoth et al., Spatial and temporal variations in ammonia emissions—a freely accessible model code for Europe. Atmos. Chem. Phys. **11**(11), 5221–5236 (2011)

Chapter 31
A High-Resolution National Emission Inventory and Dispersion Modelling—Is Population Density a Sufficient Proxy Variable?

Niko Karvosenoja, Ville-Veikko Paunu, Mikko Savolahti, Kaarle Kupiainen, Ari Karppinen, Jaakko Kukkonen and Otto Hänninen

Abstract Air quality modeling at high spatial resolution over large domains enables comprehensive health impact assessment. Spatially finely resolved emission inventories are a crucial component for reliable modeling. Spatialization of emissions from disperse emission sources (e.g. road transport) is performed using GIS-based spatial information, i.e. spatial proxies (e.g. road network and traffic volume data). For some important emission source sectors, however, it is challenging to define proxies that adequately represent the spatial distribution of emissions, and, for the lack of more representative information, population density is often used as a proxy. However, that is rarely a realistic representation and might distort the resulting assessments of their population exposure and health impacts. This study presents the impacts of the spatial allocation process and its improvements for machinery sector, by using an emission model at 250 m grid resolution in Finland. The corresponding influence on the modeled population exposure to $PM_{2.5}$ is also presented. The improvements in the gridding procedures had a substantial impact on the modeled concentrations, especially in areas with denser population. For example, the emissions in Helsinki area from the machinery sector decreased by 41% due to the improvements. We conclude that it is necessary to use more realistic spatial proxies instead of the population density for evaluating the emissions originated from various emission categories.

N. Karvosenoja (✉) · V.-V. Paunu · M. Savolahti · K. Kupiainen
Finnish Environment Institute (SYKE), POB 140, 00251 Helsinki, Finland
e-mail: niko.karvosenoja@ymparisto.fi

A. Karppinen · J. Kukkonen
Finnish Meteorological Institute (FMI), Helsinki, Finland

O. Hänninen
National Institute for Health and Welfare (THL), Helsinki, Finland

C. Mensink et al. (eds.), *Air Pollution Modeling and its Application XXVI*, Springer Proceedings in Complexity,
https://doi.org/10.1007/978-3-030-22055-6_31

31.1 Introduction

Spatially resolved emission inventories are an important input for air quality models. The quality of emission data depends on the uncertainties of several components of the inventory, e.g. activity data, emission factors and spatial representation of emissions. The uncertainties associated with activity data and emission factors in different sectors have been relatively widely assessed at regional and national level (e.g. [3, 5]). However, the impact of the accuracy of spatial disaggregation of emissions, i.e. gridding, has been less studied, but it has been identified to be potentially considerable [6].

The required spatial resolution of an emission inventory depends on the scale and impact end-point considered in an air quality model. With global to regional level models that describe mainly the long-range transboundary effects, an emission grid resolution of from 25 to 100 km is often considered to be adequate. As for regional to local level modeling that describe also the impact of local emission sources, it is commonly necessary to use spatial representation at grid resolutions down to 1 km or finer. Especially for primary pollutants that show strong concentration gradients, e.g. fine particulate matter ($PM_{2.5}$), gridding of emissions to fine horizontal resolution is required for reliable impact assessment.

Gridding of emissions from disperse emission sources (e.g. road transport) are typically performed using GIS-based spatial information, i.e. spatial proxies (e.g. road network and traffic volume data) to represent the spatial distribution of emissions. Therefore it is crucial to select appropriate proxies for different emission source sectors. Most commonly used proxies are road network data for road transport, industrial land use for industrial emissions and population density for a wide variety of emission sources [7]. However, especially population density data rarely represents adequately the spatial distribution of emissions at fine spatial resolution. Zheng et al. [9] showed that the importance of more sophisticated proxy choices than population density increased at spatial resolutions below 25 km.

Some of the more challenging sectors for emission gridding include residential wood combustion and diesel machinery [7, 8]. For example, for residential wood combustion information about the location and use of wood stoves is often scarce or lacking. Another spatially challenging emission source category is diesel machinery used in urban areas, which includes various types of equipment operating in construction and maintenance works at multitude of environments. These sources are often spatially allocated in emission inventories using population density information as a proxy. This study presents an improvement of the gridding procedure for machinery in the Finnish regional emission model FRES. The influence of using more sophisticated gridding proxies than population density was demonstrated with modeled population exposure results at 250 m grid resolution.

31.2 Methods

The Finnish Regional Emission Scenario (FRES) model [2] is an air pollution integrated assessment framework covering all anthropogenic emission sources in Finland. The dispersion of primary $PM_{2.5}$ uses source-receptor relationships at 250 m horizontal grid resolution [1] based on the urban scale multiple source Gaussian model UDM-FMI.

31.3 Population Exposure Due to Machinery Emissions—Impact of Gridding

In the context of this study, machinery refers to traffic or stationary apparatus powered with diesel or gasoline engines used in various working environments (Table 31.1). The diversity of the environments brings a challenge to the estimation of their average location distributions.

The most straightforward machinery activity types from emission gridding point of view are agricultural and forestry machinery and, to some extent, industrial machinery. Their average activities can be attributed to agricultural, forest and industrial land use types, respectively, that are typically available as GIS data to be used as gridding proxies.

However, a considerable fraction of machinery used in various types of building, construction and maintenance activities are more challenging to estimate spatially (Table 31.1). In an earlier version of the FRES model, these activities were gridded

Table 31.1 Main machinery and activity types in Finland. The improved types given in *italics*

Main machinery type	Main activity types	Activity 2015 (TJ a^{-1})
Agriculture tractors, combine harvesters	Agriculture	7160
Forest tractors, felling machines, chain saws	Forestry	3828
Snow mobiles, ATVs	Agriculture, forestry	2083
Forklift trucks, industry tractors	Industry	2890
Cranes	*Building, industry*	*532*
Road graders	*Road constr. and mainten.*	*469*
Wheel loaders, dumper trucks	*Roads construction, mines*	*5093*
Excavators	*Building, roads, industry*	*5990*
Other maintenance machinery	*Building, street mainten.*	*1434*
Lawnmowers	*Residential, greeneries*	*1172*
Diesel generators	*Road constr. and mainten.*	*1373*
Diesel compressors	*Building, street constr.*	*496*

using population density data as a proxy. This has been improved to more diverse proxy types given in Table 31.2.

The improvement in the gridding procedure moved a considerable amount of the emissions originated from construction and maintenance machinery from areas with a relatively dense population to locations that are probably more realistic. As an example, the estimate of machinery $PM_{2.5}$ emission in Helsinki area decreased from 135 to 80 Mg a^{-1} due to reallocation of spatial distribution (Fig. 31.1). Respectively, the emission estimates increased in less urban locations of Finland, e.g., mining areas and along main highways, due to the use of respective spatial proxies.

The reallocation of emissions had a considerable impact on resulting modeled population exposure to $PM_{2.5}$. Population weighted concentration due to machinery emissions over the whole Finland decreased from 0.39 to 0.24 μg m^{-3} with the improved emission gridding. These equal to 8% and 5% contribution of the total population exposure to $PM_{2.5}$ in Finland, respectively.

Table 31.2 Activity types and spatial proxies used in the improved emission gridding

Main activity type	Spatial proxy
General building and construction machinery	Population density
Road and street construction and maintenance	Roads and streets activity-weighted
Mining	Mining areas
Residential and greenery maintenance	Detached house and greenery areas

(a) **(b)**

Fig. 31.1 Spatial distribution of $PM_{2.5}$ emissions from the machinery sector in Helsinki area (**a**) before (emission in the area 135 Mg a^{-1} in 2015) and (**b**) after the improvements in emission gridding (80 Mg a^{-1})

31.4 Conclusions

The improvements in gridding implemented for $PM_{2.5}$ emissions originated from machinery had a substantial relative impact on the modeled concentrations and the resulting population exposure. Although the improvement considered less than half of the machinery activities (i.e. excluding agriculture, forestry and industry machinery), it decreased the predicted population exposure caused by all machinery by 38%, calculated over the population of Finland in 2015. Similarly, more drastic overestimation of health impacts due to flawed use of population-based proxy has been shown also for $PM_{2.5}$ emissions originated from residential wood combustion [4].

This paper has shown that it is necessary to use more realistic spatial proxies than solely the population density for evaluating the emissions originated from different emission categories.

Questions and answers

Questioner 1: Richard Menard
Question 1: Do you think we can get a quantification of the emission uncertainty as this would be important for example for top-down emissions?
Answer 1: Quantified uncertainty estimates for emission inventories widely exist concerning activity data and emission factor uncertainties in different emission source sectors and different pollutants (e.g. [11, 12]). However, the impact of assumptions for the spatial disaggregation of emissions on uncertainties is often lacking (e.g. [14]). When, e.g., a country-level top-down emission inventory is used as a basis for a city-level air quality assessment modelling, the method of spatial disaggregation of emissions within the country might increase the emission uncertainty for the area in question considerably. Thus, the Question 1 in the context of gridded emission inventories is highly relevant and important. Estimates on the uncertainties of gridded emission inventories have been presented (e.g. [10]), however, methods to quantify the uncertainties arising from spatial disaggregation are still being developed by various inventory reserch teams.

Questioner 2: Volker Matthias
Question 2: How do you know how much wood is consumed in the countryside and how much in cities [in residential wood stoves]?
A2: The question refers to the gridding procedure for residential wood combustion emissions used in the Finnish FRES model that uses house locations as a basis for gridding and takes into account differences in average wood use per household in different types of houses in different sizes of settlements [13]. The data on average wood use per household are based on questionnaires.

Acknowledgements This work has been funded by Academy of Finland in the project *Environmental impact assessment of airborne particulate matter: the effects of abatement and management strategies (BATMAN)* and NordForsk under the Nordic Programme on Health and Welfare Project #75007: *Understanding the link between air pollution and distribution of related health impacts and welfare in the Nordic countries (NordicWelfAir).*

References

1. N. Karvosenoja, L. Kangas, K. Kupiainen et al., Integrated modeling assessments of the population exposure in Finland to primary PM2.5 from traffic and domestic wood combustion on the resolutions of 1 and 10 km. Air. Qual. Atmos. Health **4**, 179–188 (2011)
2. N. Karvosenoja, Emission scenario model for regional air pollution. Monogr. Boreal Environ. Res. **32** (2008)
3. N. Karvosenoja, M. Tainio, K. Kupiainen et al., Evaluation of the emissions and uncertainties of PM2.5 originated from vehicular traffic and domestic wood combustion in Finland. Boreal Environ. Res. **13**, 465–474 (2008)
4. V.-V. Paunu, N. Karvosenoja, M. Savolahti et al., High quality spatial model for residential wood combustion emissions, in *16th IUAPPA World Clean Air Congress, Cape Town, South Africa, 29 September–4 October 2013* (2013), 4 pp
5. M. Rao, I. D'Elia, A. Piersantic, An uncertainty quantification of PM2.5 emissions from residential wood combustion in Italy. Atmos. Pollut. Res. (2018, in press)
6. P. Thunis, A. Miranda, J.M. Baldasano et al., Overview of current regional and local scale air quality modelling practices: assessment and planning tools in the EU. Environ. Sci. Policy **65**, 13–21 (2016)
7. M. Trombetti, P. Thunis, B. Bessagnet et al., Spatial inter-comparison of Top-down emission inventories in European urban areas. Atm. Environ. **173**, 142–156 (2018)
8. P. Viaene, C.A. Belis, N. Blond et al., Air quality integrated assessment modelling in the context of EU policy: a way forward. Environ. Sci. Policy **65**, 22–28 (2016)
9. B. Zheng, Q. Zhang, D. Tong et al., Resolution dependence of uncertainties in gridded emission inventories: a case study in Hebei. China. Atmos. Chem. Phys. **17**, 921–933 (2017)

References to Questions and Answers Section

10. S. Hogue, E. Marland, R.J. Andres et al., Uncertainty in gridded CO_2 emissions estimates, in *4th International Workshop on Uncertainty in Atmospheric Emissions, 7–9 October 2015*, Krakow, Poland (2015), pp. 117–124. http://www.ibspan.waw.pl/unws2015/images/publications/4thWorkshopProceedings.pdf
11. N. Karvosenoja, Emission scenario model for regional air pollution. Monogr. Boreal Environ. Res. **32** (2008)
12. J.J.P. Kuenen, A.J.H. Visschedijk, M. Jozwicka et al., TNO-MACC_II emission inventory; a multi-year (2003–2009) consistent high-resolution European emission inventory for air quality modelling. Atmos. Chem. Phys. **14**, 10963–10976 (2014)
13. V.-V. Paunu, N. Karvosenoja, M. Savolahti et al., High quality spatial model for residential wood combustion emissions, in *16th IUAPPA World Clean Air Congress, Cape Town, South Africa, 29 September–4 October 2013* (2013), 4 pp
14. E. Pisoni, D. Albrecht, T.A. Mara et al., Application of uncertainty and sensitivity analysis to the air quality SHERPA modelling tool. Atmos. Environ. **183**, 84–93 (2018)

Chapter 32
Modeling of Leisure Craft Emissions

Lasse Johansson, Jukka-Pekka Jalkanen, Erik Fridell, Ilja Maljutenko, Erik Ytreberg, Martin Eriksson, Eva Roth and Vivian Fischer

Abstract Commercial shipping fleet and its emissions can be modeled in detail, but the emission from leisure craft are often invisible for activity based, bottom-up emission inventories. A new model (FMI-BEAM) describes the emissions from the leisure craft fleet in the Baltic Sea area, complementing the existing STEAM emission modeling suite. BEAM combines information from over 3000 boat marina locations, national small boat registries, Automatic Identification System data and boat survey results to derive leisure boat emissions for over 250,000 boats around the Baltic Sea coastline. The location of marinas and boat counts were determined from satellite images and other available data. With the BEAM leisure craft simulation model the spatial and temporal distribution of air emissions can be estimated. The presented results describe our first attempt to generate fuel consumption and emission inventory for small boats which have been underrepresented in current emission inventories. Small boat activity often occurs near the coastline in vicinity of populated areas and leisure craft emissions contribute to local air quality. The emissions of carbon monoxide and hydrocarbons are high compared to the emissions of commercial shipping, because very high emission levels are allowed for old small boat engines

L. Johansson (✉) · J.-P. Jalkanen
Atmospheric Composition Research, Finnish Meteorological Institute, Erik Palmen's Square 1, 00560 Helsinki, Finland
e-mail: lasse.johansson@fmi.fi

E. Fridell
Swedish Environmental Research Institute, Aschebergsgatan 44, 41133 Göteborg, Sweden

I. Maljutenko
Tallinn University of Technology, Ehitajate Tee 5, 12616 Tallinn, Estonia

E. Ytreberg · M. Eriksson
Shipping and Marine Technology, Chalmers University of Technology, 41296 Gothenburg, Sweden

E. Roth
University of Southern Denmark, Niels Bohrs Vej 9-10, 6700 Esbjerg, Denmark

V. Fischer
Helmholtz-Zentrum Geesthacht, Max-Planck-Straße 1, Geesthacht, Germany

C. Mensink et al. (eds.), *Air Pollution Modeling and its Application XXVI*, Springer Proceedings in Complexity,
https://doi.org/10.1007/978-3-030-22055-6_32

according to current legislation. The approach described in this paper can be applied in larger scale, for example to cover European coastlines in the future.

32.1 Introduction

In contrast to commercial marine vessel traffic, small boat activities in the Baltic Sea cannot be estimated based on the Automatic Identification System (AIS), because AIS-equipment is not mandatory for boats. Since little or no information exists on the activities of boats, their contribution has all but been ignored in environmental impact studies of shipping. In many Baltic Sea countries, like in Finland for example, data of fuel sold to recreational boats are not collected and incomplete statistics exists for the number of boats and their locations in the Baltic Sea coastline. This makes assessments of small boat fuel consumption and emissions very challenging. This is in contrast with large marine vessels, for which our earlier work with Ship Traffic Emission Assessment Model (STEAM) [1, 2, 3, 4] can be applied. In this paper we present an approach which is applicable for small boats, which are not well presented in STEAM emission inventories. This approach is called the Boat Emission and Activity siMulator (BEAM) and it extends the capabilities of the waterborne emission modeling to the part of traffic which is not visible in AIS. In short: STEAM is for ships, BEAM is for boats.

32.2 Materials and Methods

The data available for small boat emission modeling varies by country. Countries like Finland do not keep track of boat marina locations or boat counts in the Baltic Sea coastline. A national boat registry is available, but it does not distinguish boats used on the coast of the Baltic Sea and inland waters. Further, no statistics exist for fuel sold to small boats or emissions and their spatio-temporal distribution.

Temporal distribution of boating activity

While AIS-data cannot be used to produce reliable pleasure boat activity data, AIS-data does contain enough small pleasure crafts for the estimation of temporal variation of activities. The resulting temporal variation of activities can be used across the Baltic Sea for all pleasure craft. It should be noted the ships that are not IMO-registered but still sending AIS-messages can also be pilot vessels, Search and Rescue (SAR) craft, small tugs and service vessels. Thus, leisure boats need to be filtered out from the AIS data.

AIS data from Helcom from period 2014-2016 and the STEAM model was used in this analysis. First, a collection of boats was isolated from the AIS-data. These were filtered out from the overall AIS data using criteria like (a) Active only during 1st May–30th September, (b) travel distances not exceeding a selected threshold value, (c) not included IMO registry and (d) vessel is not active regularly. Using the conditions above, approximately 2000 unknown vessels were filtered for further analysis. A clear majority of these ships are Swedish, Danish and German vessels.

Naturally, the start of the boating season may have large annual variation depending on the harshness of winter each year; the length of season and intensity of activity are affected by summer weather conditions. The boating activity may start at different times depending on the latitude and may vary according to summer weather conditions, but inclusion of these features is left for future work. According to the estimated profile, June-August contains approx. 75% of all activities and there is a strong diurnal variation and weekends during 10 am–2 pm have the highest activities and public holidays and summer vacation period start are visible. This temporal profile is also consistent with the distribution of voluntary SAR missions which can be indirectly linked with leisure boat activity.

Geographical distribution of boat activity

Since little or no information of the geographical distribution of pleasure boat activities exists, the approach taken in this study was to model the geographical distribution based on marina locations. The key assumption was that the geographical distribution of leisure craft activities can be associated to the vicinity of marinas. This assumption is based on the findings of the Swedish survey [5] which indicates that 90% of boats operate near their marinas.

A list of pleasure craft marinas (boat place counts, location) for each riparian state was collected based on survey data, existing national studies and satellite image analysis. The list includes more than 3000 locations for leisure boats in the Baltic Sea area and accounts for more than 250,000 boats in total. It should be noted that the list of boat mooring place counts and locations are indicative and their main purpose is to geographically distribute pre-defined national pleasure boat activities. The activity is simply spread out as a Gaussian distribution around marina, but it has been assumed that boats favor the coastline and archipelagos, which extend the range of their activity.

Technical description of the Baltic Sea boat fleet

The Swedish small boat survey describes the national vessel fleet operating along the Baltic Sea coast in detail. Data at this level of detail was not available for other countries, but since almost half of the small boats are of Swedish origin and similarities could be seen between the use and composition of Swedish and Finnish boat fleets, which together cover about 2/3 of all boats in the area, we assume that fleet description and usage of boats in the Swedish fleet apply to all boats in the Baltic Sea.

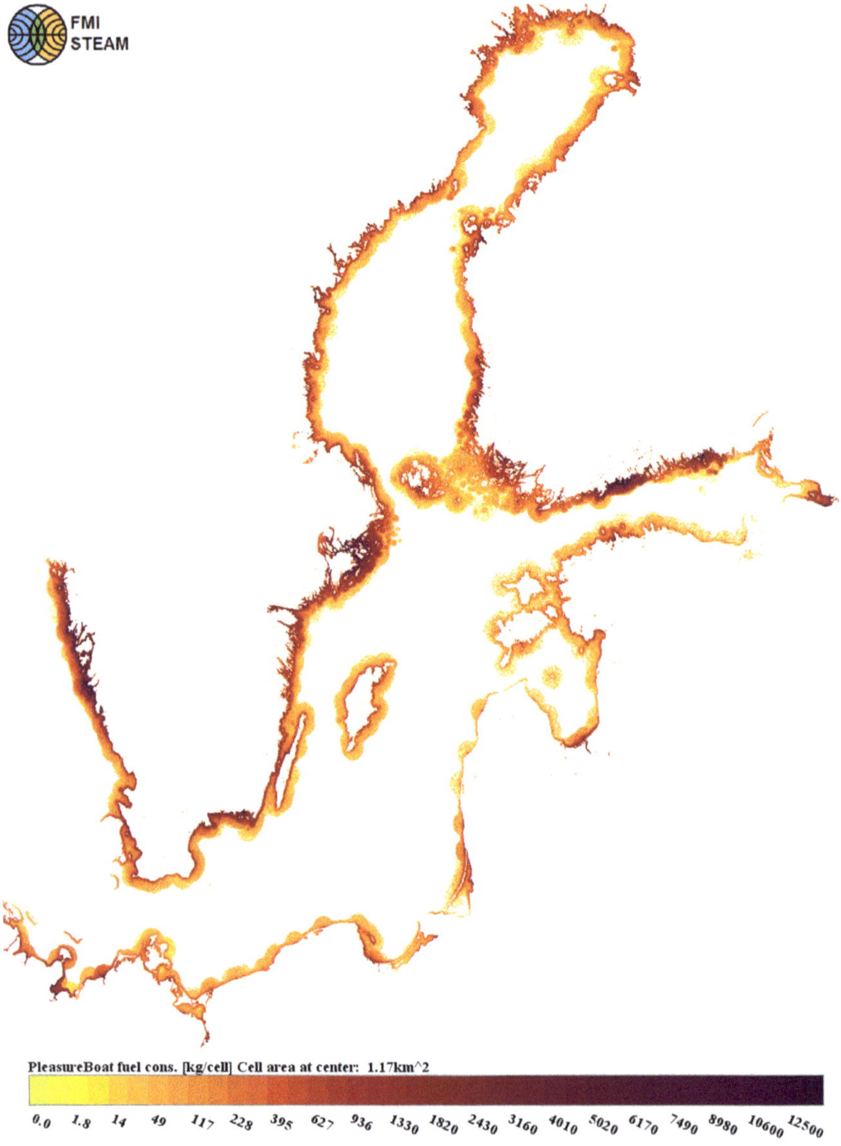

Fig. 32.1 Geographical distribution of the modeled leisure boat fuel consumption in the Baltic Sea area

32.3 Conclusion

The annual total fuel consumption of the Baltic Sea fleet is around 61,000 tons which is about 1.3% of the fuel consumed (and CO_2 emitted) by the Baltic Sea commercial shipping fleet. Figure 32.1 describes the geographical distribution of small boat fuel consumption, which is mostly concentrated along the coasts, but archipelago areas are also visible, because these are popular boating areas.

However, boats are already a more significant source of hydrocarbons than the commercial shipping fleet; the HC emissions from leisure craft in July are 6.5 times those of the large ships. Also, emissions of CO from boats during July are 2.5 times the corresponding amount from large ships. This work provides a rough estimate of the contribution of small boats to waterborne traffic emissions. The results are prone to large uncertainties concerning boat activities, but these can be reduced with further work. Our approach to small boat emission modeling is applicable to other areas, but care must be taken to adjust the activity totals and temporal distribution to fit local conditions.

One question from Kaarle Kupiainen was received after this talk in ITM2018.

"Does the PM estimate consider the influence of lubrication oil on emissions for especially the older gasoline two-stroke motors?"

Our response: The approach presented in this paper is not sophisticated enough to incorporate the PM emissions from lubrication oil. In principle, if emission factor measurements for small boat engines were conducted with enough detail, this could be determined by following typical chemical markers arising from lubrication oil. Currently, however, this is not done.

Acknowledgements This work resulted from the BONUS SHEBA project and it was supported by BONUS (Art 185), funded jointly by the EU, the Academy of Finland, Swedish Agency for Marine and Water Management, Swedish Environmental Protection Agency, FORMAS, Forschungcentrum Jülich, ICES and the Estonian Science Foundation. We are grateful to the Helcom member states for allowing the use of HELCOM AIS data in this research.

References

1. J.P. Jalkanen, A. Brink, J. Kalli, H. Pettersson, J. Kukkonen, T. Stipa, A modelling system for the exhaust emissions of marine traffic and its application in the Baltic Sea area. Atmos. Chem. Phys. **9**, 9209–9223 (2009)
2. J.P. Jalkanen, L. Johansson, J. Kukkonen, A. Brink, J. Kalli, T. Stipa, Extension of an assessment model of ship traffic exhaust emissions for particulate matter and carbon monoxide. Atmos. Chem. Phys. **12**, 2641–2659 (2012)
3. L. Johansson, J.P. Jalkanen, J. Kalli, J. Kukkonen, The evolution of shipping emissions and the costs of regulation changes in the northern EU area. Atmos. Chem. Phys. **13**, 11375–11389 (2013)

4. L. Johansson, J.P. Jalkanen, J. Kukkonen, Global assessment of shipping emissions in 2015 on a high spatial and temporal resolution. Atmos. Environ. **167**, 403–415 (2017)
5. Transportstyrelsen: Boating survey 2010—A study of Swedish small boats and how they are used. [online] Available from: https://www.transportstyrelsen.se/globalassets/global/sjofart/dokument/fritidsbatar1/batlivsundersokningen_2010.pdf

Chapter 33
Characteristics and Mitigation of Vehicular Non-exhaust Particle Emissions in Nordic Conditions

Kaarle Kupiainen, Ana Stojiljkovic, Ville-Veikko Paunu, Niko Karvosenoja, Ari Karppinen, Jaakko Kukkonen, Leena Kangas, Mari Kauhaniemi, Bruce Denby and Otto Hänninen

Abstract Assessment of air pollution health effects of road traffic should accurately characterize also the non-exhaust sources to support efficient emission mitigation. The use of the recently developed new-generation non-exhaust emission models opens new opportunities to extend and improve the traffic PM emission factors and the emission inventories. This study utilizes the non-exhaust emission model NOn-exhaust Road TRaffic Induced Particle emissions (NORTRIP) to evaluate the PM_{10}

K. Kupiainen (✉)
Ministry of the Environment, P.O. Box 35, 00023 Helsinki, Finland
e-mail: kaarle.kupiainen@ym.fi

A. Stojiljkovic · V.-V. Paunu · N. Karvosenoja
Finnish Environment Institute (SYKE), P.O. Box 140, 00251 Helsinki, Finland
e-mail: ana.stojiljkovic@ymparisto.fi

V.-V. Paunu
e-mail: ville-veikko.paunu@ymparisto.fi

N. Karvosenoja
e-mail: niko.karvosenoja@ymparisto.fi

A. Karppinen · J. Kukkonen · L. Kangas · M. Kauhaniemi
Finnish Meteorological Institute, P.O. Box 503, 00101 Helsinki, Finland
e-mail: ari.karppinen@fmi.fi

J. Kukkonen
e-mail: Jaakko.kukkonen@fmi.fi

L. Kangas
e-mail: leena.kangas@fmi.fi

M. Kauhaniemi
e-mail: mari.kauhaniemi@fmi.fi

B. Denby
Norwegian Meteorological Institute (MET), P.O. Box 43, Blindern, 0313 Oslo, Norway
e-mail: brucerd@met.no

O. Hänninen
National Institute for Health and Welfare (THL), P.O. Box 95, 70701 Kuopio, Finland
e-mail: otto.hanninen@thl.fi

C. Mensink et al. (eds.), *Air Pollution Modeling and its Application XXVI*,
Springer Proceedings in Complexity,
https://doi.org/10.1007/978-3-030-22055-6_33

and $PM_{2.5}$ emission factors separately for different climatic zones, road and street categories, and road maintenance practices in Finland. The non-exhaust emissions are influenced by climatic and weather factors, especially the road surface moisture, and have seasonal peaks particularly in winter and spring, due to enhanced formation of layers of street dust and their suspension to the air. The results demonstrated that changes in the selection of the types of winter tyres and their use, as well as road maintenance interventions could substantially reduce the non-exhaust emissions, especially in spring. Data gaps were identified in collecting the relevant inputs for the model. The methods presented in this study could be used at least in the whole Northern part of the world, i.e., the Nordic countries and Northern Central Europe, Northern Russia, North America, and parts of China and Japan.

33.1 Introduction

Modern emission abatement technologies are reducing the particle emissions from vehicular exhaust sources, but the non-exhaust sources remain a challenge. Studies and inventories of road traffic emissions should accurately characterize the non-exhaust sources to support efficient emission mitigation to the benefit of air quality and human health. In the Nordic countries key sources of non-exhaust PM emissions are wear products from pavements, tyres and brakes as well as sand and salt used for traction control [3, 4]. Additionally climatic and weather factors, especially the road surface moisture, influence the ambient emissions and result in seasonal peaks particularly in spring, due to enhanced formation of layers of street dust and their suspension to the air. With all these influencing factors the system guiding the non-exhaust PM emissions is complex and it is often difficult to identify the main sources or formulate recommendations on how to most efficiently reduce emissions in different locations. Our thinking is that a method to overcome these challenges is to use generalized, process based models for estimating non-exhaust PM emission factors that would better represent emission characteristics in broader weather, traffic, road and street conditions. However, there are not that many models available for the time being. In this work we utilized the recently developed and published non-exhaust emission model NOn-exhaust Road TRaffic Induced Particle emissions (NORTRIP) [1, 2, 4] to estimate the PM_{10} and $PM_{2.5}$ emission factors for different climatic zones, road and street categories, and road maintenance policies in Finland.

33.2 Methods

The non-exhaust emission model NORTRIP [1, 2, 4] includes a range of key processes governing road dust surface loading, PM emissions and surface moisture conditions. It uses the mass balance approach to estimate street surface moisture and dust loading and calculates the emissions of both dust and salt. The NORTRIP model has been evaluated against air quality data by [1, 2] and they found the model to predict the PM_{10} concentrations within the range of ±35% of the observed values. In this work we used the NORTRIP model to estimate and study the PM_{10} and $PM_{2.5}$ emission factors separately for different climatic zones, road and street categories, and road maintenance policies in Finland. To achieve this task, a range of input data and parameters were established. During the work we found out several missing or incomplete input data sets, especially regarding traffic and road maintenance activities.

Model input data was developed for road and street configurations and traffic data (traffic volume, vehicle types and driving speed) based on national Digiroad (https://www.liikennevirasto.fi/web/en/open-data/digiroad) and LIPASTO (http://lipasto.vtt.fi/en/index.htm) databases. In Finland winter tires are compulsory during mid-winter and studded winter tyres are very popular with traffic shares from 76 to 94% depending on the region. Studded winter tires are allowed from 1st November until the first Monday after Easter, but this period may vary depending on weather conditions. Meteorological data concerning wind speed, temperature, radiation, cloud cover, and humidity was collected from representative weather stations from the five climatic zones in Finland. Road maintenance activity data, including information about addition of salt and sand to the road surface, were developed with an automatic traction control (sand and/or salt) module of the model according to predefined rules that were set using current real-life practices in the road and street network.

We calculated hourly level emission factors with the NORTRIP model for the period January 2008 to December 2011 for each street/road category and climatic zone combination with current road maintenance practices. The model was also used to calculate emission factors in three policy scenario groups designed to simulate the impact of measures intended to reduce non-exhaust PM emissions:

(1.) "Limiting dust generation" scenario group focused on (1.1) reduction of studded tyre usage or wear effect in cars in three cases: current use (about 80%), 50% usage and 20% usage, and (1.2) limiting the use of traction sanding in different urban settings.
(2.) "Efficient and early cleaning and dust binding" scenario group applied additional extensive cleaning or extensive cleaning and dust binding with state-of-the-art machinery for highly trafficked streets.
(3.) "Maximal reduction" scenario that applies the early cleaning and dust binding criteria with a 20% studded tyres share.

33.3 Results

Annual average non-exhaust PM_{10} and $PM_{2.5}$ emission factors were estimated for the climatic zones with the currently used road maintenance practices in Finland (Table 33.1). The values in Table 33.1 are for the whole fleet; for only the light duty traffic the emission factors were approximately 60% of the presented values. The

Table 33.1 Predicted four year weighted averages of the non-exhaust PM_{10} and $PM_{2.5}$ emission factors (mg/km/vehicle) in the climatic zones with current road maintenance practices in Finland from 2008 to 2011. The minimum and maximum annual values are presented in brackets

Climatic zone	PM_{10}	$PM_{2.5}$
Northern Boreal	226 (189–253)	18 (16–20)
Middle Boreal	199 (164–234)	17 (15–19)
Southern Boreal	182 (151–217)	16 (14–18)
Hemi Boreal	179 (158–205)	15 (14–17)
Helsinki Metropolitan Area	149 (130–169)	14 (12–16)

share of $PM_{2.5}$ emissions in PM_{10} varied from 8 to 16%, depending on the policy scenario (see Table 33.1 for road maintenance practices scenario). Lower $PM_{2.5}$ shares were characteristic for the scenarios with a higher relative contribution from pavement abrasion sources. There were significant differences between the road and street categories with higher emissions on highly trafficked and high speed streets and roads due to higher abrasion.

Relatively high variability was also observed between the climatic zone and year specific emission factors (Table 33.1). Reasons for these still need further research of the data, but variability in weather conditions, reflected in surface moisture and maintenance, are possible explaining factors. Within a year the non-exhaust emission factors have seasonal peaks particularly in March and April, due to enhanced formation of layers of street dust and favorable weather conditions to allow their suspension to the air.

Study of the different road maintenance policy scenarios designed to simulate the impact of measures intended to reduce non-exhaust PM emissions indicated that the selection of winter tyre types and their use, as well as road maintenance interventions could substantially reduce the non-exhaust emissions (Table 33.2), particularly in the spring period (March-May). The reductions are pronounced for PM_{10} compared with $PM_{2.5}$. Both findings are expected because the measures target mostly road abrasion and its suspension products, which have a relatively coarse size distribution compared with the other non-exhaust sources, brakes and tyre wear that are not included in the reduction measures.

Table 33.2 Four year average differences in PM_{10} and $PM_{2.5}$ emission factors (%) compared with the current practice scenario in the Helsinki metropolitan area

Scenario	PM_{10} (%)	$PM_{2.5}$ (%)
All traction control with sand	+10	+6
50% studded tyres	−23	−12
20% studded tyres (20st)	−47	−25
Efficient cleaning	−9	−5
Efficient cleaning and dust binding (ec-bd)	−19	−11
High reduction (combination of "20st" and "ec-bd")	−54	−29

33.4 Conclusions

The non-exhaust emission factors are influenced by weather, especially the road surface moisture, and have seasonal peaks particularly in winter and spring, due to enhanced formation of layers of street dust and their suspension to the air. Results demonstrated that changes in the selection of winter tyre types and their use, as well as road maintenance interventions could substantially reduce the non-exhaust emissions, especially in spring. We experienced challenges in developing relevant input datasets for the model in many parts of the country and further work should focus on filling the data gaps. The methods presented in this study could be used at least in the whole Northern part of the world, i.e., the Nordic countries and Northern Central Europe, Northern Russia, North America, and parts of China and Japan.

Questions and answers:
Questioner 1: Peter Jackson
Question 1: Have you considered the size distribution of traction material applied to roads? Specifically the percent silt content?
Answer 1: Yes, the model includes the size distribution and the silt content of the traction material as a source of particulate matter.
Questioner 2: Cristina Guerreiro
Question 2: Does the NOTRTIP model take into account the type of asphalt for estimating asphalt wear and wear emissions?
Answer 2: Yes, the user can fill in parameters of asphalt properties related to grain size distribution and durability of the aggregate to the model and these affect the wear rate and subsequent emissions.
Questioner 3: David Niemi
Question 3: What specifically was talked about? Specifically resuspension of surface material into the air?
Answer 3: The talk and the paper is particularly about emissions of PM_{10} and $PM_{2.5}$ from non-exhaust sources to the air. The methodology includes both direct and suspension/resuspension emission pathways. However, in Finnish conditions specifically resuspension emissions are significant and key in explaining the high PM concentrations in the spring period.
Question 4: Is removal of particles through settling & impaction on snowbanks, trees, buildings included?
Answer 4: NORTRIP is an emission model. Apart from the suspension removal mechanism, particles are removed from the road surface through drainage, by vehicle spray and during snow removal. Accumulation in snow banks or trees outside the road surface is not included. Any deposition processes outside the road surface system are not part of the emission model.

Acknowledgements This work has been funded by the Academy of Finland in the project Environmental impact assessment of airborne particulate matter: the effects of abatement and management strategies (BATMAN) and NordForsk under the Nordic Programme on Health and Welfare Project #75007: Understanding the link between air pollution and distribution of related health impacts and welfare in the Nordic countries (NordicWelfAir).

References

1. B.R. Denby, I. Sundvor, C. Johansson, L. Pirjola, M. Ketzel, M. Norman, K. Kupiainen, M. Gustafsson, G. Blomqvist, G. Omstedt, A coupled road dust and surface moisture model to predict non-exhaust road traffic induced particle emissions (NORTRIP). Part 1: road dust loading and suspension modelling. Atmos. Environ. **77**, 283–300 (2013). https://doi.org/10.1016/j.atmosenv.2013.04.069
2. B.R. Denby, I. Sundvor, C. Johansson, L. Pirjola, M. Ketzel, M. Norman, K. Kupiainen, M. Gustafsson, G. Blomqvist, M. Kauhaniemi, G. Omstedt, A coupled road dust and surface moisture model to predict non-exhaust road traffic induced particle emissions (NORTRIP). Part 2: surface moisture and salt impact modelling. Atmos. Environ. **81**, 485–503 (2013). https://doi.org/10.1016/j.atmosenv.2013.09.003
3. K. Kupiainen, R. Ritola, A. Stojiljkovic, L. Pirjola, A. Malinen, J. Niemi, Contribution of mineral dust sources to street side ambient and suspension PM_{10} samples. Atmos. Environ. **147**, 178–189 (2016). https://doi.org/10.1016/j.atmosenv.2016.09.059
4. Norman, M., I. Sundvor, B.R. Denby, C. Johansson, M. Gustafsson, G. Blomqvist, S. Janhäll, (2016), "Modelling road dust emission abatement measures using the NORTRIP model: Vehicle speed and studded tyre reduction". *Atmos. Environ.*, Vol. 134, June 2016, pp. 96–108. ISSN 1352-2310, http://dx.doi.org/10.1016/j.atmosenv.2016.03.035

Part IV
Data Assimilation and Air Quality Forecasting

Chapter 34
Global CO Emission Estimates Inferred from Assimilation of MOPITT CO, Together with Observations of O_3, NO_2, HNO_3, and HCHO

Xuesong Zhang, Dylan Jones, Martin Keller, Zhe Jiang, Adam E. Bourassa, D. A. Degenstein and Cathy Clerbaux

Abstract Atmospheric carbon monoxide (CO) emissions estimated from inverse modeling analyses exhibit large uncertainties, due, in part, to discrepancies in the tropospheric chemistry in atmospheric models. We attempt to reduce the uncertainties in CO emission estimates by constraining the modeled abundance of ozone (O_3), nitrogen dioxide (NO_2), nitric acid (HNO_3), and formaldehyde (HCHO), which are constituents that play a key role in tropospheric chemistry. Using the GEOS-Chem four-dimensional variational (4D-Var) data assimilation system, we estimate CO emissions by assimilating observations of CO from the Measurement of Pollution In the Troposphere (MOPITT) and the Infrared Atmospheric Sounding Interferometer (IASI), together with observations of O_3 from the Optical Spectrograph and InfraRed Imager System (OSIRIS) and IASI, NO_2 and HCHO from the Ozone Monitoring Instrument (OMI), and HNO_3 from the Microwave Limb Sounder (MLS). Although our focus is on quantifying CO emission estimates, we also infer surface emissions of nitrogen oxides ($NO_x = NO + NO_2$) and isoprene. Our results reveal that this multiple species chemical data assimilation produces a chemical consistent state that effectively adjusts the CO–O_3–OH coupling in the model. The O_3-induced changes in OH are particularly large in the tropics. We show that the analysis results in a

X. Zhang (✉) · D. Jones · M. Keller
Department of Physics, University of Toronto, Toronto, ON, Canada
e-mail: xuesong.zhang@mail.utoronto.ca

Z. Jiang
School of Earth and Space Sciences, University of Science and Technology of China, Hefei, Anhui, China

A. E. Bourassa · D. A. Degenstein
Department of Physics and Engineering Physics, University of Saskatchewan, Saskatoon, SK, Canada

C. Clerbaux
UPMC Université Paris 6, Université Versailles St-Quentin, LATMOS-IPSL, CNRS/INSU, Paris, France

Spectroscopie de l'Atmosphére, Service de Chimie Quantique et Photophysique, Université Libre de Bruxelles, Brussels, Belgium

C. Mensink et al. (eds.), *Air Pollution Modeling and its Application XXVI*, Springer Proceedings in Complexity,
https://doi.org/10.1007/978-3-030-22055-6_34

tropospheric chemical state that is better constrained. Our experiments also evaluate the inferred CO emission estimates from major anthropogenic, biomass burning and biogenic sources.

34.1 Introduction

Surface carbon monoxide (CO) reacts through CO–hydrocarbon–NO_x chemistry cycle, affecting tropospheric ozone (O_3) and hydroxyl radical (OH) abundances. The CO emission estimates using various chemical transport models (CTMs) largely depend on the uncertainties of the employed bottom-up inventories as well as CTMs' uncertainties on chemistry and transport schemes [2]. In fact, many of the differences on regional emission estimates are due to model's chemistry scheme, especially the modelled OH and its precursor O_3. The modelled OH, serving as the cleansing agent of the troposphere, would determine the atmospheric concentrations of CO, Non-Methane Volatile Organic Compound (NMVOC), methane etc., and affect estimated CO emissions. To reduce the model's uncertainties due to chemistry, considering the significant chemical interactions of the tropospheric CO–hydrocarbon–NO_x–O_3–OH cycle, it is essential to establish strong chemical constraints on multiple tropospheric constituents. Multiple Species data Assimilation (MSA) has shown potential to better estimates CO and NO_x emissions via improving OH and its precursor O_3. Since building a chemical consistent framework throughout the assimilation would be conducive to understand both model sensitivity to certain chemicals and model uncertainties. In this paper, we use the GEOS-Chem 4D-Var data assimilation system to examine the impact of assimilating multiple chemical species to quantify CO surface emissions. We assimilated observations of O_3 from IASI and OSIRIS, NO_2 from OMI, and CO from MOPITT with two-week assimilation window running in 2016.

34.2 Methods

In our experiment, full chemistry version of GEOS-Chem adjoint model with 4D-var data assimilation scheme is used. The details of the inversion methodology are explained by Henze et al. [1]. Full chemistry version of the model together with 4D-var assimilation scheme would enable chemical feedbacks among different species throughout the assimilation window. This would allow each species to be also assimilated by observations of other related tracers. The full list of emission inventories followed v09 of the GEOS-Chem model. Our study uses GEOS-FP meteorological fields, with $4 \times 5°$ horizontal resolution. The spin-up period of all the chemical tracers is one month starting January-2016. All instrument run mentioned in the improved MSA optimizes CO emission, isoprene emissions, surface NO_x emissions as well as O_3 and HNO_3 initial conditions when all observations mentioned in Table 34.1 were

Table 34.1 Qptimized quantities with the corresponding observations used in this paper

Optimized quantities	Employed observation(s)
CO emissions	MOPITT TIR/NIR retrievals V7 J
Isoprene emissions	OMI HCHO tropospheric column v003
Surface NO_x emissions	OMI NO_2 tropospheric column v003
O_3 initial conditions	Metop-B IASI-FORLI O_3 partial column v20151001 OSIRIS O_3 mixing ratio
HNO_3 initial conditions	MLS HNO_3 mixing ratio v003
All quantities above	All instruments above

employed. In this paper, we present various runs assimilating tropospheric chemistry on August 2016 to overlap the measurement period of the ATom-1 aircraft campaign.

34.3 Results and Discussions

34.3.1 Assimilated O_3 and OH Comparing to ATom-1 Aircraft Campaign

Table 34.2 shows the comparison between the assimilated O_3 concentrations with ATom-1 aircraft observations. IASI O_3 assimilation has effectively improved the modeled O_3 concentrations over the tropics and the southern hemisphere by 30–60% within the troposphere. OSIRIS O_3, however, does not provide major improvement in lower and mid-troposphere (less than 8 km in altitude). The experiment using all instruments have shown O_3 improvements in nearly all latitude bands other than Polar Regions. The magnitude of the corrections using all instruments tend to be smaller than single instrument run except for northern mid-latitudes. Over this region, the

Table 34.2 Comparisons between the ATom O_3 observations and the assimilated runs in lower and mid troposphere (LT and MT). |Diff| is the mean absolute difference (|ATom-the a posteriori|)

|Diff|	A priori	IASI O_3	OSIRIS O_3	IASI and OSIRIS	All instruments
Global mean	15.5	8.9	13.3	9.8	8.7
90°N–60°N	11.3	16.3	11.3	11.7	10.5
60°N–30°N	13.5	12.1	13.5	13.0	11.5
30°N–30°S	18.9	5.6	18.8	8.4	7.3
30°S–60°S	14.0	8.8	14.0	9.5	8.9
60°S–90°S	8.8	6.8	9.8	7.9	6.6

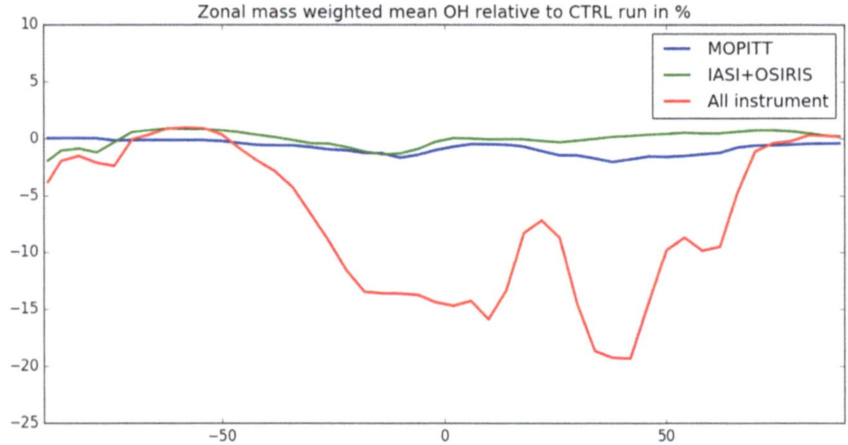

Fig. 34.1 Effect of the multi-species assimilation (MSA) on modeled zonal mass weighted OH

additional O_3 corrections are contributed by chemical feedback of the optimized O_3 precursors.

According to Fig. 34.1, the multi-species assimilation (MSA) drastically reduces the modeled OH concentration especially over the tropics as well as the northern mid-latitudes in contrast to single instrument run. The magnitude of the OH changes using MSA is larger than the aggregation of all OH changes brought by single instrument runs, suggesting strong chemistry nonlinearity in the model. The resulting global mean OH has been decreased from 13.89 to 12.38 (in 10^5 mol/cm^3), comparable to the estimated values of 12.44 from Spivakovsky et al. [3]. This shows the potential that assimilating multi-species observations could constrain tropospheric OH concentrations.

34.3.2 *CO Emission Scaling Factor in Aug-2016*

Figure 34.2 shows the difference of anthropogenic and biomass burning CO emis-

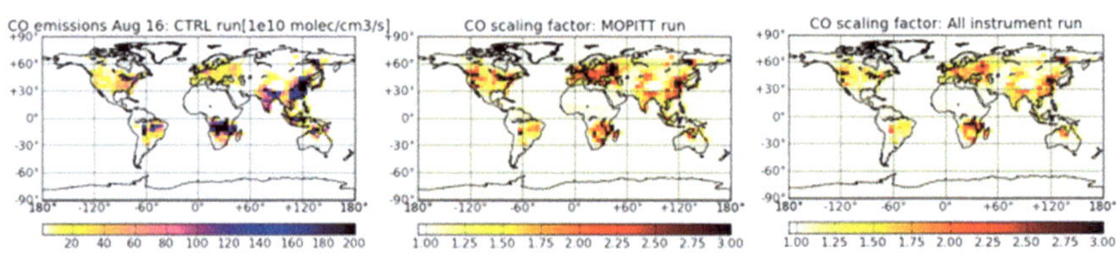

Fig. 34.2 Differences in anthropogenic and biomass burning CO emissions compared to the a priori run (left), for assimilating MOPITT only (middle) and for the multi-species assimilation MSA (right)

sions between the a priori and the a posteriori assimilating MOPITT-only, and MSA run. The single instrument run (MOPITT CO assimilation in this case) showed increased emissions more than factor over 2.5 in Europe, the eastern and western United States, and East Asia. The biomass burning emissions assimilating MOPITT CO showed that the emissions over southern Africa, northern Australia, and northwestern North America were also increased around by a factor of 2. When assimilating all chemical species, the optimized emission changes over both anthropogenic and biomass burning emission regions were consistent with MOPITT assimilation. The all instrument assimilation showed larger emissions in Europe but lower emissions in East Asia, comparing to the single instrument run. These differences on CO emissions in MSA were due to other observations affecting the modeled OH budget.

34.4 Conclusion

In this study, we found that the 4D-Var scheme seeks to obtain a model trajectory in state space that best matches all the available observations over the assimilation period. It thus provides consistent chemical state with all available observations over the assimilation period. The multi-species assimilation reduced the absolute mean bias in modeled O_3, relative to Atom-1 data. We also found the global mean tropospheric OH has been decreased from 13.89 to 12.38 (in 10^5 mol/cm^3). The resulting MOPITT CO-only and multi-species assimilation produced similar CO emission estimates for the major source regions (except for southern Africa), suggesting large increases in the emissions compared to the a priori. The emission was increased by 50–120% in both major anthropogenic emission hotspots such as North America, Europe, East Asia, and biomass burning regions such as southern Africa.

Questions

Pablo Saide:

Question: Can we use the emission statistics from these assimilation studies to avoid biases from CO surface flux?

Answer: CO data assimilation can quantify the biases of the bottom-up emission inventories. In our case, we showed that the bottom up emission inventories over the northern extratropics were still underestimated. However, the surface flux biases cannot be fully removed even we use the optimized emission statistics since they were still affected by model and observation error.

Shuzhan Ren:

Question: How much confidence do you have on your constrained OH concentrations?

Answer: Our comparison between our assimilated results and Spivakovsky et al. [3] showed that the modeled global mean tropospheric OH has been reduced by 10%

using MSA. If we do not assume significant trend of tropospheric OH since 2000, then our assimilated OH looks more realistic than the CTRL run.

Acknowledgements This work was supported by funding from Environment and Climate Change Canada and the Natural Sciences and Engineering Research Council (NSERC).

References

1. D.K. Henze, A. Hakami, J.H. Seinfeld, Development of the adjoint of GEOS-Chem. Atmos. Chem. Phys. **7**, 2413–2433 (2007)
2. K. Miyazaki, H.J. Eskes, K. Sudo, M. Takigawa, M. van Weele, K.F. Boersma, Simultaneous assimilation of satellite NO_2, O_3, CO, and HNO_3 data for the analysis of tropospheric chemical composition and emissions. Atmos. Chem. Phys. **12**, 9545–9579 (2012)
3. C.M. Spivakovsky, J.A. Logan, S.A. Montzka, Y.J. Balkanski, M. Foreman-Fowler, D.B.A. Jones, L.W. Horowitz, A.C. Fusco, C.A.M. Brenninkmeijer, M.J. Prather, S.C. Wofsy, M.B. McElroy, Three-dimensional climatological distribution of tropospheric OH: update and evaluation. J. Geophys. Res. Atmos. **105**(D7), 8931–8980 (2000)

Chapter 35
Experimental Forecasting Using the High-Resolution Research Configuration of GEM-MACH

Paul Makar, Ayodeji Akingunola, Balbir Pabla, Craig Stroud, Jack Chen, Philip Cheung, Michael Moran, Wanmin Gong, Qiong Zheng and S. M. Li

Abstract Experimental air-quality forecasts for the Canadian provinces of Alberta and Saskatchewan have been carried out since 2012, using a 10 km/2.5 km nested resolution version of Environment and Climate Change Canada's Global Environmental Multiscale-Modelling Air-quality and Chemistry (GEM-MACH) on-line air-quality model. We describe here some of the main results of that work, and a major upgrade of this forecasting system (based on work carried out following a 2013 monitoring intensive field campaign in the Athabasca oil sands region of Canada). The new

P. Makar (✉) · A. Akingunola · B. Pabla · C. Stroud · P. Cheung · M. Moran · W. Gong · Q. Zheng · S. M. Li
Air Quality Research Division, Science and Technology Branch, Environment and Climate Change Canada, Toronto, ON, Canada
e-mail: paul.makar@canada.ca

A. Akingunola
e-mail: ayodeji.akingunola@canada.ca

B. Pabla
e-mail: balbir.pabla@canada.ca

C. Stroud
e-mail: craig.stroud@canada.ca

P. Cheung
e-mail: philip.cheung@canada.ca

M. Moran
e-mail: mike.moran@canada.ca

W. Gong
e-mail: wanmin.gong@canada.ca

Q. Zheng
e-mail: qiong.zheng@canada.ca1

S. M. Li
e-mail: shao-meng.li@canada.ca

J. Chen
Air Quality Research Division, Science and Technology Branch, Environment and Climate Change Canada, Ottawa, ON, Canada
e-mail: jack.chen@canada.ca

C. Mensink et al. (eds.), *Air Pollution Modeling and its Application XXVI*, Springer Proceedings in Complexity, https://doi.org/10.1007/978-3-030-22055-6_35

forecasting system has been designed in preparation for a follow-up field campaign, taking place during April and June of 2018.

35.1 GEM-MACH

Version 2 of the Global Environmental Multiscale—Modelling Air-quality and CHemistry (GEM-MACH) model [5, 6, 8, 10] was used for the simulations described here. The model uses a 2 or 12-bin sectional representation of aerosol microphysics, with eight particle species (sulphate, nitrate, ammonium, secondary organic aerosol, primary organic aerosol, elemental carbon, sea-salt and crustal material). The 2-bin configuration of the model has been used for ongoing daily experimental forecasts of pollutants and their deposition since 2012, while the 12-bin configuration has been used for research simulations. The model configuration employs three levels of nesting:—analysis-driven meteorological output from the Canadian operational Regional Deterministic Prediction System is used to simultaneously drive both a GEM-MACH 10 km resolution grid cell size North American domain simulation and a Canadian High Resolution Deterministic Prediction System 2.5 km grid cell size simulation over Canada and the USA. The former and latter provide chemical and meteorological boundary and initial conditions respectively to a 2.5 km grid cell size simulation domain encompassing the Canadian provinces of Alberta and Saskatchewan (Fig. 35.1).

The emissions used in the simulations are the result of a multi-year effort to improve emissions inventories in the region and include efforts to incorporate emissions data estimated from 2013 research flights around individual oil sands facilities [13].

The use of these emissions data as well as the 2 and 12-bin configurations of the model resulted in several important findings:

(1) Simulations employing model-measurement fusion predicted future acidifying aquatic ecosystem damage at 2013 emissions levels, for an area greater than 3.8×10^5 km^2 [9].
(2) A parameterization of bi-directional fluxes of ammonia was required in order to simulate observed ammonia concentrations [12].
(3) Improved estimates of emitted mass and speciation of volatile organic compounds resulted in significant performance improvements for predicted VOC chemistry and secondary organic aerosol formation [11].
(4) The standard algorithms for estimating the rise of industrial plumes underestimated plume height, when driven using meteorological observations [4]. However, a revised algorithm showed much greater skill in estimating this key parameter [1].
(5) The 12-bin size distribution was shown to improve the model's aerosol performance for all statistical metrics [1].

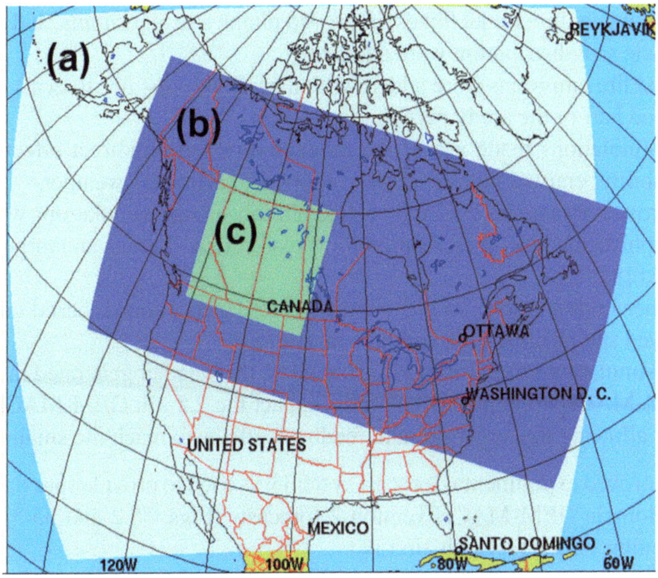

Fig. 35.1 Model domains for GEM-MACH simulations for experimental forecasts in Alberta and Saskatchewan (after [1]). **a** North American 10 km grid cell resolution GEM-MACH domain. **b** 2.5 km grid cell resolution high resolution deterministic prediction system domain. **c** 2.5 km grid cell resolution GEM-MACH domain

(6) Feedbacks between chemistry and meteorology were shown to improve air-quality model performance, and change the local weather [7].

35.2 Spring/Summer 2018 Aircraft Campaign

The above and additional observational studies identified further sources of uncertainty in accurately simulating transformation, transport and fate of emissions from sources in Alberta: (1) What is the identity and emission rates of intermediate volatility organic compounds? (2) What is the seasonality of the emissions and chemical processes? (3) What are the processes controlling organic aerosol formation? (4) What are the emission levels and sizes of particles containing base cations in the region? (5) What are the relative impacts of different source types in the region? These questions prompted a second field campaign in the region, taking place in two periods in 2018: April 3rd to April 20th, and May 31st to July 6th. GEM-MACH at 2.5 km resolution was the forecast tool for directing a research aircraft during this study.

Several improvements to the simulation system were included for consideration in the setup of the new experimental forecast:

(A) Ammonia bidirectional fluxes, improved plume rise algorithms, and the 12-bin particle size distribution were adopted.
(B) Forest fire emissions were added, with a parallel forest fire forecast during the second half of the study period.
(C) All simulations made use of the model's direct and indirect effect feedback capability; emitted pollutants were allowed to affect the weather.
(D) Improved organic aerosol and on-line photolysis rate calculations were added.
(E) A sophisticated parameterization for forest canopy shading and turbulence was added [8].
(F) Further improvements to the emissions were incorporated, based on satellite-derived spatial allocation and new inventories.
(G) The simulation setup was modified to allow the weather generated in the 10 km GEM-MACH simulation to directly impact the 2.5 km GEM-MACH simulation, allowing the larger domain feedback effects to reach the smaller scales.

The improved experimental forecast system was used to provide four daily 2.5 km grid cell domain GEM-MACH version 2 forecasts (Figs. 35.2 and 35.3), to guide flight planning for the research aircraft.

The second phase of the experiment (May 28th to July 5th, 2018) included an additional set of parallel forecasts at 0 and 12Z, which included forest fire emissions generated using the new Canadian Forest Fire Emissions Processing System (CFFEPS; see [2], these proceedings). These additional simulations were used to direct the research aircraft during forest fire events in the region, aiding in flight planning for direct sampling of forest fire emissions passing through the oil sands industrial area during a forest fire event on June 25th, 2018, (Fig. 35.3).

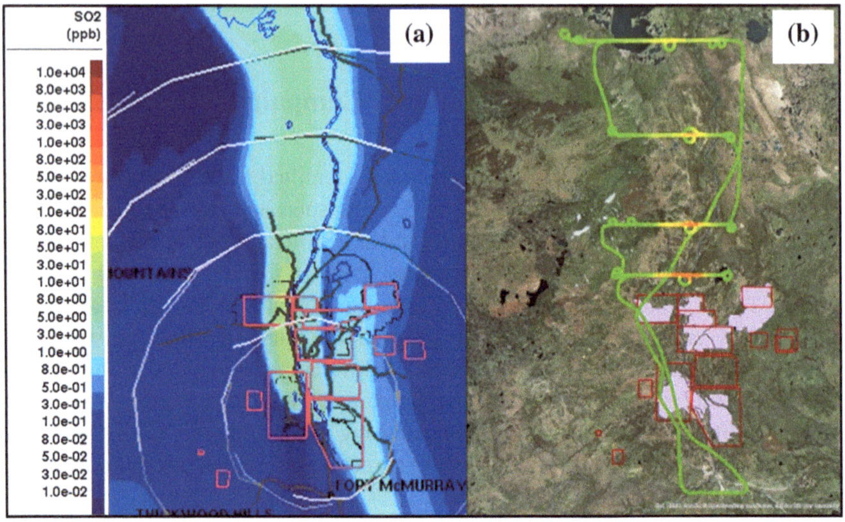

Fig. 35.2 Example comparison for Flight 5, April, 2018 of **a** field forecast of SO_2 concentrations (ppbv) with **b** aircraft- observed SO_2 plume location

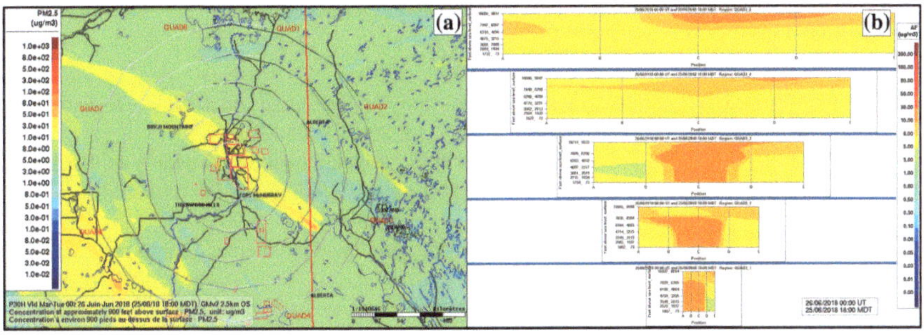

Fig. 35.3 Example forecast for June 25th, 2018. **a** Forecast 330 m agl $PM_{2.5}$ concentrations showing forest fire plume originating to the SE of oil sands facilities, exiting to NW. **b** Cross-sections of $PM_{2.5}$ concentrations including forest fire plume

Question and Answer
Questioner: Greg Yarwood
Question:
Can you comment on what flight planning strategies are effective for emissions characterization, and how much variability in emissions is seen between flights for the same facility?
Answer:
Emissions characterization flights are "box" flights around an individual facility, flown under wind speeds and direction which imply relatively clean upwind boundary conditions. The agreement between SO_2 emissions estimated from repeat flights around the facilities in 2013 showed a very good agreement with CEMS observations. Part of the reason for returning to the facilities in 2018 is to determine the extent of variability in different years and seasons (2013 was a summer-only study, while 2018 captures both winter—snow covered conditions were observed in winter—and summer conditions).
Questioner: Rohit Mathur
Questions:

(1) Will the aircraft campaign observation data set (which appears to be an exciting and extensive data set) be available to others for use in model assessment?
(2) Could you briefly describe how Hg emissions from wildfires was estimated?

Answer to (1): Yes. The 2013 data are available publicly (and can be downloaded from the Oil Sands Monitoring program website). The 2018 data will be made publicly available once quality assurance and control are completed.
Answer to (2): Briefly, the initial estimates were based on the Fire Inventory from NCAR (FINN), with an inversion technique used to optimize emission factors for gaseous elemental Hg for five vegetation types represented in North American fires. The Hg data shown in the presentation were for Canadian domain results [3], which goes into the construction of the emissions data in detail—readers interested in this data should consult that reference, included below.

References

1. Akingunola et al., A chemical transport model study of plume rise and particle size distribution for the Athabasca oil sands. Atm. Chem. Phys. **18**, 8667–8688 (2018)
2. J. Chen et al., Impact of urban land use and anthropogenic heat on air quality in urban environments, in *Proceeding of 36th ITM on Air Pollution Modelling and Its Application*, this vol., (2018)
3. A. Fraser et al., How important is biomass burning in Canada to mercury contamination? Atm. Chem. Phys. **18**, 7263–7286 (2018)
4. Gordon et al., A comparison of plume rise algorithms to stack plume measurements in the Athabasca oil sands. Atm. Chem Phys. Discuss., in review (2017). https://doi.org/10.5194/acp-2017-1093
5. P.A. Makar et al., Feedbacks between air pollution and weather, part 1: effects on weather. Atm. Env. **115**, 442–469 (2015)
6. P.A. Makar et al., Feedbacks between air pollution and weather, part 2: effects on chemistry. Atm. Env. **115**, 499–526 (2015)
7. Makar et al., Coupled chemistry-meteorology: simulations at 2.5 km resolution, in GAW report, No 226 (2016), https://www.wmo.int/pages/prog/arep/gaw/documents/Final_GAW_226_10_May.pdf (Baklanov et al.)
8. P.A. Makar et al., The effects of forest canopy shading and turbulence on boundary layer ozone. Nat. Comm. (2017). https://doi.org/10.1038/ncomms15243
9. P.A. Makar et al., Estimates of exceedances of critical loads for acidifying deposition in Alberta and Saskatchwan. Atm. Chem. Phys. **18**, 9897–9927 (2018)
10. M.D. Moran et al., Particulate-matter forecasting with GEM-MACH15, a new Canadian air-quality forecast model, in *Air Pollution Modelling and Its Application XX*, ed. by D.G. Steyn, S.T. Rao (Springer, Dordrecht, 2010), pp. 289–292
11. C.A. Stroud et al., Impact of aircraft-derived organic gaseous and particle emissions on air quality predictions: an Athabasca oil sands case study. Atm. Chem. Phys. **18**, 13531–13545 (2018)
12. Whaley et al., Contributions of natural and anthropogenic sources to ambient ammonia in the Athabasca oil sands and north-western Canada, Atm. Chem. Phys. 2011–2034 (2018)
13. Zhang et al., Emissions preparation and analysis for multiscale air quality modelling over the Athabasca oil sands region of Alberta, Canada. Atm. Chem. Phys. **18**, 10459–10481 (2018)

Chapter 36
An Air Quality Modeling System Providing Smoke Impact Forecasts for Health Protection in Southeastern USA

M. Talat Odman, Ha Ai, Yongtao Hu, Armistead G. Russell, Ambarish Vaidyanathan and Scott L. Goodrick

Abstract The HiRes-2 system, operational since 2015, provides daily forecasts of potential prescribed fire impacts on air quality. The system uses the WRF and CMAQ models for meteorology and air quality computations. A decision tree model predicts prescribed fire activity based on the weather forecast by analyzing historical burning patterns under similar meteorological conditions. Prescribed fire emissions are estimated using land-based but satellite-enhanced fuel load maps, consumption forecasts, and emission factors derived from laboratory and field measurements. The DDM-3D feature of CMAQ is used to calculate the impacts of prescribed fire emissions on pollutant concentrations. The expanded new system, HiRes-X, has many extensions for dynamic air quality and human exposure management. Daily forecasts of air quality and prescribed fire impacts are disseminated through an online interactive tool, which uses webGIS technologies to display the forecasting products in real time and retrospectively, along with relevant earth observation datasets. These datasets can be overlapped with geographical and population information to visualize the air pollution impacts on sensitive places and vulnerable communities.

M. T. Odman (✉) · H. Ai · Y. Hu · A. G. Russell
School of Civil and Environmental Engineering, Georgia Institute of Technology, Atlanta, GA 30332, USA
e-mail: odman@gatech.edu

H. Ai
e-mail: hai8@gatech.edu

Y. Hu
e-mail: yh29@mail.gatech.edu

A. G. Russell
e-mail: ted.russell@gatech.edu

A. Vaidyanathan
Centers for Disease Control and Prevention, 4770 Buford Highway, Atlanta, GA 30341, USA
e-mail: dvq3@cdc.gov

S. L. Goodrick
Southern Research Station, US Forest Service, 320 Green Street, Athens, GA 30602, USA
e-mail: sgoodrick@fs.fed.us

C. Mensink et al. (eds.), *Air Pollution Modeling and its Application XXVI*,
Springer Proceedings in Complexity,
https://doi.org/10.1007/978-3-030-22055-6_36

36.1 Introduction

Air quality forecasts provide useful information for avoiding exposure to high levels of pollutants. However, additional information is needed about the contribution of specific emission sources to develop mitigation strategies and avoid repeated pollution episodes. Dynamic air quality management is a paradigm where, depending on the episode, different sources are targeted with varying levels of control. Prescribed burning is a major source of air pollution in Southeastern USA. According to the 2014 emissions inventory, ¼ of all primary $PM_{2.5}$ emissions in Southeastern USA originate from the treatment of approximately 2 million hectares of land by prescribed fire each year. The use of prescribed burning for forest and crop health has economical and ecosystem-related benefits; however, these benefits must be weighed and managed with respect to the potential health and welfare impacts. Prescribed burning is also very suitable source for dynamic management. Unlike most other sources that cannot be curtailed without significant economic, social and political consequences, prescribed burns can be easily deferred to another day.

There is an increasing need to forecast prescribed burn activity and its impact on air quality and human health in Southeastern USA, and to disseminate these forecasts, along with other useful burn information, through a convenient online tool for analysis and management purposes. WebGIS-based online tools have already been developed for emissions inventory spatial analysis [1], marine pollution monitoring and forecasting [2], and urban air quality monitoring data visualization [3]. NOAA operates a website (https://www.star.nesdis.noaa.gov/smcd/spb/aq/eidea) that brings together satellite-based fire detection, cloud, smoke and aerosol optical thickness data with ground-level $PM_{2.5}$ observations for research and application communities to explore the impacts of fires. Centers for Disease Control and Prevention (CDC) is also developing an online tool for identifying populations vulnerable to wildfire smoke hazards [4]. Until now, a system that combines fire and air quality observations with forecasts, focusing specifically on prescribed fires, did not exist.

Here, HiRes-X is introduced as the first modeling system and online interactive tool (https://sipc.ce.gatech.edu/SIPFIS/map/) for forecasting and analysis of prescribed burn impacts in Southeastern USA. First, the weather-based burn activity forecasting capability, the fire emission prediction component, and the air quality and burn impact prediction methods are reviewed. Then, efforts to expand and extend the system are discussed. Expansion efforts include broader coverage over the Southeastern USA and using fire-related data from a wide range of earth observations. Extensions include providing the forecasts to local and state public health agencies and to CDC on a real-time basis. Finally, examples from recent applications of HiRes-X to the prediction of the prescribed fire impacts are presented.

36.2 Extended Forecasting System: HiRes-X

The HiRes-2 system is operational since 2015, forecasting air quality in Southeastern USA over a domain centered on Georgia. The system uses WRF (version 3.6) and CMAQ (version 5.0.2) models for meteorology and air quality computations. In addition, using the DDM-3D feature of CMAQ, it provides daily forecasts of the potential contribution of prescribed fires to air quality. In HiRes-X, the burn activity forecast coverage, formerly limited with Georgia, has been expanded to other states in the Southeastern USA. A classification and regression tree (CART) was built using meteorological data and either burn permit data in Georgia or satellite-based burn location and area data in other states, for recent years. The satellites cannot always detect the prescribed fires in Southeastern USA due to their small size and low heat intensity, overlaying canopy in understory burns, and frequent cloud cover. On a given day, if more than a threshold amount of land (e.g., 40 hectares), which varies by state, was permitted or detected to burn in any county, that day is defined as a "burn day". The CART branches out at specific meteorological parameter values with each branch leading to a "burn day" or a "no-burn day".

If a burn day is forecasted, burns of sizes typical for lands owned by institutional, large commercial and small private burners totaling the daily average burn area in the region are randomly distributed to the respective lands. The Fuel Characteristic Classification System (FCCS) maps are used to determine the fuel loads at each burn location. Then, the fraction of the fuel that will be combusted is calculated using the CONSUME model together with forecasted meteorological parameters such as fuel moisture. Next, fire emissions are calculated by multiplying the amount of fuel mass consumed with the emission factors (EFs) for different pollutants. The EFs used here are specific to the fuels in the Southeastern USA and were derived from prior field measurements. Finally, the plume height is calculated from the heat released and the emissions are vertically distributed and injected into the corresponding layers of CMAQ.

Other than the above forecasting products generated in-house, the system is also fetching datasets from third parties. These datasets are ozone and $PM_{2.5}$ observations from the national air quality monitoring network (https://www.airnow.gov) and fire locations and burned areas from Hazard Mapping System (HMS) Fire and Smoke Product (http://www.ospo.noaa.gov/Products/land/hms.html), both downloaded on a daily basis, and burn locations and areas from Florida and Georgia's burn permit records, updated on an annual basis.

Air quality observational data are ingested into a MySQL database tables immediately after being downloaded onto the local server. HMS detection datasets are transformed to a simple text format before archival. Permit data undergo quality checks, during which missing items are filled, if possible. For example, latitude and longitude of the burns, if not provided, are derived from provided addresses by utilizing Google Earth services. Finally, the checked records are ingested into the MySQL database while inaccurate or incomplete records are discarded.

The HiRes-X online tool can display the data as point markers or contours on a map. The background map is provided by OpenStreetMap, which supports spatial navigation and zooming in for very detailed local geographical information. While air quality observation sites, HMS fire detections, forecast or permitted burn locations are presented on the map as markers through Leaflet's pointToLayer function, spatial distributions of air quality index grades or burn impact levels are plotted through another Leaflet function for polygons in GeoJSON format. As a third data display method, site specific observations and forecasts are presented using time-series plotting functions from the D3.js library. All the above display features are realized by using Javascript functions and support interactive capability to display according to date, pollutant, region, site, or item selection.

36.3 Applications of HiRes-X

Several types of analyses can be performed with HiRes-X. These include fire analyses comparing burn permit records to satellite detected fires and satellite derived burned areas as well as to forecasts of burn activity. HiRes-X can be used as a screening tool for fire-related exceptional events that lead to exceedances and violations of air quality standards. Forest and air quality managers can use HiRes-X to regulate prescribed burning operations with minimal human exposure to fire smoke. HiRes-X can also be used to access various forecasting products for personal use such as residential community level or mobility based air quality forecasts. Here, application to exceptional event screening will be illustrated although the selected event did not lead to an exceedance or violation, but only to moderate air pollution.

On 13 March, 2018, the recorded daily average $PM_{2.5}$ concentration at St. Marks Wildlife Refuge and Tallahassee in Florida were 17.2 and 13.7 $\mu g\ m^{-3}$, respectively, while the HiRes-2 forecasts were only 4.95 and 7.20 $\mu g\ m^{-3}$. On that day, HMS detected a lot of fires in the region (Fig. 36.1). Hires-X burn forecast predicted about the same amount of prescribed fires around St. Marks. Both HMS fire and smoke analysis (not shown) and HiRes-X burn impact forecast suggested heavy smoke presence near St. Marks under northwesterly winds. The burn impact forecast predicted a prescribed fire contribution of 5–10 $\mu g\ m^{-3}$ to the daily average $PM_{2.5}$ concentrations at St. Marks and 2–5 $\mu g\ m^{-3}$ at Tallahassee. This level of impact was sufficient to turn the forecast from green (good) to yellow (moderate or >12 $\mu g\ m^{-3}$) for St. Marks but not for Tallahassee.

36.4 Conclusion

The expanded version of the HiRes-2 air quality forecasting system incorporates new elements targeted for dynamic air quality management in Southeastern USA. Forecasts of the impacts of prescribed burn emissions can be used for averting air

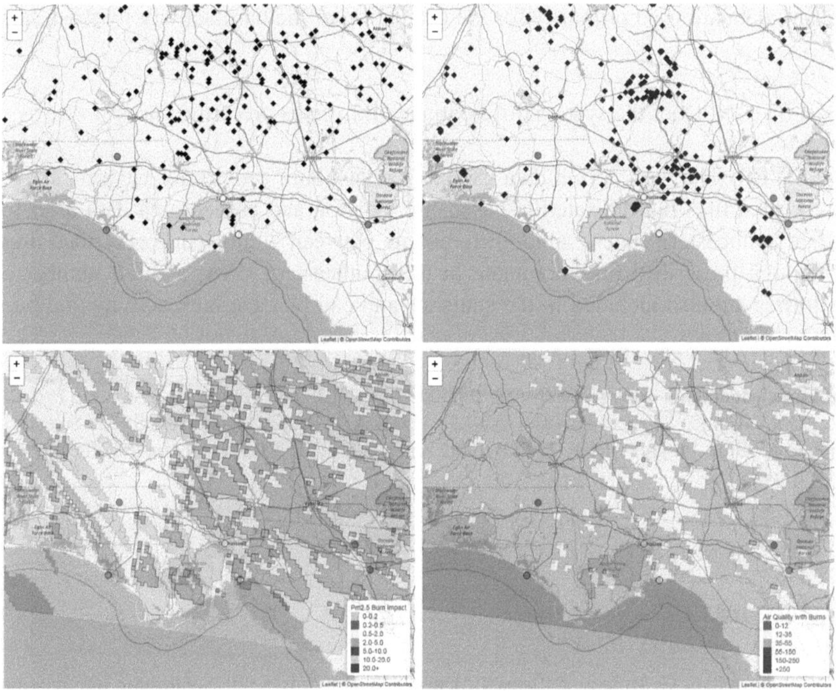

Fig. 36.1 Locations of forecast (top left) and satellite detected burns (top right) along with forecast air quality (bottom right) and contributions of burns (bottom left) on 13 March 2018. St. Marks along the Gulf Coast of Florida is marked with a light circle indicative of daily average $PM_{2.5}$ >12 $\mu g\ m^{-3}$. Directly to its north-northwest is Tallahassee, also marked with a light circle (https://sipc.ce.gatech.edu/SIPFIS/map/)

pollution episodes while maximizing the ecological and economic benefits of the burns through cohesive air quality and burn management. Extension efforts include providing the forecasts to local and state public health agencies in Southeastern USA and to CDC, on a real-time basis.

QUESTIONER: Jack Chen

QUESTION: Does the system take into account actual permitting info from local, state, federal agency to adjust forecast burning activity?

ANSWER: A large number of applications for prescribed burning are accepted online or over the phone in the morning of the burn day. Permitted burns are typically ignited before noon. This does not leave sufficient time for using the actual permit information in the burn activity and burn impact forecasts. For this reason, the activity forecast is based on the weather forecast and prior burn statistics. The actual burn permit data are used primarily for the evaluation of the burn activity forecast, later on.

QUESTION: Does the system differentiate wild versus prescribed fires?

ANSWER: The burn permit data used to forecast burn activity for Georgia are for prescribed fires but the satellite-based fire location and burned area data used for other southeastern states cannot differentiate between wild versus prescribed fires. Although prescribed fires dominate in the southeastern USA, a small number of wildfires are also included in the statistics and can bias the burn activity and burn impact forecasts.

Acknowledgements This research was supported by the Joint Fire Science Program (16-1-08-1), NASA Applied Sciences Program (NNX11AI55G & NNX16AQ29G) and US EPA Science to Achieve Results Program (RD83521701). The statements made do not represent the official views of the sponsors.

References

1. D. Gkatzoflias, G. Mellios, Z. Samaras, Development of a web GIS application for emissions inventory spatial allocation based on open source software tools. Comput. Geosci. **52**, 21–33 (2013)
2. M. Kulawiak, A. Prospathopoulos, L. Perivoliotis, M. łuba, S. Kioroglou, A. Stepnowski, Interactive visualization of marine pollution monitoring and forecasting data via a Web-based GIS. Comput. Geosci. **36**, 1069–1080 (2010)
3. W. Lu, T.H. Ai, X. Zhang, Y.K. He, An interactive web mapping visualization of urban air quality monitoring data of China. Atmosphere **8** (2017)
4. A. Vaidyanathan, F. Yip, P. Garbe, Developing an online tool for identifying at-risk populations to wildfire smoke hazards. Sci. Total Environ. **619–620**, 376–383 (2018)

Chapter 37
Evaluation of Air Quality Maps Using Cross-Validation: Metrics, Diagnostics and Optimization

Richard Ménard and Martin Deshaies-Jacques

Abstract Since 2002, Environment and Climate Change Canada (ECCC) has been producing hourly surface analyses of pollutants covering North America which have been used in numerous health impact studies. The analyses are produced using an optimum interpolation scheme that combines the operational air quality forecast model GEM-MACH outputs (CHRONOS model outputs were used prior to 2009) with real-time hourly observations of O_3, $PM_{2.5}$, PM_{10}, NO_2, and SO_2. We examine how passive observations, that are not used to create the analysis (i.e. cross-validation), can be used to evaluate the analysis error and to optimize the input error statistics.

37.1 Introduction

Maps (or *analyses* as called in data assimilation) of surface pollutants over Canada and USA have been produced hourly since 2002 using an optimum interpolation scheme combining the operational air quality model output of GEM-MACH (or CHRONOS prior to 2009) with real-time AirNow hourly observations [4, 6, 10]. Maps of surface O_3 and $PM_{2.5}$ were thus produced, and PM_{10}, NO_2, NO and SO_2 were added in April 2015 along with a 3-hourly averaged air quality health index (AQHI) map [11]. Collected over a long period of time, i.e. since 2002, this multiyear analysis was examined by Robichaud and Ménard [10] and O_3 maps have been used in several health impact studies [1–3, 12]. What we will discuss here is how to quantify the error of these analyses.

It is important to remark that to produce these analyses requires the knowledge of the statistics of the observation and model errors. Furthermore, the observation error is not simply the instrument error but should also include; errors due to the mismatch

R. Ménard (✉) · M. Deshaies-Jacques
Air Quality Research Division, Environment and Climate Change Canada, Dorval H9P 1J3, Canada
e-mail: Richard.menard@canada.ca

M. Deshaies-Jacques
e-mail: martin.deshaies-jacques@canada.ca

C. Mensink et al. (eds.), *Air Pollution Modeling and its Application XXVI*,
Springer Proceedings in Complexity, https://doi.org/10.1007/978-3-030-22055-6_37

in scales being sampled in a single observation vs that of the model grid box, subgrid scale variability that may be captured by the observation but not by the model, missing modeling processes, etc., that collectively we call representativeness error. The observation error is thus not well known and need to be estimated. Although from the analysis algorithm we can compute an analysis error variance, this value depends only on the input observation and model error statistics used in the algorithm and may not represent the true analysis error if they are incorrectly specified. Furthermore, misspecified error statistics generally leads to a degradation of the analysis accuracy. One way to evaluate the true analysis error is with a forecast, but the forecast accuracy depends on many other things than the analysis accuracy. It depends on the accuracy of emissions, the coherence between observed and unobserved species, the boundary layer mixing and ventilation, etc. The impact of observations may also be limited as the AirNow observations are confined to the surface. Furthermore, to estimate and tune the input errors statistics based on forecasts is computationally expensive. Thus, our objective is to develop a method to evaluate the true analysis error but without relying on a model forecast. Doing so, we also contribute in maximizing the information content provided by the observations.

Our method to evaluate the analysis error variance is based on cross-validation. To perform cross-validation we separate the observation data set into N spatially random-distributed observation sites (here we use $N = 3$), performing an analysis using $N-1$ subsets and verifying the analysis with the remaining single subset. This leads to an objective measure of analysis accuracy [8, 9] where the process is done N times by permutation of all subsets used for evaluation. An example of three subsets of $PM_{2.5}$ for cross-validation is illustrated in Fig. 37.1 of Ménard and Deshaies-Jacques [8].

One of the main sources of the true observation and model errors comes in observation-minus-model residuals, often noted as *O-B* (*B* for background). As the observations are compared with the (interpolated) model values, they thus contains errors of representativeness. The variance of the *O-B* can also be considered as the sum of observation and model error variances (this is evident when observation and model errors are uncorrelated). We could obtain an estimate of observation error variance and an estimate of model error variance, if we would know their ratio and that is exactly what optimization using cross-validation is capable of obtaining, i.e. the ratio of error variances that minimizes the analysis error variance (see Ménard and Deshaies-Jacques [9] for a full discussion). An example of this for $PM_{2.5}$ is provided in Fig. 37.1.

37.2 Main Results of Cross-Validation

In Fig. 37.1, the variance of the independent observation-minus-analysis using the (N-1) subsets, i.e. var(O-A), for each subsets is displayed with circles. The mean variance over the 3 subsets is displayed with a solid black line. Verifying against the same set of observation as those used for the analysis is also displayed along the

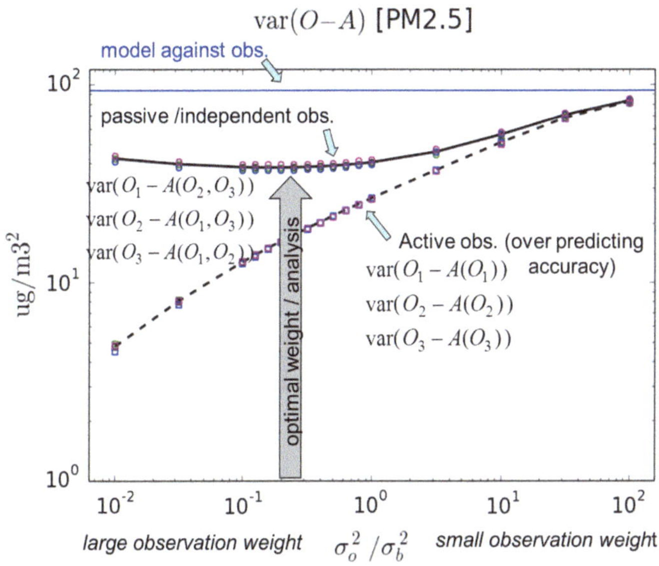

Fig. 37.1 Variance of observation-minus-analysis residuals of $PM_{2.5}$ for both active and independent observations

dash line. A series of analyses were performed with different observation weight, as controlled by the ratio σ_o^2/σ_b^2 while maintaining the sum of observation and model error variance constant and equal to the var (O-B) [5, 7].

To illustrate the different diagnostics from which an estimate of analysis error variance can be obtained, we use a geometrical representation of random variables, where uncorrelated variables are orthogonal, and the norm (or length) of a side represents the standard deviation. In Fig. 37.2 is displayed the truth T, O the observations used for the analysis, B the background or model value, and O_c the independent observations used to evaluate the analysis. Since the analysis is a linear combination of the observations and the model (background) the analysis A lies on the line joining O and B. As usual, we assume that observation error, i.e. the vector in the direction (T, O), is uncorrelated with the model error, i.e. vector in the direction (T, B). Thus the triangle ΔOTB is a right triangle. The analysis is optimal when the analysis error, i.e. the norm of the vector (T, A), is minimal, which actually occur when (T, A) is orthogonal to (O, B).

For an optimal analysis, the triangle ΔOAT and ΔTAB are right triangles. Assuming that the independent observations O_c have uncorrelated errors with active observation errors (T, O) and have also uncorrelated errors with the background error (T, B), the independent observation O_c thus lie perpendicular to the analysis plane formed by T, O, B, and A. It is then evident that the var(O_c-A) which is plotted in Fig. 37.1, which is the length of (O_c, A) reaches a minimum when the analysis A is closest to T, that is when the analysis is optimal. Different points along the line (O, B) correspond to different ratio σ_o^2/σ_b^2, so we conclude that the minimum observed in

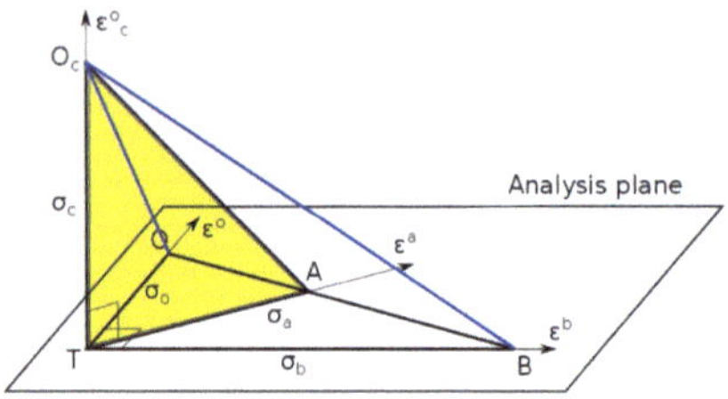

(a) Marseille *et al.* diagnostic

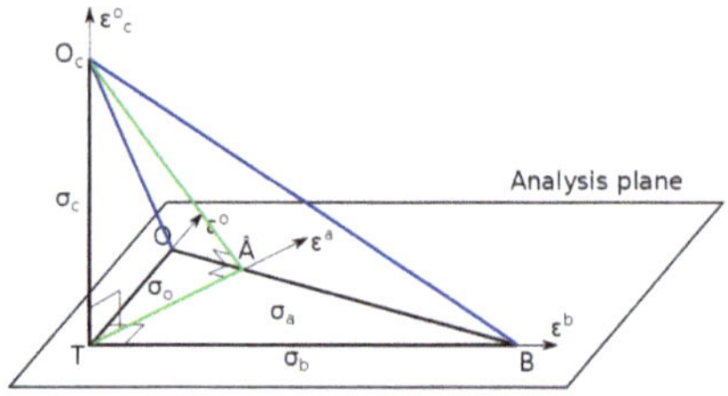

(b) Optimal analysis

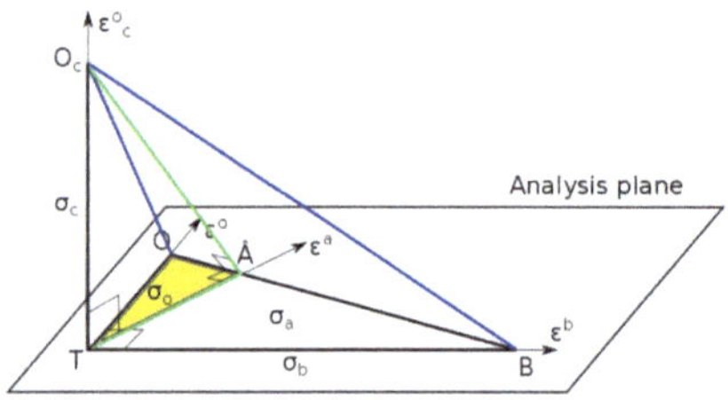

(c) Hollingsworth-Lönnberg (1989)

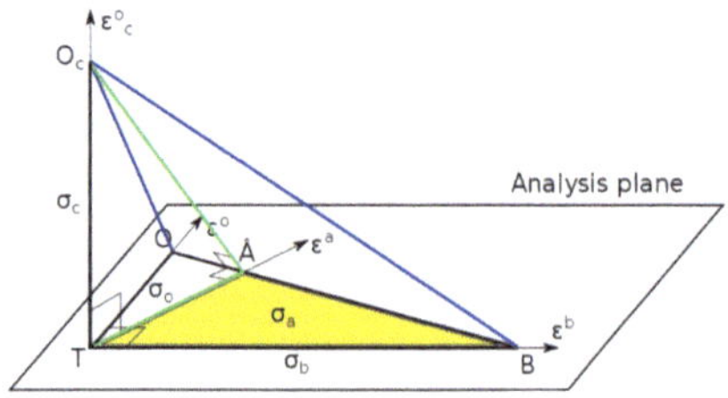

(d) MDJ diagnostic in active space

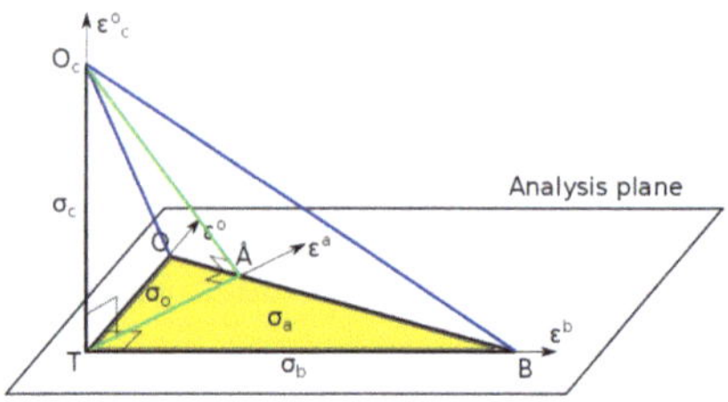

(e) Desroziers *et al.* diagnostics

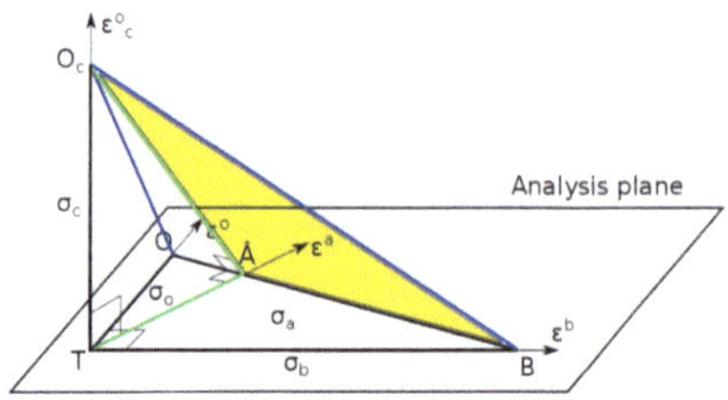

(f) MDJ diagnostics independent/passive space

Fig. 37.2 Geometric (i.e. Hilbert space) representation of a scalar analysis and cross-validation. The yellow right triangles indicate the sides (i.e. variances) from which different diagnostics of analysis error variance can be deduced using different methods in (**a**), (**c**), (**d**), (**e**), and (**f**)

Fig. 37.1, correspond to the ratio of error variances that minimizes the analysis error variance. Since var(O-B) was kept constant, we thus establish the value of σ_o^2 and of σ_b^2. This is how cross-validation can be used to determine the input error statistics that minimize the analysis error variance.

Different diagnostics to evaluate the analysis error variance actually correspond to the different right triangles illustrated in panels (a)–(f) of Fig. 37.2. A complete discussion and calculation of the analysis error variance for O_3 and $PM_{2.5}$ has been presented in Ménard and Deshaies-Jacques [9], where it is shown that when the background error correlation length is estimated and provide a chi-square diagnostic value close to one, then the different estimates of analysis error variance are very close to another.

With such an optimization procedure, we evaluate that the error standard deviation of the model is 9 ppbv for O_3 and 8.7 μg/m^3 for $PM_{2.5}$ at 21 UTC for the summer 2014, while the optimal analysis reduces the error to 5.4 ppbv for O_3 and 4.6 $\mu g/m^3$ for $PM_{2.5}$. Correspondingly, the fractional bias for the model O_3 and $PM_{2.5}$ is respectively 6.5% and 22% and with an optimal analysis reduces to 0.3% for O_3 and 2% for $PM_{2.5}$ (see [8, 9]).

37.3 Conclusions

We have developed a framework using cross-validation to evaluate the analysis error variance and optimize the analysis by obtaining the optimal observational weights. This method does not require a forecast initialized by the analysis. The method is quite general and could be used in many different contexts of off-line analysis and assimilation.

Questions and Answers

Questioner: Shuzhan Ren

Questions: Can this optimal ratio of σ_o^2/σ_b^2 be put into an analytical form? Are there any plans to do data assimilation of concentrations and surface emissions at the same time?

Answer. Conducting a series of cross-validation analyses for each ratio σ_o^2/σ_b^2 is costly, but we precompute this off-line using monthly statistics. We have not yet developed a scheme to evaluate the minimum of var(O_c-A) by a gradient search, but this would be useful when we will conduct this procedure in assimilation mode. In the future, we plan to perform data assimilation to adjust concentrations as well as emissions. The cross-validation procedure could then be used to obtain the optimal analysis, while the best forecast could be obtained by adjusting, in addition, the emissions.

References

1. D.L. Crouse, P.A. Peters, P. Hystad, J.R. Brook, A. van Donkelaar, R.V. Martin, P.J. Villeneuve, M. Jerrett, M.S. Goldberg, C.A. Pope III, M. Brauer, R.D. Brook, A. Robichaud, R. Ménard, R.T. Burnett, Associations between ambient $PM_{2.5}$, O_3, and NO_2 and mortality in the Canadian Census Health and Environment Cohort (CanCHEC) over a 16-year follow-up. Environ. Health Perspect. (Open Access) **123**, 1180–1186 (2015). https://doi.org/10.1289/ehp.1409276
2. D.L. Crouse, L. Pinault, A. Balram, P. Hystad, P.A. Peters, A. van Donkelaar, R.V. Martin, R. Ménard, A. Robichaud, P.J. Villeneuve, Urban greenness and mortality in Canada's largest cities: a national cohort study. The Lancet, Planet Earth **1**, e289–e297 (2017), www.thelancet.com/planetary-health

3. D.L. Crouse, A. Balram, P. Hystad, L. Pinault, M. van den Boshm, H. Chen, D. Rainham, E.M. Thomson, C.H. Close, A. van Donkelaar, R.V. Martin, R. Ménard, A. Robichaud, P.J. Villeneuve, Associations between living near water and risk of mortality among urban Canadians. Environ. Health Perspect. **126**(7) (2018). https://doi.org/10.1289/EHP3397
4. R. Ménard, A. Robichaud, The chemistry-forecast system at the Meteorological Service of Canada. In: The ECMWF Seminar Proceedings on Global Earth-System Monitoring, Reading, UK, 5–9 September 2005; pp. 297–308 (2005)
5. R. Ménard, Error covariance estimation methods based on analysis residuals: theoretical foundation and convergence properties derived from simplified observation networks. Q. J. R. Meteorol. Soc. **142**, 257–273 (2016). https://doi.org/10.1002/qj.2650
6. R. Ménard, M. Deshaies-Jacques, N. Gassett, A comparison of correlation-length estimation methods for the objective analysis of surface pollutants at Environment and Climate Change Canada. J. Air Waste Manag. Assoc. **66**, 874–895 (2016). https://doi.org/10.1080/10962247.2016.1177620
7. R. Ménard, M. Deshaies-Jacques, Error covariance estimation methods based on analysis residuals and its application to air quality surface observation networks. In: C. Mensink, G. Kallos (Eds.) Air Pollution and Its Application XXV. Springer International AG, 249-253 (2017). https://doi.org/10.1007/978-3-319-57645-9_40
8. R. Ménard, M. Deshaies-Jacques, Evaluation of analysis by cross-validation. Part I: Using verification metrics. Atmosphere **9**, 86 (2018a). https://doi.org/10.3390/atmos9030086, www.mdpi.com/journal/atmosphere
9. R. Ménard, M. Deshaies-Jacques, Evaluation of analysis by cross-validation. Part II: Diagnostic and optimization of analysis error covariance. Atmosphere **9**, 70 (2018b). https://doi.org/10.3390/atmos9020070, www.mdpi.com/journal/atmosphere
10. A. Robichaud, R. Ménard, Multi-year objective analysis of warm season ground-level ozone and PM2.5 over North-America using real-time observations and Canadian operational air quality models. Atmos. Chem. Phys. **14**, 1769–1800 (2014). https://doi.org/10.5194/acp-14-1769-2014
11. A. Robichaud, R. Ménard, Y. Zaïtseva, D. Anselmo, Multi-pollutant surface objective analyses and mapping of air quality health index over North America. Air Qual. Atmos. Health **9**, 743–759 (2016). https://doi.org/10.1007/s11869-015-0385-9
12. D. Stieb, R. Ménard, M. Smith-Doiron, J. Yao, S. Henderson, L. Pinault, A. Robichaud, A. van Donkelaar, R.V. Martin, J. Brook, Spatial variability in ambient ozone and fine particles concentrations and population susceptibility among Canadian health regions. Can. J. Public Health (in print) (2019)

Chapter 38
Ensemble-Based Data Assimilation and Forecasting of Volcanic Ash

Andreas Uppstu, Julius Vira and Mikhail Sofiev

Abstract Volcanic ash and other aerosols such as desert dust form significant hazards for aviation and can cause both direct safety threats and significant economic losses. However, forecasts of aviation hazards have generally been deterministic, although the available computational resources would easily allow for them to be ensemble-based. In principle, ensemble-based forecasts can enable more accurate error estimates and thus an improved risk management framework. Advanced data assimilation methods, such the Ensemble Kalman Filter, coupled with a meteorological forecast ensemble, provide increased accuracy and the possibility to estimate the source term by taking into account its correlation with the observed ash concentration.

38.1 Introduction

Real-time data assimilation of observations of volcanic ash has potential both to improve flight safety and reduce the aviation related costs of volcanic eruptions. Within the EUNADICS-AV project, ensemble-based forecasting is suggested as the base for future products aimed to improve flight safety and to optimize flight routing and scheduling. In addition to volcanic eruptions, hazards included in the project are releases of radioactive matter, dust storms and vegetation fires. The Finnish Meteorological Institute utilizes the chemical transport model SILAM [3], coupled with the Ensemble Kalman Filter (EnKF), which is a data-assimilation method naturally suited for ensemble-based forecasting, to assimilate satellite- and ground-based

---

A. Uppstu (✉) · J. Vira · M. Sofiev
Atmospheric Composition Research, Finnish Meteorological Institute, Helsinki, Finland
e-mail: Andreas.Uppstu@fmi.fi

M. Sofiev
e-mail: Mikhail.Sofiev@fmi.fi

J. Vira
Cornell University, Ithaca, NY, USA

C. Mensink et al. (eds.), *Air Pollution Modeling and its Application XXVI*, Springer Proceedings in Complexity,
https://doi.org/10.1007/978-3-030-22055-6_38

observations of both volcanic ash and SO_2. We demonstrate through a simulated eruption of the Etna volcano how the EnKF can be utilized to deliver assimilated forecasts of the dispersion of volcanic ash and how the forecast could be visualized.

38.2 Methods

In order to provide assimilated forecasts, an atmospheric transport model, a volcano eruption model, a data-assimilation method and a set of observations that are obtainable in near real-time are needed. The ensemble is formed through perturbing the time stamp of the meteorological forecast, with the standard deviation of the perturbations being 3.0 h, and through perturbing the emission source.

38.2.1 A Volcanic Eruption Model

The volcanic eruption model is based on an empirical relationship found between the height of the eruption plume and the eruption rate [2]. However, methods for determining the vertical distribution of ash within the plume as well as for determining the particle size distribution of the eruption require more study. In this study, it is assumed that 75% of the erupting mass is distributed uniformly over the top 25% of the plume, with the remaining mass being uniformly distributed over the bottom part of the plume. Within the ensemble, the eruption rate is assumed to be log-normally distributed. Compared to a normal distribution, this allows for a significantly larger range of different eruptions to be included within the ensemble, which is essential for a proper description of the a priori probabilities.

In order to estimate the strength of the eruption, the logarithm of the eruption rate is included as a part of the control vector of the data assimilation process. To avoid a collapse of the ensemble and to take into account temporal changes in the a priori probability distribution, the distribution of the eruption rates is set to constantly evolve towards a pre-defined distribution within a correlation time of 24 h. Thus, if no observations of the erupting plume are made, the eruption rate distribution reverts to the original one within a time scale of one day.

38.2.2 A Simulated Eruption of Etna

In order to test the methodology, we applied it on a simulated eruption of the Etna volcano. For the simulation, we utilized the emission term estimated for the 2010 Eyjafjallajökull eruption [4]. Simulated retrievals of volcanic ash loadings from the satellite-borne IASI instrument as well as of ash concentrations from the satellite-borne CALIOP lidar and the EARLINET ground-based lidar network were created using readily available data of the satellite orbital parameters and lidar station locations. The simulated retrievals were perturbed with a normally distributed relative random error of 0.2 times the retrieved value and additionally a normally distributed random error of 0.04 mg/m^3 for the concentrations and of 20 mg/m^3 for the integrated columns.

The eruption was simulated using meteorological data based on the European Centre for Medium-Range Weather Forecasts (ECMWF) ERA-Interim reanalysis dataset [1], whereas in the assimilation of the simulated observations, the meteorological data was obtained from the ECMWF operational forecasts. Both the simulation and the assimilation were performed using a 0.25° × 0.25° grid with 28 vertical levels up to an altitude of 12,250 m.

38.3 Results

In order to quantify the performance of the forecast, we have analyzed how well the data assimilation is able to predict simulation cells with a concentration of greater than 200 mg/m^3, which is the threshold for the Enhanced Procedures Zone (EPZ) defined by UK Civil Aviation Authority [5]. We thus classify cells based on whether an ash concentration exceeding the EPZ threshold is present or not in the simulation, and correspondingly, whether it is present or not in the forecast (Table 38.1). From the values presented in Table 38.1 one can compute the following statistical measures:

$$\begin{aligned} Model\,accuracy &= (a+d)/(a+b+c+d) \\ Probability\,of\,detection &= a/(a+c) \\ Probability\,of\,false\,detection &= b/(b+d) \\ False\,alarm\,rate &= b/(a+b) \\ Bias &= (a+b)/(a+c)-1 \\ Odds\,ratio &= ad\big/bc \end{aligned} \tag{38.1}$$

Our statistical values are based on a grid having six cells in the vertical direction (thus each of them is approximately 2000 m thick), whereas the horizontal cell size is the same as in the actual computations, i.e. 0.25° × 0.25°. Statistical figures for the ensemble average and the 90th and the 98th percentile are presented in Table 38.2. Figure 38.1 shows the vertically integrated ash concentration at a specific simulation

Table 38.1 Contingency matrix for the discretized outcomes of the assimilation. The variables *a–d* indicate numbers of cells

	Present	Not present
Predicted	*a*	*b*
Not predicted	*c*	*d*

Table 38.2 Quantitative performance of the forecast based on Eq. (38.1)

	Average	90th percentile	98th percentile
Model accuracy (%)	99.5	99.4	99.0
Probability of detection (%)	55	74	84
False alarm rate (%)	43	54	72
Probability of false detection (%)	0.24	0.50	1.25
Bias (%)	−3	60	199
Odds ratio	504	572	419

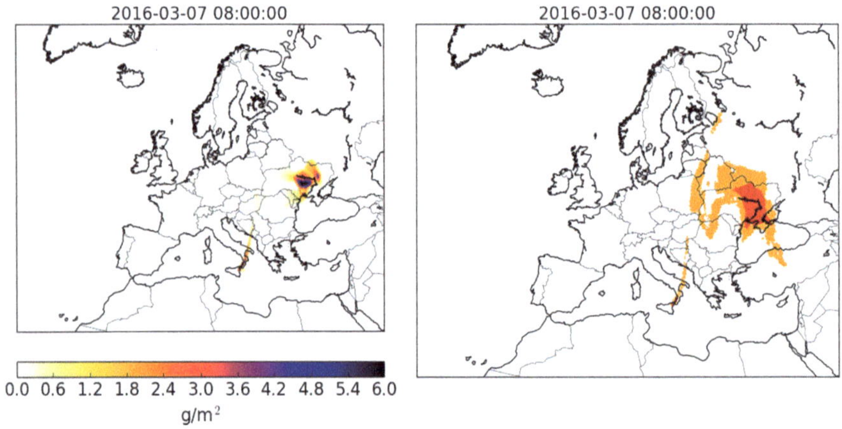

Fig. 38.1 Left: Vertically integrated ash concentration at a specific time of the simulated eruption. Right: Example of a warning that could be issued for FL 220 based on the forecast ensemble and simulated observations. In the red area the average value of the ensemble contains ash exceeding the EPZ threshold, whereas in the orange area at least 2% of the ensemble members exceed the threshold

time and an image illustrating the distribution of ash within the forecast ensemble at a specific flight level. With such an approach, one can define areas of interest based on combinations of ensemble fractions and threshold values. Higher level utilization of the ensemble could involve computing path integrals of the ensemble-member specific ash concentrations along aircraft routes in order to estimate the distribution of total doses of ash that can be encountered.

38.4 Discussion and Conclusions

In the case of a simulated volcano eruption, EnKF-based data-assimilation is able to provide reasonably good forecasts of the dispersion of volcanic ash. In our test scenario, the model accuracy is very good, but this is obviously due to the vast majority of cells being free of volcanic ash. The probability of detection increases from 55 to 84% when the average value is replaced by the 98th percentile, at the cost of the forecast area growing in size by a factor of three. However, even in such a case, only 1.25% of the total number of cells within the simulation domain are wrongly predicted to contain an amount of ash exceeding the EPZ threshold.

As the initial estimate of the eruption rate is assumed to have a large uncertainty and as the transport model used in the assimilation run is the same as in the simulation, the inability of the ensemble to cover all areas is presumably partly due to the perturbation of the meteorological data being too limited. In principle better results could be obtained using an actual meteorological forecast ensemble instead of a perturbed single forecast, but in a real time application such an approach could be limited by the data transfer capacity.

The simulation results help to verify the functionality of the EnKF despite that all of its mathematical requirements are not met. For example, the forward model is generally not linear and the relationship between the eruption rate and the in-air ash concentration can in principle be very complex. It is worth noting that our test does not take into account all important sources of uncertainty, such as inaccuracies and nonlinearities of the observation operator. Additionally, real measurements are in general not fully capable of distinguishing different types of aerosols from each other and can also be affected by water clouds.

QUESTIONER: Pieter De Meutter

QUESTION: Which criteria were used to choose the perturbations in the meteorological ensemble, and did you validate the spread in the meteorological ensemble?

ANSWER: The meteorological ensemble was created from a single ECMWF forecast by perturbing only the forecast time. Thus, the ensemble members are valid meteorological fields from a physical point of view, barring issues that may rise from the diurnal cycle being misaligned. The spread of the meteorological ensemble was tested in the actual data assimilation run, which showed that perturbing the forecast time of the ECMWF forecast with a standard deviation of 3 h was not sufficient to fully cover the actual dispersion of ash acquired with the ERA Interim dataset (given an ensemble size of 80).

QUESTIONER: Richard Menard

QUESTION: Taking into account that the correlation structures you get with this problem are non-isotropic and elongated and that you need to used localization which makes the correlation structures more isotropic, have you considered using a much large number of ensemble members?

ANSWER: Yes, larger ensembles were considered, but the available storage size prevented testing those. Since the presentation, more space efficient methods for generating larger ensembles have been implemented in our transport model (SILAM), and the effect increasing the size of the ensemble will be studied.

Acknowledgements The authors acknowledge funding from the EUNADICS-AV and the Nordic Centre of Excellence EmblA projects. Initial volcano model development was performed with support from the VAST and SMASH projects of the European Space Agency (ESA).

References

1. D.P. Dee et al., The ERA-Interim reanalysis: configuration and performance of the data assimilation system. Q. J. R. Meteorol. Soc. **137**, 553–597 (2011)
2. L.G. Mastin et al., A multidisciplinary effort to assign realistic source parameters to models of volcanic ash-cloud transport and dispersion during eruptions. J. Volcanol. Geoth. Res. **186**, 10–21 (2009)
3. M. Sofiev, J. Vira, R. Kouznetsov, M. Prank, J. Soares, E. Genikhovich, Construction of the SILAM Eulerian atmospheric dispersion model based on the advection algorithm of Michael Galperin. Geosci. Model Dev. **8**, 3497–3522 (2015)
4. A. Stohl et al., Determination of time- and height-resolved volcanic ash emissions and their use for quantitative ash dispersion modeling: the 2010 Eyjafjallajokull eruption. Atmos. Chem. Phys. **11**, 4333–4351 (2011)
5. C. Zehner et al., Monitoring volcanic ash from space, in *ESA-EUMETSAT Workshop on the 14 April to 23 May 2010 eruption at the Eyjafjöll Volcano*, South Iceland (ESA/ESRIN, 26–27 May 2010) (ESA Publication STM-280, 2010)

Chapter 39
Performance Differences of the National Air Quality Forecasting Capability When There is a Major Upgrade in the Chemistry Modules

Pius Lee, Li Pan, Youhua Tang, Daniel Tong, Barry Baker, Hyuncheol Kim and Rick Saylor

Abstract There have been large advancement in modeling science and chemical constituent measurement of air pollutants harmful to the public health. The National Oceanic and Atmospheric Administration (NOAA) National Air Quality Forecasting Capability (NAQFC) is a vital service that NOAA provides to the general public to help safeguarding the public health as well as the environmental resilience through announcement of information-driver mitigation and adaptation action. NAQFC is poised to upgrade from using the Community Multiscale Air Quality Model (CMAQ) version 5.0.2 to version 5.2. This is noticeable a multiple sub-version number leaping forward corresponding to major upgrades in chemistry and emission sciences. The following lists the major science upgrade: (a) upgrade gas chemistry for the Carbon-Bond Mechanism version 5 (CB05) to version 5 Revision1 (CB05R1); (b) Inclusion of Halogen chemistry; (c) Employed more explicit speciation for isoprene and monoterpenes from biogenic sources; (d) Upgraded the aerosol module using a more sophisticated secondary aerosol production suite of multi-generational oxidation mechanism; and (e) Application of a fuller set of National Emission Inventory (NEI) from base year 2014. We tested the new system for a summer case retrospectively and compared its forecast performance with the real-time operational NAQFC. The U.S. Environmental Protection Agency (EPA) AIRNow monitoring network was used to verify the forecast accuracy. We noticed considerable discrepancies in the performance of the two realization of forecasting simulations.

P. Lee (✉) · L. Pan · Y. Tang · D. Tong · B. Baker · H. Kim
NOAA/Air Resources Laboratory, College Park, MD, USA
e-mail: pius.lee@noaa.gov

L. Pan · Y. Tang · D. Tong · B. Baker · H. Kim
UMD/Cooperative Institute for Climate and Satellites, College Park, MD, USA

D. Tong · R. Saylor
NOAA/ARL/Atmospheric Turbulence and Diffusion Division, Oak Ridge, TN, USA

P. Lee · L. Pan · Y. Tang · D. Tong · B. Baker · H. Kim · R. Saylor
Center for Spatial Information Science and Systems, George Mason University, Fairfax, VA, USA

C. Mensink et al. (eds.), *Air Pollution Modeling and its Application XXVI*,
Springer Proceedings in Complexity, https://doi.org/10.1007/978-3-030-22055-6_39

39.1 Introduction

Perennial over-estimate on surface ozone concentration in urban centers prompted the U.S. National Air Quality Forecasting Capability (NAQFC) [1] examines the possible impact of the new halogen chemistry module introduced in the US EPA Community Multi-scale Air Quality Model (CMAQ) version 5.2 carbon Bond 05 Mechanism (CB05)-eh module [2]. This is a potentially a major upgrade to include halogen chemistry in air quality forecasting in a national scale in the USA. Its impetus stemmed from the possible simultaneous improvement in reduction in the perennial positive bias in O_3 and PM2.5 forecast in coastal urban locations. The US EPA has tested the halogen chemistry module in the CB05-eh version by studying sensitivity cases based on turning on or off of the halogen chemistry when simulating the air composition over the conterminous USA (CONUS) as a stand-alone limited area domain with Sea-Salt halogen as the primary source for halogen gases. Two additional sensitivity cases were also tested with the CB05-eh used in a hemispheric CMAQ simulation at roughly $1^\circ \times 1^\circ$ spatial resolution. This rather coarse spatial resolution simulation predicted the chemical constituent concentrations at the lateral boundaries of the conterminous US. It was revealed that the CB05-eh module did a variable emission label.

CB05R1 with halogen chemistry has a net effect of reducing surface ozone concentration where halogen gaseous constituents represent a significant vapor pressure in relation to that of NOx and VOC. In particular due to changes in the oxidant the overall activity of atmospheric primary. Basically the dry deposition velocity for ozone was increased at least 4 times the current NAQFC dry deposition capability. Primarily, the halogen chemistry enabled the halogen gases to catalytically react with PM2.5 giving the initialization of the CMAQ modeling work. We chose a summer period for the testing.

39.2 The Testing of CMAQ5.2 with CB05-eh

Summer 2017 was chosen to be tested. It is important for the esusus The mechanism that impacts ozone dry deposition by the existence of significant amount of halogen gases comparable to those of NOx and VOC modifies O_3 deposition speed.

The phenomenon of competition for hydroxyl and hydro-peroxyl radicals depletes nitrogen-oxides. Table 39.1 showed the statistics about the performance next box in common sense a sanctity.

Table 39.1 Performance metric for NAQFC (base) versus Halogen chemistry included sensitivity (Cl and Br)

MD8A O_3	Case	Obs	Mean	Bias	RMSE	Correlation
CONUS (52,200)	Base	41.7	42.8	1.1	9.00	0.75
	Cl and Br		43.5	1.8	9.11	0.76
Northeast (7900)	Base	38.7	41.6	2.9	8.62	0.78
	Cl and Br		42.4	3.7	9.01	0.78
Southeast (8500)	Base	38.8	43.7	4.89	8.95	0.75
	Cl and Br		44.4	5.58	9.21	0.77
Up Middle (11,180)	Base	41.5	42.5	0.96	7.10	0.81
	Cl and Br		43.5	1.93	7.31	0.82
Low Middle (5145)	Base	41.8	45.7	3.91	9.82	0.71
	Cl and Br		46.4	4.64	10.02	0.72
Rocky M. (7600)	Base	48.4	46.0	− 2.43	8.93	0.71
	O_n_G		46.7	− 1.78	8.56	0.73
Pacific Coast (7260)	Base	47.5	43.8	− 3.63	11.48	0.75
	Cl and Br		44.0	− 3.48	11.48	0.75

39.3 Test Case and Evaluation

The evaluation covered the usual performance statistics. It is rather straight forward to see how the coastal area specific urban area associated high positive bias to be targeted using the new chlorine and bromine chemistry upgrade to decrease coastal and water surface resistance for dry deposition velocity.

Overall the daily eight hour average maximum ozone concentration was reduced across the coastal areas in general and where there are strong NOx emission there corresponds a larger reduction in surface ozone concentration.

39.4 Results and Summary

The general agreement with our test result of a summer cast in 2017 with that reported for a 2014 study be the US EPA also over the CONUS gave confidence that the inclusion of the halogen chemistry in ozone production (or loss) was consistent with the coastal area targeted phenomenon. Namely the overall effect of increase hydroperoxyl radical depletion due to chlorine oxidation and bromine oxidation increases the production of ozone. However the increased dry deposition loss of ozone due to the correction of rate of dissolvable ozone over moist surfaces depleted ozone to the surface by depositional loss is larger. The phenomenon was strongest during the highest ozone concentration during its diurnal maximum concentrations. It was contribution most to the reduction of high positive bias in daily 8 h maximum ozone concentration (Table 39.1). Also the time series for depicting the time evolution of the surface ozone concentration over the month of July 2017 was shown over CONUS (Fig. 39.1) and the Gulf of Mexico (Fig. 39.2).

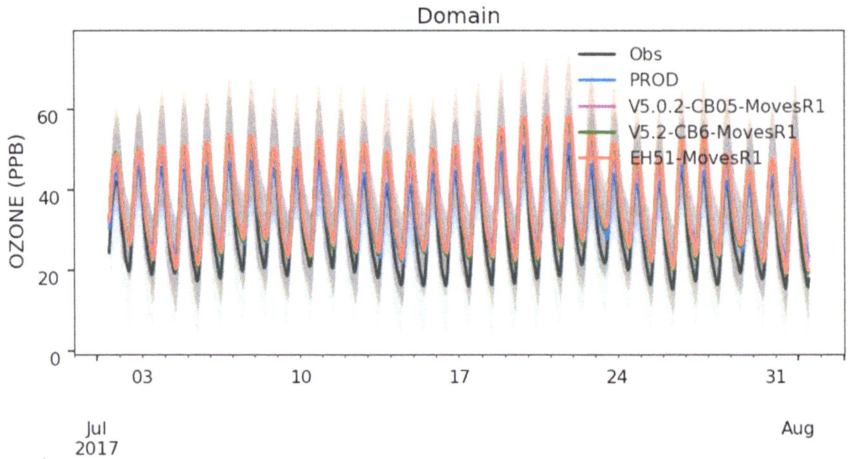

Fig. 39.1 Time series of surface ozone concentrations over the CONUS: observation in black, current NAQFC prediction in blue and red depicted that after halogen chemistry has been included

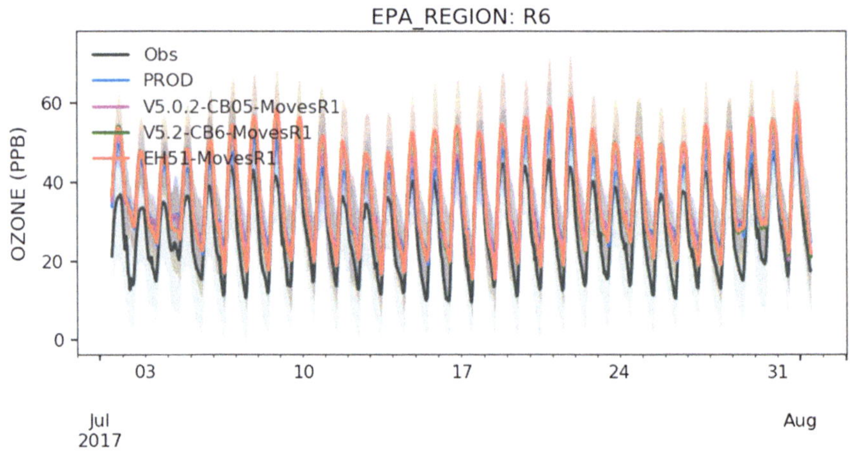

Fig. 39.2 Time series of surface ozone concentrations over the Gulf Coast: observation in black, current NAQFC prediction in blue and red depicted that after halogen chemistry has been included

References

1. P. Lee, J. McQueen, I. Stajner, J. Huang, L. Pan, D. Tong, H. Kim, Y. Tang, P. Shafran, H.-C. Huang, J. Gorline, S. Upadhayay, R. Artz, NAQFC developmental forecast guidance for fine particulate matter (PM2.5). Weather Forecast. **32**(1), 343–360 (2017). https://doi.org/10.1175/WAF-D-15-0163.1
2. G. Sarwar, D. Kang, K. Foley, D. Schwede, B. Gantt, R. Mathur, Technical note: Examining ozone deposition over seawater, in Atmospheric Environment, vol. 141 (Elsevier Science Ltd, New York, NY, 2016), pp. 255–262

Chapter 40
Total Deposition Maps Evaluated from Measurement-Model Fusion in North America (ADAGIO Project)

Alain Robichaud, Amanda Cole, Michael Moran, Alexandru Lupu, M. Shaw, G. Roy, M. Beauchemin, V. Fortin and R. Vet

Abstract Environment and Climate Change Canada's ADAGIO project (Atmospheric Deposition Analysis Generated by Optimal Interpolation using Observations) produces maps of wet, dry and total annual deposition of oxidized and reduced nitrogen (N) and sulphur (S) and ozone in Canada and the United States by combining in an optimal way observed and modeled data. Optimal interpolation methods are used to provide the best objective analyses of seasonally-averaged surface concentrations of gaseous, particulate, and precipitation species predicted by Environment and Climate Change Canada's in-line regional air quality model GEM-MACH (Global Environmental Multiscale model—Modelling Air quality and Chemistry) based on the difference between the modeled and measured values at network observation sites. The resulting objective analyses (OA) for gas and particulate species concentration fields are then combined with effective deposition velocities from GEM-MACH to calculate dry deposition. Concentrations of precipitation ions are combined with precipitation amounts from the Canadian Precipitation Analysis (CaPA), in which all available precipitation data sets are used to adjust precipitation amounts predicted by GEM, to calculate wet deposition. Results from the 2010 development year are compared with previously-generated wet deposition kriging maps, results from the

A. Robichaud (✉)
Modelling and Integration Research Section, Air Quality Research Division, Environment and Climate Change Canada, Dorval, Canada
e-mail: alain.robichaud@canada.ca

M. Moran · A. Lupu
Modelling and Integration Research Section, Air Quality Research Division, Environment and Climate Change Canada, Downsview, Canada

A. Cole · M. Shaw · R. Vet
Measurements and Analysis Research Section, Air Quality Research Division, Environment and Climate Change Canada, Downsview, Canada

G. Roy · M. Beauchemin
High Impact Weather National Laboratory, Meteorological Service of Canada, Environment and Climate Change Canada, Montreal, Canada

V. Fortin
Environmental Numerical Weather Prediction Research, Meteorological Research Division, Environment and Climate Change Canada, Dorval, Canada

C. Mensink et al. (eds.), *Air Pollution Modeling and its Application XXVI*, Springer Proceedings in Complexity, https://doi.org/10.1007/978-3-030-22055-6_40

USEPA's Total Deposition (TDEP) method, and surface measurements not used in the analysis where available. It was found that the biggest sources of uncertainties are the dry deposition velocities and error statistics (weight matrix used to produce OA). Therefore, more work is needed to reduce these uncertainties.

40.1 Introduction

This paper describes a new method to obtain atmospheric dry and wet deposition maps in Canada and US. The methodology is based on optimal interpolation for air quality [3] and sensitivity tests for error statistics. Deposition maps use GEM-MACH, v2 [2] output combined with quality-controlled and screened observations of different monitoring networks in Canada and the US (CAPMON, NAPS, AMON, AQS, and CASTNET). Wet, dry and total deposition from the ADAGIO project will replace krigged maps of wet-only deposition currently used to evaluate N and S deposition to Canadian and US emissions [5].

40.2 Methods

Objective analysis (OA): model fusion with observations (data assimilation techniques) (Figs. 40.1 and 40.2).

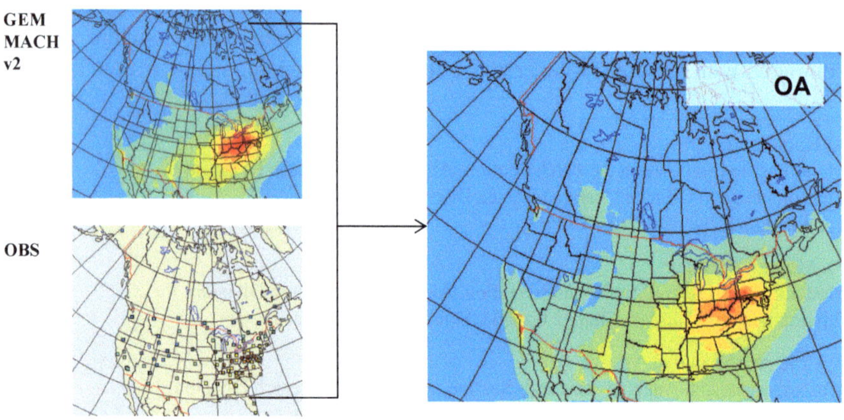

Fig. 40.1 Model (top left)-data (bottom left) fusion using DA techniques to get OA

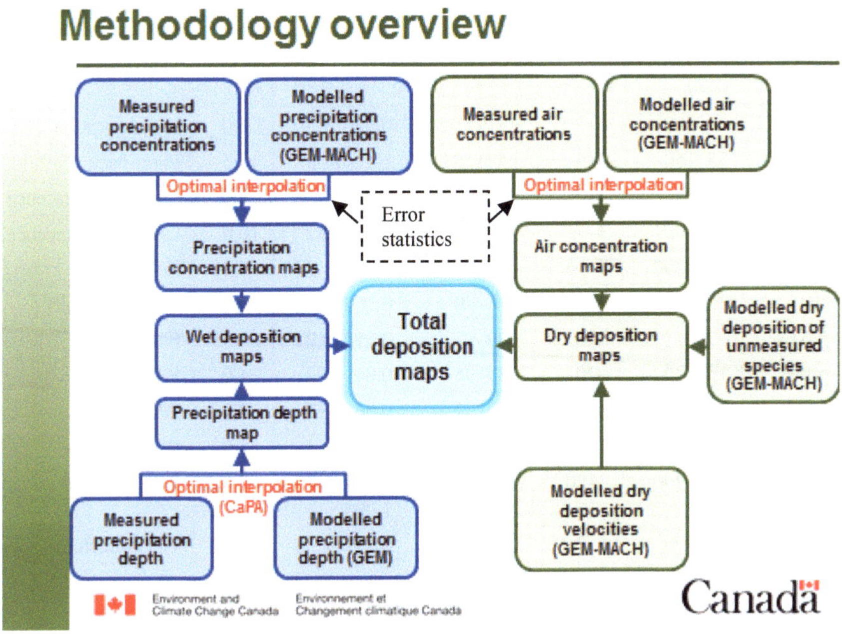

Fig. 40.2 Methodology to produce deposition maps over Canada and US

40.3 Results

First, OA gridded concentrations are obtained for all species according to data assimilation techniques [1] adapted to air quality [3]. One of the most difficult steps is to obtain appropriate error statistics. Sensitivity tests over a wide range of correlation lengths and variance ratios have been done to obtain optimum values. Results are given in Table 40.1. Dry deposition is then computed by multiplying weighted dry deposition velocity by OA concentration and wet deposition obtained by multiplying precipitation amount from the Canadian Precipitation Analysis (CaPA).

The comparison between ADAGIO and US/TDEP initiative [4] is remarkable and of interest since the two projects use a different model and methodology while results are fairly similar for N total deposition for year 2010 (Fig. 40.3) indicating rather high certainty of the results.

40.4 Summary and Conclusions

Maps of N and S deposition for North America have been produced using objective analyses based on data assimilation techniques and sensitivity tests based on 5 ensembles to find error statistics. It was found that the correlation length is the

Table 40.1 Results from sensitivity tests for OA (range corresponds to different seasons)

Species	Corr. length (km)	Variance ratio	N	Xhi^2	Obs. error (%)	Observation data methods
O_3	70–200	1–5	837–1473	0.986–1.50	5.0	UV absorption
SO_2	55–150	2–6	521–543	1.19–1.38	16.5	Filter pack
NO_2	60–73	3–21	465–484	1.005–1.53	10.0	Chemiluminescence
NO	20–100	3–15	410–464	0.92–1.82	10.0	Chemiluminescence
HNO_3	150–200	2–13.5	89–94	1.096–1.37	6.0	Filter pack
NH_3	30–320	2–25	18–20	0.164–1.02	10.0	Passive sampler
p-NO_3	~200	~20	89–95	0.592–0.92	5.5	Filter pack
p-NH_4	155–200	~20	89–95	0.81–0.92	4.0	Filter pack
p-SO_4	155–200	2–20	89–94	0.56–0.91	7.0	Filter pack
w-NO_3	200	1–3	225–244	1.04–1.41	5.0	Wet collector
w-NH_4	190–200	2–20	225–248	0.821–1.093	2.0	Wet collector
w-SO_4	80–200	1–3	216–248	0.797–1.00	5.0	Wet collector

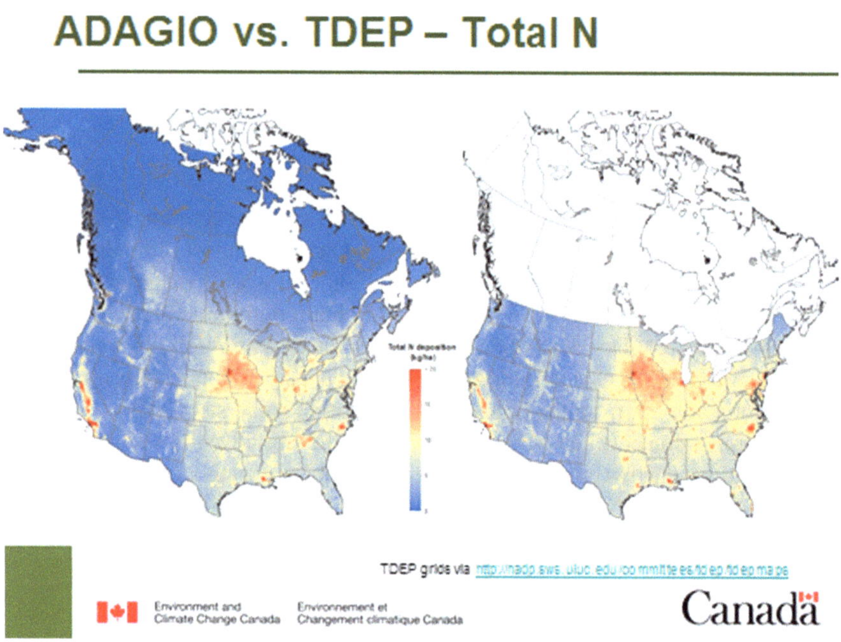

Fig. 40.3 Comparison between ADAGIO total nitrogen deposition for 2010 with US/TDEP initiative

most sensitive parameter for OA production, and dry deposition velocity is the highest source of uncertainly for deposition maps. Preliminary inter-comparison with deposition maps produced by US EPA (Schwede and Lear [4], TDEP project) shows reasonable agreement despite different methodologies used. A new version of deposition maps is currently underway using the techniques of sensitivity tests. Finally, the results of ADAGIO are also compared to previous maps obtained using kriging algorithm (e.g., [6]) and gives consistent results.

Acknowledgements The first author wishes to acknowledge helpful discussions with Martin Deshaies and Richard Ménard concerning some aspects of error statistics and cross-validation.

References

1. E. Kalnay, *Atmospheric Modeling, Data Assimilation and Predictability* (Cambridge University Press, Cambridge, 2003)
2. M.D. Moran, S. Ménard, R. Pavlovic, D. Anselmo, S. Antonopoulos, A. Robichaud, S. Gravel, P.A. Makar, W. Gong, C. Stroud, J. Zhang, Q. Zheng, H. Landry, P.A. Beaulieu, S. Gilbert, J. Chen, A. Kallaur, Recent advances in Canada's national operational air quality forecasting system, in *32nd NATO-SPS ITM*, 7–11 May 2012 Utrecht, NL (2012)
3. A. Robichaud, R. Ménard, Y. Zaïtseva, D. Anselmo, Multi-pollutant surface objective analyses and mapping of air quality health index over North America. Air Qual. Atmos. Health (2016). https://doi.org/10.1007/s11869-015-0385-9
4. B.D. Schwede, G.G. Lear, A novel hybrid approach for estimating total deposition in the United States. Atmos. Environ. **92**, 207–220 (2014)
5. R. Vet, C.-U. Ro, Contribution of Canada–United States transboundary transport to wet deposition of Sulphur and nitrogen oxides—a mass balance approach. Atmos. Environ. **42** (2008). https://doi.org/10.1016/j.atmosenv.2007.12.034
6. R. Vet et al., A global assessment of precipitation chemistry and deposition of sulfur, nitrogen, sea salt, base cations, organic acids, acidity and pH, and phosphorus. Atmos. Env. **13**, 3–100 (2014)

Chapter 41
Importance of Inventory Representativeness for Air Quality Forecasting: A Recent North American Example

Michael D. Moran, Qiong Zheng, Junhua Zhang, Radenko Pavlovic and Mourad Sassi

Abstract North American air quality (AQ) forecasts made by the Environment and Climate Change Canada (ECCC) operational regional AQ prediction system since 2015 have used input emissions files based on Canadian, U.S., and Mexican national emissions inventories for base years 2010, 2011, and 1999, respectively. Since 2010, however, emissions of many criteria air pollutants have declined in both Canada and the U.S.. We recently tested new input emissions files based on a 2013 Canadian inventory, a projected 2017 U.S. inventory, and a 2008 Mexican inventory in the ECCC regional AQ prediction system. For Canada, the switch from the 2010 inventory to the 2013 inventory reduced SO_2, NO_x, and VOC annual anthropogenic emissions by 12%, 2%, and 4%, respectively. For the continental U.S., adoption of the projected 2017 inventory reduced SO_2, NO_x, and VOC annual anthropogenic emissions relative to the 2011 inventory by 65%, 33%, and 11%, respectively, suggesting the importance of emissions base-year representativeness for AQ forecasting. Moreover, the use of these new input emissions fields for 2016 and 2017 test periods improved AQ forecasts in comparison to the operational model for Canada and the U.S., in particular for summertime ozone forecasts over the eastern U.S.. A new version of the ECCC forecast system that uses these updated input emissions files was accepted for operational implementation in mid 2018.

M. D. Moran (✉) · Q. Zheng · J. Zhang
Air Quality Research Division, Environment and Climate Change Canada, Toronto, ON, Canada
e-mail: mike.moran@canada.ca

R. Pavlovic · M. Sassi
Air Quality Policy-Issue Response Section, Environment and Climate Change Canada, Montreal, QC, Canada

C. Mensink et al. (eds.), *Air Pollution Modeling and its Application XXVI*,
Springer Proceedings in Complexity, https://doi.org/10.1007/978-3-030-22055-6_41

41.1 Introduction

Although current air quality is determined by current pollutant emissions, operational AQ forecasts are usually performed based on retrospective emissions inventories, that is, on inventories that are representative of a past base year. When emissions levels are changing rapidly, however, retrospective inventories may not accurately represent current emission levels, as is now the case in North America due to the continuing decrease of emissions of many criteria air contaminants (CACs) in both Canada and the U.S. since the 1990s (Fig. 41.1). One way to deal with this lack of representativeness for AQ forecasting is to use a projected, future-year emissions inventory that is closer to the forecast year. Such inventories are sometimes created for AQ policy development applications by incorporating assumptions about projected changes to the economy, to population and housing, to the on-road and off-road vehicle fleets, etc. along with any emissions changes expected to result from the phased implementation of existing emissions control legislation.

Since 2015 ECCC's operational Regional Air Quality Deterministic Prediction System (RAQDPS), which produces 48 h AQ forecasts on a North American domain twice daily, has used input emissions files (named SET2.1.1) based on version 1 of the 2010 Canadian Air Pollutant Emission Inventory (APEI), version 1 of the 2011 U.S. Environmental Protection Agency's (EPA) National Emissions Inventory (NEI), and version 1 of the 1999 Mexican emissions inventory [6]. The decision to use these particular inventories was not based on their representativeness of 2015 (or later) emissions but rather was because they were the most recent Canadian, U.S., and Mexican emissions inventories available at the time. This paper describes the impact of adopting newer Canadian and Mexican inventories and a projected U.S. inventory on 2016 and 2017 North American AQ forecasts made by the RAQDPS.

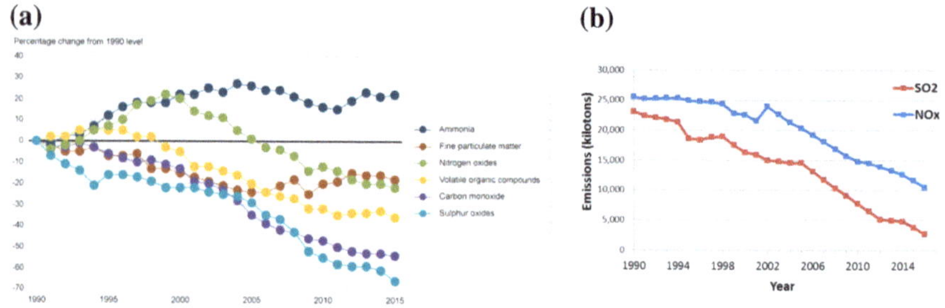

Fig. 41.1 Trends in annual **a** Canadian CAC emissions (normalized to 1990) for the 1990–2015 period and **b** U.S. SO_2 and NO_x emissions for the 1990–2016 period (*Source* [2, 3])

41.2 Description of Updated Emissions

By 2017 a 2013 Canadian APEI and a 2008 Mexican inventory for AQ modelling had become available as well as the first version of the 2014 U.S. NEI [1, 4]. In addition, the U.S. EPA had also developed several versions of a projected 2017 NEI for its 2011 Air Emissions Modeling Platform for policy development applications [5]. Given the continuing decline in Canadian and U.S. air pollutant emissions shown in Fig. 41.1, there was a strong argument for AQ forecasting in 2017 to move to inventories whose base years were more recent than 2010, 2011, and 1999. As well, the significant decrease in U.S. SO_2 and NO_x emissions that had occurred even between 2014 and 2016 suggested that an inventory for base year 2017 might be more appropriate for forecasting in 2017 and beyond than one for base year 2014, even though the former was a projected inventory whereas the latter was an actual inventory. One additional consideration was that a second version of the 2014 U.S, NEI was expected to be released at the end of 2017, so that the available version of the 2014 U.S. NEI was only a preliminary version.

Figure 41.2a compares the magnitude of total anthropogenic emissions (summed over point, area, on-road, and off-road source sectors) for six CAC species for version 1 of the 2013 Canadian APEI versus version 1 of the 2010 Canadian APEI. SO_2, NO_x, and VOC emissions were estimated to have decreased by 12%, 2%, and 4%, respectively, between 2010 and 2013 whereas NH_3, $PM_{2.5}$, and PM_{10} emissions increased by 5, 30, and 15%. Figure 41.2b shows a similar comparison for the magnitude of total anthropogenic CAC emissions for the projected 2017 inventory in the U.S. EPA 2011v6.3 emissions modeling platform versus the 2011 inventory in the EPA 2011v6.0 emissions modeling platform [5]. SO_2, NO_x, VOC, and $PM_{2.5}$ emissions were estimated to have decreased by 65%, 33%, 11%, and 1%, respectively, between 2011 and 2017, whereas PM_{10} emissions increased by 18% and NH_3 emissions were roughly unchanged.

The SMOKE emissions processing system was used to process these three new inventories to create a new set of input emissions files (named SET3.1) for the RAQDPS 10 km North American grid (for more details, see [8]). Figure 41.3 shows

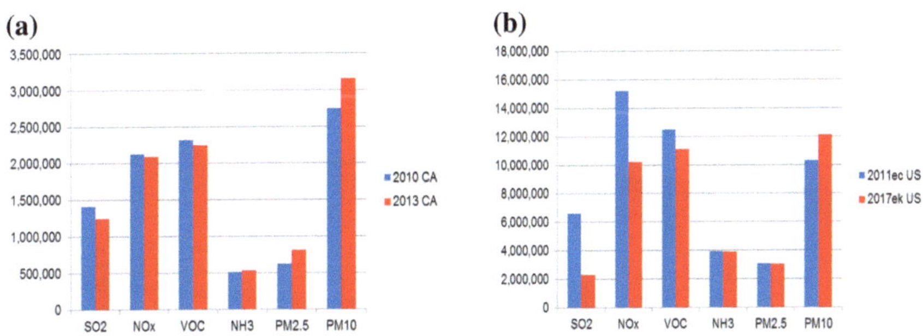

Fig. 41.2 Comparison of old and new **a** Canadian and **b** U.S. inventory emissions (tons/year)

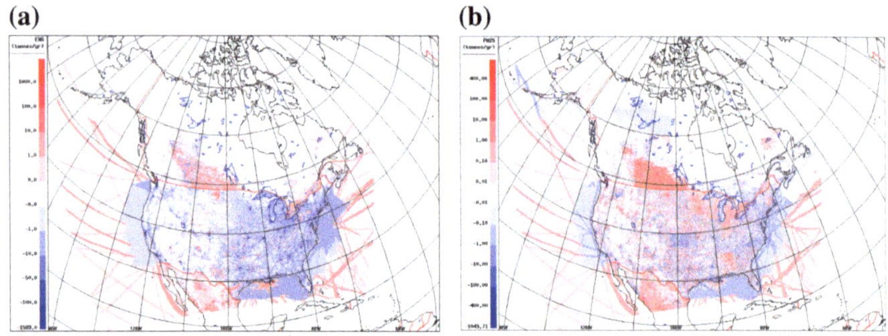

Fig. 41.3 Maps of differences in July emissions distribution (new–old; tonnes/grid cell/month) of **a** NO and **b** $PM_{2.5}$

two examples of the spatial distribution of differences between the SET2.1.1 and SET3.1 emissions files. In particular, NO (and NO_x) emissions decreased almost everywhere in the contiguous U.S. but not in Canada.

41.3 Impact of Updated Emissions

To evaluate the impact of using the new input emissions files versus the existing operational emissions files, the operational version of RAQDPS was re-run with the SET3.1 emissions files for two 6-week periods in summer 2016 (July 1–August 16) and winter 2017 (January 1–February 18).

Figure 41.4 shows the mean difference fields for NO_2, O_3, and $PM_{2.5}$ surface abundances between the RAQDPS runs using the two sets of emissions for the winter period (Moran et al. [7] for the summer plots). NO_2 levels decreased as a direct result of decreases in NO_x emissions (Fig. 41.3a) while O_3 levels increased due to decreased NO titration. $PM_{2.5}$ concentrations decreased over the continental U.S. due to decreased SO_2 and NO_x emissions (i.e., PM precursors) but increased over western Canada due to increased primary $PM_{2.5}$ emissions (Fig. 41.3b). Savic-Jovcic

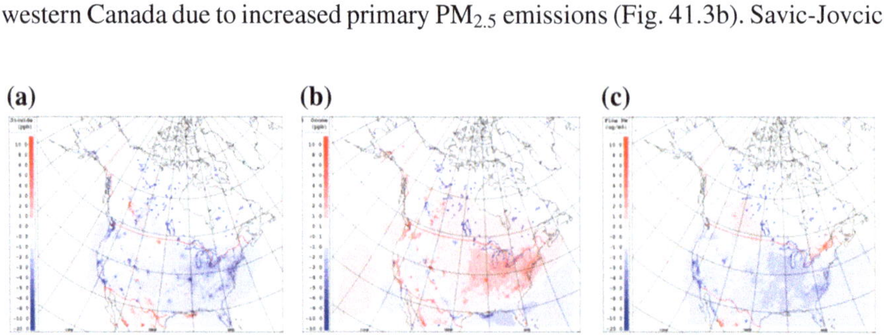

Fig. 41.4 Differences (new–old) of mean **a** NO_2 (ppbv), **b** O_3 (ppbv), and **c** $PM_{2.5}$ ($\mu g\ m^{-3}$) surface abundance fields for 45-day winter 2017 period

et al. [9] show mean 48 h observed and predicted NO_2 time series for eastern Canada (EC) and the eastern U.S. (EUS) for both the summer 2016 and winter 2017 periods and O_3 and $PM_{2.5}$ time series for summer 2016. NO_2 concentration bias is reduced over the EUS for both periods, suggesting that the new NO_x emissions from the 2017 projected U.S. NEI are more representative for 2017 than the 2011 NEI emissions, consistent with Fig. 41.1. Unlike the winter period shown in Fig. 41.4b, O_3 levels over the EUS are reduced in summer 2017 due to decreased production from lower NO_x emissions [7].

41.4 Conclusion

To make AQ forecasts we should ideally use current emissions, but national emissions inventories are generally retrospective. This becomes an issue when emissions are changing significantly from year to year, as has been the case in North America for the past decade or more.

Projected emissions inventories may be less accurate than actual inventories for the same base year, but they have the advantage of being available years sooner and they may also be quite accurate if projected emissions changes are largely the result of known emission control legislation. This study shows one example where a projected 2017 emissions inventory gave more representative estimates of emissions levels in 2016 and 2017 than did a retrospective inventory for 2011.

A new version of the ECCC forecast system that uses these updated input emissions files was accepted for operational implementation in mid 2018 [8].

References

1. AQMAS, Documentation for SMOKE-Ready 2013 and 2025 Air Pollutant Emission Inventory (APEI) Package version 1. Report number AQSC-17-003, Air Quality Modelling Applications Section, Environment and Climate Change Canada, Montreal, Quebec, April, 15 pp. (2017)
2. ECCC (2018), http://ec.gc.ca/indicateurs-indicators/default.asp?lang=en&n=E79F4C12-1
3. M. Moran, Q. Zheng, J. Zhang, R. Pavlovic, RAQDPS Version 013: upgrades to the CMC Operational Regional Air Quality Deterministic Prediction System released in June 2015. Canadian Meteorological Centre Technical Note, Montreal, November, 54 pp. (2015), http://collaboration.cmc.ec.gc.ca/cmc/cmoi/product_guide/docs/lib/op_systems/doc_opchanges/Technical_Note_GEM-MACH10_v1.5.3+SET2.1.1_Emissions_9Nov2015.pdf
4. M.D. Moran, Q. Zheng, J. Zhang, R. Pavlovic, M. Sassi, Impact on recent North American AQ forecasts of replacing a retrospective U.S. emissions inventory with a projected inventory, in *16th CMAS Conference*, 23–25 October 2017, Chapel Hill, North Carolina (2017), https://www.cmascenter.org/conference//2017/slides/moran_impact_recent_2017.pptx

5. M. Moran, R. Pavlovic, Regional Air Quality Deterministic Prediction System (RAQDPS): update from version 019 to version 020. Canadian Meteorological Centre Technical Note, September, Montreal, Quebec, 48 pp. (2018), http://collaboration.cmc.ec.gc.ca/cmc/cmoi/product_guide/docs/changes_e.html
6. U.S. EPA (2018), https://www.epa.gov/air-emissions-inventories/air-pollutant-emissions-trends-data
7. U.S. EPA (2018), https://www.epa.gov/air-emissions-inventories/2014-national-emissions-inventory-nei-data
8. U.S. EPA (2018), https://www.epa.gov/air-emissions-modeling/2011-version-6-air-emissions-modeling-platforms
9. V. Savic-Jovcic, M.D. Moran, R. Pavlovic, H. Landry, Q. Zheng, J. Zhang, A. Lupu, S. Ménard, A. Akingunola, S. Gravel, M. Sassi, D. Davignon, Overview of the 2018 Canadian operational regional air quality deterministic prediction system: new features and performance improvements. This volume (2019)

Chapter 42
Can Assimilation of Ground Particulate Matter Observations Improve Air Pollution Forecasts for Highly Polluted Area of Europe?

Małgorzata Werner, Maciej Kryza and Jakub Guzikowski

Abstract In this study we present the influence of assimilation of ground PM2.5 measurements on forecasted concentrations of particulate matters for low air quality episode observed in the year 2017 over Poland. The episode was not reproduced by a standard forecasting system, based on the Weather Research and Forecasting with Chemistry model (WRF-Chem), working operationally without data assimilation. Here, we used Grid point Statistical Interpolation (GSI) system to assimilate ground observations from 42 stations measuring PM2.5 concentrations. The results show that the assimilation of PM2.5 concentrations has a positive impact on modelled concentration of PM2.5 and PM10. The greatest positive impact is noticed for the period with the high measured concentrations of pollutants. The results also show that for some stations the assimilation of PM2.5 and PM10 may lead to overestimation of concentrations at the warmer period characterised by lower overall PM concentrations. Further study with an application of degree-day factors for residential emissions and GSI assimilation is planned as the next step.

42.1 Introduction

Central and Eastern European countries suffer from high concentrations of particulate matter (PM) in winter periods, which is mainly related to high emissions from SNAP sector 2 (residential combustion). Meteorological conditions i.e. low temperatures and anticyclone systems make favourable conditions to accumulation of air pollutants

M. Werner (✉) · M. Kryza · J. Guzikowski
Department of Climatology and Atmosphere Protection,
University of Wroclaw, Wroclaw, Poland
e-mail: malgorzata.werner@uwr.edu.pl

M. Kryza
e-mail: maciej.kryza@uwr.edu.pl

J. Guzikowski
e-mail: jakub.guzikowski@uwr.edu.pl

C. Mensink et al. (eds.), *Air Pollution Modeling and its Application XXVI*,
Springer Proceedings in Complexity,
https://doi.org/10.1007/978-3-030-22055-6_42

and deterioration of air quality. In such conditions it is very difficult to correctly predict air pollution concentrations.

Many operational air-quality models are initialized using concentrations of chemical species obtained from the previous day's forecast, without considering observations. Data assimilation has not been extensively applied to air quality models because chemical observations are relatively sparse compared to meteorological observations [5]. Data assimilation in chemical models may improve initial conditions and provides tools for better estimates of emission intensity.

In this study we aim to present the influence of assimilation of ground PM2.5 measurements on forecasted concentrations of particulate matter for low air quality episode in the year 2017 over Poland. The episode was not reproduced by forecasting system working operationally without data assimilation. The system is based on the Weather Research and Forecasting with Chemistry model (WRF-Chem) and produces forecasts daily, starting at 00 UTC and with 72 h lead time and uses chemistry cycling. We used Grid point Statistical Interpolation (GSI) system to assimilate ground observations from 42 stations measuring PM2.5 concentrations at hourly basis. The modelled concentrations of PM2.5 and PM10 were compared with observations and with the standard forecasts calculated without data assimilation.

42.2 Data and Methods

42.2.1 The WRF-Chem Model and the GSI System

In this study we used version 3.9 of WRF-Chem. We applied two one way nested domains—the outer domain covers Europe at a 12 km $\times$ 12 km grid and the inner domain is focused on Poland at 4 km $\times$ 4 km resolution. The mother domain has 285 and 332 points in the west-east and south-north direction, respectively. We used 35 vertical levels with the lowest layer top at about 30 m. The simulations are driven by the GFS meteorological data, available every 3 h, at a $0.5^\circ \times 0.5^\circ$ spatial resolution. The gas phase chemistry model used in this study was the Regional Acid Deposition Model, version 2 (RADM2, [4]) and the aerosol module included Goddard Chemistry Aerosol Radiation and Transport scheme (GOCART, [1]). Two emission databases were used in the simulations. For the outer domain we included TNO MACC III database at $1/8^\circ \times 1/16^\circ$, for the inner domain for SNAP sector 2 and SNAP sector 7 up to date emissions with improved emission factors were provided by the Chief Inspectorate of Environmental Protection.

We used the 3D-Var component of the GSI system. The system is described in details in [3] and here only the basis is provided. Analyses are obtained by minimizing a cost function:

$$J(x) = (x - x_b)^T B^{-1}(x - x_b) + (y - H(x))^T R^{-1}(y - H(x))$$

where x is a vector of analysis, x_b is the background (forecast) vector, y is an observation vector, B is the background error covariance matrix, H is an observation operator, and R is the observation error covariance matrix. The background error covariance matrix B is separated into vertical and horizontal components, and is represented as a product of error variances and spatial correlation matrices. The correlation matrices simulate Gaussian shapes in space and in the GSI are modelled with recursive filters. The observational data of PM2.5 concentrations, available at 1 h resolution, are provided by the Chief Inspectorate of Environmental Protection (http://powietrze.gios.gov.pl). There were 42 stations available for the study period.

42.2.2 The Simulation Period and Verification

The study period covers episode of a very low air quality and high concentrations of particulate matter in Poland on 11–25 February 2017. The episode was not reproduced by a standard forecasting system. More details are given in [2]. The most recent forecasts are available here: powietrze.uni.wroc.pl.

In this study we run and compared two simulations. The first simulation was initialised each day at 00 UTC, using the chemistry cycling from a previous forecast (GOCART). The second run the simulation was also initialised at 00 but the initial conditions were modified by incorporation of observational data by the GSI system (GOCART_GSI). Both forecasts were run for the 72 h (as a standard procedure in our system), however only first 24 h were analysed in this study. To verify the results we used factor of two (FAC2), mean bias (MB), mean gross error (MGE), normalised mean bias (NMB), and correlation coefficient (R).

42.3 Results

Mean observed concentrations of PM10 and PM2.5 were respectively 57.2 and 46.4 $\mu g\ m^{-3}$ during the period of 11–25 February 2017. Very high concentrations were observed in the first part of the analysed episode. PM2.5 concentrations reached or exceeded 150 $\mu g\ m^{-3}$ at many Polish stations between 13th and 18th of February (Fig. 42.1).

In general the WRF-Chem model underestimated observed PM2.5 concentrations. However, all statistics were improved after assimilation of ground based observations of PM2.5. MB decreased from -11.4 to -4.8 $\mu g\ m^{-3}$, FAC2 increased from 0.68 to 0.74 and R from 0.63 to 0.73 (Table 42.1). Time series in Fig. 42.1 shows that GOCART_GSI better fits in the observations during the peak of concentrations, however for the warmer period it overestimates concentrations at the Warsaw station. An improvement between GOCART_GSI and GOCART is also observed for PM10 but the changes are smaller than for PM2.5. FAC2 and R increased from 0.63 to 0.65 and from 0.56 to 0.63, respectively. The GOCART simulation underestimated observed

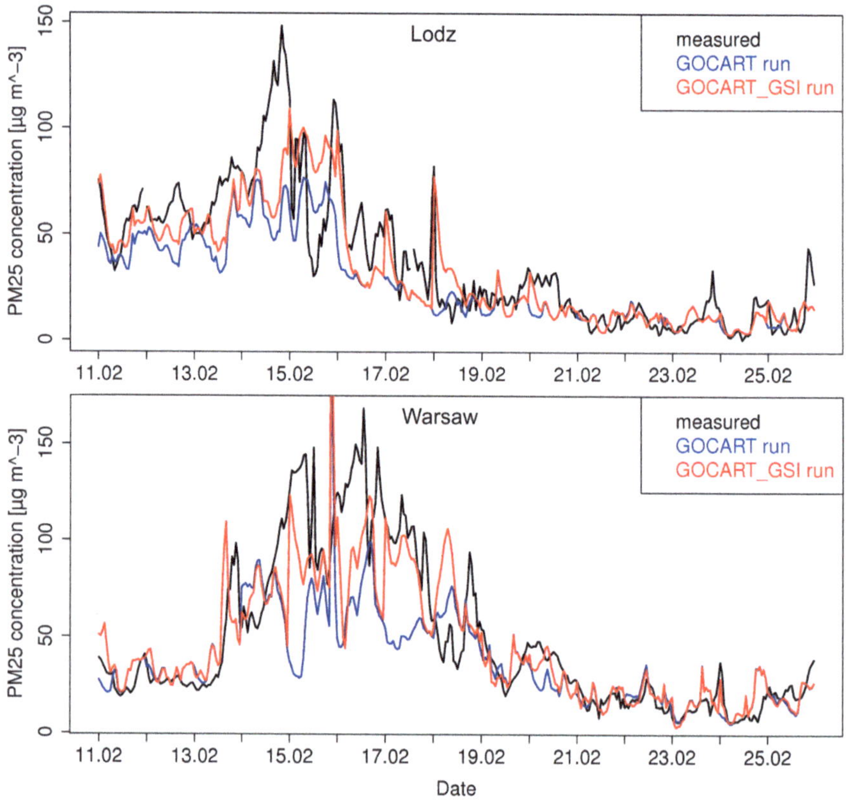

Fig. 42.1 Time series of modelled and measured PM2.5 concentrations at two selected stations—Lodz and Warsaw

Table 42.1 Statistics based on hourly observations for the GOCART and GOCART_GSI simulation. N-number of pairs model-observation

	N	FAC2	MB	MGE	NMB	R
PM2.5						
GOCART	15,038	0.68	− 11.36	24.14	− 0.23	0.63
GOCART_GSI	15,038	0.74	− 4.80	20.75	− 0.10	0.73
PM10						
GOCART	41,309	0.63	− 6.15	33.42	− 0.10	0.56
GOCART_GSI	41,309	0.65	1.38	32.01	0.02	0.63

PM10 concentrations with MB equal to −6.15 μg m^{-3}, while the GOCART_GSI slightly overestimated with MB equal to 1.38 μg m^{-3}.

Analysis of temporal changes of MB for GOCART_GSI for all stations indicates that the model underestimated peaks of very high concentrations of PM and slightly overestimated concentrations in the second (warmer) part of the episode (the figures not shown here).

42.4 Summary and Conclusions

The study with the WRF-Chem model and the GSI assimilation tool was carried out for the episode of very high concentrations of particulate matter followed by the period with lower PM concentrations in Poland on 11–25 February 2017. The results show that the assimilation of ground based measurements of PM2.5 concentrations has a positive impact on modelled concentration of PM2.5 and PM10. The greatest positive impact is noticed for the period with the high measured air pollution concentrations. The results also show that for some stations the GOCART_GSI simulation overestimates concentrations at the warmer period characterised by lower observed PM concentrations. Previous analysis done by [2] showed that overestimation of PM concentrations at warm winter periods can be related to too high emission factors for the residential emissions. Thus, further study with an application of degree-day factors for residential emissions and GSI assimilation is planned as the next step.

Acknowledgements The study was supported by the National Science Centre, Poland project no. UMO-2016/23/B/ST10/01797. We are grateful to the CIEP for providing the emissions data from the project "Supporting the air quality assessment system with application of modelling of PM10, PM2.5, SO_2, NO_2, B(a)P".

References

1. M. Chin, P. Ginoux, S. Kinne, O. Torres, B.N. Holben, B.N. Duncan, R.V. Martin, J.A. Logan, A. Higurashi, T. Nakajima, Tropospheric aerosol optical thickness from the GOCART model and comparisons with satellite and sun photometer measurements. J. Atmos. Sci. **59**, 461–483 (2002)
2. M. Kryza, M. Werner, M. Dore, Application of degree-day factors for residential emission estimate and air quality forecasting. Int. J. Environ. Pollut. In Review (2018)
3. M. Pagowski, G.A. Grell, S.A. McKeen, S.E. Peckham, D. Devenyi, Three-dimensional variational data assimilation of ozone and fine particulate matter observations: some results using the weather research and forecasting-chemistry model and grid-point statistical interpolation. Q. J. R. Meteorol. Soc. **136** (2010)
4. W.R. Stockwell, P. Middleton, J.S. Chang, X. Tang, The second generation regional acid deposition model chemical mechanism for regional air quality modeling. J. Geophys. Res. **95** (1990)
5. J.B. Wu, J. Xu, M. Pagowski, F. Geng, S. Gu, G. Zhou, Y. Xie, Z. Yu, Modeling study of a severe aerosol pollution event in December 2013 over Shanghai China: an application of chemical data assimilation. Particuology **20**, 41–51 (2015)

Chapter 43
Assimilation of Meteorological Data in Online Integrated Atmospheric Transport Model—Example of Air Quality Forecasts for Poland

Maciej Kryza, Małgorzata Werner and Jakub Guzikowski

Abstract In this work we analyse the impact of meteorological data assimilation on the performance of the air quality forecasting system for the area of Poland (Central Europe). The forecasting system uses the WRF-Chem model, which is online integrated meteorology and air chemistry transport model. The forecasts are run each day for the next 48 h, using two nested domains of 12 km × 12 km (Europe) and 4 km × 4 km (Poland) and 35 vertical levels. In this work we analyse the period of 11–25 February 2017, during which poor air quality was observed at the beginning, followed by unusually warm days with low concentrations of pollutants. Two sets of forecasts are compared. In the first group, we use the forecasts with no data assimilation. Secondly, we use the community Gridpoint Statistical Interpolation system (GSI) to assimilate surface and radiosonde meteorological data. Both sets of forecasts are compared with hourly measurements of PM10 and PM2.5 for Poland. Assimilation of meteorological data overall improves the air quality forecasts, but not always leads to better representation of high-concentration episode.

43.1 Introduction

Poland is among the several European countries that have severe problems with air quality. The number of days with exceeded threshold value for 24 h mean concentrations of PM10 each year is larger than allowed 35 at many locations [2]. Long-term exposure to atmospheric aerosols is often linked with health problem, including

M. Kryza (✉) · M. Werner (✉) · J. Guzikowski (✉)
Department of Climatology and Atmosphere Protection, Institute of Geography and Regional Development, Wrocław University, Wrocław, Poland
e-mail: Maciej.kryza@uwr.edu.pl

M. Werner
e-mail: malgorzata.werner@uwr.edu.pl

J. Guzikowski
e-mail: jakub.guzikowski@uwr.edu.pl

C. Mensink et al. (eds.), *Air Pollution Modeling and its Application XXVI*,
Springer Proceedings in Complexity,
https://doi.org/10.1007/978-3-030-22055-6_43

respiratory and cardiovascular diseases and increased mortality [4], not mentioning deterioration of life quality. The air quality forecasts are becoming more and more frequently used to support local authorities in reducing the population exposure.

In this work we present the results of the air quality forecasting system developed within the LIFE-APIS/PL project [6] for the area of Poland using the WRF-Chem model. The study period covers 11–25 February 2017, when high PM2.5 and PM10 concentrations were observed, but not always accurately forecasted by the system. The main aim of this work is to test if we can improve the performance of the forecasting system by assimilation of the conventional surface and radiosonde meteorological data using the Grid point Statistical Interpolation (GSI) system.

43.2 Data and Methods

43.2.1 Study Period

In this study we analyse the period of 11–25 February 2017. During the first half of this period, central Europe was under the influence of a high-pressure system, with the centre located over Poland and Belarus. Calm wind conditions were favourable for accumulation of atmospheric pollutants. Moreover, low air temperatures, observed especially over the night, lead to increased emission from low elevated residential sources. Hourly concentrations of PM2.5 and PM10 frequently exceeded 100 $\mu g\ m^{-3}$ at number of measuring sites. After 17th of February, advection of warm air masses from the south of Europe was observed. This lead to higher wind speeds and air temperatures exceeded the long term mean values for February. This resulted in a rapid decrease of PM2.5 and PM10 concentrations in Poland.

43.2.2 WRF-Chem Model Configuration

The Weather Research and Forecasting (WRF) model with chemistry (WRF-Chem), version 3.9, was used in this study [1]. The model was configured using two one-way nested domains with spatial resolutions of 12 km × 12 km and 4 km × 4 km. Both domains had 35 vertical levels. RADM2 chemical mechanism with the Kinetic Pre-Processor (KPP) for gas chemistry and GOCART scheme for aerosols were applied. For the outer domain covering Europe, the model uses the TNO MACC III emission inventory [3]. The same emission database was also used for the nested domain except for Poland, where high-resolution emission inventory, developed by the Chief Inspectorate of Environmental Protection (CIEP) was applied.

Two sets of forecasts were calculated. They both were initialised each day at 00 UTC. The base model runs were using the data from the Global Forecast System (GFS) as meteorological initial and boundary conditions. For the second set of the model forecasts, the initial conditions were modified by assimilation of the conventional surface land and radiosonde meteorological data, using the 3D-Var component of the Grid point Statistical Interpolation (GSI) tool [5]. For chemistry, the initial conditions came from chemistry cycling from a previous day forecast.

43.2.3 Evaluation of the Model Results

Two sets of PM2.5 and PM10 forecasts were compared with the hourly measurements gathered at a measuring network operated by CIEP in Poland. For PM2.5, measurements from 43 stations were available. For PM10, information from 116 sites were used for forecasts evaluation. Three error statistics were calculated to summarize the model performance: mean error (ME), index of agreement (IOA) and factor of two (FAC2).

43.3 Results

The model performance for the two sets of PM2.5 and PM10 forecasts is summarized in Table 43.1. Both model runs underestimate the observed concentrations of PM2.5 and PM10 for this period. Application of the data assimilation improves the forecasts both for PM2.5 and PM10. ME decreased, and IOA increased for the GSI run. The FAC2 statistic is similar for both model runs.

Figure 43.1 shows the details of the model performance for the entire period and two selected stations. For Wrocław, there are only small differences between the base and GSI runs. The GSI run is slightly better at the beginning of the high-concentration period (14–15 February), where the base run shows rapid decrease in PM2.5 concentrations during the day. However, both the base and GSI simulations underestimate PM2.5 (and PM10, not presented) for these days. For Kraków, assimilation of conventional meteorological data improved the model performance for PM2.5 and PM10 significantly for the beginning of the high-concentration episode (14th and 15th of

Table 43.1 Performance of the WRF-Chem model forecasts for the base and GSI runs

	ME [$\mu g\ m^{-3}$]	IOA [−]	FAC2 [−]
PM2.5 concentrations			
Base run	−14.53	0.68	0.81
GSI run	−13.81	0.70	0.81
PM10 concentrations			
Base run	−11.55	0.66	0.56
GSI run	−10.18	0.68	0.57

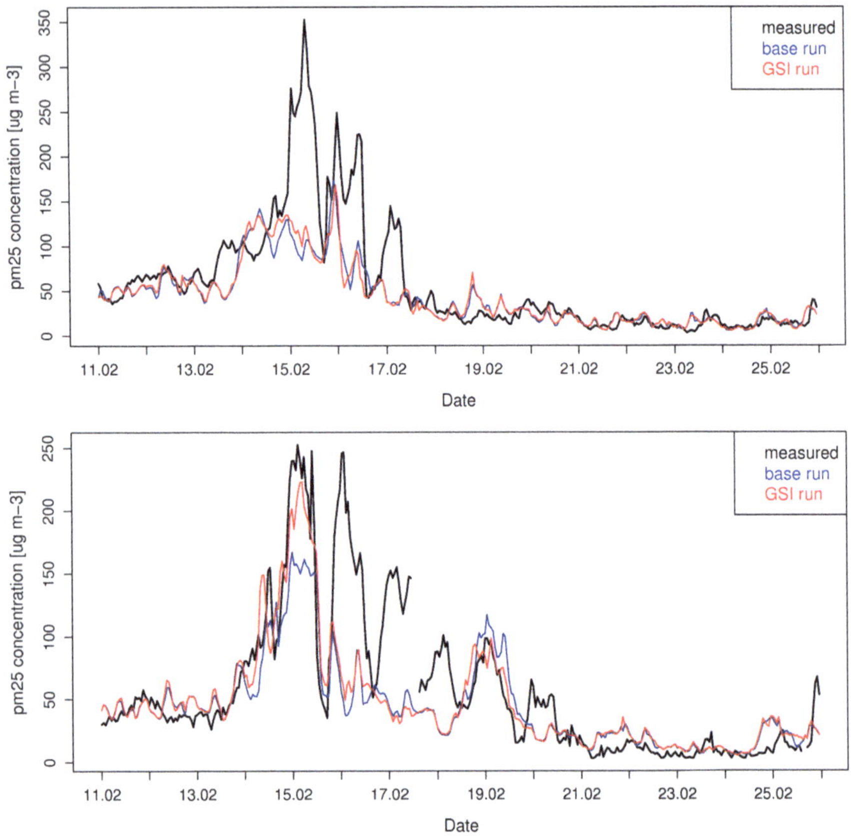

Fig. 43.1 Observed and modelled hourly PM2.5 concentrations at Wrocław (top) and Kraków (bottom)

February). However, the pollutants concentrations are still underestimated for 16–18 of February.

If the model performance is summarized for all the stations according to the forecast lead time, it is visible that the impact of the data assimilation is the strongest at the beginning of the simulation (Fig. 43.2). For IOA, the base and GSI runs become very similar after 12 h of the simulations, but again, after 24 h IOA is larger for the GSI run. Similar pattern is observed for ME—the largest decrease for the GSI run is observed for the first 12 h. After that both simulations show similar performance until 24 h lead time. Then ME is higher for the GSI run.

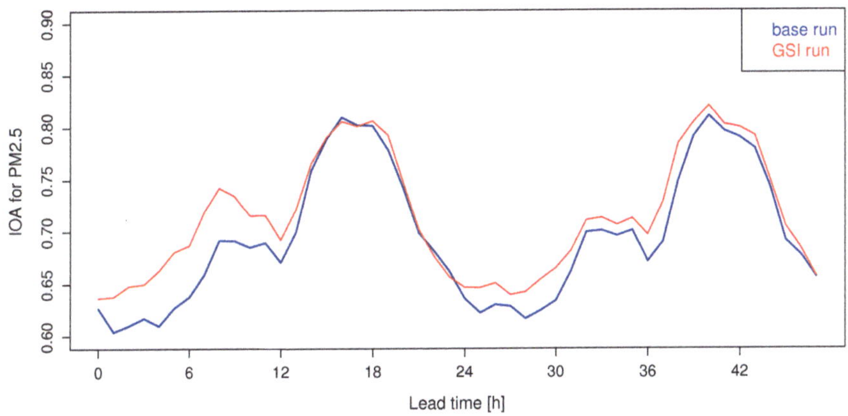

Fig. 43.2 Model performance according to the forecast lead time (all stations)

43.4 Summary and Conclusions

In this work we have shown that the application of meteorological data assimilation may improve the performance of the air quality forecasts of PM2.5 and PM10. Despite the overall improvement, the model simulation with data assimilation still failed to reproduce some peak values of observed PM2.5 and PM10. The results presented here are preliminary and should be further tested for the longer study periods.

Acknowledgements The study was supported by the National Science Centre, Poland project no. UMO-2016/23/B/ST10/01797. We are grateful to the CIEP for providing the emissions data from the project "Supporting the air quality assessment system with application of modelling of PM10, PM2.5, SO_2, NO_2, B(a)P".

References

1. G.A. Grell, S.E. Peckham, R. Schmitz, S.A. McKeen, G. Frost, W.C. Skamarock, B. Eder, Fully coupled "online" chemistry within the WRF model. Atmos. Environ. **39**, 6957–6975 (2005). https://doi.org/10.1016/j.atmosenv.2005.04.027
2. J. Krynicka, A. Drzeniecka-Osiadacz, Analysis of variability in PM 10 concentration in the Wrocław agglomeration. Polish J. Environ. Stud. **22**(4), 1091–1099 (2013)
3. J.J.P. Kuenen, A.J.H. Visschedijk, M. Jozwicka, H.A.C. Denier van der Gon, TNO-MACC_II emission inventory; a multi-year (2003–2009) consistent high-resolution European emission inventory for air quality modelling. Atmos. Chem. Phys. **14**(20), 10963–10976 (2014). https://doi.org/10.5194/acp-14-10963-2014
4. C.A. Pope, D.W. Dockery, Health effects of fine particulate air pollution: lines that connect. J. Air Waste Manag. Assoc. **56**(6), 709–742 (2006)
5. H. Shao, J. Derber, X.-Y. Huang, M. Hu, K. Newman, D. Stark, M. Lueken, C. Zhou, L. Nance, Y.-H. Kuo, B. Brown, Bridging research to operations transitions status and plans of community GSI. Bull. Am. Meteorol. Soc. **97**(8) (2016). https://doi.org/10.1175/bams-d-13-00245.1

6. M. Werner, M. Kryza, H. Ojrzyńska, C.A. Skjøth, K. Wałaszek, A.J. Dore, Application of WRF-Chem to forecasting PM10 concentration over Poland. Int. J. Environ. Pollut. **58**(4) (2015). https://doi.org/10.1504/ijep.2015.077458

Chapter 44
Distinguishing Between Remote and Local Air Pollution Over Taiwan: An Approach Based on Pollution Homogeneity Analysis

Pavel Kishcha, Sheng-Hsiang Wang, Neng-Huei Lin, Arlindo da Silva, Tang-Huang Lin, Po-Hsiung Lin, Gin-Rong Liu, Boris Starobinets and Pinhas Alpert

Abstract An analysis of pollution homogeneity has been conducted to distinguish between remote and local pollution which contributes to month to month changes in aerosol optical depth (AOD) over the Taiwan area. This was carried out using both AERONET measurements at six monitoring sites in Taiwan and NASA MERRA-2 reanalysis, over the 15-year period from 2002 to 2017. As a measure of air pollution homogeneity we used the AOD standard deviation: the more homogeneous the spatial distribution of air pollution, the lower the AOD standard deviation is. Using this approach, we found that, over Taiwan, in autumn, inhomogeneous local air pollution is predominant, while, in spring, homogeneous remote air pollution from south-east Asia dominates. In autumn, when inhomogeneous aerosols from local sources dominate, the AOD standard deviation is essentially higher than that in spring, when homogeneous aerosols from remote sources dominate. Our approach allowed us to distinguish between homogeneous remote and inhomogeneous local sulfate air pollution of similar optical properties and chemical composition.

P. Kishcha (✉) · B. Starobinets · P. Alpert
Department of Geophysics, Tel-Aviv University, Tel-Aviv, Israel
e-mail: pavel@cyclone.tau.ac.il

S.-H. Wang · N.-H. Lin · T.-H. Lin · G.-R. Liu
Department of Atmospheric Sciences, National Central University, Taoyuan, Taiwan

A. da Silva
NASA Goddard Space Flight Center, Greenbelt, MD, USA

P.-H. Lin
National Taiwan University, Taipei, Taiwan

C. Mensink et al. (eds.), *Air Pollution Modeling and its Application XXVI*,
Springer Proceedings in Complexity,
https://doi.org/10.1007/978-3-030-22055-6_44

44.1 Introduction

Taiwan is an island located at approximately 200 km eastward from the Pacific coast of South East Asia. Both remote air pollution from highly polluted south-east Asia (including emissions of bio-mass burning) and local air pollution from Taiwanese industries and transportation contribute to the distribution of aerosol optical depth (AOD) over the Taiwan area. Distinguishing between remote and local air pollutants is important for raising awareness about their separate effects on regional air quality, weather and climate. To this end, an approach has been developed, based on the homogeneity of both AOD data from NASA MERRA-2 aerosol reanalysis and AERONET AOD measurements [2].

44.2 Method

According to Kishcha et al. [2], AOD and its standard deviation can be used as a measure of air pollution homogeneity. To estimate month to month changes in air pollution homogeneity over Taiwan, we used AOD monthly data from six AERONET stations during the 15-year period from 2002 to 2017 (Fig. 44.1a). The analysis of AERONET AOD was complemented with the analysis of MERRA-2 reanalysis AOD data during the 15-year study period. MERRA-2 reanalysis includes AOD of the following aerosol species: black and organic carbon, sulfate, desert dust, and sea salt [1]. MERRA-2 is based on the up-to-date NASA GEOS-5 model with aerosol data assimilation. According to Randles et al. [3] and Buchard et al. [1], for the

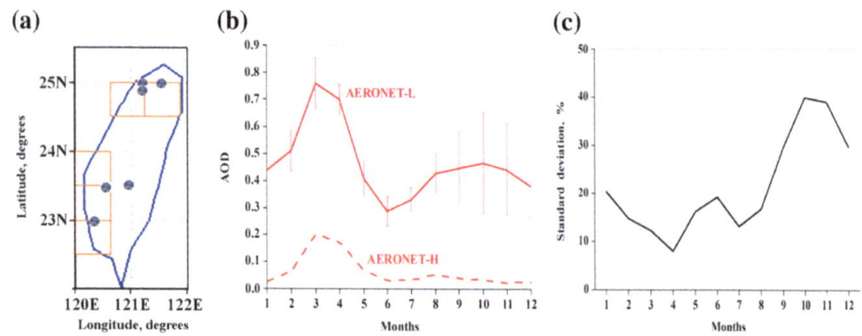

Fig. 44.1 **a** The Taiwan area, where the closed circles show the location of AERONET sites, while the orange open boxes represent GEOS-5 model grid boxes located in the vicinity of AERONET sites. Each GEOS-5 model grid box is 0.5° latitude by 0.625° longitude; **b** month to moth changes in 15-year mean AERONET AOD (2002–2017). AERONET-L designates AERONET AOD averaged over the five low-elevated AERONET sites, while AERONET-H designates AOD from the high-elevated (2800 m a.s.l.) Lulin site. The AOD standard deviation is shown by the vertical red lines; **c** month to month changes in the AERONET AOD standard deviation from the five low-elevated sites, in percentage to the 15-year mean AOD (Updated from Kishcha et al. [2])

aerosol data assimilation GEOS-5 uses AOD retrievals from the Multiangle Imaging SpectroRadiometer (MISR), the Moderate Resolution Imaging Spectroradiometer (MODIS) on both NASA Terra and Aqua satellites; the Advanced Very High Resolution Radiometer (AVHRR) instruments over bright surfaces, and ground-based AERONET direct measurements. Month to month changes in MERRA AOD were analyzed over the following two areas: the open ocean area in the vicinity of Taiwan (25.5 N–29 N; 122.5E–124.5E) and the Taiwan area (22 N–25.5 N; 120E–122E). In addition, a spatial distribution of wind vectors in the 700–800 hPa layer was analyzed, using 15-year monthly mean MERRA data.

44.3 Results and Discussion

Month to month changes in AERONET AOD measurements in Taiwan show the primary maximum (AOD $\approx$ 0.75) in spring (March–April) and the secondary maximum (AOD $\approx$ 0.45) in autumn (September–October) (Fig. 44.1b). AOD from the high-elevated site in Lulin is essentially lower than AOD from the low-elevated AERONET monitoring sites. Therefore, air pollution in Taiwan is mainly vertically distributed below 2800 m a.s.l., in line with Kishcha et al. [2]. As shown in Fig. 44.1b, the standard deviation of AERONET AOD in spring is significantly lower than that in autumn, indicating the more homogeneous AOD spatial distribution in spring than in autumn. Therefore, aerosols originating from remote sources were more homogeneously distributed over Taiwan than local aerosols [2]. As a measure of air pollution homogeneity, we used the AOD standard deviation: the more homogeneous the spatial distribution of air pollution is, the lower the AOD standard deviation is. To quantify changes in aerosol homogeneity over Taiwan over the year, we analyzed month to month changes in the AERONET AOD standard deviation in percentage to the 15-year mean AOD (Fig. 44.1c). One can see that, in autumn (October), the AOD standard deviation is essentially higher than that in spring (March–April) (Fig. 44.1c): this fact points out that, in autumn, inhomogeneous aerosols from local sources are predominant, while, in spring, homogeneous aerosols from remote sources dominate.

The above mentioned finding is supported by the analysis of spatial distributions of 15-year monthly mean MERRA wind vectors in the 700–800 hPa layer [2], their Fig. 7). In particular, Kishcha et al. [2] showed that, in the spring season (March), prevailing strong west and south-west winds blow mainly from land to sea, causing transport of anthropogenic air pollution (including biomass burning) from its sources in continental Asia towards Taiwan. Therefore, in spring, aerosols from remote sources are likely to be predominant. By contrast, in autumn (October), weak east winds are observed over Taiwan, indicating insignificant transport of continental air pollution towards Taiwan [2]. Therefore, in autumn, aerosols from local sources are likely to dominate.

To support our findings about the origin of air pollution responsible for the spring and autumn AOD maxima, we conducted a comparison of month to month changes in MERRA AOD over the open ocean area in the vicinity of Taiwan (25.5 N–29 N;

122.5E–124.5E), where there are no local sources of anthropogenic aerosols, and over the Taiwan area (22 N–25.5 N; 120E–122E), where there are local sources of anthropogenic aerosols (Fig. 44.2). For sulfate aerosols, which are the major pollutant in Taiwan, MERRA AOD peaked in both spring and autumn over Taiwan, but not over the open ocean, where only the spring maximum exists (Fig. 44.2, the magenta lines). This is evidence of the contribution of local sulfate aerosols to the maximum in autumn over Taiwan. By contrast to sulfates, there are no local sources of carbonaceous aerosols (OC & BC) neither in the Taiwan area no in the open ocean area in the vicinity of Taiwan. MERRA AOD for carbonaceous aerosols (OC & BC) shows similar seasonal variations with only one maximum in spring over the two areas (Fig. 44.2, green lines). This indicates that remote carbonaceous aerosols dominate in spring over Taiwan.

Furthermore, our approach allowed us to distinguish between homogeneous remote and inhomogeneous local sulfate air pollution of similar optical properties and chemical composition. To this end we analyzed sulfate aerosol homogeneity over Taiwan using month to month changes in MERRA sulfate AOD and its standard deviation. MERRA sulfate AOD was averaged over the five model grid boxes of 0.5° latitude by 0.625° longitude each, located in the vicinity of the low-elevated AERONET sites (Fig. 44.3). One can see that the standard deviation of MERRA sulfate AOD in autumn was higher than that in spring (Fig. 44.3). This is similar to AERONET AOD (Fig. 44.1). Therefore, inhomogeneous sulfate aerosols from local sources are predominant in autumn, while homogeneous sulfate aerosols from remote sources dominate in spring. The same approach can be applied to distinguish between remote and local air pollution over different land areas on a global scale.

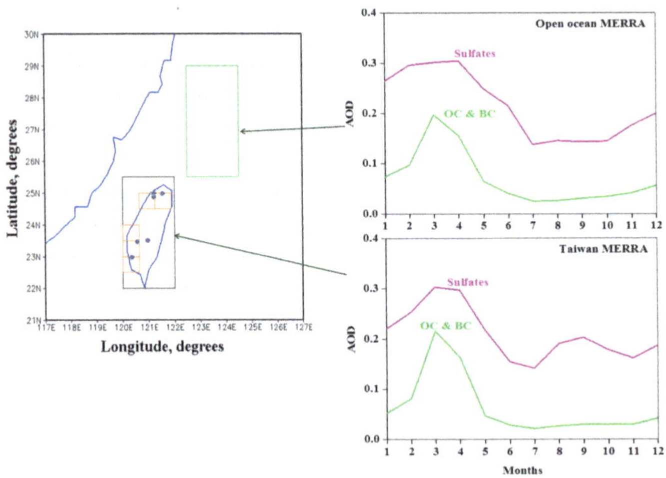

Fig. 44.2 Month to month changes in MERRA AOD over Taiwan and over the open sea area (the green open rectangle in the left panel) in the vicinity of Taiwan, over the 15-year period from 2002–2017. The green lines designate organic and black carbon (OC & BC), while the magenta lines designate sulfate aerosols (Updated from Kishcha et al. [2])

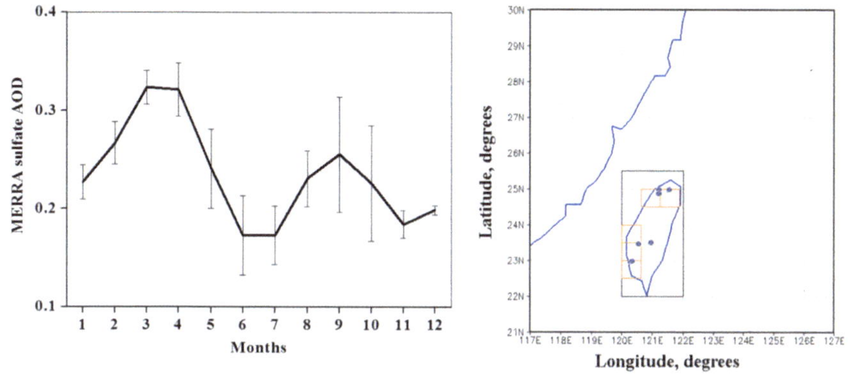

Fig. 44.3 The left panel shows month to month changes in 15-year mean MERRA sulfate AOD over Taiwan. MERRA sulfate AOD was averaged for the five specified GEOS-5 grid boxes of 0.5° latitude by 0.625° longitude each, located in the vicinity of AERONET sites (the orange open boxes on the right panel). The standard deviation of AOD is shown by the black vertical lines. The right panel represents the study area (Updated from Kishcha et al. [2])

References

1. V. Buchard, A. Randles, A.M. da Silva, A. Darmenov, P.R. Colarco, R. Ggovindaraju, R. Ferrare, J. Hair, A.J. Beyersdorf, L.D. Ziemba, H. Yu, The MERRA-2 Aerosol Reanalysis, 1980 Onward. Part II: evaluation and case studies. J. Clim. **30**, 6851–6872 (2017). https://doi.org/10.1175/JCLI-D-16-0613.1
2. P. Kishcha, S.-H. Wang, N.-H. Lin, A.M. da Silva, T.-H. Lin, P.-H. Lin, G.-R. Liu, B. Starobinets, P. Alpert, Differentiating between local and remote pollution over Taiwan. Aerosol Air Qual. Res. **18**, 1788–1798 (2018). https://doi.org/10.4209/aaqr.2017.10.0378
3. C.A. Randles, A.M. da Silva, V. Buchard, P.R. Colarco, A. Darmenov, R. Govindaraju, A. Smirnov, B. Holben, R. Ferrare, J. Hair, Y. Shinozuka, C.J. Flynn, The MERRA-2 Aerosol Reanalysis, 1980 Onward. Part I: system description and data assimilation evaluation. J. Clim. **30**, 6823–6850 (2017). https://doi.org/10.1175/JCLI-D-16-0609.1

Chapter 45
Overview of the 2018 Canadian Operational Regional Air Quality Deterministic Prediction System: New Features and Performance Improvements

Verica Savic-Jovcic, Michael Moran, Radenko Pavlovic, Hugo Landry, Qiong Zheng, Junhua Zhang, Alexandru Lupu, Sylvain Ménard, Ayodeji Akingunola, Sylvie Gravel, Mourad Sassi and Didier Davignon

Abstract Since the last ITM in October 2016, the Canadian operational Regional Air Quality Deterministic Prediction System (RAQDPS) has been ported to a new high-performance computing system and has been updated to use a new meteorological initialization method, a new meteorological "piloting" model, a new and faster version of the GEM-MACH code, and a new set of input emissions files and to produce an expanded set of output fields. These updates are briefly described and some examples are given of their impact on RAQDPS forecast performance, including improved NO_2 forecasts and a large reduction (~10 ppbv) in summertime ozone overpredictions for the eastern United States.

45.1 Introduction

Environment and Climate Change Canada's (ECCC) Regional Air Quality Deterministic Prediction System (RAQDPS) became operational in 2001. Since then, the RAQDPS has switched to a completely new chemical transport model (CTM) and undergone frequent updates to improve its capabilities and performance. The current system is built on a limited-area version of the GEM-MACH (Global Environmental Multi-scale—Modeling Air quality and Chemistry) CTM, which consists of an in-line chemistry model embedded in ECCC's multi-scale meteorological forecast model, GEM. GEM-MACH's forecast grid covers North America with horizontal

V. Savic-Jovcic (✉) · M. Moran · Q. Zheng · J. Zhang · A. Lupu · A. Akingunola
Air Quality Research Division, Environment and Climate Change Canada, Toronto, ON, Canada
e-mail: verica.savic-jovcic@canada.ca

R. Pavlovic · H. Landry · S. Ménard · M. Sassi · D. Davignon
Air Quality Modelling Applications Section, ECCC, Montreal, QC, Canada

S. Gravel
Air Quality Research Division, ECCC, Montreal, QC, Canada

C. Mensink et al. (eds.), *Air Pollution Modeling and its Application XXVI*,
Springer Proceedings in Complexity, https://doi.org/10.1007/978-3-030-22055-6_45

grid spacing of 10 km and 80 hybrid vertical levels. The RAQDPS is run twice per day to provide 48 h forecasts of Canada's national Air Quality Health Index (AQHI).

The AQHI has a range from 0 to 10+ and is a year-round, additive, no-threshold, hourly, health-based AQ index. It was developed from daily time-series analysis of air pollutant concentrations and mortality data and is based on a weighted sum of NO_2, O_3, and $PM_{2.5}$ concentrations [7]. RAQDPS forecasts provide point-specific values of NO_2, O_3, and $PM_{2.5}$ concentrations to UMOS-AQ, a statistical post-processing package for bias correction that can both compensate for systematic model errors and account for unresolved subgrid-scale phenomena at selected points, namely AQ station locations. RAQDPS forecasts are also used as first-guess fields for operational objective analyses of hourly North American surface concentration fields for O_3, NO_2, $PM_{2.5}$, PM_{10}, and SO_2 [5].

45.2 Recent Updates to the RAQDPS

A new high-performance computing (HPC) system to support ECCC forecasting was installed in 2017. The heart of the new HPC system consists of a pair of Cray XC40 computers, each with 30,780 cores, which replaced two IBM Power7-based massively parallel supercomputers. Before this major computing system transition a decision was made that no new forecast model versions should be introduced into operations during the transition. The version of the RAQDPS that was operational at the beginning of 2017, version 018, went into service in autumn 2016 [2]. It thus continued to run on the IBM supercomputers until autumn 2017, when it was replaced by equivalent code (version 019) that had been ported to the Cray supercomputers.

During 2017, work also continued in parallel to develop a new version of the RAQDPS for implementation after the HPC transition. This new version, version 020, has many changes and upgrades, including the following five major improvements:

(1) The RAQDPS follows the same meteorological initialization procedure as the regional configuration of the GEM meteorological forecast model, the Regional Deterministic Prediction System (RDPS; [1]). RDPS5, the pre-transition operational RDPS version, employs an intermittent data assimilation cycle and a digital filter, but RDPS6, the new RDPS version, employs a continuous data assimilation cycle and an incremental analysis update (IAU) approach. RAQDPS020 has also adopted this new initialization procedure.
(2) Meteorological lateral boundary conditions (LBCs), sometimes referred to as "piloting files", are provided to the limited-area RAQDPS every hour by the RDPS. RAQDPS020 has adopted the RDPS6 as its piloting model in place of RDPS5.
(3) The GEM-MACH code used in the RAQDPS has been updated from v2.2.0 to v2.3.1 for RAQDPS020. Several features have been added: (a) modifications to the gas-phase dry deposition module for snow-covered surfaces and sea ice; (b) a new capability to turn off chemistry calculations at upper levels; and (c)

addition of particle ammonium and nitrate emissions. Many code changes were also made to fix minor bugs and to make the code faster (modified inorganic heterogeneous chemistry module, improved input of chemical LBCs, turning off most chemistry in stratosphere), more robust (e.g., improved thread safety), and cleaner (e.g., removal of unused variables).

(4) New input emissions files were generated from three more recent national emissions inventories: a 2013 Canadian inventory; a projected 2017 U.S. inventory; and a 2008 Mexican inventory. Due to this change, SO_2, NO_x, and VOC emissions are lower in both Canada and the U.S. Details are given in Moran et al. [3].

(5) Nineteen hourly surface dry deposition fields and eleven hourly wet deposition fields were added to RAQDPS020 as routine output fields. The availability of these output fields will allow seasonal and annual wet deposition fields to be calculated at the end of each forecast year, and these fields can in turn be used by the new ECCC ADAGIO tool (Atmospheric Deposition Analysis Generated by optimal Interpolation from Observations; [6]) that is being developed for objective analysis of dry, wet, and total deposition.

More details can be found in Moran and Pavlovic [4].

45.3 Impact of Recent Updates

The new RAQDPS020 was run for a six-week summer period in 2016 (July 1–Aug. 14) and a six-week winter period in 2017 (Jan. 1–Feb. 14) for which RAQDPS019 results were already available. Figures 45.1 and 45.2 show mean 48-h variations of composite surface NO_2 volume mixing ratios (VMR) at measurement stations in eastern Canada (EC) and the eastern U.S. (EUS) for the summer and winter periods, respectively. NO_2 emissions were 2% lower in Canada and 33% lower in the U.S. for the RAQDPS020 runs [3]. The impact of these emissions decreases can be seen in both figures, and the NO_2 overprediction is reduced in both EC and EUS in the winter and in the EUS in the summer.

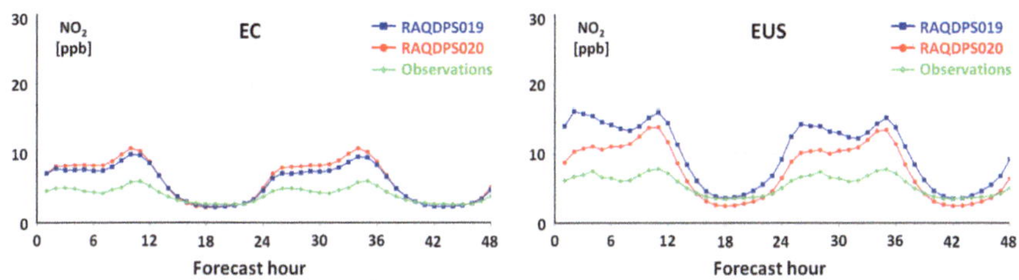

Fig. 45.1 Mean 48-h NO_2 VMR time series (ppbv) for the 45-day summer 2016 period for EC and EUS

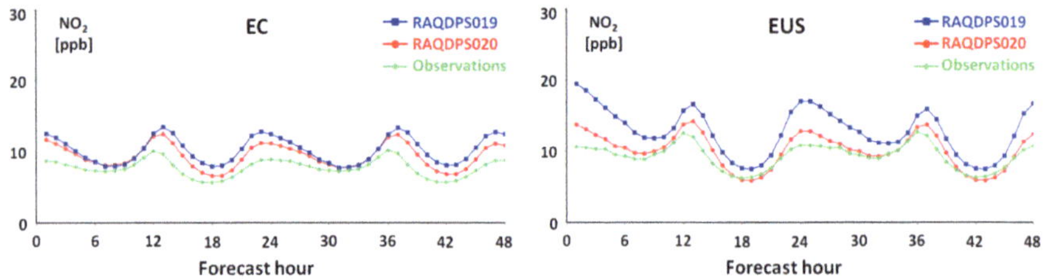

Fig. 45.2 Mean 48-h NO_2 VMR time series (ppbv) for the 45-day winter 2017 period for EC and EUS

The decreases in surface NO_2 levels are accompanied in summer 2016 by decreases in surface O_3 levels, especially in the EUS where mean daytime O_3 VMR decreased by as much as 10 ppbv (Fig. 45.3).

Figure 45.4 shows comparable composite time series for summer 2016 for surface $PM_{2.5}$ concentration. This is a puzzling result at first glance because mean $PM_{2.5}$ levels decreased in both EC and EUS in summer 2016 for RAQDPS020 even though primary $PM_{2.5}$ emissions were higher in Canada and roughly unchanged in the U.S. [3]. This reduction results from lower levels of secondary particle sulphate (p-SO_4) due to the large decreases in SO_2 emissions in the new input emissions files in both Canada (12%) and the U.S. (65%). Interestingly, these decreases in p-SO_4 were much smaller in the winter 2017 period (not shown) when gas-phase conversion of SO_2 to p-SO_4 is much slower.

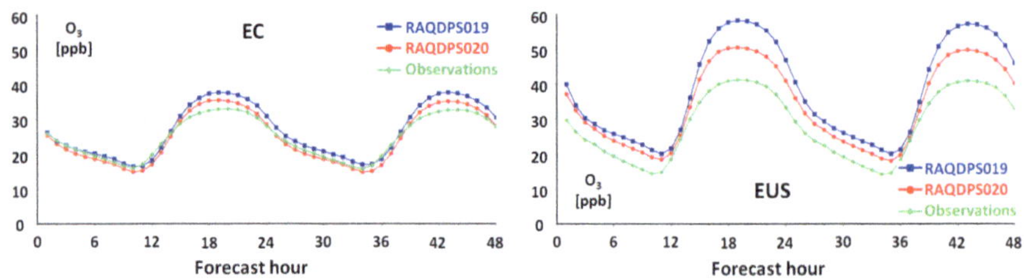

Fig. 45.3 Mean 48-h O_3 VMR time series (ppbv) for the 45-day summer 2016 period for EC and EUS

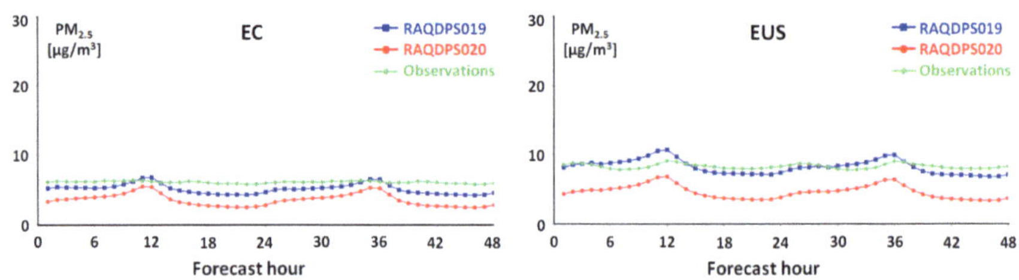

Fig. 45.4 Mean 48-h $PM_{2.5}$ concentration time series ($\mu g\ m^{-3}$) for the 45-day summer 2016 period for EC and EUS

Analysis of additional sensitivity tests (not shown) in which only one major change was introduced at a time (e.g., new RDPS6 only, new GEM-MACH code only, new input emissions files only) suggests that the differences seen in Figs. 45.1, 45.2, 45.3 and 45.4 arise primarily from the emissions changes. Changes to the GEM-MACH code had some impact, particularly for surface O_3 levels in the winter, whereas switching to the new RDPS version had only minor impacts.

45.4 Summary and Conclusion

A number of significant changes were made to the ECCC RAQDPS in 2018. Test runs for summer 2016 and winter 2017 periods suggested that the new version (RAQDPS020) outperforms the previous operational version (RAQDPS019), including improved NO_2 and O_3 forecasts. The performance of this new version was also evaluated for a 2018 period, and the RAQDPS020 subsequently became the new ECCC operational AQ forecast system in September 2018 [4].

References

1. J.-F. Caron, T. Milewski, M. Buehner, L. Fillion, M. Reszka, S. Macpherson, J. St-James, Implementation of deterministic weather forecasting systems based on ensemble-variational data assimilation at Environment Canada. Part II: the regional system. Mon. Wea Rev. **143**, 2560–2580 (2015)
2. M.D. Moran, A. Lupu, J. Zhang, V. Savic-Jovcic, S. Gravel, A comprehensive performance evaluation of the next generation of the Canadian operational regional air quality deterministic prediction system. In: Air Pollution Modeling and its Application XXV, Springer, pp. 75–81 (2017). https://doi.org/10.1007/978-3-319-57645-9
3. M.D. Moran, Q. Zheng, J. Zhang, R. Pavlovic, M. Sassi, Importance of inventory representativeness for air quality forecasting: a recent North American example. This volume (2019)
4. M.D. Moran, R. Pavlovic, Regional Air Quality Deterministic Prediction System (RAQDPS): Update from version 019 to version 020. Canadian Meteorological Centre Technical Note, September, Montreal, 48 pp (2018). http://collaboration.cmc.ec.gc.ca/cmc/cmoi/product_guide/submenus/raqdps_e.html
5. A. Robichaud, R. Ménard, Y. Zaitseva, D. Anselmo, Multi-pollutant surface objective analyses and mapping of air quality health index over North America. Air Qual. Atmos. Health **9**, 743–759 (2016)
6. A. Robichaud, A. Cole, M.D. Moran, A. Lupu, M. Shaw, G. Roy, M. Beauchemin, V. Fortin, R. Vet, Total deposition maps evaluated from measurement-model fusion in North America (ADAGIO project). This volume (2019)
7. D.M. Stieb, R.T. Burnett, M. Smith-Dorion, B. Orly, H.H. Shin, V. Economou, A new multipollutant, no-threshold air quality health index based on short term associations observed in daily time-series analyses. J Air Waste Manage Assoc **58**, 435–450 (2008). https://doi.org/10.3155/1047-3289.58.3.435

Chapter 46
PROGNOS: A Meteorological Service of Canada (MSC) Initiative to Renew the Operational Statistical Post-processing Infrastructure

Stavros Antonopoulos, Christian Saad, Jacques Montpetit, Andrew Teakles and Jonathan Baik

Abstract A new MSC initiative, named PROGNOS, aims to provide a more versatile, modular and innovative weather and air quality post-processing system to replace the current operational system (UMOS). PROGNOS has extensible statistical modeling capabilities. Currently in development, it issues real-time experimental air quality and temperature forecasts for cities across Canada and will eventually be applied to other meteorological fields and numerical models. The batch updates of the statistical models occur weekly using parallel processing in a cluster computing environment. Less flexible but more computationally efficient, online updating methods are also being evaluated. Several statistical modeling approaches have been explored including multiple linear regression, random forest, and Kalman filter prototypes for air quality forecasts. Logging, parameterisation, diagnostic and visualization features are also being explored. Medium to long term milestones include integrating seasonal and other transitional schemes as well as gridded post-processing.

S. Antonopoulos (✉) · C. Saad · J. Montpetit
Meteorological Service of Canada (MSC), Environment and Climate Change Canada (ECCC), Montreal, QC, Canada
e-mail: stavros.antonopoulos@canada.ca

C. Saad
e-mail: christian.saad@canada.ca

J. Montpetit
e-mail: jacques.montpetit@canada.ca

A. Teakles
Meteorological Service of Canada, ECCC, Dartmouth, NS, Canada
e-mail: rew.teakles@canada.ca

J. Baik
Meteorological Service of Canada, ECCC, Vancouver, BC, Canada
e-mail: jonathan.baik@canada.ca

C. Mensink et al. (eds.), *Air Pollution Modeling and its Application XXVI*, Springer Proceedings in Complexity,
https://doi.org/10.1007/978-3-030-22055-6_46

46.1 Introduction

Research and development of the PROGNOS post-processing system started as an answer to various operational needs that the current system (UMOS; [10]) has difficulty addressing. For example, adapting and validating UMOS to very frequent numerical model (NWP) updates has become tedious and time consuming. Furthermore, UMOS has a limited ability to evaluate various statistical modeling and data pre-processing methods as well as to post-process new predictands. In addition, accounting for new observation datasets in the current operational system has proven difficult. Therefore, UMOS is limiting our ability to accommodate evolving forecasting program requirements such as gridded post-processing. The proposed modular and flexible design of the PROGNOS system will address these challenges while enhanced portability and transparency in the underlying code will help mitigate some of the ongoing support requirements.

46.2 Architecture

46.2.1 Technologies

PROGNOS is built on open source technologies in Python, R and Bash languages. The data pre-processing and statistical post-processing components of the system are programmed in R, which makes use of the extensive statistical R libraries and support from the online community. The data flow (I/O) managing modules mostly use R with a few Python-based tools, but a gradual and complete move towards Python tools is in progress. Daily experimental runs of the system are conducted with Maestro, an ECCC sequencer [5], which is an application that submits sets of tasks (job scripts) in a user-defined order. To enable portability, usability and flexibility of the system, a centralized configuration approach is currently under development using YAML (YAML Ain't Markup Language), a human-readable, structured data serialization syntax.

To ensure that the system is isolated, portable and reproducible, the following library management technologies are used: the R Packrat package [8] and the Simple Software Manager (SSM), an ECCC-developed packaging solution [4].

Furthermore, point based observation, predictor and forecast data are currently stored and managed using SQLite, a serverless relational database management system, with plans to move to PostgreSQL hosted on a dedicated server. Adopting SQL technologies is an improvement to the proprietary standards used in the current system. It enhances I/O flexibility and versatility in both assimilating and selecting predictors as well as in stratifying observations according to station metadata.

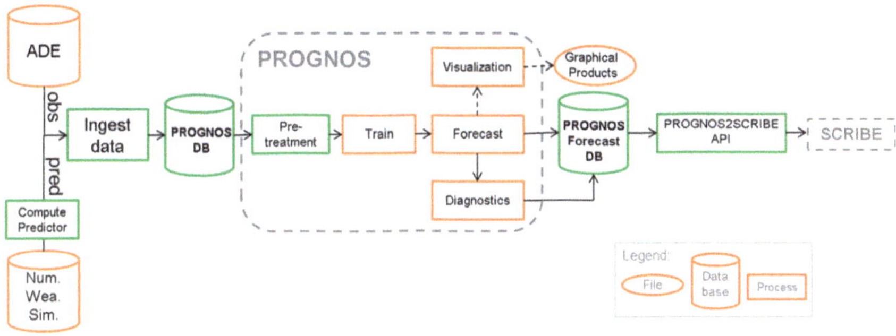

Fig. 46.1 Representation of the PROGNOS flow chart (future developments in green)

46.2.2 Infrastructure and Flow Chart

As shown below (Fig. 46.1), the proposed system consists of three main components: (1) Data ingestion modules that fill the PROGNOS database with observations as well as calculated and/or interpolated predictors; (2) the PROGNOS core consisting of various modules responsible in part for the pre-processing of the input data, the training of the forecast models and the production of diagnostics and forecasts; and (3) the output module, which holds the forecast database and is also responsible for converting the forecasts to a format compatible with SCRIBE, the current public forecast dissemination system [9].

46.3 System Features

Several post-processing approaches such as LASSO [7], random forest [1] and Kalman filter [3] have been explored in previous versions of the system; however, most of the recent development has been centered on multiple linear regression (MLR) techniques. A batch calibration approach is currently implemented and executed weekly to update model coefficients. Moreover, capabilities for updatable model output statistics are being considered. The predictor selection scheme is a stepwise approach, which consists of sequential replacement [2] using a weighted Bayesian information criterion (BIC; [6]). The BIC allows for the comparison between candidate MLR equations in order to select the best fit model for each observation station and forecast hour. The possibility of adding weighted transition schemes to adjust for seasonal changes and numerical model updates is accounted for.

The training, data pre-processing and the forecasting modules have the capability of running in parallel to make use of the full potential computational resources available. The experimental runs are using a simple multi-node approach to distribute the processing of each data block to different CPUs. Additional code optimisation

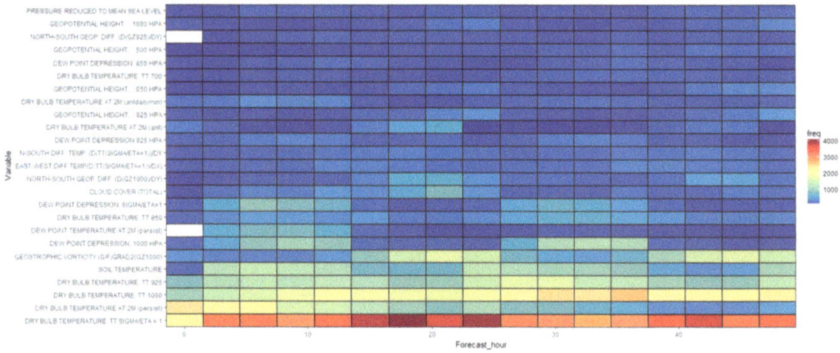

Fig. 46.2 Predictor usage heat map representing the frequency (in color) of predictors selected (y-axis) per forecast hour (x-axis) for the MLR models trained over a network of meteorological stations to predict surface temperature

and parallel computing is planned to facilitate the transition from station (point) post-processing to gridded post-processing in support of the next generation weather prediction systems.

Basic diagnostic tools used to extract the MLR model coefficients and model statistics per station, run hour and forecast hour are available. These diagnostic outputs are interpreted by developers amongst other experts to monitor the system. They can also be used to generate graphical products such as predictor usage heat maps for groups of stations (Fig. 46.2); a product often requested by meteorologists to best understand the nature and number of variables that explain the behavior of the predictand.

To improve the traceability and to facilitate debugging for development and monitoring purposes, a three-level logging feature is integrated in PROGNOS: info, warning and errors. Each log includes the logging level, the time-stamp and the associated message.

46.4 Conclusion

Numerous forecast comparisons with the current operational system have shown that PROGNOS attains the same or better skill for ambient concentration of ozone (O_3), fine particulate matter (PM_{25}) and nitrogen dioxide (NO_2) forecasts. The same holds true for surface temperature forecasts.

PROGNOS shows great potential and represents a significant renewal effort to the operational post-processing systems within ECCC. Development efforts are underway to mature PROGNOS to satisfy all operational requirements and to replace the current UMOS system when ready. The current focus is on ensuring that the design of the system and each of its components is modular and flexible while the code remains portable, transparent and resource efficient. Parallel work includes improvements to the diagnostic and visualization features for enhanced monitoring and consultation of experimental forecasts.

An iterative development approach is adopted to expand the PROGNOS system in order to better serve research and development projects. The next major milestone consists of adapting the data ingestion and pre-processing routines in PROGNOS to use historical records from the PROGNOS database. From there, PROGNOS will extend its modeling to consider additional predictands and NWP sources as well as season and model transition schemes. Over the medium to long term, PROGNOS will be integrating gridded post-processing methods to meet the evolving air quality and meteorological forecast program requirements.

References

1. L. Breiman, Random forests. Mach. Learn. **45**, 5–32 (2001)
2. M. Cassotti, F. Grisoni, R. Todeschini, Reshaped sequential replacement algorithm: an efficient approach to variable selection. Chemometr. Intell. Lab. Syst. **133**, 136–148 (2014)
3. R.E. Kalman, A new approach to linear filtering and prediction problems. Trans. ASME, Ser. D, J. Basic Eng. **82**(1), 35–45 (1960)
4. J. Marshall, SSM v7.8 documentation (2006), https://goo.gl/xyi4ER
5. D. Racette, ECCC Maestro sequencer code on Github (2017), https://github.com/racetted/maestro
6. G. Schwarz, Estimating the dimension of a model. Ann. Stat. **6**, 461–464 (1978)
7. R. Tibshirani, Regression shrinkage and selection via the lasso. J. Royal Stat. Soc. Ser. B, **58**(1), 267–288 (1996)
8. K. Ushey, J. McPherson, J. Cheng, A. Atkins, J.J. Allaire, Packrat: a dependency management system for projects and their r package dependencies. R package version 0.4.8-1 (2016), https://CRAN.R-project.org/package=packrat
9. R. Verret, D. Vigneux, J. Marcoux, F. Petrucci, C. Landry, L. Pelletier, G. Hardy, Scribe 3.0 a product generator, in *Preprints 13th International Conference on Interactive Information and Processing Systems for Meteorology, Oceanography and Hydrology*, 2–7 February 1997 (AMS, Long Beach, California, 1997), pp. 392–395
10. L.J. Wilson, M. Vallée, The Canadian updateable model output statistics (UMOS) system: Design and development tests. Weather Forecast. **17**, 206–222 (2002)

Part V
Model Assessment and Verification

Chapter 47
Hierarchical Clustering for Optimizing Air Quality Monitoring Networks

Joana Soares, Paul Makar, Yayne-Abeba Aklilu and Ayodeji Akingunola

Abstract Hierarchical clustering (HC) analysis groups datasets into clusters based on their (dis)similarity, and can be used to assess air-quality monitoring networks representability. The methodology describe here is a new approach to designing optimized air-quality monitoring networks by combining Kolmogorov-Zurbenko filtering (KZ) and HC of observed and modelled time series. Here we present the optimization of the air quality network in the province of Alberta, Canada, for NO_2, SO_2, $PM_{2.5}$ and O_3. The study suggests that network optimization will vary depending on chemical species due to different emissions sources and/or the results of secondary chemistry. Making use of hourly and time-filtered time series allows identifying emission sources, with much of the signal identifying sources emissions residing in shorter time scales (hourly to daily) due to short-term variation of concentrations, and background concentrations can be identified by larger time scales (monthly or over). The methodology is also capable of generating maps of sub-regions within which a single station will represent the entire sub-region, to a given level of dissimilarity, when applied to gridded datasets such as chemical transport modelling output.

47.1 Introduction

Canada is home to one of the largest deposits of oil sands in the world. The Governments of Canada and Alberta have set-up the Oil Sands Monitoring (OSM) Plan to integrate the existing monitoring networks across the Province, for better understanding the air-quality within and downwind of the oil sands region. A potentially powerful tool for assessing the consistency and spatial representativeness of the exist-

J. Soares (✉) · P. Makar · A. Akingunola
Air Quality Modelling and Integration Section, Air Quality Research Division, Environment and Climate Change, Toronto, ON M3H 5T4, Canada
e-mail: joana.soares@canada.ca

Y.-A. Aklilu
Environmental Monitoring and Science Division, Alberta Environment and Parks, Edmonton, AL T5J 5C6, Canada

C. Mensink et al. (eds.), *Air Pollution Modeling and its Application XXVI*, Springer Proceedings in Complexity,
https://doi.org/10.1007/978-3-030-22055-6_47

ing air quality network or design a new one is the combined use of the model and clustering analysis [1]. HC is a well-established methodology used to determine the inherent or natural groupings of objects [2]. The similarity among members is determined by a distance metric, which is used to create a similarity matrix in which data are cross-compared. This is followed by operations on the similarity matrix which group data according to their degree of (dis)similarity with respect to that metric.

47.2 Methodology

The methodology presented here is based on the associativity analysis described in the work of Solazzo and Galmarini [4] and references therein. The HC analysis uses the hourly or time-filtered time series of observations at different monitoring stations in Alberta, and analyses this data based on two dissimilarity metrics, 1-R (R being the Pearson correlation coefficient), and Euclidean distance to understand the similarity between time series in terms of temporal and magnitude variation, respectively. Dissimilarity analysis results may be used to rank stations in terms of potential redundancy, where the lowest levels of dissimilarity may point to stations being potentially redundant.

The methodology can be applied to hourly resolution or to time-filtered time series. A Kolmogorov-Zurbenko filter [6] was applied to filter the time series to remove daily, weekly and monthly or shorter time scales signal from the hourly time series.

The methodology was first applied to hourly observations, and to hourly model results extracted at station locations, to assess the model's ability to recreate the associations between observed records. We then apply the same methodology to modelled gridded-data and assess the extent to which model output can be used as a potential surrogate for observations in clustering analysis.

For this study, observational hourly data was collected for NO_2, SO_2, $PM_{2.5}$ and O_3 from the monitoring networks of Alberta for the period between August 2013 and July 2014 was collected. There are nine networks within Alberta (Fig. 47.1b): Alberta Capital Airshed Alliance (ACAA), Calgary Regional Airshed Zone (CRAZ), Lakeland Industrial Community Association (LICA), Fort Air Partnership (FAP), Peace Airshed Zone Association (PAZA), Palliser Airshed Society (PAS), Parkland Airshed Management Zone (PAMZ), West Central Airshed Society (WCAS), and Wood Buffalo Environmental Association (WBEA). The data was subjected to additional quality assurance and control procedures to avoid gaps in the time series of observations: continuous station data should be rejected if their hourly data records for the analysis period have more than 10% of the total data for the year missing, or contain data gaps of more than 168 consecutive hours in duration [4].

Model simulations were carried out for the same time period, over a domain centred over North America with 10 km grid spacing with the Global Environmental Multiscale—Modelling Air-quality and Chemistry modelling system [3]. The resulting outputs were used as initial and boundary conditions for a nested set of

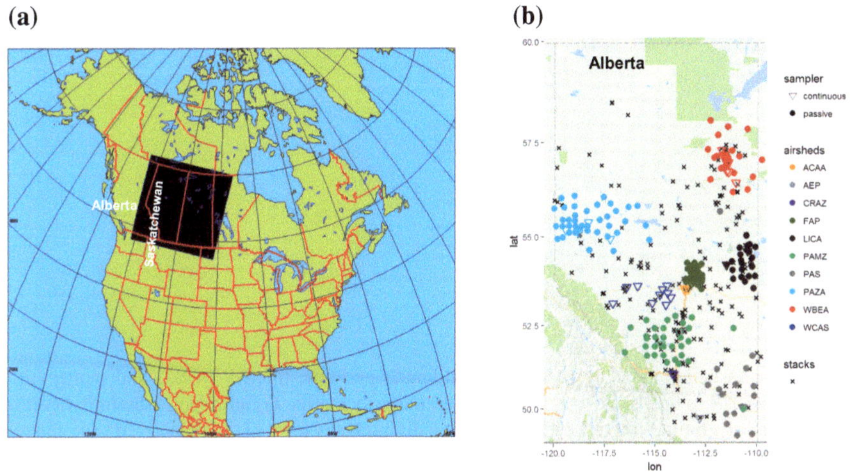

Fig. 47.1 Study area: **a** model domain covering the provinces of Alberta and Saskatchewan, and **b** continuous and passive monitors and main stacks in the Province of Alberta. Stations are colour-coded according to networks and plotted with different polygons to distinguish between passive (circle) and continuous (inverted triangle) samplers

simulations at 2.5 km resolution, for a domain covering the provinces of Alberta and Saskatchewan (Fig. 47.1a). The model was driven by regulatory reported emissions and additional emissions data emissions developed for the model simulations of OSM [5] to better simulate Athabasca oil sand surface mining and processing facilities.

47.3 Results

The dissimilarity analysis was applied observational time series data for all the stations complying with the QA/QC criteria described in Sect. 47.2. Figure 47.2 shows the spatial distribution of the clusters of O_3 and NO_2 continuous monitors for hourly (a) O_3 and (c) NO_2) and time-filtered time series for weekly and shorter time scales (b) O_3 and (d) NO_2). The results shows that for similar number of clusters, the level of dissimilarity for O_3 is lower than NO_2, showing that the stations measuring O3 are clearly more similar than the stations measuring NO_2. This methodology shows how results can be very different between species that are primary emitted and secondarily formed. It also shows that when shorter time scales are removed, the analysis loses information about local sources, as stations start to cluster across airsheds.

Figure 47.3 depicts the resulting mapped 1-R cluster analysis over the main facility in the Oil Sand area, when each model grid-cell is considered as a potential monitoring station location This is a specific sub-section of the model grid; namely a 72 × 72 block of model grid-squares centred on the Athabasca Oil Sands.

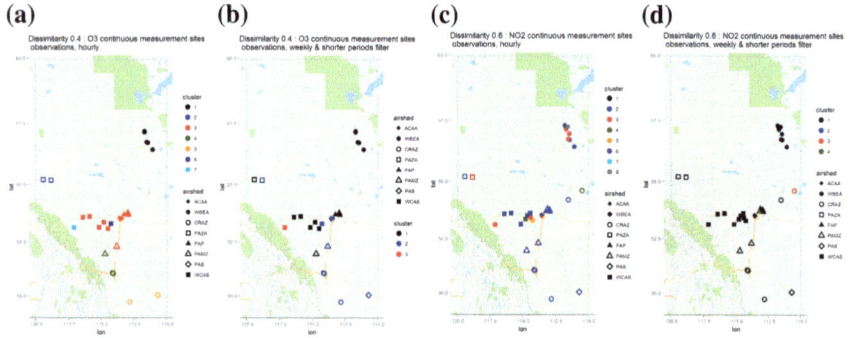

Fig. 47.2 Associativity analysis for observed hourly and time-filtered (weekly and shorter) time series using 1-R as the metric to compute the dissimilarity matrix: O_3 assuming a dissimilarity level of 0.4 **a** hourly, **b** time-filtered; NO_2 assuming a dissimilarity level of 0.6 **c** hourly, **d** time-filtered. Stations are colour-coded by cluster, and networks are plotted with different polygons

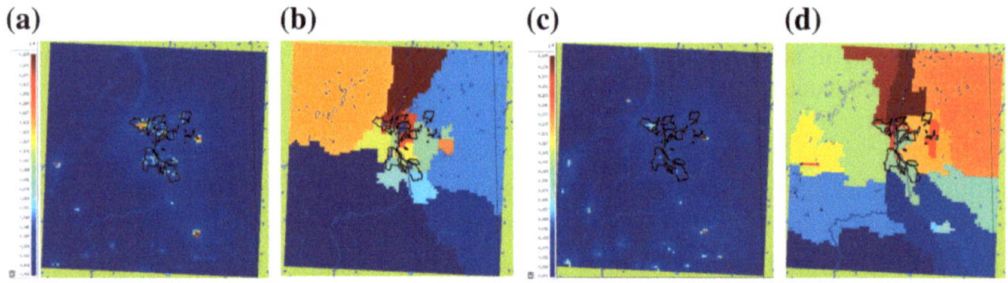

Fig. 47.3 Dissimilarity maps based on 1-R metric for **a** SO_2 and **c** $PM_{2.5}$ modelled hourly output at each GEM-MACH grid-cell. Associativity analysis maps for modelled NO_2 and SO_2, generated using a 1-R dissimilarity level of **b** 0.65, and **d**) 0.8, respectively. Main mining facilities operating in the Athabasca oil sands region are shown in black contours enclosing transparent light grey shading

Each of the coloured areas reflects the area of representativeness of a single station. Overlaying georeferenced information such as road maps and orography will provide a good basis for a assessing the potential location of stations. This example shows that gridded model output may be used to generate an optimized monitoring network.

47.4 Conclusion

We have assessed Alberta monitoring stations by means of a methodology based on filtering and hierarchical clustering. The methodology identifies stations influenced by unique sources but also identifies stations that contain in their observations records outliers, large gaps of data and even measurement errors. The methodology also identifies different measurement technologies (not shown here). The analysis shows a lower tendency of the data to cluster according to sources as the shorter time scales are removed, with more "signal" of a specific source be present in the hourly data. This is more visible in species such as SO_2, primarily emitted from high stacks, and the least in O_3, secondarily produced. We have assessed the model capability to mimic the observation, by taking model results extracted at station locations and apply the same methodology. We have shown that the model is able to reproduce the observed clusters particularly well, especially for secondary pollutants such O_3. We treated every grid cell of the model output as a potential station location, for different chemicals, and apply the methodology to generate maps of dissimilarity metric and of clustering of that metric to provide information for network optimization. These maps will give areas of representativeness for single stations. Note that this methodology is not taking into consideration road access or power availability, but the maps generated can be layered with other information to assist in decision making.

Acknowledgements This project was jointly supported by the Climate Change and Air Quality Program of Environment and Climate Change Canada, Alberta Environment and Parks, and the JOSM program.

References

1. B.S. Everitt, S. Landau, M. Leese, D. Stahl. *Cluster Analysis*, 5th Edition, Wiley Series in Probability and Statistics, p. 330 (2011)
2. R.A. Johnson, D.W. Wichern. (2007), *Applied Multivariate Statistical Analysis*. Pearson Prentice Hall, Pearson Education Inc. Upper Saddle River, NJ, USA
3. M.D. Moran, S. Menard, D. Talbot, P. Huang, P.A. Makar, W. Gong, H. Landry, S. Gravel, S. Gong, L.-P. Crevier, A. Kallaur, M. Sassi, Particulate-matter forecasting with GEM-MACH15, a new Canadian air-quality forecast model. In: D.G. Steyn, S.T. Rao (Eds.), *Air Pollution Modelling and its Application XX*, Springer, Dordrecht, pp. 2890–292 (2010)
4. E. Solazzo, S. Galmarini, Comparing apples with apples: using spatially distributed time series of monitoring data for model evaluation. Atmos. Environ. **112**, 234–245 (2015)
5. J. Zhang, Q. Zheng, M.D. Moran, P.A. Makar, A. Akingunola, S.-M. Li, G. Marson, M. Gordon, R. Melick, S. Cho, Emissions preparation for high-resolution air quality modelling over the Athabasca oil sands region of Alberta, Canada. 21st Intern. Emissions Inventory Conference, 13–17 April, San Diego (2015). http://www.epa.gov/ttn/chief/conference/ei21/session1/zhang_emissions.pdf
6. I.G. Zurbenko, *The Spectral Analysis of Time Series* (North-Holland, Amsterdam, 1986), p. 236

Chapter 48
Continental-Scale Analysis of Atmospheric Deposition Over North America and Europe Using the AQMEII Database

Christian Hogrefe, Stefano Galmarini, Efisio Solazzo, Roberto Bianconi, Roberto Bellasio, Peng Liu and Rohit Mathur

Abstract Participants in the Air Quality Model Evaluation International Initiative (AQMEII) have conducted three rounds of model evaluation and intercomparison activities over Europe and North America since 2010, resulting in a large dataset of modeled meteorology, air quality and deposition fields for 2006–2010 that is available to the community for ongoing research on model evaluation. This study presents a brief analysis of some of the deposition fields generated during these past phases of AQMEII, quantifying both model-to-model variability and the level of agreement with available wet deposition measurements. We also discuss potential future AQMEII work focused on evaluating deposition and using modeled deposition fields for applications such as producing maps of estimated total deposition.

48.1 Introduction

The Air Quality Model Evaluation International Initiative (AQMEII) has conducted three phases of regional air quality model evaluation and intercomparison over Europe and North America [3–5], hereafter referred to as AQMEII1, AQMEII2 and AQMEII3. Most publications resulting from these activities focused on air concentrations of trace gases and particulate matter. Aiming to eventually contribute to the

C. Hogrefe (✉) · R. Mathur
Atmospheric and Environmental Systems Modeling Division, Center for Environmental Measurement and Modeling, U.S. Environmental Protection Agency, Research Triangle Park, NC 27711, USA
e-mail: hogrefe.christian@epa.gov

S. Galmarini · E. Solazzo
European Commission Joint Research Centre, Ispra, Italy

R. Bianconi · R. Bellasio
Enviroware srl, Concorezzo, MB, Italy

P. Liu
NRC Research Associate, Computational Exposure Division, National Exposure Research Laboratory, U.S. Environmental Protection Agency, Research Triangle Park, NC 27711, USA

C. Mensink et al. (eds.), *Air Pollution Modeling and its Application XXVI*,
Springer Proceedings in Complexity, https://doi.org/10.1007/978-3-030-22055-6_48

Table 48.1 Number of models used in the analysis. The numbers in parentheses show the number of models with gridded deposition data available from the ENSEMBLE database, some of these models were not used in the analysis due to missing variables or clearly erroneous data values

	AQMEII1 2006	AQMEII2 2006	AQMEII2 2010	AQMEII3 2010
North America	6 (6)	3 (4)	4 (6)	2 (2)
Europe	9 (9)	–	9 (14)	11 (12)

body of work stemming from atmospheric deposition model intercomparison studies conducted on both the global [2, 10] and regional scale [1, 12, 13], this study presents a brief overview and analysis of existing AQMEII deposition data sets and discusses goals for a potential future AQMEII activity.

48.2 Model Simulations and Observations

The modeled deposition fields analyzed in this study were extracted from the ENSEMBLE database that stores all model data submitted by AQMEII1, AQMEII2, and AQMEII3 participants. These fields are provided on common grids defined over North America and Europe and contain monthly total deposition fluxes for 2006 and 2010. While roughly 20 different wet and dry deposition fields were requested from participants, not all participants submitted all fields. In this short overview, we focus on wet deposition of SO_4^{2-}, NO_3^- and NH_4^+ and dry deposition of ozone. Out of the 18 (35) model simulations for which gridded deposition fields for at least some species are stored on ENSEMBLE over North America (Europe), 15 (29) were used in this study while the others were discarded due to missing variables or clearly erroneous data values. Further information on these simulations is provided in Table 48.1. Observations of SO_4^{2-}, NO_3^-, and NH_4^+ wet deposition were obtained from National Acid Deposition Program (NADP) monitors over the U.S. and European Monitoring and Evaluation Programme (EMEP) monitors over Europe. Monthly total observed values were calculated from the weekly NADP and daily EMEP measurements and corresponding monthly model values were matched to each observation site. No data completeness criteria were applied in this study except that for the calculations of annual total values, months with missing observations were also excluded from the calculation of modeled annual totals.

48.3 Results and Discussion

Figure 48.1 shows distributions of simulated annual total ozone dry deposition over land cells in Europe and North America. Median values vary between 1680 and 7650 kg/km^2/yr for different simulations over Europe and between 1150 and 7950 kg/km^2/yr for different simulations over North America. Some AQMEII2 sim-

Table 48.2 Comparison of observed and modeled annual total wet deposition. Annual totals were computed at each station and then spatially averaged across all sites for observations and each simulation. The median and range of these values across models is shown for each AQMEII phase. The units are kgS/km^2/yr for $SO_4{}^{2-}$ and kgN/km^2/yr for $NO_3{}^-$ and $NH_4{}^+$

		NADP			EMEP		
		$SO_4{}^{2-}$	$NO_3{}^-$	$NH_4{}^+$	$SO_4{}^{2-}$	$NO_3{}^-$	$NH_4{}^+$
2006	# Sites	255			68		
	Observed	290	140	156	276	253	301
	AQMEII1 Median	271	132	132	204	129	157
	AQMEII1 Range	239–485	9–236	35–171	143–473	45–364	62–431
	AQMEII2 Median	318	124	53	–	–	–
	AQMEII2 Range	316–362	124–140	48–121	–	–	–
2010	# Sites	241			67		
	Observed	182	108	143	294	237	308
	AQMEII2 Median	209	92	48	61	41	72
	AQMEII2 Range	68–242	8–105	17–97	19–186	9–193	26–181
	AQMEII3 Median	189	162	206	177	138	224
	AQMEII3 Range	171–189	86–162	143–206	19–327	33–313	97–1008

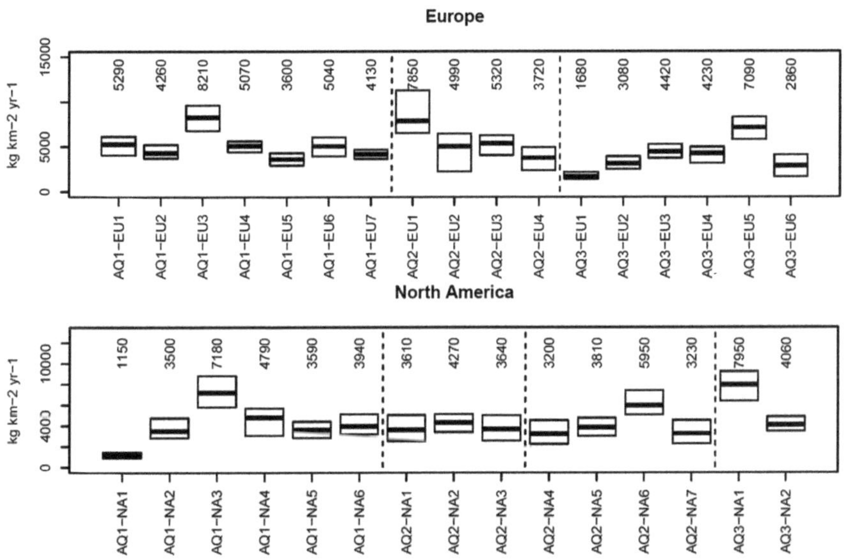

Fig. 48.1 Distribution of modeled annual total ozone dry deposition fluxes across all land cells. The lines and vertical numbers above each box indicate the median of the distributions while the boxes indicate the interquartile range. Each box represents a different simulation with "AQ1", "AQ2", and "AQ3" denoting the three phases of AQMEII and the remaining characters randomly labeling individual simulations

ulations over Europe were excluded from this plot since their values were a factor of 10 higher than the maximum AQMEII2 model displayed (AQ2-EU1). Given the importance of ozone dry deposition for model performance [9], future work needs to be directed at understanding the drivers for these model differences at the process level, following step-wise approaches such as those employed in Schwede et al. [6].

Table 48.2 provides a comparison of observed and modeled annual wet deposition. Based on the comparison of the median model results and observations, there is a tendency to underestimate SO_4^{2-}, NO_3^-, and NH_4^+ wet deposition over both continents in 2006 and 2010. However, the range of simulated deposition values across models is substantial, varying between species, years, and AQMEII phases from a factor of 2 to a factor of 20. The range of variability is consistent with the AQMEII1 model-to-model analysis presented in Solazzo et al. [8] and highlights that—besides better understanding process drivers for these differences—future work is also needed for developing and applying formal criteria to construct deposition ensembles for impact assessments, refining ad hoc criteria used in earlier studies [11, 13]. It is envisioned that future AQMEII work could contribute to efforts to construct regional and global maps of total deposition based on fusing measurements and model simulations [7, 11, 14].

Acknowledgements and Disclaimer We gratefully acknowledge NADP for providing wet deposition measurements over the U.S. (http://nadp.slh.wisc.edu/) and NILU for providing EMEP wet deposition measurements over Europe through EBAS (http://ebas.nilu.no/). The views expressed in this article are those of the authors and do not necessarily represent the views or policies of the U.S. Environmental Protection Agency.

References

1. G.R. Carmichael et al., Atmos. Env. **36**, 175–199 (2002). https://doi.org/10.1016/S1352-2310(01)00448-4
2. F. Dentener et al., Glob. Biogeochem. Cycl., 20, GB4003 (2006). https://doi.org/10.1029/2005gb002672
3. S. Galmarini, S. T. Rao, Atmos. Env., 45 (2011). https://doi.org/10.1016/j.atmosenv.2011.03.025
4. S. Galmarini et al., Atmos. Env., 115 (2015). https://doi.org/10.1016/j.atmosenv.2015.06.009
5. S. Galmarini et al., Atmos. Chem. Phys. **17**, 1543–1555 (2017). https://doi.org/10.5194/acp-17-1543-2017
6. D.B. Schwede et al., Atmos. Env., 45 (2011). https://doi.org/10.1016/j.atmosenv.2010.11.050
7. D.B. Schwede, G.G. Lear, Atmos. Env. (2014). https://doi.org/10.1016/j.atmosenv.2014.04.008
8. Solazzo et al., Atmos. Env., 53 (2012), https://doi.org/10.1016/j.atmosenv.2012.02.045
9. E. Solazzo et al., Atmos. Chem. Phys. **17**, 10435–10465 (2017). https://doi.org/10.5194/acp-17-10435-2017
10. J. Tan et al., Atmos. Chem. Phys. **18**, 6847–6866 (2018). https://doi.org/10.5194/acp-18-6847-2018
11. R. Vet et al., Atmos. Env., 93 (2014). https://doi.org/10.1016/j.atmosenv.2013.10.060
12. M.G. Vivanco et al., Atmos. Env., 151 (2017). https://doi.org/10.1016/j.atmosenv.2016.11.042
13. M.G. Vivanco et al., Atmos. Chem. Phys. **18**, 10199–10218 (2018). https://doi.org/10.5194/acp-18-10199-2018
14. WMO, WMO Report #234, S. Carou and R. Vet, World Meteorological Organization, Geneva, Switzerland (2017)

Chapter 49
Multi Model Study on the Impact of Emissions on CTMs

Johannes Bieser, Martin Otto Paul Ramacher, Marje Prank, Efisio Solazzo and Andreas Uppstu

Abstract Emission data are a key driver and a major source of uncertainty to atmospheric chemistry transport models. As part of the international model-intercomparison study AQMEII chemistry transport models (CTMs) with harmonized input data have been used to evaluate the impact of emission datasets on different species and compare it to the effect of meteorology and parametrization of the CTM.

49.1 Introduction

Emission data are often cited as a key source for model uncertainty in CTM studies. Albeit an intuitively convincing argument, the exact impact of emission data on modeled concentration and deposition fields has not been quantified in the scientific literature. Moreover, to assess the impact of the emission dataset one needs to investigate the relative impact of emissions compared to other drivers of CTMs, namely the meteorological data and the practical implementation of physical and chemical processes. The presented study is a first step towards a comprehensive understanding of the impact of emission data on CTM results.

J. Bieser (✉) · M. O. P. . Ramacher
Helmholtz-Zentrum Geesthacht, Institute of Coastal Research, Max-Planck-Strasse 1, 21502 Geesthacht, Germany
e-mail: johannes.bieser@hzg.de

M. Prank · A. Uppstu
Finnish Meteorological Institute, Helsinki, Finland

E. Solazzo
European Commission, Joint Research Centre, Ispra, Italy

C. Mensink et al. (eds.), *Air Pollution Modeling and its Application XXVI*,
Springer Proceedings in Complexity,
https://doi.org/10.1007/978-3-030-22055-6_49

49.2 Methodology

This study is based on regional European CTM results from the Air Quality Modeling Evaluation International Initiative (AQMEII3) [1], Solazzo and Galmarini [9]. In the course of AQMEII3, input data for the regional CTMs were harmonized in a way that allows for the investigation of the impact of the various CTMs. Based on identical input data each model calculated concentration

and deposition fields for the year 2010 on a regional domain over Europe. Subsequently, the CTM results were uploaded into the ENSEMBLE system for evaluation and comparison with measurement data [2]. In order to evaluate the impact of emissions we supplied AQMEII3 participants with an alternative emission dataset. The key issue was to create an emission dataset that is actually independent from the default dataset based on MACC/EDGAR emissions [3]. The problem is that all emission datasets are based on the same or at least sufficiently similar national total emission estimates for criteria pollutants [4]. Moreover, the spatial disaggregation of national and sectoral emission totals is mostly identical as it is nowadays typically based on the official European population density maps land use classifications.

Here, we created what we call a maximum feasible independent emission dataset based on the emission model SMOKE for Europe [4]. The difference to other European emission models is the implementation of bottom up modules for emissions from shipping [5], on-road transport [6], and agriculture [7] as well as explicit plume rise calculations [8]. The employed bottom-up models force a dynamic redistribution of emissions and thus impact the temporal and spatial distribution as well as annual emission totals. This dataset, further referred to as SMOKE-EU, was distributed among groups participating in the AQMEII3 exercise. So far, there are only results from two models (SILAM and CCLM-CMAQ) available for the analysis. A detailed model description can be found in Solazzo and Galmarini [9]. The harmonized input data is described in Galmarini et al. [1].

49.3 Analysis

Depending on species (e.g. CO, NO, NO_2, SO_2), the ENSEMBLE database contains up to 20 million validated observations for a single air pollutant in 2010. We calculated the normalized mean bias (NMB), correlation coefficient (CORR), and root mean squared deviation (RMSD) for each data point and CTM run. Based on NMB and RMSD we determined the difference between CTM runs with the same model and CTM runs with the same emission dataset (Fig. 49.1). The difference in NMB and RMSD is then calculated and used to determine the absolute impact on CTM results (Fig. 49.2). We do not dwell into the question whether a single model configuration performs better or worse. The focus of the analysis is to compare the impact of emissions to other drivers on CTM results. Moreover, the question is whether we can identify common behavior for different CTMs when using identical emission datasets. The main goal of the study is to elucidate the role of emission data in a way that allows for an informed evaluation of the impact of emissions unlike the often general identification of undetermined issues with 'the emissions'.

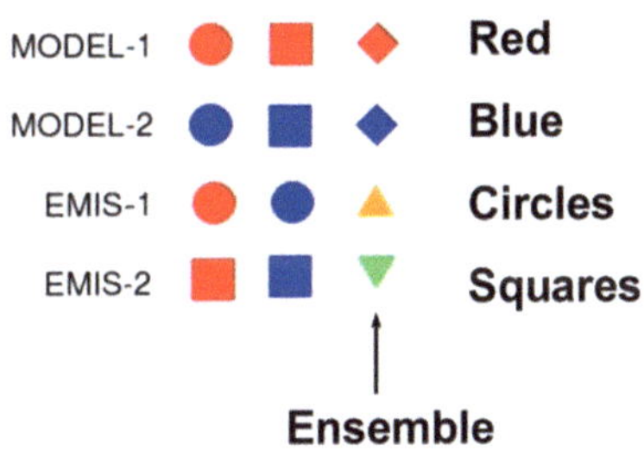

Fig. 49.1 Icons used to identify CTM/emission combinations. Circles indicate same emission with different CTM and vice versa for squares

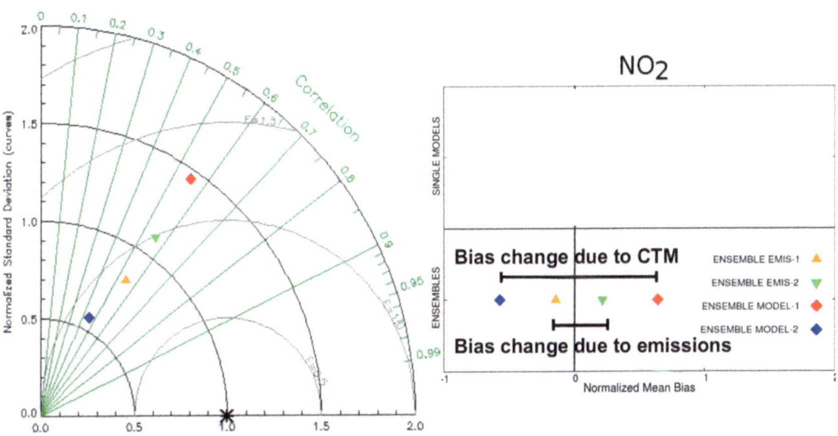

Fig. 49.2 RMSD and NMB analysis for NO_2 concentrations in air for 2010

49.4 Results

We investigated the model ensemble for six major air pollutants: NO_2, NO, CO, O_3, SO_2, $PM_{2.5}$. For these species we calculated two aggregated measures to determine the impact of emissions (Fig. 49.3). Firstly the relative impact which is defined as the fraction of RMSD change due to emission dataset divided by the total RMSD change due to all input parameters (Eq. 49.1). This value is defined between 0 and 1 and defines the fraction of change due to the emissions. Secondly, we calculated an absolute measure which is defined as the total RMSD change divided by the annual observation mean (Eq. 49.2). By definition this value lies between 0 and infinity, however it is to be expected that changes in CTM results should realistically be lower than the annual mean concentration which is why we expect, and find, values below 1.

$$\frac{\Delta RMSD_{emission}}{\Delta RMSD_{emission} + \Delta RMSD_{CTM} + \Delta RMSD_{meteorology}} \tag{49.1}$$

$$\frac{\Delta RMSD_{emission} + \Delta RMSD_{CTM} + \Delta RMSD_{meteorology}}{Observation\, mean} \tag{49.2}$$

We find that, independent of the CTM, the emissions have a large impact on SO_2 (70%), $PM_{2.5}$ (43%), NO (41%) concentrations. For CO (12%) and ozone (19%) the impact of emissions is much lower than that of meteorology and CTM. Moreover, including the absolute impact (Eq. 49.2) we find that nitrogen oxides exhibit the

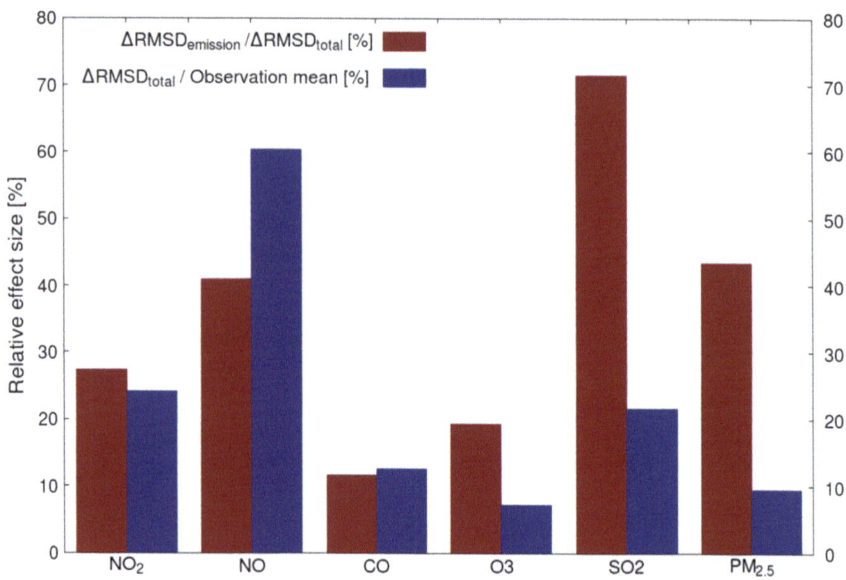

Fig. 49.3 Relative (red, Eq. 49.1) and absolute (blue, Eq. 2) impact on CTM results

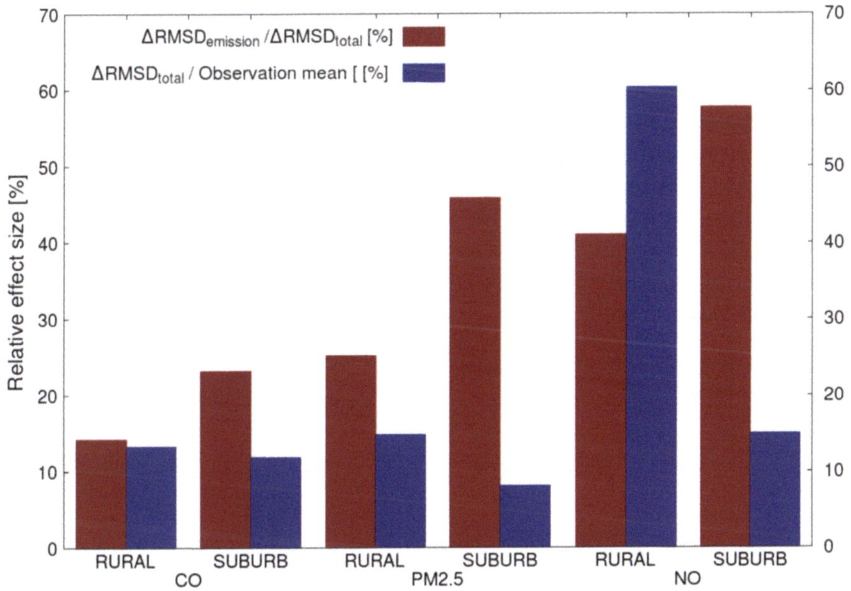

Fig. 49.4 Distinct evaluation of rural and suburban measurement stations

strongest change compared to annual mean concentrations with 60% for NO and 26% for NO_2. For $PM_{2.5}$ on the other hand, where we saw a large relative contribution of the emissions, we find that the change is low compared to the mean observations. We are hypothesizing that secondary particle production is probably dominating the CTM results for this species.

Finally, we compared results for rural and suburban station (Fig. 49.4). It can be seen, that the absolute (blue) impact on CTM results is larger in rural compared to suburban locations. This can be explained by the vicinity of the measurement stations to primary sources of air pollutants. One can see that the long lived species like CO and species with secondary sources such as $PM_{2.5}$ show a much lower absolute impact compared to NO which is strongly emitted in the urban environment. However, it is interesting to note that the relative impact of the emission model increases in suburban areas. This effect cannot be easily explained and might be an artifact of low model resolution (12×12 km^2).

49.5 Discussion

We find that the impact of the emission dataset on modeled concentrations is of similar size as the effect of meteorology and CTM for NO_2, NO, SO_2, $PM_{2.5}$. We hypothesize that this is mostly due to emissions of NO_x and NH_3 which influence secondary inorganic aerosol formation. Due to the bottom-up modeling of emissions from agriculture and on-road transport these are the species with the highest expected difference in the SMOKE-EU emission dataset. For CO and ozone, however we find only a comparably small impact of the emission dataset. Especially for CO this result is surprising as the SMOKE-EU dataset uses a distinct spatial and temporal disaggregation of the emissions based on heating demand compared to fixed annual profiles for the MACC/EDGAR dataset. As expected, the impact of emissions increases near sources. Thus, with increasing CTM resolution it will become more important to correctly model emission hot-spots such as traffic and residential emissions [10].

Finally, this study is only based on two CTMs and the findings need to be verified by a larger model ensemble.

Question and Answers

Questioneer: C. Geels

Q: You talked about changed emissions—but actually it was "only" the temporal profiles?

A: This is a very important question indeed, thus I will answer it in detail:

(1) For emission from shipping and on-road transport we use bottom-up models that result in independently calculated annual total emissions. Moreover, we gain high resolution spatial distribution data. This means the exact location of each ship at each time step and vehicle road densities which are based on vehicle counting and traffic demand models. Finally, we use dynamic emission factors depending on vehicle type, fuel, and temperature. This has, for example, a large impact on the NO/NO_2 split.
(2) For emission from residential heating we calculate individual temporal profiles for each grid cell based on a heating demand formula. We then normalize only the annual total emissions which leads to a spatial redistribution of emissions to cold regions with a different distribution for each day.
(3) For the agricultural emissions we use a similar approach for emissions from animal husbandry and fertilizer application. Here, the redistribution is based on temperature and wind speed. But also legal aspects such as national and European legislation on fertilizer application are implemented into the model. Moreover, we use a plant growth model that calculates additional NH_3 emissions which are not covered by the national totals.

Q: For some components like NH_3 the annual emission might change from year to year just because of temperature changes. So warm year more NH_3 emissions—do you think we should include that.

A: We already do that to a certain extend but I think NH_3 emissions exhibit the highest uncertainty and variability of the standard (criteria) pollutants. Especially because of the impact on secondary inorganic particle conversion I would strongly support the development of more sophisticated emission models for agricultural emissions.

References

1. S. Galmarini, B. Koffi, E. Solazzo, T. Keating, C. Hogrefe, M. Schulz, A. Benedictow, J.J. Griesfeller, G. Janssens-Meanhout, G. Carmicheal, J. Fu, F. Dentener, Technical note: coordination and harmonization of the multi-scale, multi-model activities HTAP2, AQMEII3, and MICS-Asia3: simulations, emission inventories, boundary conditions, and model output formats. Atmos. Chem. Phys. **17**, 1543–1555 (2017)
2. S. Galmarini, R. Bianconi, W. Appel, E. Solazzo, S. Mosca, P. Grossi, M. Moran, K. Schere, S.T. Rao, ENSEMBLE and AMET: two systems and approaches to a harmonized, simplified and efficient facility for air quality models development and evaluation. Atmos. Environ. **53**, 51–59 (2012)
3. J.J.P. Kuenen, A.J.H. Visschedijk, M. Jozwicka, H.A.C. Denier van der Gon, TNO-MACC_II emission inventory; a multi-year (2003–2009) consistent high-resolution European emission inventory for air quality modelling. Atmos. Chem. Phys. **14**, 10963–10976 (2014)
4. J. Bieser, A. Aulinger, V. Matthias, M. Quante, P. Builtjes, SMOKE for Europe—adaptation, modification and evaluation of a comprehensive emission model for Europe. Geosci. Model Dev. **4**(1), 47–68 (2011)
5. A. Aulinger, V. Matthias, M. Zeretzke, J. Bieser, M. Quante, A. Backes, The impact of shipping emissions on air pollution in the greater North Sea region—part 1: current emissions and concentrations. Atmos. Chem. Phys. **16**(2), 739–758 (2016)
6. S. Stefan, J. Bieser, S. Ehrenberger, U. Kugler, German and European ground-transport emissions in three different scenarios until 2040. Transport Research Arena, TRA 2018, 16–19 April 2018, Wien, Österreich (2018)
7. A. Backes, A. Aulinger, J. Bieser, V. Matthias, M. Quante, Ammonia emissions in Europe, part I: development of a dynamical ammonia emission inventory. Atmos. Environ. **131**, 55–66 (2016)
8. J. Bieser, A. Aulinger, V. Matthias, M. Quante, H.A.C. Denier van der Gon, Vertical emission profiles for Europe based on plume rise calculations. Environ. Pollut. **159**(10), 2935–2946 (2011)
9. E. Solazzo, S. Galmarini, Error apportionment for atmospheric chemistry-transport models—a new approach to model evaluation. Atmos. Chem. Phys. **16**, 6263–6282 (2016)
10. V. Matthias, J.A. Arndt, A. Aulinger, J. Bieser, H.A.C. Denier van der Gon, R. Kranenburg, J. Kuenen, D. Neumann, G. Pouliot, M. Quante, Modeling emissions for three-dimensional atmospheric chemistry transport models. JAWM **68**(8), 2018 (2018)

Chapter 50
Evaluation of the New Version of Stratospheric Chemistry Module of the SILAM CTM

Risto Hänninen, Mikhail Sofiev, Rostislav Kouznetsov and Viktoria Sofieva

Abstract The effect of different bromine and chlorine species on ozone is here evaluated using a System for Integrated modeLling of Atmospheric coMposition, SILAM. The new stratospheric chemistry module includes 15 chlorine and 9 bromine species together with their gas-phase and heterogeneous reactions with other species. Evaluation of ozone concentrations is obtained by using the steady-state, which is obtained after spin-up time of several decades. The amount of halogens is estimated by including bromine emissions from sea (e.g. sea-salt), from biomass burning and from terrestrial sources. For chlorine we used the emissions from the GEIA inventory, with additional yearly scaling of the different CFC emissions.

50.1 Introduction

It is well-known that the amount of ozone is affected by halogens. Especially the heterogeneous reactions in polar stratospheric clouds (PSC) are expected to be the main cause for the ozone depletion in polar regions: the reactions in PSC convert bromine and chlorine species to their active form which eventually react with ozone. The concentrations of different gases in stratosphere also affect their concentrations in troposphere. Therefore, a better model for stratospheric chemistry reactions will also improve the modelling of, e.g. ozone, NO_2, and NO_3 in the lower troposphere where these different species may have important health effects.

R. Hänninen (✉) · M. Sofiev · R. Kouznetsov · V. Sofieva
Finnish Meteorological Institute, Helsinki, Finland
e-mail: Risto.Hanninen@fmi.fi

M. Sofiev
e-mail: Mikhail.Sofiev@fmi.fi

R. Kouznetsov
e-mail: Rostislav.Kouznetsov@fmi.fi

V. Sofieva
e-mail: Viktoria.Sofieva@fmi.fi

C. Mensink et al. (eds.), *Air Pollution Modeling and its Application XXVI*,
Springer Proceedings in Complexity,
https://doi.org/10.1007/978-3-030-22055-6_50

Compared with the previous version of the stratospheric module in SILAM [7] we have now included the effect of inorganic bromine species. Additionally, the model now includes the effect of CFC-11, CFC-12, CCl_4, together with methyl chloride and bromide, plus CH_3CCl_3. The simulation results will later be compared with satellite measurements of ozone, NO_2, NO_3, and PM profiles by the GOMOS satellite, and total ozone column by OMI and TOMS satellite observations.

50.2 Model

Our simulations are based on the SILAM model [7], where the updated stratospheric module has been further expanded and now includes reactions for both the chlorine and bromine species. The gas-phase and photolysis reactions are essentially same than what are present, e.g., in the FinROSE model [4], except with the inclusion of the 5 organic chlorine species and methyl bromide.

For heterogeneous reactions occurring in polar stratospheric clouds we have expanded (compared with the previous version of SILAM) the number of reactions from 5 to 13, which reflects the inclusion of the bromine species. The full list of the heterogeneous reactions is illustrated in Table 50.1. The updated reaction rate coefficients for different reactions (gas-phase and heterogeneous) are taken from IUPAC database, or from JPL publication 15–10 [2]. The model takes into account (i) supercooled ternary solutions (STS) consisting of sulphuric acid, nitric acid and water, (ii) particles consisting of nitric acid trihydrate (NAT) and (iii) particles composed of water ice.

Table 50.1 Heterogeneous reactions included in the model. Symbol (g) indicates that the compound is in the gas phase and (s) that the compound is in the condensed phase

Reaction	Number
$ClONO_2(g) + H_2O(s) \rightarrow HOCl(g) + HNO_3(s)$	(HR1)
$BrONO_2(g) + H_2O(s) \rightarrow HOBr(g) + HNO_3(s)$	(HR2)
$N_2O_5(g) + H_2O(s) \rightarrow 2\ HNO_3(s)$	(HR3)
$ClONO_2(g) + HCl(s) \rightarrow Cl_2(g) + HNO_3(s)$	(HR4)
$HOCl(g) + HCl(s) \rightarrow Cl_2(g) + H_2O(s)$	(HR5)
$BrONO_2(g) + HCl(s) \rightarrow BrCl(g) + HNO_3(s)$	(HR6)
$HOBr(g) + HCl(s) \rightarrow BrCl(g) + H_2O(s)$	(HR7)
$N_2O_5(g) + HCl(s) \rightarrow ClNO_2(g) + HNO_3(s)$	(HR8)
$ClONO_2(g) + HBr(s) \rightarrow BrCl(g) + HNO_3(s)$	(HR9)
$HOCl(g) + HBr(s) \rightarrow BrCl(g) + H_2O(s)$	(HR10)
$BrONO_2(g) + HBr(s) \rightarrow Br_2(g) + HNO_3(s)$	(HR11)
$HOBr(g) + HBr(s) \rightarrow Br_2(g) + H_2O(s)$	(HR12)
$N_2O_5(g) + HBr(s) \rightarrow BrNO_2(g) + HNO_3(s)$	(HR13)

In our updated module the composition of aqueous HNO_3–H_2SO_4 stratospheric aerosol is calculated non-iteratively by using the model by [3]. Additionally, the solubility of HCl, HOCl, and HBr in STS is calculated using the effective Henry's law constants [3, 5, 6].

In addition to new bromine species, we have also included the halogen emissions in order to be able to study a steady state situation. For chlorine species we used the GEIA inventory, and included the emissions of CH_3Cl, HCl, Cl, CH_3CCl_3, CFC-11, and CFC-12. In order to cover the most important ozone depletion substances, we also included CCl_4 (CFC-10). Also, since the GEIA inventory covers only one year, we scaled the CFC and CH_3CCl_3 (methyl chloroform) emissions using the known yearly estimates based, e.g., on the AFEAS data [1]. These emission estimates typically extended up to year 2000. For the following years the emissions of CFC-12, CCl_4, and methyl chloroform was decreased to zero by year 2010, in accordance with the Montreal Protocol. For CFC-11, whose production was essentially zero already in 2000, we release yearly 7.5% of the still existing "banks" of CFC-11 (mainly in closed cell foams), which was about 950 Gg in 1999 [1]. In addition to CH_3Cl emissions from GEIA, which covers emissions from sea and biomass burning, we included terrestrial emissions of 2330 Gg [9].

We included bromine emissions from sea-salt, which contains approximately 0.223% of bromine [8], and should cover at least 90% of all the bromine emissions to the atmosphere. Currently the bromine sea-salt emissions are in the form of Br_2. Additionally, we release CH_3Br due to biomass burning (23 Gg), terrestial (18 Gg), and from the sea (32 Gg) [9]. Other standard emissions are taken from the MACCity emission inventory.

50.3 Results

The preliminary simulations that cover yearly changing CFC-, and CH_3CCl_3-emissions, extend from year 1942–2017. Any emissions prior to 1942 is released during the first year of simulation. These simulations have a spatial resolution of 2° and they use ERA_interim for meteorology, which is simply copied from years 1980–2017 to cover also the years 1942–1979. CFC emissions included in the model produce reasonable concentrations of these gases. For example, the average mixing ratios for CFC-11, CFC-12, and CCl_4 in 2011 are 253, 522, and 88 ppt, respectively. These are in line with the tropospheric measurements [9].

Our simulations clearly indicate that the ozone depletion above Antarctic is sensitive to the amount of active chlorine that originates mainly from CFCs and methyl chloroform due to the photolysis. With current emissions the total ozone column above the Antarctic has a thickness that is consistent with the observations. This is illustrated in Fig. 50.1, where monthly averaged total ozone column is shown for October 2011 (output of the zoomed run with 1° resolution).

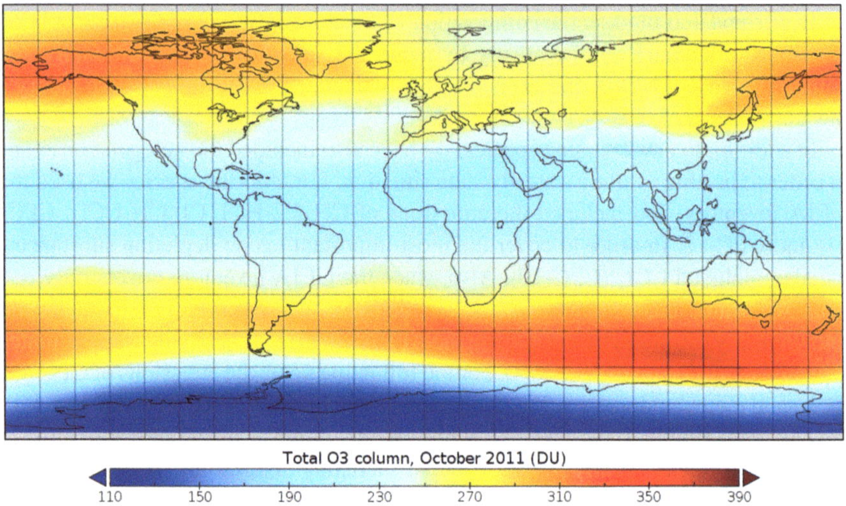

Fig. 50.1 Calculated total ozone column for October 2011 (monthly mean)

50.4 Conclusion

Our updated stratospheric module in SILAM is now capable of modelling the stratospheric chemistry where both chlorine and bromine reactions are included, together with their heterogeneous reactions. We have also included the main CFC species and their emissions. Now the model provides a reasonable estimate for the ozone, also above the Antarctic, where the ozone hole is sensitive to the amount of ozone depleting substances. In the future, a proper verification of the model against satellite measurements is planned.

Questions and Answers

Questioner: Kirill Semeniuk.

Question: What altitude range does ozone hole form in?

Answer: The altitude where the ozone concentration is decreased due to ozone depleting substances (ODS) depends slightly on the latitude. Above Antarctic, where the ozone hole appears, the maximum depletion occurs around 18–19 km altitude, and the dominant depletion occurs between 14 and 23 km. Figure 50.2 illustrates this. There we have averaged the ozone concentration during the year 2011, and compared it with a situation where there are no CFCs (CFC-11, CFC-12, CFC-10), neither methyl chloroform (CH_3CCl_3). The figure shows the difference between these two cases.

Question: Are PSCs prognostic or diagnostic?

Answer: In SILAM the ice and NAT particles are prognostic, while STS are diagnostic.

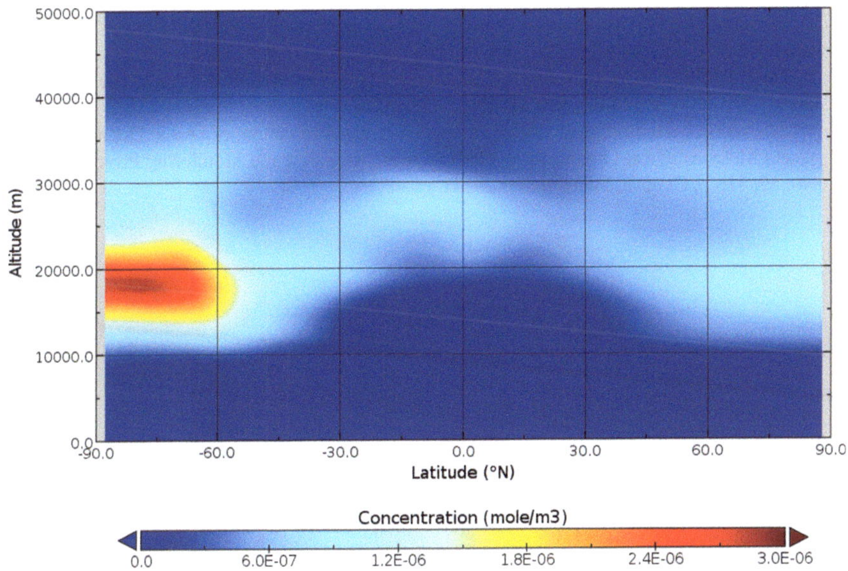

Fig. 50.2 Ozone destroyed by ODS (CFC-11, CFC-12, CFC-10 and CH_3CCl_3) in 2011 (yearly average)

Acknowledgements The SILAM stratospheric modules were developed within the Academy of Finland project ASTREX. Support of the Academy of Finland project GLORIA and NordForsk project NordicWelfAir is kindly appreciated.

References

1. AFEAS, The Alternative Fluorocarbons Environmental Acceptability Study. Production, Sales and Atmospheric Release of Fluorocarbons through 2003 (2005), https://agage.mit.edu/data/afeas-data
2. J.B. Burkholder, S.P. Sander, J. Abbatt, J.R. Barker, R.E. Huie, C.E. Kolb, M.J. Kurylo, V.L. Orkin, D.M. Wilmouth, P.H. Wine, Chemical kinetics and photochemical data for use in atmospheric studies, Evaluation No. 18 JPL Publication 15–10, Jet Propulsion Laboratory, Pasadena (2015), http://jpldataeval.jpl.nasa.gov
3. K.S. Carslaw, B. Luo, T. Peter, An analytic expression for the composition of aqueous HNO_3–H_2SO_4 stratospheric aerosols including gas phase removal of HNO_3. Geophys. Res. Letters **22**(14), 1877–1880 (1995)
4. J. Damski, L. Thölix, L. Backman, P. Taalas, M. Kulmala, FinROSE—middle atmospheric chemistry transport model. Boreal. Environ. Res. **12**, 535–550 (2007)
5. T. Huthwelker, B.P. Luo, S.L. Clegg, K.S. Carslaw, P. Brimblecombe, Solubility of HOCl in water and aqueous H_2SO_4 to stratospheric temperatures. J. Atmos. Chem. **21**, 81–95 (1995)
6. B. Luo, K.S. Carslaw, T. Peter, S.L. Clegg, Vapor pressures of $H_2SO_4/HNO_3/HCl/HBr/H_2O$ solutions to low stratospheric temperatures. Geophys. Res. Lett. **22**(3), 247–250 (1995)
7. M. Sofiev, R. Kouznetsov, M. Prank, J. Soares Alves Antunes, J. Vira, V. Tarvainen, A long-term re-analysis of atmospheric composition and air quality, in *Air Pollution Modeling and its*

Application XXV ITM 2016, ed. by C. Mensink, G. Kallos, pp. 55–59 (2017). https://doi.org/10.1007/978-3-319-57645-9_9

8. X. Yang, J.A. Pyle, R.A. Cox, Sea salt aerosol production and bromine release: role of snow on sea ice. Geophys. Res. Lett. **35**(L16815), 1–5 (2008)
9. WMO, World Meteorological Organization. Scientific Assessment of Ozone Depletion: 2014. World Meteorological Organization, Global Ozone Research and Monitoring Project—Report No. 55, 416 p, Geneva, Switzerland (2014)

Chapter 51
Lightning NO_X Distribution and Its Impact on Ozone Over the Contiguous United States During 2011

Daiwen Kang, Rohit Mathur, Limei Ran, George Pouliot, David Wong, Kristen Foley, Wyat Appel and Shawn Roselle

Nitrogen oxides (NO_X: NO + NO_2) play a critical role in controlling atmospheric chemistry, especially for the tropospheric ozone (O_3) formation and distribution. As one of the major natural NO_X sources, lightning NO_X (LNO_X) production is estimated to be in the range of 10–15% of the total global NO_X emissions budget [6]. In addition, lightning activity and the tropospheric distribution of LNO_X exhibit strong spatial and temporal variations [5]. To estimate LNO_X impact on air quality accurately, the spatial and temporal distributions of LNO_X need to be quantified using robust LNO_X production and distribution schemes in air quality models.

Beginning with the version (v5.2) of the Community Multiscale Air Quality (CMAQ, Appel et al. [2]) model, A LNO_X production scheme is implemented based on hourly gridded lightning flashes from the National Lightning Detection Network (NLDN) to estimate gridded hourly total LNO_X across the contiguous US for retrospective model applications. After the column total LNO_X is calculated, it is distributed vertically through the model layers based on the double-peak algorithm described in Allen et al. [1]. In this study, the relative contributions of LNO_X to the total NO_X emissions budget in time and space for April to September 2011 over the contiguous United States are quantified using the 2011 National Emissions Inventory (NEI) for anthropogenic NO_X emissions and soil NO emissions estimated by CMAQ inline biogenic emission model [3, 4]. Model simulations with LNO_X (NLDN) and without LNO_X (Base) were performed. The impact of LNO_X on O_3 air quality at the surface were assessed by region and month against gas phase measurements from the EPA's Air Quality System (AQS; https://www.epa.gov/aqs) for hourly gas species, accounting for spatial differences in the relative importance of LNO_X relative to other sources of NO_X emissions. Vertical profiles were also examined against ozone-sonde data measured from the Deriving Information on Surface

D. Kang (✉) · R. Mathur · L. Ran · G. Pouliot · D. Wong · K. Foley · W. Appel · S. Roselle
National Exposure Research Laboratory, U.S. Environmental Protection Agency, Research Triangle Park, Durham, NC 27711, USA
e-mail: kang.daiwen@epa.gov

C. Mensink et al. (eds.), *Air Pollution Modeling and its Application XXVI*,
Springer Proceedings in Complexity,
https://doi.org/10.1007/978-3-030-22055-6_51

Conditions from Column and Vertically Resolved Observations Relevant to Air Quality (DISCOVER-AQ; https://www-air.larc.nasa.gov/missions/discover-aq/discover-aq.html) 2011 campaign.

51.1 NO_X Emissions and the Relative Contributions of LNO_X

The spatial distributions of anthropogenic NO_X (a), soil NO (b), LNO_X generated by CMAQ with NLDN lightning flash data (c), and the relative contribution ratio of LNO_X to total NO_X (a + b + c) during July 2011 are shown in Fig. 51.1. As indicated by Fig. 51.1d, the LNO_X ratios are largest over the Rocky Mountain (RM) area (Fig. 51.2a) followed by the Southeast (SE) region; the LNO_X ratios up to 90% of the total NO_X emissions were observed in some locations signifying that LNO_X emissions were the primary NO_X sources over these areas during that particular month.

To examine the differences in LNO_X contributions over space and time, the monthly LNO_X ratios from April to September 2011 over the entire domain and for several sub regions are shown in Fig. 51.2b. Contributions of LNO_X emissions to the total NO_X emissions ranging from 10% (September) to 22% (July) are observed over the contiguous US. Collectively over the six-month period, the largest LNO_X ratios were observed in SE and that was followed by RM. The largest and smallest

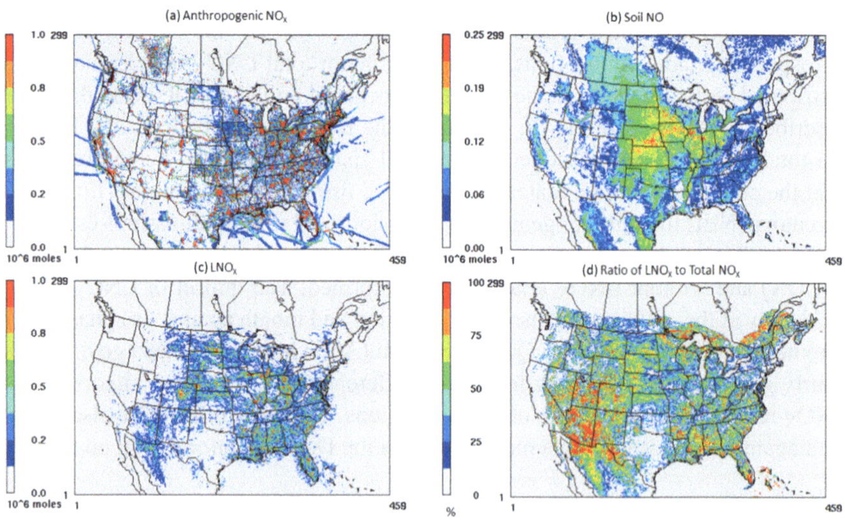

Fig. 51.1 Total monthly emissions (in 10^6 mol) and the LNO_X ratios during July, 2011: **a** anthropogenic NO_X, **b** soil NO, **c** LNO_X, and **d** the ratio of LNO_X to total NO_X emissions

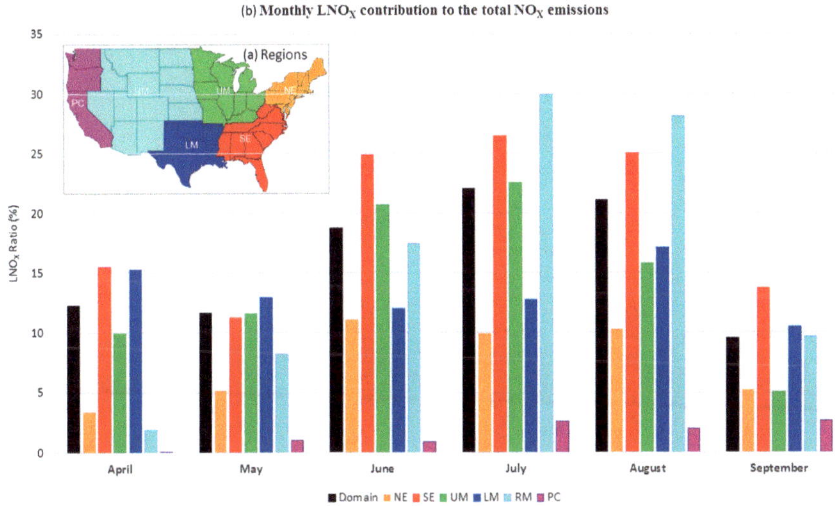

Fig. 51.2 **a** Regions, **b** The monthly ratio of LNO_X to the total NO_X emissions over the domain and each of the regions as shown in (**a**)

LNO_X ratios were observed at the RM region during July and August and Pacific Coast (PC) region (less than 3%) across all months, respectively.

51.2 Assessing the Impact on Air Quality

To assess the LNO_X impact on air quality, simulations with LNO_X (NLDN) and without LNO_X (Base) using CMAQv5.2 over the 12 km contiguous US domain were conducted. The gridded daily lightning flash rate from NLDN and the corresponding changes (NLDN—Base) of the maximum surface hourly O_3 mixing ratios on July 11th, 2011 are presented in Fig. 51.3. While the maximum change of hourly O_3 mixing ratio can be as large as 70 ppb, it ranges mostly from 6 ppb to 30 ppb along the path of lightning strikes.

The ozone-sonde data at Beltsville and Edgewood, MD measured during 2011 DISCOVER-AQ campaign on days when significant lightning impact was observed in the model simulations at each location were used to evaluate the impact of LNO_X on vertical O_3 profiles. The observed ozone-sonde data were paired with model prediction in time and space and the average values over all the ozone-sonde launches (one or two measurements per day) on the selected days at each location were taken. As indicated in Fig. 51.4, the O_3 mixing ratios aloft are significantly underestimated but overestimated near the ground-level by Base simulation, while with LNO_X in the NLDN simulation both the underestimation aloft and the near-surface overestimation are mitigated. And even though the benefit in model performance from the addition of LNO_X emissions was not revealed by the overall statistics over the regions such as SE

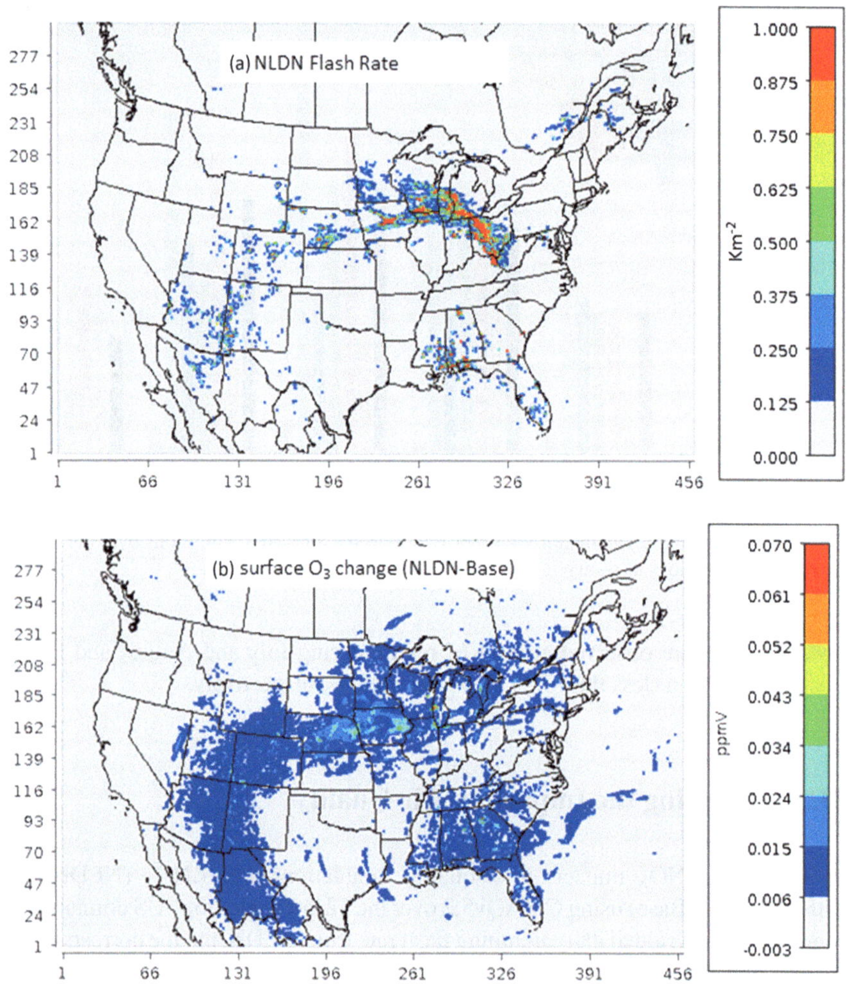

Fig. 51.3 The daily NLDN lightning flash rate (**a**) and the maximum change (NLDN—Base) of hourly surface O_3 (**b**) on July 11, 2011

where surface O_3 mixing ratios were already over estimated by the Base simulation, the increase of performance skill is obvious from the improved vertical structure of O_3, at least for the two locations shown here. A potentially significant contribution of LNO_X emissions to the total NO_X emissions budget across much of the US during the

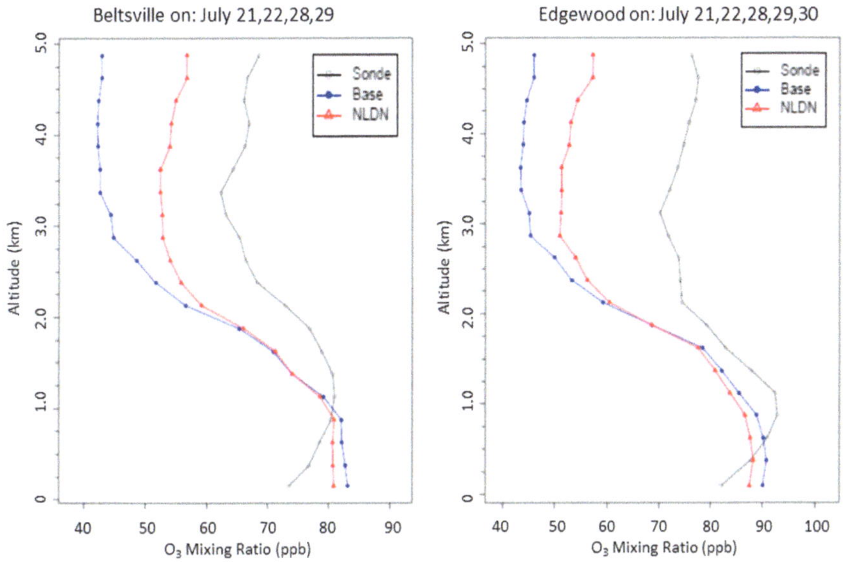

Fig. 51.4 Mean vertical O_3 profiles from ozone-sonde observations and model simulations at Beltsville and Edgewood, MD on days when significant lightning impact was observed near each location

warmer months is present, thus it is important to count LNO_X emissions in regional air quality model simulations to improve the accuracy of air quality predictions.

Question and Answer

Questioner: Jack Chen

Question: Does the NLDN observations of lightning flash data have different flash type by in-cloud, cloud-ground? If so, are they handled differently in the model?

Answer: No. the NLDN observations only contain cloud-to-ground flashes. We used the satellite-based climatological intracloud to cloud-to-ground flash ratios to estimate gridded hourly total LNO_x.

Acknowledgements We sincerely thank Vaisala for making the NLDN data available to us.

Disclaimer The views expressed in this paper are those of the authors and do not necessarily represent the views or policies of the U.S. EPA. Mention of trade names or commercial products does not constitute endorsement or recommendation for use.

References

1. D.J. Allen, K.E. Pickering, R.W. Pinder, B.H. Henderson, K.W. Appel, A. Prados, Impact of lightning-NO on eastern United States photochemistry during the summer of 2006 as determined using CMAQ model. Atmos. Chem. Phys. **12**, 1737–1758 (2012). https://doi.org/10.5194/acp-12-1737-2012
2. K.W. Appel, Coauthors, Description and evaluation of the Community Multiscale Air Quality (CMAQ) model version 5.1, Geosci. Model Dev. **10**, 1703–1732 (2017). https://doi.org/10.5194/gmd-1703-2017
3. E. Kinnee, C.D. Geron, T. Pierce, United States land use inventory for estimating biogenic ozone precursor emissions. Ecol. Appl. **7**(1), 46–58 (1997). Ecological Society of America, NY
4. J. Novak, T. Pierce, Natural emissions of oxidant precursors. Water Air Soil Pollut. **67**, 57–77 (1993)
5. G. Medici, K.L. Cummins, D.J. Cecil, W.J. Koshak, S.D. Rudlosky, The intracloud lightning fraction in the contiguous United States. Mon. Wea. Rev. **145**, 4481–4499 (2017). https://doi.org/10.1175/MWR-D-16-0426.s1
6. L.T. Murray, Lightning NO_x and impacts on air quality. Curr. Pollut. Rep. (2016). https://doi.org/10.1007/s40726-016-0031-7

Chapter 52
Is a Model's Scatter Really "Very Small" or Is Model A Really "Performing Better" Than Model B?

Steven Hanna and Joseph Chang

Abstract Many papers are published in which a dispersion model's predictions are compared with field observations and/or with other models' predictions. Standard model performance measures are used such as Fractional Bias (FB). Many times, subjective statements are made such as "The model has very small scatter" or "Model A is performing better than Model B". About 30 years ago, we developed the BOOT model evaluation software, which has two main components: 1. Calculation of model performance measures such as FB; and 2. Calculation of confidence limits (e.g., 95%) on performance measures and on the difference in a performance measure between two models. Bootstrap or Jackknife resampling methods are employed. We briefly review the methodology in BOOT's Component 2, which is seldom used by researchers. We present an example from a project where several urban puff models' predictions are compared with JU2003 field data, and where assessments are carried out regarding whether, for example, it can be concluded, with 95% confidence, that the difference in FB for two models is not significantly different from zero.

52.1 Introduction

Dispersion models are often used in decision-making regarding pollutant impacts on the public and on the environment. Among other considerations, the decision-maker is interested in the uncertainties in the dispersion model outputs, and whether one model is better than another. The purpose of this paper is to describe a quantitative method to assess these uncertainties and determine whether a performance measure or a difference in performance measures between models are statistically significant.

S. Hanna (✉)
Hanna Consultants, 7 Crescent Ave, Kennebunkport, ME 04046-7235, USA
e-mail: hannaconsult@roadrunner.com

J. Chang
RAND Corporation, Arlington, VA 22202-5050, USA
e-mail: jchang@rand.org

C. Mensink et al. (eds.), *Air Pollution Modeling and its Application XXVI*,
Springer Proceedings in Complexity,
https://doi.org/10.1007/978-3-030-22055-6_52

To assess dispersion model performance and uncertainties, the models are evaluated by comparison with several sets of field data. The BOOT model evaluation software [2, 4] is widely used to evaluate the performance of individual models. The software also allows the differences in the statistical performance measures between several models to be assessed to determine whether they are significant at some confidence level (usually 95%). The BOOT methodology for assessing confidence limits is demonstrated here using several urban puff dispersion models applied to the Oklahoma City Joint Urban 2003 (JU2003) field data [1, 3] as part of the Urban Dispersion International Evaluation Exercise (UDINEE) [5].

A formal statistical test such as that in BOOT is needed to compare the performance measures for two models. BOOT automatically can account for the possibility that the two models' predictions may be correlated. For example, if the concentrations predicted by model B are always exactly two times those predicted by model A, then there will always be a significant difference between the performance measures calculated for the two models. This can be shown by expanding the formula for the variance between fluctuations in two variables X and Y, and finding that a term "-2R" is present, where R is the correlation between fluctuations in X and in Y.

52.2 Statistical Methods

The following equations define the statistical performance measures that are used in the BOOT evaluation software [2]. These include the fractional bias (FB), the geometric mean bias (MG), the normalized mean square error (NMSE), the geometric variance (VG), and the fraction of predictions within a factor of two of observations (FAC2).

$$\mathrm{FB} = 2(\langle \mathrm{Co}\rangle - \langle \mathrm{Cp}\rangle)/(\langle \mathrm{Co}\rangle + \langle \mathrm{Cp}\rangle) \quad (52.1)$$

$$\mathrm{MG} = \exp(\langle \ln \mathrm{Co}\rangle - \langle \ln \mathrm{Cp}\rangle) \quad (52.2)$$

$$\mathrm{NMSE} = (\langle (\mathrm{Co}-\mathrm{Cp})^2\rangle/\langle \mathrm{Co}\rangle\langle \mathrm{Cp}\rangle \quad (52.3)$$

$$\mathrm{VG} = \exp\left(\langle(\ln \mathrm{Co}-\ln \mathrm{Cp})^2\rangle\right) \quad (52.4)$$

$$\mathrm{FAC2} = \text{Fraction of predictions that are within a factor of two of observations} \quad (52.5)$$

where Cp is model prediction and Co is observation of concentration; and $\langle\ \rangle$ is average over the dataset. A perfect model has MG, VG, and FAC2 = 1.0; and FB and NMSE = 0.0.

To estimate confidence limits on differences in a performance measure (PM) between two or more models, we start with a master table containing N rows, each of which represents one time period and one concentration sampler. Bootstrap or jackknife resampling methods are used to determine the difference ΔPM (i, j) between two models (i and j). The values of ΔPM (i, j) are ranked and used to define 95% confidence limits. If the 95% confidence limits on ΔPM (i, j) overlap 0.0, then it can be concluded, with 95% confidence, that the difference in PM between the two models is not significantly different from 0.0.

52.3 JU2003 Data Set and Models Evaluated

The JU2003 field data involved releases of puffs or plumes of SF_6 in the Oklahoma City domain in 2003 [1]. Three to six puffs were released on each of the ten IOP days [3, 6]. Observations of SF_6 concentration were made by 10 real-time samplers (resolution of 0.5 s) at distances of a few hundred meters. In the UDINEE model comparison study [5], predictions of ten models from several countries were compared with the puff observations. In this paper, the dosage/Q outputs from the UDINEE data set are compared, where Q is the total mass of SF_6 in a puff. Puff-sampler pairs are included in the analysis only if both predicted and observed max 0.5 s concentrations exceed 400 ppm. The UDINEE model comparison reports and output files use anonymous designations (e.g., Model M1). In the comparisons discussed in this paper, three of the models are compared for two of the IOPs (3 and 8). BOOT was applied to these models and data and the performance measures listed above were calculated. In addition, confidence limits on the performance measures and the differences in performance measures were calculated by BOOT.

52.4 Results

Of the many tables and figures that show the statistical results, we select one figure and one table for discussion. There were four puffs released during each of IOP3 and IOP8 and, for each puff, six or seven samplers produced valid data. A total of 48 puff-sampler data pairs during IOP3 and IOP8 are analyzed. We consider only Models M3, M5, and M11, since they produced model predictions for all 48 puff-sampler pairs. Here we present results for dosages (concentrations integrated over time, with units ppt-s) divided by mass emitted, Q.

As an aid to understanding the quantitative calculations by BOOT, it is useful to look at scatter plots such as Fig. 52.1, where predicted and observed dosage/Q are plotted for models M3, M5, and M11. The viewer's first impression is that the points are a "shotgun blast pattern" with little skill by the models. The scatter covers a range of plus and minus about a factor of ten. Although models M3 and M5 appear to overpredict more often than model M11, the points for the three models overlap.

Quantitative performance measures (PMs) for the data points in Fig. 52.1 are listed in Table 52.1. Model M11 appears to be doing "better" for most PMs, while models M3 and M5 are indicating a tendency to overpredict. But are these differences significant? It was mentioned earlier that the significance of differences between two models depends partly on how well the two models are correlated. When we plotted the predictions of M3 versus M11, and of M5 versus M11, there was no obvious trend or correlation seen. However, a clear correlation could be seen between the predictions of M3 and M5.

As an example of significance tests with a single performance measure (FB), it is found that M11 is the only model of the three tested whose FB value (−0.18) is not significantly different from 0.0 at the 95% confidence limit. Looking at the model-to-model difference in FB values, the three combinations of model pairs all have differences ΔFB (>0.71) that are significantly different from zero at the 95% confidence limit. Thus, based just on the performance measure FB and this specific

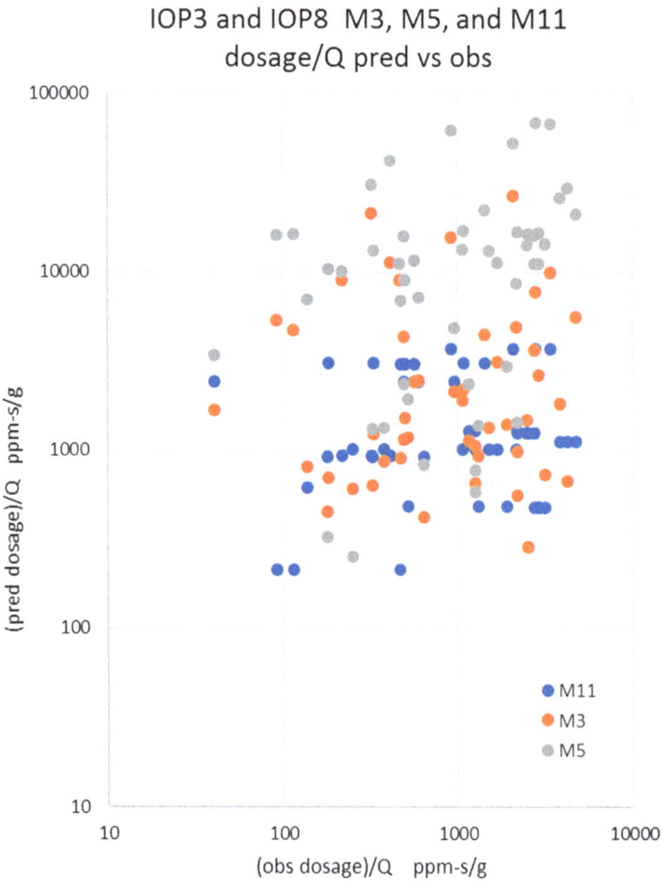

Fig. 52.1 Scatter plot of predictions versus observations of dosage/Q

Table 52.1 Model performance measures (PMs) for puffs and samplers in IOP3 and IOP8

Dosage/Q IOP3 & IOP8	FB	NMSE	FAC2	MG	VG	HI1	HI2
Obs.						4702	4233
M3	−0.97	6.92	0.31	0.40	24	26660	21322
M5	−1.68	22.74	0.15	0.11	1300	68615	68222
M11	−0.18	1.27	0.31	0.66	5.9	3709	3709

data set, it is possible, from a statistical confidence point of view, to say that (1) Model M11's fractional mean bias (FB) is not significantly different from 0.0, and (2) model M11's FB is significantly better (at the 95% confidence level) than the FB for models M3 and M5.

52.5 Further Comments

The above model performance evaluation exercise is intended as an example of application of the BOOT statistical confidence assessment method. The UDINEE exercise has involved many more models and endpoints and JU2003 IOPs, which will be described in an upcoming special issue of *Boundary Layer Meteorology*.

Acknowledgements Steven Hanna acknowledges the support of the U.S. Defense Threat Reduction Agency (DTRA), with Dr. Ronald Meris as Project Manager. Thomas Mazzola of Engility Corp provided much guidance. Dr. Miguel Hernández-Ceballos of JRC, Ispra, Italy provided the UDINEE data.

References

1. K.J. Allwine, J. Flaherty, Joint Urban 2003: study overview and instrument locations. Pacific Northwest National Laboratory, Richland, WA. PNNL-15967 (2006)
2. J.C. Chang, S.R. Hanna, Air quality model performance evaluation. Meteorol. Atmos. Phys. **87**, 167–196 (2004)
3. K.L. Clawson, R.G. Carter, D.J. Lacroix, C.J. Biltoft, N.F. Hukari, R.C. Johnson, J.D. Rich, S.A. Beard, T. Strong, Joint Urban 2003 (JU2003) SF_6 Atmospheric Tracer Field Tests, NOAA Tech Memo OAR ARL-254. Air Resources Laboratory, Silver Spring, MD, pp. 162 + Appendices (2005)
4. S.R. Hanna, Confidence limits for air quality models, as estimated by Bootstrap and Jackknife resampling methods. Atmos. Environ. **23**, 1385–1395 (1989)

5. M.A. Hernández-Ceballos, S. Galmarini, S.R. Hanna, T. Mazzola, J.C. Chang, R. Bianconi, R. Bellasio, UDINEE Project: international platform to evaluate urban dispersion models' capabilities to simulate radiological dispersion device, in *Proceedings of HARMO17 Conference*, Budapest Hungary, 28 p (2016)
6. Y. Zhou, S.R. Hanna, Along-wind dispersion of puffs released in built-up urban areas. Bound.-Layer Meteorol. **125**, 469–486 (2007)

Chapter 53
Sensitivity of Atmospheric Composition Mesoscale Simulations in the Mediterranean to the Meteorological Data and Chemical Boundary Conditions

D. G. Amanatidis, S. Myriokefalitakis, Georgios Fanourgakis, N. Daskalakis and M. Kanakidou

Abstract Limited area models applied at higher resolution than global models and using datasets of higher resolution are generally expected to more accurately represent the spatiotemporal variability of key meteorological and climate parameters such as near surface temperature, pressure, wind speed and atmospheric composition. However, limited area models require boundary conditions and the accuracy of such datasets reflects on the accuracy of the mesoscale simulations of atmospheric composition, in particular of the longer-lived species. The effects of various resolution meteorological data and of different chemical boundary and initial conditions on the simulated concentrations of the chemical gases and aerosols in the Mediterranean have been here investigated. Two different simulations in three domains of progressively increasing horizontal resolution were performed for the year 2016 using the mesoscale Weather Research and Forecasting (WRF) meteorological model with the chemistry module (WRF-Chem) and an additional one with different chemical boundary and initial conditions. Meteorology from the Global Forecast System (GFS) at two different horizontal resolutions ($1 \times 1°$ and $0.25 \times 0.25°$) available

D. G. Amanatidis · G. Fanourgakis · M. Kanakidou (✉)
Environmental Chemical Processes Laboratory, Department of Chemistry, University of Crete, Voutes Campus, P.O.Box 2208, 70013 Heraklion, Greece
e-mail: mariak@uoc.gr

D. G. Amanatidis
e-mail: dimaman25@gmail.com

G. Fanourgakis
e-mail: fanourg@uoc.gr

S. Myriokefalitakis
IERSD, National Observatory of Athens, Lofos Koufou, Penteli, Greece
e-mail: steliosm@noa.gr

S. Myriokefalitakis · N. Daskalakis
LAMOS, Institute of Environmental Physics (IUP), University of Bremen, Bremen, Germany
e-mail: daskalakis@uni-bremen.de

C. Mensink et al. (eds.), *Air Pollution Modeling and its Application XXVI*,
Springer Proceedings in Complexity,
https://doi.org/10.1007/978-3-030-22055-6_53

from NOAA has been used in the model to investigate their impact on the simulated air pollutants. The higher resolution meteorological input improves the comparison of model results to observations. The results are found sensitive to the chemical boundary and initial conditions.

53.1 Introduction

The impact of air pollution on human health, ecosystems, and climate is uncontestable. The continuous growth of urbanization, transportation and industrialization recorded all over the Mediterranean during the last century led to a rise in the anthropogenic emissions of several key trace gases and aerosols. Modern society is becoming increasingly vulnerable to changes in weather, climate and air quality, supporting the need of models for reliable projections on regional to local scales. Trace gases and aerosols affect climate both directly and indirectly and in turn regional air quality is strongly influenced by weather and climate. Several studies [1, 2] have shown strong correlation between surface ozone concentrations and surface temperature. Also, temperature increase leads to increases of biogenic emissions [5]. Local meteorological conditions (e.g. high solar radiation, wind speeds, humidity) favor the formation of secondary pollutants [6]. This study investigates the sensitivity of mesoscale model simulations of meteorology and air quality for July 2016 on the different meteorological and chemical boundary and initial conditions used by the model for the Mediterranean focusing on the Eastern Mediterranean.

To study the impact of distinct sources on air pollution and atmospheric deposition in the South Eastern Mediterranean the mesoscale Weather Research and Forecasting model version 3.8 [4] coupled with chemistry (WRF-CHEM) was setup for the area and evaluated. The present study focuses on evaluating the model by comparing simulation results with surface observations at ground-based stations of Finokalia on the island of Crete in Greece and Ayia Marina in Cyprus. It also investigates the importance of the model spatial resolution for the simulated pollutant concentrations and in particular for gases O_3, CO and NO_2 and for aerosols BC, OC, SO_4^{2-}, NO_3^-.

53.2 Data and Methodology

The WRF-CHEM simulations of the spatiotemporal variability of key meteorological and climate parameters such as near surface temperature, wind speed, surface pressure and atmospheric composition, are first evaluated. Three model domains extended vertically up to 50 hPa with 30 hybrid levels are used: a coarse (36 × 36 km), an intermediate (12 × 12 km) and a higher horizontal resolution (7.2 × 7.2 km) (Fig. 53.1a). The intermediate and finest resolutions were run on one-way nested setup. The National Centers for Environmental Predictions (NCEP) meteorology analyses, available every 6 h in 1° × 1° (GFS-FNL) and 0.25° × 0.25°

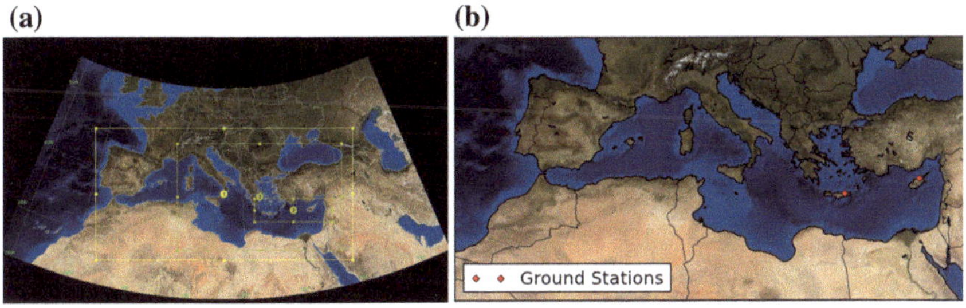

Fig. 53.1 **a** Chosen model domains; **b** Ground stations used for model evaluation

(GDAS) horizontal resolution and 27 vertical levels from the surface up to 10 hPa were used to compute the meteorological fields for the mesoscale model. Six-hourly varying chemical initial and boundary conditions have been extracted from 2 different 3-dimensional global chemistry-transport models simulations: the TM4-ECPL (Daskalakis et al. [7] at 3° × 2° and the MOZART-4 [3] at 2.5° by 1.9°, longitude x latitude horizontal resolution.

All tracers are initialized in the beginning of the simulation in the model domain and every 6 h at the lateral and upper boundaries of the coarse domain. Hourly model output is used to compare with hourly and 24 h in some cases, observed concentrations. All our simulations are conducted with the same model physics configuration to facilitate intercomparison.

Hourly observations of O_3, CO and NO_2 and daily observations of BC, OC, SO_4^{2-}, NO_3^- are taken from two surface monitoring stations with available data for the studied period (Fig. 53.1b): Finokalia in north east of Crete in Greece and Ayia Marina in north west of Cyprus, and compared with model results to evaluate simulated trace gases and aerosols and how their levels are effected by the meteorological factors.

53.3 Results

53.3.1 Sensitivity to Meteorological Input Data

Statistical metrics are derived by comparing the two model simulations (1° × 1° and 0.25° × 0.25° meteorology) to hourly observations for gases and to daily for aerosols. The correlation coefficient (R) and Normalized Mean Bias (NMB) for temperature at 2 m, surface pressure and wind speed at 10 m, variables that are calculated by the model, are also computed. High R (0.88–0.95) especially for Ayia Marina with low NMB (−11 to −9%) are calculated for temperature. Surface pressure shows a good correlation (0.91–0.94) and low NMB reaching −1 to 5% for both stations. For wind speed for both resolutions, low correlations and over-predictions by the WRF model are found especially for Ayia Marina, which shows the effect of an unresolved topography of mountains, hills and valleys and other smaller scale terrain features.

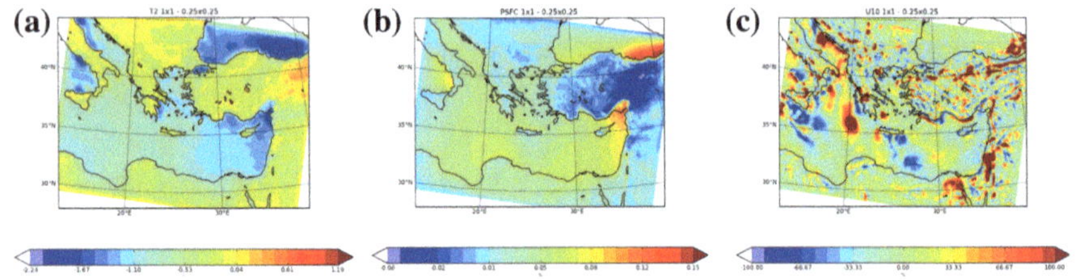

Fig. 53.2 Monthly mean percentage differences of simulated distributions of **a** Temperature at 2 m, **b** Surface Pressure, **c** Wind Speed at 10 m between intermediate domain simulations using 1° × 1° and 0.25° × 0.25° meteorological fields

The percent differences in temperature, surface pressure and wind speed between the two simulations, shown in Fig. 53.2 for the intermediate model domain, range between −2% and 1.2% for the temperature, −0.06% and 0.15% for surface pressure and locally over 100% for the wind speed which indicates aforementioned issues with the topography.

The results obtained using the high resolution meteorology dataset (0.25° × 0.25°) show better performance when compared with observations. As a consequence the model simulates the meteorology in higher precision than when using the lower resolution meteorological input. This improvement is expected to also affect the chemistry simulations.

For O_3, and for Finokalia, relatively good correlation coefficients of 0.5–0.6 are found between model results and observations, associated with low NMB reaching −7%. The use of meteorological datasets of different resolution does not result in any significant improvement in the comparison of simulations with observations. For CO low correlation coefficients (0.4) for both stations are found with NMB of 17–38%. The use of higher resolution meteorological input is improving the correlation of simulated CO concentrations with observations at Ayia Marina (R = 0.5). In particular, by increasing the resolution of the meteorological input, simulated NO_2 concentrations are significantly improved (R = 0.7 and NMB = 10% for Ayia Marina and R = 0.4 and NMB = 73% for Finokalia). For aerosol species the calculated correlations between observations and model results are: 0.5 for Black Carbon (BC) at both stations (NMB of −2 to 13%) and 0.6 (NMB −72%) and 0.7 (NMB −70%) for Organic Carbon (OC) at Ayia Marina and Finokalia, respectively. Similar correlation is found for sulfate (R = 0.6 and NMB around −30%) for both stations; while for nitrate low correlations are found around 0.4 for Ayia Marina and 0.5 for Finokalia with underestimations of −42% to −54%, respectively. For most aerosol species significant improvements are calculated when using higher resolution meteorological input.

When comparing the results of the simulations using the two different meteorological datasets changes between −38 and 7% are found for O_3, that are justified by the low differences in the simulated temperatures. For CO the computed differences are smaller than for O_3 (−5 to 23%) with the highest differences mainly downwind urban areas and near the coasts. NO_2 shows larger divergence of −14 to 102% mostly

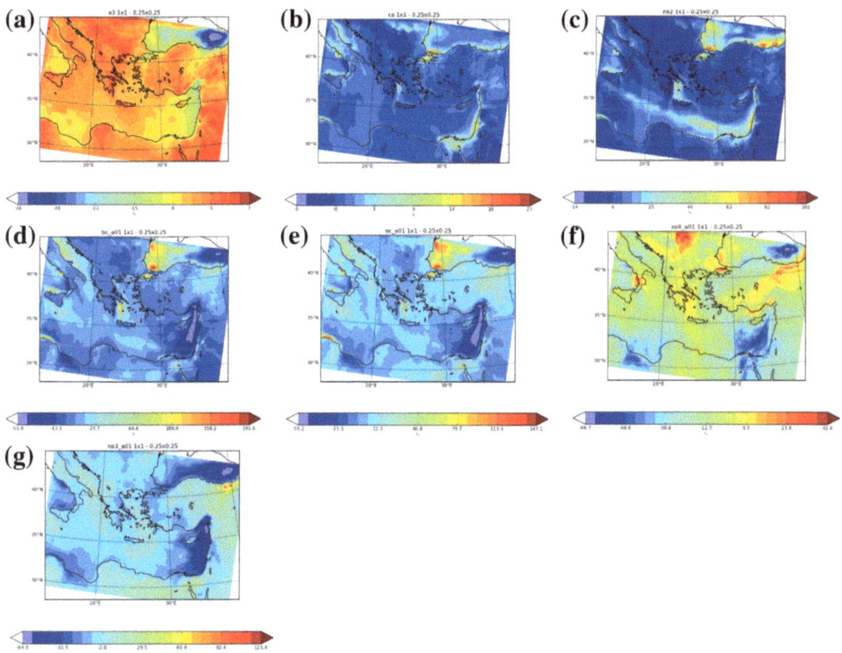

Fig. 53.3 Mean monthly percentage differences of WRF-Chem simulations of **a** Ozone, **b** CO, **c** NO_2, **d** BC, **e** OC, **f** SO_4^{2-}, **g** NO_3^-, at surface using 1° × 1° and 0.25° × 0.25° meteorological fields

in coastal areas and close to the shipping lanes, showing the importance of meteorology in the simulations of plume chemistry in the model. Large changes in the computed concentrations of all aerosols are imprinted on the distributions ranging from −53.9% to 191% for BC, −55% to 147% for OC, −67% to 41.6% for SO_4^{-2} and −65 to 124% for NO_3^- (Fig. 53.3), which are most probably associated with the computed large changes of wind speed in the meteorological forecasts (Fig. 53.2c).

53.3.2 Sensitivity to Chemical Boundary Conditions

The results shown in Sect. 53.3.1 have been performed with chemical boundary conditions from MOZART-4 model. To investigate the effect of using different chemical boundary conditions an additional simulation with WRF-Chem with meteorological input at, 0.25 × 0.25°, resolution was performed using chemical boundary and initial conditions from the TM4-ECPL model. Figure 53.4 shows the mean percentage difference ([(TM4-ECPL)-(MOZART-4)]/(MOZART-4)) of gases and aerosols for July 2016. Over the entire domain higher ozone concentrations are computed with the TM4-ECPL chemical boundary and initial conditions (positive monthly mean difference reaching about 80%), the correlation with the observations is improved (0.48)

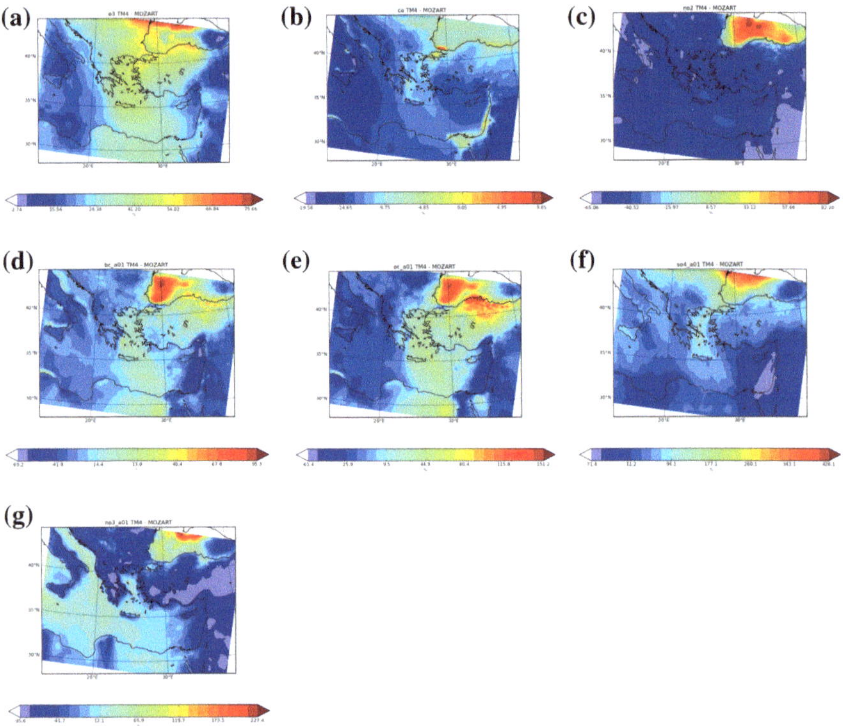

Fig. 53.4 Mean monthly percentage differences of WRF-Chem simulations of **a** Ozone, **b** CO, **c** NO_2, **d** BC, **e** OC, **f** SO_4^{2-}, **g** NO_3^-, at surface due to different chemical boundary and initial conditions input ([(TM4-ECPL)-(MOZART-4)]/(MOZART-4))

and the positive NMB of 21% shows a slight overestimation of O_3 by the model. For CO concentrations small changes up to 10% are found over the hotspot areas such as Istanbul, Cairo and over Lebanon. As for O_3 the correlation with observations has improved, reaching 0.51 for the high resolution domain.

NO_2 concentrations show large differences (reaching about 80%) over the Black Sea which might be due to difference in the emissions from shipping in the two global models during the simulated period. Larger difference near the boundaries and over the Black Sea are computed for $SO_4{}^{2-}$, OC and BC aerosols, reaching 400%, 150% and 95% respectively, showing the importance of long range transport and chemistry for these species. High correlations reaching 0.8 and lower NMB (−29%) for OC are found when using TM4-ECPL boundaries. These results show the importance of chemical boundary conditions for the oxidant and aerosol levels computed by WRF-Chem model.

53.4 Conclusions

This study using the WRF-Chem models shows a reasonable agreement between computed and observed surface $O_{3,}$ CO, BC, OC and SO_4^{2-} concentrations in both stations in the East Mediterranean with available data; while the model shows lower performance for NO_2 and $NO_3{}^-$. Increases in the horizontal resolution of the model lead to improvements in the model performance, which is most noticeable for aerosols. Meteorological input of higher resolution seems to improve the simulations of both atmospheric transport and chemical composition. Use of different chemical boundary conditions (model results from TM4-ECPL instead of MOZART-4) also improves the correlation with observations, in particular for ozone and OC, and shows the large sensitivity of the model results to this input.

References

1. B.J. Bloomer, J.W. Stehr, C.A. Piety, R.J. Salawitch, Observed relationships of ozone air pollution with temperature and emissions. Geophys. Res. Lett. **36**, L09803 (2009). https://doi.org/10.1029/2009GL037308
2. L. Camalier, W. Cox, P. Dolwick, The effects of meteorology on ozone in urban areas and their use in assessing ozone trends. Atmos. Environ. **41**, 7127–7137 (2007)
3. L.K. Emmons et al., Description and evaluation of the model for ozone and related chemical Tracers, version 4 (MOZART-4). Geosci. Model Dev. **3**, 43–67 (2010)
4. G.A. Grell et al., Fully coupled "online" chemistry within the WRF model. Atmos. Environ. **39**, 6957–6975 (2005). https://doi.org/10.5194/acp-14-10119-2014
5. U. Im et al., The impact of temperature changes on summer time ozone and its precursors in the Eastern Mediterranean. Atmos. Chem. Phys. **11**, 3847–3864 (2011)
6. M. Kanakidou et al., Megacities as hot spots of air pollution in the East Mediterranean Atmospheric Environment (2011). https://doi.org/10.1016/j.atmosenv.2010.11.048
7. N. Daskalakis et al., Large gain in air quality compared to an alternative anthropogenic emissions scenario. Atmos. Chem. Phys. **16**, 9771–9784 (2016). https://doi.org/10.5194/acp-16-9771-2016

Chapter 54
Quantification of Uncertainty in Lagrangian Dispersion Modelling, Using ECMWF's New ERA5 Ensemble

Andy Delcloo and Pieter De Meutter

Abstract FLEXPART is a Lagrangian particle dispersion transport model which is originally designed for calculating the long-range and mesoscale dispersion of air pollutants from point sources. Through the years, these type of models have proven to be a very useful tool in an operational context for the protection of the population in case of accidents in a nuclear power plant. In the meantime, FLEXPART has evolved into a more comprehensive tool for atmospheric transport modelling and analysis, and it can be used for a wide range of applications. The model can be used in a forward or backward mode, making it possible to trace back the source pollution contribution of a certain pollutant. To perform the FLEXPART dispersion simulations under consideration, we will use meteorological data from the European Centre for Medium Range Forecasts (ECMWF), more specifically the new ERA5 10-member climate data reanalysis at a 63 km resolution. We will explore how good we can access the model uncertainty in an objective way by taking advantage of ensemble weather forecasts.

54.1 Introduction

The International Monitoring System (IMS) is being setup to verify compliance with the Comprehensive Nuclear-Test-Ban Treaty. Part of the IMS will consist of forty stations that can measure certain airborne radioactive xenon isotopes, which are created after a nuclear explosion. However, radioactive xenon isotopes are regularly detected by IMS stations and it has been shown that these originate from civilian sources such as medical isotope production facilities and nuclear power plants [2]. This xenon background could mask the signatures of a nuclear explosion.

A. Delcloo (✉) · P. De Meutter
Royal Meteorological Institute of Belgium, Brussels, Belgium
e-mail: Andy.Delcloo@meteo.be

Department of Physics and Astronomy, Ghent University, Ghent, Belgium

P. De Meutter
SCK CEN, Belgian Nuclear Research Centre, Boeretang 200, 2400 Mol, Belgium

C. Mensink et al. (eds.), *Air Pollution Modeling and its Application XXVI*, Springer Proceedings in Complexity,
https://doi.org/10.1007/978-3-030-22055-6_54

One of the possible ways to deal with this issue is to simulate the contribution of civilian sources to the IMS station detections explicitly. To ascertain the level of agreement that can be made between such atmospheric transport modelling and IMS observations, two international challenges have been set up, of which the results of the latest challenge have been published recently [1]. During the “blind phase” of this second challenge, participants were asked to simulate the Xe-133 contribution from the nuclear facility Ansto (near Sidney, Australia) to six IMS stations in the southern hemisphere. Real emission data were not made available then and participants were asked to perform simulations with unit releases. In the current “open phase” of the challenge, participants are encouraged to perform additional in-depth research while having available the emission data and IMS observations for research purposes.

In this paper, we redo this second international ATM challenge using the new ERA5 dataset from ECMWF. ERA5 is unique compared to previous reanalyses since it consists of 10 members. The spread between the members represents the meteorological uncertainty.

54.2 Method

We have used detailed time-resolved emission data from the nuclear facility Ansto. Flexpart [3] has been run multiple times (using each of the ERA5 ensemble members) in forward mode to simulate Xe-133 activity concentrations at six IMS stations in the southern hemisphere for which observations were available for the period 11 May 2013 until 10 June 2013.

The (dis)agreement between the observed and simulated Xe-133 activity concentration is quantified as in Maurer et al. [1], which is the following rank score (Eq. (54.1)):

$$\text{Rank} = \text{R}^2 + (1 - |\text{FB}|/2) + \text{F5} + \text{ACC} \tag{54.1}$$

with R the correlation coefficient, FB the fractional bias, F5 the factor of simulated activity concentrations within a factor 5 of the observations and ACC the accuracy with the associated threshold being the sample-specific minimum detectable concentration (roughly the accuracy of discriminating detections from non-detection) (Fig. 54.1).

54.3 Results

See Fig. 54.2.

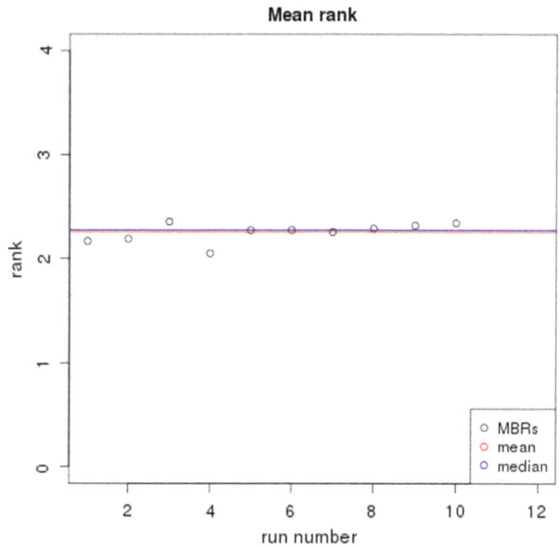

Fig. 54.1 Mean rank for the 10 ensemble members. The mean rank over all six IMS stations is shown

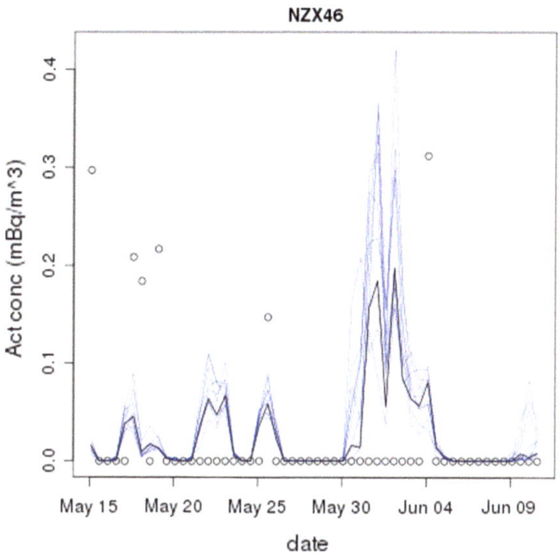

Fig. 54.2 Observed (black dots) and simulated (blue lines) activity concentration (mBq/m^3) for the 10 members at the IMS station NZX46. Observations below the detection limit were set to zero

54.4 Discussion

When we compare these results with the scores obtained during the Maurer et al. [1] intercomparison exercise, it is found that the setup using the ERA5 data results in a good rank (the median rank for all participants in Maurer et al. [1] was 2.07).

Each member of the ERA5 ensemble has a roughly equivalent rank. This is a feature of a good ensemble, where all members are drawn from the same probability distribution.

It is also shown that there are other sources of uncertainty which could not be quantified. In this exercise, this was mainly related to the emission uncertainty from other sources of radioxenon which were not taken into account in the challenge, which results in an ensemble spread that does not represent therefore the full uncertainty.

Acknowledgements The authors acknowledge ANSTO stack emission data obtained from the CTBTO Preparatory Commission in the framework of a vDEC agreement (https://www.ctbto.org/specials/vdec/).

Disclaimer The views expressed herein are those of the authors and not necessarily reflect the views of the CTBTO Preparatory Commission.

References

1. C. Maurer, P. Eslinger, J. Bar, J. Kusmierczyk-Michulec, P. Seibert, B. Orr, A. Philipp, O. Ross, S. Generoso, P. Achim, M. Schoeppner, A. Malo, A. Ringbom, O. Saunier, D. Quelo, A. Mathieu, Y. Kijima, A. Crawford, A. Stein, T. Chai, F. Ngan, S. Leadbetter, P. De Meutter, A. Delcloo, R. Britton, A. Davies, L. Glascoe, D. Lucas, M. Simpson, P. Vogt, M. Kalinowski, T. Bowyer, International challenge to model the long-range transport of radioxenon released from medical isotope production to six comprehensive nuclear-test-ban treaty monitoring stations. J. Environ. Radioact. (Accepted for publication)
2. P.R. Saey, The influence of radiopharmaceutical isotope production on the global radioxenon background. J. Environ. Radioact. **100**(5), 396–406 (2009)
3. A. Stohl, C. Forster, A. Frank, P. Seibert, G. Wotawa, The Lagrangian particle dispersion model FLEXPART version 6.2. Atmos. Chem. Phys. **5**(9), 2461–2474 (2005)

Chapter 55
Assessment of Fine-Scale Dispersion Modelling for Near-Road Exposure Applications

Jennifer L. Moutinho, Donghai Liang, Jeremy Sarnat and Armistead G. Russell

Abstract Detailed measurements and dispersion modeling were conducted to develop more accurate integrated metrics to assess exposure to potentially high pollutant levels of primary traffic emissions. A 13-week intensive sampling campaign was conducted at six monitoring sites surrounding one of the busiest highway segment in the US with the study area focusing on the Georgia Institute of Technology campus to capture the heterogeneity in pollutant concentrations related to primary traffic emissions. A dispersion model (RLINE) was used to develop spatial concentration fields at a fine-spatial resolution over the area of primary exposures. Initial RLINE results were highly biased, due either to errors in the emissions or the model. Analysis suggests that both may be important, depending upon species, though the largest errors were due to how the model represents near-source dispersion, especially when the wind aligns with the road segments. To correct for high near-road bias, the RLINE results were calibrated using measurement observations after the urban background was removed. Performing the calibration hourly also reduced the bias observed in the diurnal profile. Both the measurement observations and dispersion modeling results show that the highway has a substantial impact on primary traffic pollutant (particularly elemental carbon and carbon monoxide) concentrations and captures the prominent spatial gradients across the campus domain, though the gradients were highly species dependent. These improved concentration fields were used to enhance the characterization of pollutant spatial distribution around a traffic hotspot and to quantify personal exposure to primary traffic emissions.

J. L. Moutinho (✉) · A. G. Russell
School of Civil and Environmental Engineering, Georgia Institute of Technology, Atlanta, GA, USA
e-mail: jennifer.moutinho@gatech.edu

A. G. Russell
e-mail: ted.russell@gatech.edu

D. Liang · J. Sarnat
Rollins School of Public Health, Emory University, Atlanta, GA, USA
e-mail: donghai.liang@emory.edu

J. Sarnat
e-mail: jsarnat@emory.edu

C. Mensink et al. (eds.), *Air Pollution Modeling and its Application XXVI*,
Springer Proceedings in Complexity,
https://doi.org/10.1007/978-3-030-22055-6_55

55.1 Introduction

On-road mobile emissions lead to elevated air pollutant concentrations in near-road environments and steep concentration gradients exist with distance from major roads. Exposure to traffic related air pollutants (TRAP), including carbon monoxide (CO), nitrogen oxides (NOx), and fine particulate matter ($PM_{2.5}$), has been linked to many adverse health effects [2]. While vehicles are a significant source of pollutants in the near-road environment, improved vehicle engine technology, emissions control systems, and fuel regulations have reduced emissions leading to a decrease in traffic related air pollutants (TRAP) near heavily trafficked roads [2]. Quantifying the pollutant concentration attributed to traffic pollution from regional pollution through measurement becomes more difficult as vehicle emissions decrease. In addition, near-road measurements are often limited within a city leading to a limited understand of the spatial gradients. Dispersion modeling can provide spatially and temporally resolved concentration fields of TRAPs for assessing exposure.

55.2 Methods

Hourly-modeled primary traffic-related CO, NOx, and $PM_{2.5}$ concentrations at a 25 × 25 m grid resolution were generated using RLINE (Research Line Source Model), a steady-state, Gaussian plume dispersion model [3]. The Atlanta Regional Commission (ARC) provided link-based, on-road mobile source emissions for the 20-country region surrounding metro Atlanta using a traffic demand and mobile source emissions model. ARC estimated 2010 CO, NOx, and PM2.5 emissions for 43,712 links based on modeled traffic volume, vehicle speed, and fleet demographics. The surface meteorological data was from the National Weather Service at the Hartsfield-Jackson Atlanta International Airport (ATL) and preprocessed using AERMINUTE. The upper air data was from the Peachtree City Falcon Field Airport (FFC).

Initial RLINE results were corrected by applying ARC monthly and hourly diurnal emissions variability, and hours with wind speeds less than 1 ms^{-1} were replaced. The 2010 ARC link-based emissions were scaled to 2014 emissions levels using the mobile emissions ratio from the Motor Vehicle Emissions Simulator [4]. Corrected RLINE results were calibrated to the hourly surface concentrations measurements collected as part of the Dorm Room Inhalation to Vehicle Emissions (DRIVE) study. A linear regression between the hourly-corrected RLINE concentrations and the hourly measurements at the six monitoring locations scaled the corrected results and the intercept was removed to exclude the local background. In a second process, the hourly-corrected RLINE results were calibrated to the hourly DRIVE surface measurements using 24 linear regressions to account for differences in the diurnal profile. A third process averaged three hourly RLINE concentrations at each grid with wind direction varying five degrees in either direction to help reduce bias when the wind aligns with the road links.

The DRIVE study was a 13-week sampling campaign (September 2014–December 2014) focusing on traffic related air pollutants (TRAPs) surrounding the busiest highway in Atlanta, GA with an annual average daily traffic of 320,000 vehicles [1]. The study included six monitoring sites and focused on the Georgia Institute of Technology campus in the geographical core of Atlanta with its eastern edge bordering the major highway. The sites included: a roadside location 10 m from the highway (RD), an EPA Near-road Monitoring Network site located 70 m north of the RD site and 5 m from the highway (EPD), a dormitory room 20 m from the highway (ND), a dormitory room 1.4 km from the highway (FD), and an urban background site located 2.3 km from the highway.

55.3 Results and Discussion

Simulated concentrations from raw RLINE outputs show the spatial gradients, however concentrations in the near-road environment were unreasonably high (Fig. 55.1a). The corrected RLINE output removed all low wind speed events, which led to bias high concentrations due to stagnation events trapping emissions at the road, and reintroduced monthly and hourly diurnal emissions variability (Fig. 55.1b). The single linear regression applied to the corrected RLINE output (Fig. 55.1c) simulate realistic average species concentrations and captures the spatial gradients. The additional processes described will be developed and presented. The average simulated impact of on-road mobile sources on primary $PM_{2.5}$, NOx, and CO concentrations at

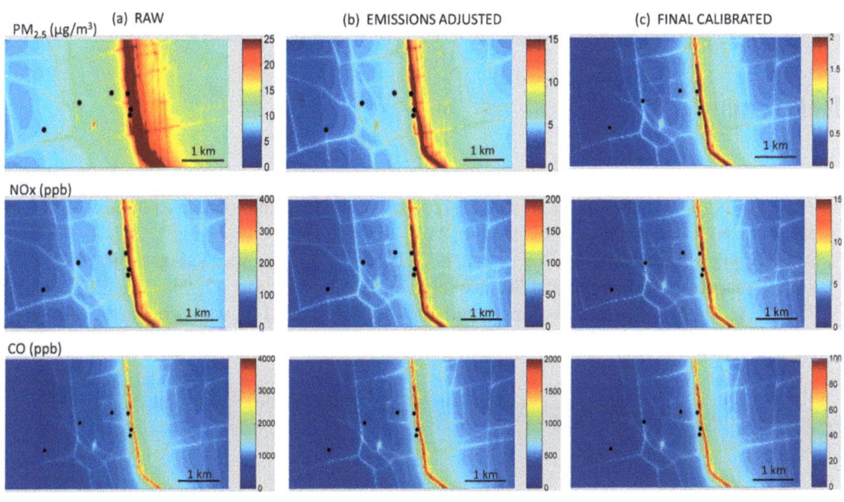

Fig. 55.1 RLINE fine scale dispersion modeling for simulated PM2.5, NOx, and CO concentrations (September–December 2014). **a** Raw RLINE output; **b** Corrected RLINE output; **c** Calibrated by single linear calibration. Black dots represent the sampling locations. Resolution: 25 × 25 m

the RD site was 1.3 μg m^{-3}, 8.6 ppb, and 51.5 ppb respectively, while the concentration at the urban background (JST) site was 0.4 μg m^{-3}, 2.8 ppb, and 16.9 ppb. The maximum hourly concentrations were simulated when wind speeds were low and therefore dispersion was low which often occurring at night when emissions were also low. The wind direction measured at the RS site measured easterly winds more than 70% of the time leading to direct influence from the highway emissions on the monitor sites. Even though the major emissions source in the domain is the highway, the surface streets (AADT 20,000) are an observable source [1].

55.4 Conclusion

These processes reduced unrealistically high concentrations that were observed on and near the main highway emissions source. The linear calibration helped decrease the bias high levels improving the average concentration fields, however hourly bias was observed. The diurnal calibration improved both the bias high concentrations and the distribution of the concentrations.

Acknowledgements Research described in this article was conducted under contract to the Health Effects Institute (HEI), an organization jointly funded by the United States Environmental Protection Agency (EPA) (Assistance Award No. R-82811201) and certain motor vehicle and engine manufacturers. The contents of this article do not necessarily reflect the views of HEI, or its sponsors, nor do they necessarily reflect the views and policies of the EPA or motor vehicle and engine manufacturers. Atmospheric Research & Analysis, Inc. and the Georgia Department of Natural Resources provided additional measurement data.

References

1. GA DOT, Traffic data (2012), www.dot.ga.gov/
2. HEI, Traffic-related air pollution: a critical review of the literature on emissions, exposure, and health effects. Special Report 17 (Boston, MA, 2010)
3. M.G. Snyder, A. Venkatram, D.K. Heist, S.G. Perry, W.B. Petersen, V. Isakov, RLINE: a line source dispersion model for near-surface releases. Atmos. Environ. **77**, 748–756 (2013)
4. US EPA, Motor vehicle emissions simulator (MOVES) (2014), www.epa.gov/otaq/models/moves/

Chapter 56
Detailed Assessment of a Smog Situation Detected in the Sajó Valley, Hungary

Zita Ferenczi, Emese Homolya and László Bozó

Abstract The impacts of air pollution on the environment and health is a very actual topic in Hungary. By understanding the connections between the pollution that humanity produces and the meteorological situations, we can make the changes that are necessary to prevent smog episodes and to pass on a more sustainable world to future generations. Primarily during winter and fall seasons, episodes of poor air quality related to high concentrations of particulate matter are frequent, especially in the eastern part of Hungary. These situations are often connected to special meteorological conditions—such as cold air cushion—which do not help the mixing and dilution of air pollutants. Usually this type of meteorological condition is coupled with very low ambient air temperature, which can urge the usage of more solid fuel (wood and coal), therefore result in an increase of PM_{10} emission from domestic heating. Such a situation developed at the end of January 2017 in the Sajó Valley, Hungary. The aim of the present study is to examine the meteorological background of the extreme high PM_{10} concentrations which could contribute to the developing of the smog situation, and to estimate the growth rate of the emission from residential combustion due to the low ambient air temperature.

56.1 Introduction

Nowadays particulate matter (PM) is one of the most frequently cited pollutants since it has many negative impacts on the ecosystem, built environment and especially on human health. Local and regional meteorology, including wind speed, wind direction, atmospheric stability, long-range transport and pollution dispersion are all

Z. Ferenczi (✉) · E. Homolya · L. Bozó
Hungarian Meteorological Service, Kitaibel Pál u. 1, Budapest 1024, Hungary
e-mail: ferenczi.z@met.hu

E. Homolya
e-mail: homolya.e@met.hu

L. Bozó
e-mail: bozo.l@met.hu

C. Mensink et al. (eds.), *Air Pollution Modeling and its Application XXVI*,
Springer Proceedings in Complexity,
https://doi.org/10.1007/978-3-030-22055-6_56

factors that play important roles in PM concentration reduction strategies. Analysis of local and regional meteorology is important to completely understand the processes responsible for the spatial and temporal distribution of PM in all geographic regions [1, 4].

In Hungary, one key issue in air quality research is to find the sources and the influencing meteorological factors for aerosol particles, which cause elevated levels of particulate matter concentrations in the Carpathian Basin. Although the quality of the emission databases have been improved in the latest years in Europe, the determination of the quantitative contributions of different sources (e.g. domestic heating and automobile exhaust) to PM_{10} and $PM_{2.5}$ concentrations remains a difficult task. The difficulty is not only due to the uncertainties in the emission data, we do not have an accurate knowledge about how the meteorological situations influence the air pollutant levels either. These two facts may result in the air quality forecasts for this region not always being successful, especially in the case of particulate matter. The situation is further complicated by the fact that PM can not only be emitted directly but it can also form in the atmosphere when gaseous pollutants (SO_2 and NO_x) undergo transformation to form secondary inorganic particles [2]. This complexity demonstrates, that meteorology plays a key issue in the air pollution analyses. This is also the reason why the smog situations have to be examined in depth in Hungary.

56.2 A Special Meteorological Situation in Hungary: The Cold Air Cushion

The cold air cushion is a special meteorological situation that is related to inversion in the upper atmosphere. It most frequently evolves in areas that are surrounded by chains of mountains. Anticyclonic events trigger the development of cold air cushions as they foster downward motions in the air. By serving as a barrier for mixing motions, inversion causes the air to stabilize and it hinders the movement of the air mass out of the basin. This effect is enhanced by the presence of the surrounding mountains.

A meteorological situation like the cold air cushion has environmental consequences. The lack of mixing processes, together with the increased rate of emission caused by residential combustion, due to the cold, pollutants of anthropogenic origin accumulate rapidly. Concentrations of pollutants in the air can easily reach and exceed levels that are considered harmful for the health.

56.3 Influence of Meteorological Parameters on PM_{10}

Such an anticyclonic situation occurred at the end of January 2017 in the Carpathian Basin, which caused temperatures in the Sajó Valley, located in the north-eastern part of Hungary, to drop well below −10 °C for a period of time covering the last days of the month. The height of the planetary boundary layer remained below 200 m and the winds were light throughout the same period. The stagnation-index calculated based on model forecasts also indicated unfavourable meteorological conditions.

At the same time a smog situation was detected in Hungary. The concentrations of PM_{10} were above the limit values all over the country. Extreme high concentrations were measured in the eastern part of Hungary in the morning of 31 January 2017. In this period, the highest one-hour average concentration was above 500 $\mu g/m^3$ (Fig. 56.1).

The PM_{10} concentrations started to rise on 18 January, and they reached the highest value during the last 3 days of January. Our analysis demonstrated that the constellation of the unfavourable meteorological conditions resulted in the formation of the smog situation in the Carpathian Basin. Figure 56.2 shows the time series of the meteorological parameters (air temperature, wind speed, boundary layer height) which affect PM_{10} concentrations. During these days, wind speed was very low and the boundary layer height remained below 150 m, which is extreme. In the cases of very low wind speed and boundary layer height, the mixing of the air pollutants with the ambient air is quite poor. This was one of the reasons behind the formation of the smog situation. Another reason was the increasing emissions due to the cold weather. Good correlation value was found between PM_{10} concentrations and temperatures as well. The air temperature below minus 10 °C induces an enhanced usage of solid

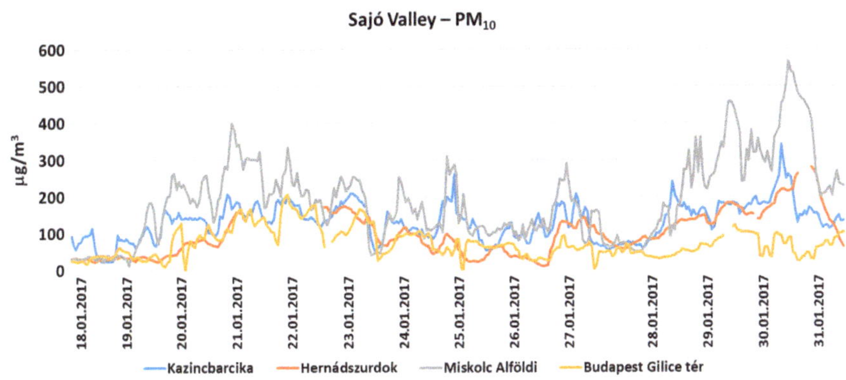

Fig. 56.1 PM_{10} concentration values detected at different monitoring stations in the Sajó Valley

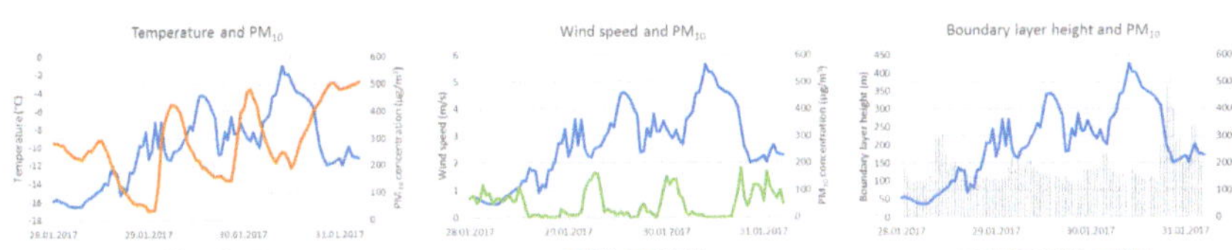

Fig. 56.2 Relationship between local meteorological parameters and PM_{10} concentration (Miskolc, Hungary)

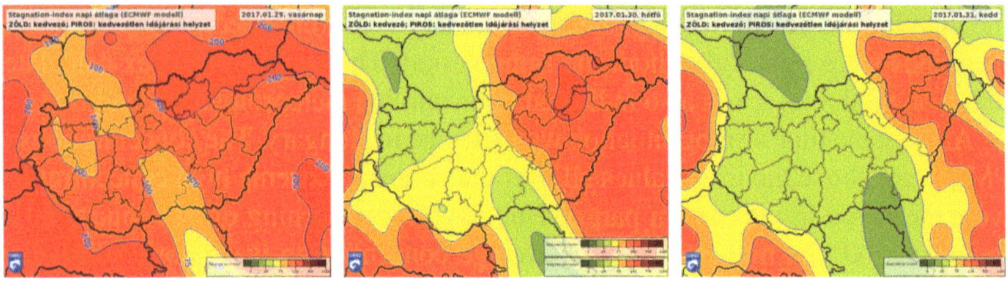

Fig. 56.3 Calculated SI index values (29/01/2017, 30/01/2017, 31/01/2017)

fuel, which means that the emission of PM_{10} from domestic heating becomes higher than usual.

There is a widely used SI index in Hungary which characterizes the vertical and horizontal mixing of the lower atmosphere [3]. When the value of SI is above 100, it is most likely that a smog situation will occur in Hungary.

$$SI = \sqrt{\frac{10^6}{MLHx|v|}} \tag{56.1}$$

where MLH is the mixing layer height (m) and v is the wind speed at 10 m height (m/s). On 29 January, 2017, the SI index was above 100 throughout Hungary and reached 200 in the eastern parts of the country, and this value predetermined the formation of the smog situation in this region (Fig. 56.3).

56.4 Chemical Transport Model Simulation

The CHIMERE chemical transport model was applied to simulate the transport and chemical transformations of air pollutants in the Sajó Valley for the 4-day smog episode in January 2017. The AROME numerical weather prediction model provided the gridded meteorological inputs for the chemical model calculations. Two different EMEP anthropogenic emission datasets were used to a defined grid, covering the Carpathian Basin with a 0.1° × 0.1° spatial resolution. In the first case the original 0.1° × 0.1° EMEP database was applied, and in the second case the 50 × 50 km EMEP database was downscaled to 0.1° × 0.1°, using the emission preprocessor of the CHIMERE model package. The vertical domain included 8 layers and extended to 500 mbar.

Figure 56.4 shows the results of the model simulations between 28 and 31 January. All in all, we found that the PM_{10} concentration values that we got as the result of the model simulations were significantly lower than the measurements, leading us to a conclusion that the model could not detect the smog situation over the territory in question appropriately.

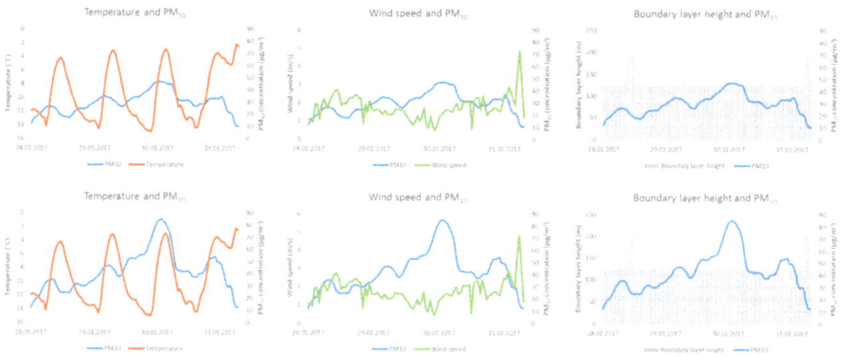

Fig. 56.4 Time series of model calculations for PM_{10}, temperature, wind speed and boundary layer height

56.5 Conclusion

The effects of the meteorological parameters on high level PM_{10} concentrations were examined in depth. Special attention was given to the parameters that influence the vertical and horizontal mixing of air pollutants, as they describe all the important processes in the lower atmosphere in terms of air quality. Chemical transport model calculations were carried out using the CHIMERE model. By comparing the model simulation with the measured data we found that the model could not detect the smog situation with high PM_{10} values properly. PM_{10} concentration values that we got as the result of the model simulations were significantly lower than the measurements. Depending on the spatial resolution of the initial emission database, model results show a considerable difference in the magnitude of PM_{10} values in the environment. This analysis showed that in areas where residential heating is a determinant source of PM_{10} emissions during winter, further development is needed in order to integrate a temperature dependent emission density in the model system.

Acknowledgements This work has been supported by GINOP-2.3.2-15-2016-00055 Project through the National Research, Development and Innovation Office, Hungary.

References

1. Z. Chen, X. Xie, J. Cai, D. Chen, B. Gao, B. He, N. Cheng, B. Xu, Understanding meteorological influences on PM2. 5 concentrations across China: a temporal and spatial perspective. Atmos. Chem. Phys. 1–30 (2017)
2. Z. Ferenczi, K. Imre, L. Bozó, Application of trajectory clustering for determining the source regions of secondary inorganic aerosols measured at K-puszta background monitoring station, Hungary, in *International Technical Meeting on Air Pollution Modelling and its Application XXV* (2018), pp. 593–597

3. J. Holst, H. Mayer, T. Holst, Effect of meteorological exchange conditions on PM_{10} concentration. Meteorol. Z. **17**(3), 273–282 (2008)
4. J.L. Pearce, J. Beringer, N. Nicholls, R.J. Hyndman, N.J. Tapper, Quantifying the influence of local meteorology on air quality using generalized additive models. Atmos. Environ. **45**, 1328–1336 (2011)

Chapter 57
Comparison of the Performance of AERMOD and CALPUFF Dispersion Model Outputs to Monitored Data

Jackson Mak, Camille Taylor, Melanie Fillingham and Jamie McEvoy

Abstract AERMOD and CALPUFF are two Ontario approved regulatory air dispersion models used to assess air quality compliance to provincial standards. Modelled results from these dispersion models may not always be representative of actual concentrations due to their inherent assumptions and atmospheric simplifications. This research aims to assess the near field performance of these two models by comparing modelled concentrations predictions with monitored observations. The case study analyzed was for 24 h average nickel (Ni) from a Facility located in Northern Ontario. The Facility has a monitoring program set up to measure dust, from which the monitored metal concentrations were speciated. Statistical analysis of the modelled results demonstrated that CALPUFF 24 h average modelled concentrations showed better agreement with monitored results than those modelled using AERMOD, however, even these CALPUFF results were often 2x (or 0.5x) the monitored value.

57.1 Introduction

In Ontario, local air quality is regulated under O. Reg. 419/05. Within this regulation, two dispersion models are approved for assessing compliance with provincial standards, AERMOD v14134 (a steady-state Gaussian dispersion model) and CALPUFF v.6.263 level 080827 (a Lagrangian puff model). In general, models tend to be conservative. When compared to maximum monitored results, research has found that

J. Mak (✉) · C. Taylor (✉) · M. Fillingham (✉) · J. McEvoy (✉)
Golder Associates Ltd., Ottawa, ON K2H 5B7, Canada
e-mail: Jackson_Mak@golder.com

C. Taylor
e-mail: Camille_Taylor@golder.com

M. Fillingham
e-mail: Melanie_Fillingham@golder.com

J. McEvoy
e-mail: Jamie_McEvoy@golder.com

C. Mensink et al. (eds.), *Air Pollution Modeling and its Application XXVI*,
Springer Proceedings in Complexity,
https://doi.org/10.1007/978-3-030-22055-6_57

AERMOD will not underestimate maximum concentrations but will sometimes over-predict as high as by a factor of 5 [5, 10]. The occurrence of maximum concentrations is often at times of low-wind speeds (when dispersion is poor). During these stagnation conditions, model errors have been over-emphasised utilizing both AERMOD [8] and CALPUFF [3].

Dispersion models are more adequate at predicting longer-timed averaged concentrations (e.g. annual) than they are shorter-timed averages (e.g. hourly) [1]. They are also not accurate at estimating specific spatial or temporal occurrences, which can pose issues for pollutants that range drastically with time and location [9]. As a result, it is a common practice to evaluate the monitored and modelled results based on unpaired basis. The treatment of volume sources in AERMOD is another area of conservatism, as they do not include upward velocity [11].

Dispersion models are sensitive to the inputted meteorological data, with variability in the meteorological dataset carrying through to the dispersion model results. These arise due to assumptions that simplify the model but are not always well-representative of actual atmospheric conditions during each modelling hour. One example of this that occurs in both the meteorological pre-processors AERMET and CALMET (preprocessors of regulatory dispersion models AERMOD and CALPUFF, respectively) is the Bowen Ratio, which is estimated by month or season and land-use type. Assuming consistent monthly or seasonal values does not represent actual conditions. Similar assumptions occur for parameters such as Albedo and surface roughness.

The objective of this study is to assess the model performance of both AERMOD and CALPUFF by comparing modelled concentrations to measured concentrations at two Facilities within Ontario. The bias from each of these dispersion models were analyzed to see whether these models produce inconsistent results.

57.2 Methodology

AERMOD v14134 and CALPUFF v.6.263 level 080827, the two Ontario approved regulatory models at the time of the experiment, were used to model 24 h average Nickel (Ni). For AERMOD, the modelling was performed as per the Ontario Ministry of the Environment, Conservation and Parks' (Ministry) "Guideline A-11: Air Dispersion Modelling Guideline for Ontario," [6]. The Ministry's "Technical Bulletin: Using Combined Assessment of Modelled and Monitored (CAMM) Results to Refine Emission-Rate Estimates" [7], was followed to conduct the assessment. The CALPUFF modelling was performed as per the British Columbia Ministry of Environment's "Air Quality Dispersion Modelling Guideline" [2].

Emissions rates and source characterization for these Facilities were obtained from confidential Facility specific data. The emission rate sources include: direct measurements, estimates from emission factors (such as those from the U.S. EPA's AP-42), mass balances, and engineering calculations. The Facilities' emission data

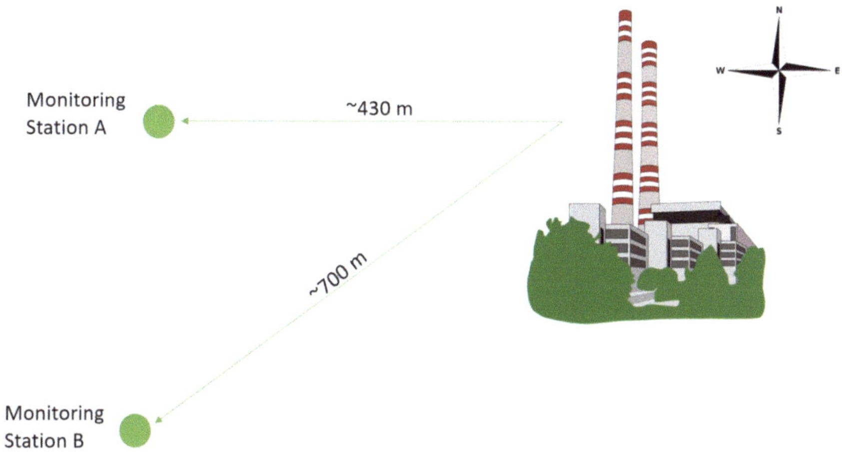

Fig. 57.1 Schematic representing relative locations of monitoring stations with respect to the main sources at the facility

will remain constant between both model assessments and as such, was not the focus of the assessment.

The Facility has five years of data at two dust monitoring stations (Fig. 57.1). The dust was speciated out to Ni based on performed analytical analyses. Only data that was downwind from the Facility and that exceeded a wind speed threshold of 1 m/s were considered "hits" and used in the analysis. According to the Ministry's CAMM technical bulletin, it is recommended that a Facility have at least 30 "hits." Both monitoring stations at the Facility had over 30 "hits" and therefore met the criteria. Background data was not taken into consideration in modelling.

The meteorological data for AERMOD was prepared from the CALMET data by extracting the ground-level meteorological conditions for the grid cell where the facility is situated. The data was also extracted to match the monitoring period. CALMET data used to drive CALPUFF dispersion modelling was initialized by Rapid Update Cycle (RUC) model output and surface meteorological fields recorded at the nearby airport station. The RUC model is an operational weather prediction system employed by National Oceanic and Atmospheric Administration (NOAA) with approximately 20 km horizontal resolution and 50 vertical layers.

57.3 Results and Discussion

Within this extended abstract, the comparisons between monitored concentrations and modelled concentrations using AERMOD and CALPUFF are presented separately for each of the two monitoring stations. All of the results presented are unpaired in time and space, due to the known inability of dispersion models to accurately predict the time and location of concentrations.

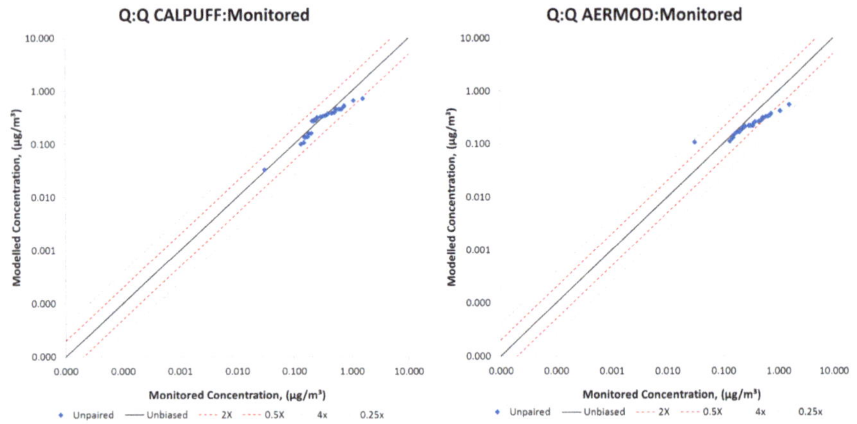

Fig. 57.2 Q:Q plot comparing CALPUFF (left) and AERMOD (right) modelled concentrations to monitored concentrations at station A

57.3.1 Monitoring Station a Results

Both AERMOD and CALPUFF models show poor agreeance with monitored data at station A, with points falling outside of the 0.5 to 2x range (Fig. 57.2). CALPUFF results were mainly over-predicted, whereas AERMOD results were mainly under-predicted. At low concentrations (<0.1 $\mu g/m^3$), both models over-predicted. Both models underestimated the maximum concentrations.

57.3.2 Monitoring Station B

CALPUFF had better agreeance for Ni concentrations with the monitored data at station B compared to AERMOD (Fig. 57.3), with majority of the points falling within the 0.5 to 2x range. AERMOD results were mostly under-predicted, whereas CALPUFF results fluctuated between over and under predicting. Both models under-estimated the maximum concentration.

57.3.3 AERMOD Versus CALPUFF

At both monitoring stations, results generated using CALPUFF were in better agreeance with monitored data than those generated using AERMOD. However, both models tended to under predict maximum concentrations.

Background data was not included in the modelling assessments, which results in concentration predictions less than the expected actual values. If background had

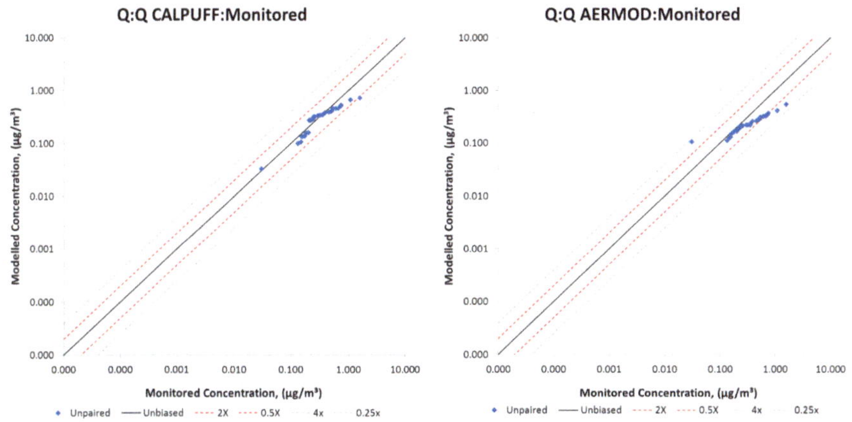

Fig. 57.3 Q:Q plot comparing CALPUFF (left) and AERMOD (right) modelled concentrations to monitored concentrations at station B

been considered, the modelled values would be higher than what has been displayed in this study. There was a source modelled within CALPUFF using variable temperatures (buoyancy), a method not supported by AERMOD. The average temperature was used in the AERMOD, creating a slight discrepancy between modelling variables.

It would be expected that if pollutants with shorter averaging times, such as 1 h, were analyzed with a similar approach that the results may have worse correlation. This has been shown to occur in other studies [4], as dispersion models are notorious for poorly predicting concentrations of short-averaging periods.

57.4 Conclusion

From the results displayed in this study, CALPUFF 24 h average modelled concentrations showed better agreement with monitored results than those modelled using AERMOD, however, even these CALPUFF results were often 2x (or 0.5x) the monitored value.

References

1. N.E. Bowne, R.J. Londergan, *Overview, Results, and Conclusions for the EPRI Plume Model Validation and Development Project: Plains Site*. EPRI EA–3074. Electric Power Research Institute, Palo Alto, CA (1983)
2. British Columbia Ministry of Environment, *Air Quality Dispersion Modelling Guideline* (2015)
3. R.M. Cox, J. Sontowski, C.M. Dougherty, An evaluation of three diagnostic wind models (CALMET, MCSCIPUF, and SWIFT) with wind data from the dipole pride 26 field experiments.

Meteorol. Appl. **12**, 329–341 (2005)
4. M.D. Gibson, S. Kundu, M. Satish, Dispersion model evaluation of $PM_{2.5}$, NO_x and SO_2 from point and major line sources in Nova Scotia, Canada using AERMOD gaussian plume air dispersion model. Atmos. Pollut. Res. **5**, 157–167 (2013)
5. S.A. Guerra, Innovative dispersion modeling: practices to achieve a reasonable level of conservatism in AERMOD modeling demonstrations. EM Mag. 25–29 (2014)
6. Ontario Ministry of the Environment, Parks, and Conservation (MEPC), *Guideline A-11: Air Dispersion Modelling Guideline for Ontario* (2017)
7. Ontario Ministry of the Environment, Parks, and Conservation (MEPC), *Technical Bulletin: Using Combined Assessment of Modelled and Monitored (CAMM) Results to Refine Emission-Rate Estimates* (2011)
8. W. Qian, A. Venkatram, Performance of steady-state dispersion models under low wind-speed conditions. Bound.-Layer Meteorol. **138**, 475–491 (2010)
9. R.G. Rhoads, *Accuracy of Air Quality Models: Staff Report.* U.S. Environmental Protection Agency, Research Triangle Park, NC (Docket No. A–80–46, II–G–6) (1981)
10. A.S. Rood, Performance evaluation of AERMOD, CALPUFF, and legacy air dispersion models using the winter validation tracer study dataset. Atmos. Environ. 707–720 (2014)
11. K. Silverman, E. Sargent, Q. Zeyuan, Comparison of the industrial source complex and AERMOD dispersion models: case study for human health risk assessment. J. Air Waste Manag. Assoc. 1439–1446 (2007)

Chapter 58
Model of Arrival Time for Gas Clouds in Urban Canopy

Hana Chaloupecká, Zbyněk Jaňour, Klára Jurčáková and Radka Kellnerová

Abstract The aim of this paper is to present a new model of arrival time for gas clouds. To create such a model, simulations of short-term gas leakages were conducted in a wind tunnel with a neutrally stratified boundary layer. Into the tunnel, a model of an idealized urban canopy in scale 1:400 was placed. For simulations of the short-term gas discharges, ethane was utilized. Concentration time series were measured by a fast flame ionisation detector. The experiments were repeated about 400 times to get statistically representative datasets. The ensembles of concentration time series were measured at about 50 individual positions. From these data, puff arrival times were computed. The results showed that a suitable probability distribution to describe the variability in values at individual positions for arrival time is lognormal. Moreover, the parameters of this distribution do not change randomly with the change in the measurement position but their change can be described by functions. Utilizing them, probability density functions of arrival time can be constructed and whatever quantile of arrival time at a chosen position can be computed. Such a model could help emergency services to estimate how the situation could look like during the accident not only in the most frequently occurred but also in the extreme cases.

58.1 Introduction

Leakages of gases during an accident last very often less than one hour (e.g., [2]). The dispersion of these gas clouds depends on the actual turbulent flow present in the time and the place. Hence, wide range of dispersion scenarios can happen under the same mean ambient conditions. To be able to react effectively to an accident,

H. Chaloupecká (✉) · Z. Jaňour · K. Jurčáková · R. Kellnerová
Institute of Thermomechanics, Czech Academy of Sciences, Prague, Czech Republic
e-mail: hana.chaloupecka@it.cas.cz

H. Chaloupecká
Faculty of Mathematics and Physics, Charles University, Prague, Czech Republic

C. Mensink et al. (eds.), *Air Pollution Modeling and its Application XXVI*,
Springer Proceedings in Complexity,
https://doi.org/10.1007/978-3-030-22055-6_58

emergency services need to know how the situation after the leakage can look like in mean as well as in extreme cases.

One of the crucial information for emergency services is the time when a gas cloud gets into individual places affected by the incident. Chaloupecká et al. [2] compared arrival times as found by different definitions and proposed a new method applicable also in the operative stage. Lübcke et al. [5] found out that the mean value of arrival time increases with the increasing distance from the gas source but this increase seems to be higher close to the source. Chaloupecká et al. [1] discussed the mean arrival time in an urban canopy in which flow was parallel with a street. Within this parallel street, the mean value of arrival time increases linearly with the increasing distance from the source. But a range of values of an arrival time ensemble at each measurement position is very wide and therefore a change of values of different quantiles at individual sampling positions is also an important aspect. Hence, we propose a new model describing an evolution of probability density functions of arrival time in an idealized urban canopy.

58.2 Methods and Experimental Set-up

The experiments were conducted in a wind tunnel with a neutrally stratified boundary layer. Its characteristics agreed with the recommendations of VDI [7] for flows found in towns (see [2] for more details). Into the tunnel, a model of an idealized urban canopy with pitched roofs in scale 1:400 was placed (Fig. 58.1). This type of urban canopy is typical for European cities (e.g., [3]). For simulations of the short-term gas discharges, ethane was utilized. The discharges of 1 s set on a programmer logic controller were created by an electromagnetic valve. Concentration time series were

(a) **(b)**

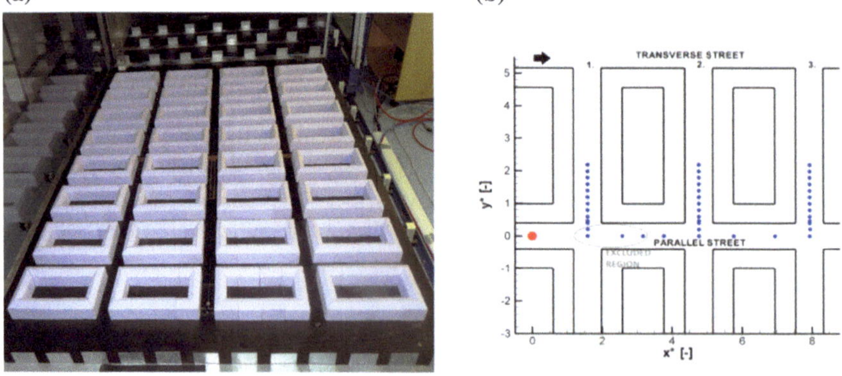

Fig. 58.1 Model of idealized urban canopy: placement in wind tunnel (**a**), scheme of investigated area (**b**)—big circle—position of ground-level point source, small circles—sampling positions, arrow—flow direction

measured by a fast flame ionization detector. The experiments were repeated about 400 times to get statistically representative datasets. The ensembles of concentration time series were measured at about 50 individual positions within human breathing zone at the model (Fig. 58.1b). From these data, puff arrival times (at*) were computed. We utilized a threshold method utilizing residual concentrations for finding arrival time, which is described in [2].

In the paper, we use dimensionless quantities (e.g., [7]).

$$x^* = \frac{x}{H}, y^* = \frac{y}{H}, t^* = \frac{tU_{ref}}{H} \tag{58.1}$$

In these relations, x, y stand for horizontal coordinates, H the characteristic height (the height of modelled buildings), t time, U_{ref} reference speed (measured in the middle of the wind tunnel). The probability density function (pdf) of lognormal distribution (e.g., [8]) with parameters μ, σ is defined as

$$f(x|\mu, \sigma) = \frac{1}{x\sigma\sqrt{2\pi}} exp\left\{\frac{-(lnx - \mu)^2}{2\sigma^2}\right\}; x > 0 \tag{58.2}$$

A quantile function is defined as

$$Q(p) = inf\{x \in R; p \le F(x)\} \tag{58.3}$$

for a probability $0 < p < 1$ and a cumulative distribution function F of a random variable X. The sample quantile $Q(p)$ is approximately the value expected to exceed a randomly chosen member of the dataset with probability p, as suggested [8].

58.3 Results

The results showed that the ensembles of arrival time usually follow a lognormal distribution according to Chi-square-goodness-of-fit-test (e.g., [4]) with significance level 0.05. Along the parallel street, all datasets can be fitted by a lognormal distribution. But within the transverse streets, some measurement positions in which the data do not seem to follow the lognormal distribution exist. These positions are highlighted by star markers in Figs. 58.2 and 58.3.

The parameters of the fitted distributions do not change randomly with the increasing distance from the gas source, but seem to chance according to some rules and therefore could be fitted by functions. For the analysis, we used weighted least square method (e.g., [4]). The weights were chosen according to p-values derived from the chi-square-goodness-of-fit-test.

The value of the parameter μ increases with the increasing distance from the source in the parallel street. This increase is approximately linear. An approximately linear increase of the parameter μ can also be seen with the increasing distance from

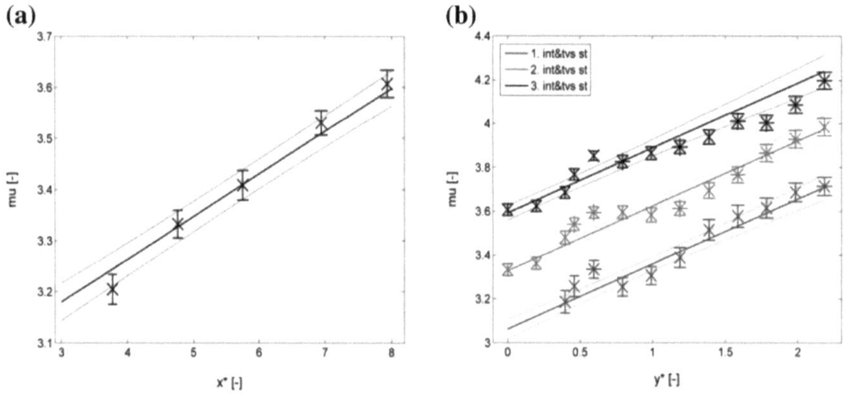

Fig. 58.2 Parameter μ of arrival time with 95% CI in the parallel street (**a**) and in the intersections (int) and transverse streets (tvs st) (**b**) with fitted model (thick line) and its 95% CI (thin line)

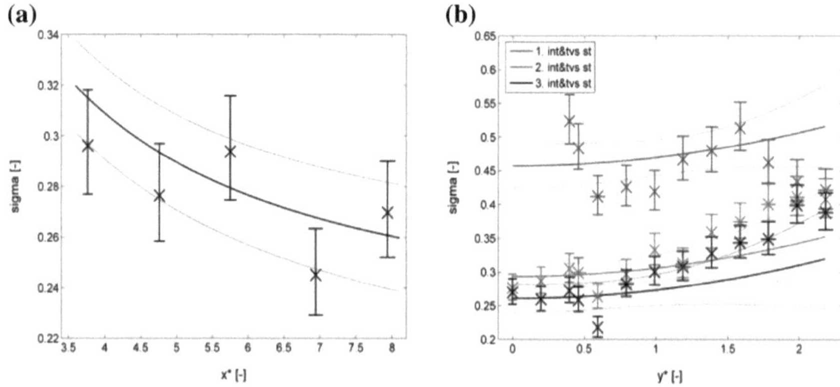

Fig. 58.3 Parameter σ of arrival time with 95% CI in the parallel street (**a**) and in the intersections (int) and transverse streets (tvs st) (**b**) with fitted model (thick line) and its 95% CI (thin line)

the middle of the parallel street into the transverse street in all three investigated transverse streets. But one can spot a variation of this behaviour just behind the intersection. In this region, a "hump" (a bigger increase and after a few sampling positions a decrease of the parameter value) can be seen. This behaviour is apparently connected with the presence of a recirculation vortex (e.g., [6]) in this region. Comparing the transverse streets, the slope of the fitted linear function decreases from the first to the third transverse street. But this change is not statistically significant; the values of the slope overlap in the 95% confidence intervals (CIs). Hence, we did not distinguish the slopes during creation of the model. Utilizing the linear approximation, the overall model describing an evolution of μ in the investigated area is

$$\mu = 0.083x^* + 0.30y^* + 2.93 \tag{58.4}$$

The model (Fig. 58.2) is valid for all the sampling positions except for an excluded region (Fig. 58.1b). The coefficient of determination R^2 for the model is 0.97.

The second parameter of the lognormal distribution σ decreases with the increasing distance from the source in the parallel street. In contrast, the value of the parameter σ has an increasing tendency looking from the middle of the parallel street into the second or the third transverse street. In the first transverse street, such a tendency is not noticeable. The appropriate model describing an evolution of σ in the investigated area is

$$\sigma = \frac{0.39}{x^*} + 0.01y^{*2} + 0.21 \tag{58.5}$$

The model (Fig. 58.3) is valid for the same sampling positions as the parameter μ above. The coefficient of determination R^2 for the model is 0.89.

We utilized the proposed model to calculate the 0.05, 0.25, 0.5, 0.75 and 0.95 quantiles of at* and compared them with the same quantiles derived from the data from the experiments. The values matched in the 95% CIs except for a few quantiles for positions close to the source (not showed).

58.4 Conclusion

The paper dealt with arrival time of gas clouds. It proposed the new model for construction of the probability density function of arrival time within the idealized urban canopy. Arrival time ensembles in our experiments usually follow a lognormal distribution. The evolution of its parameters within our model can be described by the equations $\mu = 0.083x^* + 0.30y^* + 2.93$, $\sigma = 0.39/x^* + 0.01y^{*2} + 0.21$. Such models could be utilized to predict the situation after the short-term gas leakage not only in the most frequently occurred (mode of pdf) but also in extreme cases (e.g., 0.95 quantile of pdf).

Acknowledgements The authors would like to thank the Technical Agency of the Czech Republic—TA CR (TJ01000383) and the Institute of Thermomechanics (RVO 61388998) for their financial support.

References

1. H. Chaloupecká, Z. Jaňour, Š. Nosek, Short-term gas dispersion in idealised urban canopy in street parallel with flow direction, in *EPJ Web of Conference* (2016)
2. H. Chaloupecká, Z. Jaňour, J. Mikšovský, K. Jurčáková, K. Kellnerová, Evaluation of a new method for puff arrival time as assessed through wind tunnel modelling. PSEP **111**, 194–210 (2017)
3. E. Heathcote, *The Historic Mixed-Use Courtyard Buildings of Central Europe*. Global Property Insight, Financial Times (2014)
4. P.G. Hoel, *Introduction to Mathematical Statistics* (Wiley, USA, 1966), p. 427
5. L. Lubcke, H. Harms, E. Berbekar, B. Leitl, Puff dispersion in a simplified central-European city, in *International Workshop on Physical Modelling and Dispersion Phenomena* (2013)
6. L. Soulhac, V. Garbero, P. Salizzoni, P. Mejean, R.J. Perkins, Flow and dispersion in street intersections. Atmos. Environ. **43**, 2981–2996 (2009)
7. VDI, Environmental meteorology, physical modelling of flow and dispersion processes in the atmospheric boundary layer, Application of wind tunnels. VDI-Standard: VDI 3783 Blatt 12 (2000)
8. D.S. Wilks, *Statistical Methods in the Atmospheric Sciences* (Academic Press, USA, 2006), p. 626

Part VI
Aerosols in the Atmosphere

Chapter 59
Evaluation of Seven Chemical Aging Modeling Schemes with the 2D-VBS Framework Against Ground and Airborne PEGASOS Campaign Measurements

Eleni Karnezi, Benjamin N. Murphy and Spyros N. Pandis

Abstract The 2D-VBS framework describing the organic aerosol (OA) distribution as a function of its volatility and O:C is used in PMCAMx-Trj, a one–dimensional Lagrangian chemical transport model. The model is used to simulate the atmospheric OA during two PEGASOS campaigns in the Po Valley in Italy during 2012 and in Hyytiälä, Finland during 2013. Po Valley is an area with many industrial and agricultural sources and Hyytiälä is characterized by high biogenic secondary OA (SOA) levels. The simulations are evaluated with AMS measurements both at the ground and aloft with a Zeppelin airship. There were seven aging schemes with different assumptions about functionalization, biogenic SOA aging, and fragmentation that succeeded in reproducing the AMS measurements within measurement error. For all seven schemes, the assumed enthalpy of evaporation had a surprisingly small effect on the diurnal and vertical profiles of O:C and OA concentration. Even if the seven schemes have relatively different characteristics and assumptions, all provide a rather similar picture about the different sources and processes contributing to the total OA in these two very different areas.

E. Karnezi (✉) · S. N. Pandis
Department of Chemical Engineering, Carnegie Mellon University,
Pittsburgh, PA 15213, USA
e-mail: ekarnezi@andrew.cmu.edu

S. N. Pandis
e-mail: spyros@chemeng.upatras.gr

B. N. Murphy
National Exposure Research Laboratory, United States Environmental Protection Agency,
Research Triangle Park, Durham, NC 27709, USA
e-mail: ben.n.murphy@gmail.com

S. N. Pandis
Department of Chemical Engineering, University of Patras, 26504 Patra, Greece

C. Mensink et al. (eds.), *Air Pollution Modeling and its Application XXVI*,
Springer Proceedings in Complexity,
https://doi.org/10.1007/978-3-030-22055-6_59

59.1 Introduction

Organic aerosol is an important component of fine atmospheric particulate matter. The physical properties of the OA and its formation mechanisms are still highly uncertain. Using lumped species for the representation of the hundreds or even thousands of different OA components in atmospheric chemical transport models is computationally efficient and has been used in almost all previous studies. The two-dimensional volatility basis set (2D-VBS) framework [1] lumps organic compounds into surrogates along two axes, the axis of volatility and the axis of oxygen content (in this work expressed as O:C ratio), for the description of the OA chemical aging reactions. Murphy et al. [2, 3] used a 2D-VBS module for the first time in the one-dimensional Langrangian chemical transport model PMCAMx-Trj, with 12 volatility bins (effective saturation concentrations C^* varying from 10^{-5} up to 10^6 $\mu g\ m^{-3}$ at 298 K) and 13 O:C bins (from 0 to 1.2 with a step equal to 0.1). They evaluated various parameterizations with ground measurements at three European sites and their results showed that the simplest parameterization for the chemical aging reactions was the most successful.

59.2 Model Description

In this work, we use the one-dimensional PMCAMx-Trj model with the 2D-VBS module [2, 3] simulating the air parcels that arrived at each PEGASOS monitoring location: San Pietro Capofiume in Italy during 2012 and Hyytiälä during 2013.

59.2.1 The Seven OA Chemical Aging Parameterizations

In the simulations performed, we evaluated a series of OA chemical aging parameterizations. Two simple functionalization schemes, the 1-bin and 2-bin case, in which there is either one or two bin volatility reduction for every reaction respectively with 50% probability for an increase of 1 or 2 oxygen atoms and a more rigorous scheme, the detailed functionalization scheme, initially suggested by [1]. Two biogenic SOA aging schemes and different fragmentation probabilities varying from zero (no fragmentation) to unity are explored. Seven aging schemes (out of approximately 200 tested) reproduced well the AMS ground and aloft O:C and OA measurements. Table 59.1 summarizes the 7 parameterizations that were found to reproduce the OA observations in Italy.

Table 59.1 Parameters of the 7 optimized 2D-VBS schemes used in the simulations

Functionalization scheme	bSOA increase during aging	Fragmentation probability	Parameterization name
1-bin	No	$b = 0$	1-bin
1-bin	Yes	$b = 0.15$	1-bin/bSOA/$b = 0.15$
2-bin	No	$b = 0$	2-bin
2-bin	No	$b = 0.1$	2-bin/$b = 0.1$
2-bin	Yes	$b = 0.4$	2-bin/bSOA/$b = 0.4$
Det	No	$b = 0.3$	Det/$b = 0.3$
Det	Yes	$b = 0.7$	Det/bSOA/$b = 0.7$

59.3 Results

59.3.1 Simple Functionalization (1-bin Case)

The base case as described in [3] or else the simple functionalization scheme (1-bin) was used for the first set of evaluations. Figure 59.1 shows the predicted average diurnal O:C variation and the corresponding ground AMS measurements. There is encouraging agreement between model and measurements, with differences noticed mostly in the afternoon where there seems to be overprediction by the model to some extent. The main conclusion is that the OA in Po Valley is oxidized with almost little average diurnal O:C variation. The average modeled O:C is equal to 0.64 comparable to the average AMS measurement that is equal to 0.58. The average vertical O:C profile was also relatively flat especially in the lowest 1 km of the atmosphere and both the model and the Zeppelin measurements suggested that the OA was highly oxidized. The 1-bin case was also successful in reproducing the corresponding observations in the boreal forest of Hyytiälä.

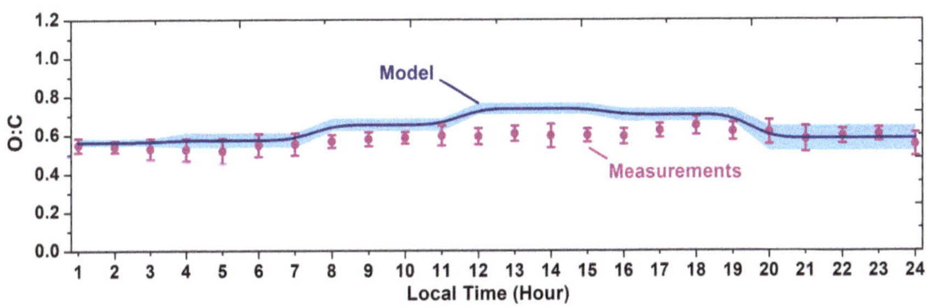

Fig. 59.1 Average ground diurnal O:C evolution in Po Valley, Italy for the simple functionalization (1-bin case). The blue line represents the model simulation results and the shaded area one standard deviation from all the daily simulations. The pink symbols show the stationary AMS measurements and the error bars correspond to one standard deviation from all the days modeled

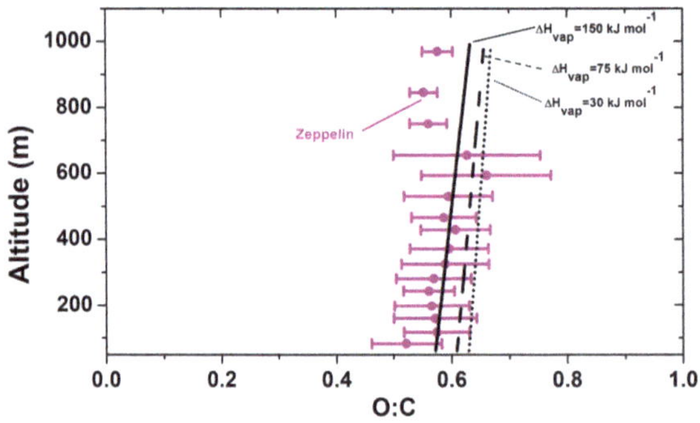

Fig. 59.2 Average vertical profile of O:C assuming various vaporization enthalpies in the model as $\Delta H_{vap} = 30$ kJ mol^{-1} (black short dot line), $\Delta H_{vap} = 75$ kJ mol^{-1} (black dashed line), and $\Delta H_{vap} = 150$ kJ mol^{-1} (black line) for the Po Valley in Italy during the PEGASOS flights. Once again, the pink symbols show the Zeppelin measurements and the error bars correspond to one standard deviation from all the days modeled

59.3.2 *The Role of ΔH_{vap}*

The OA effective enthalpy of evaporation is a property that is considerably uncertain with different values assumed in various CTMs. In our simulations, we tested the simple functionalization scheme (1-bin case) assuming three different enthalpies of evaporation with values of 30, 75, and 150 kJ mol^{-1}. We evaluated the model predictions of the vertical profiles of O:C and OA concentration against the Zeppelin measurements.

The modeled vertical profiles of O:C were surprisingly similar for the various assumed effective ΔH_{vap} value (Fig. 59.2). The predicted OA concentrations were also quite similar. This lack of sensitivity can be explained by the intricate interactions and feedbacks between partitioning of the semi-volatile OA components and their gas-phase chemical aging reactions. All aging mechanisms were tested for the sensitivity of ΔH_{vap} and the conclusions were the same.

59.3.3 *Predicted OA Composition*

The seven aging schemes were successful in reproducing the measurements in Po Valley in Italy (Fig. 59.3a) and all predicted relatively similar source contributions. This was also the case for the boreal forest in Finland where all seven aging schemes once again predicted quite similar source contributions (Fig. 59.3b).

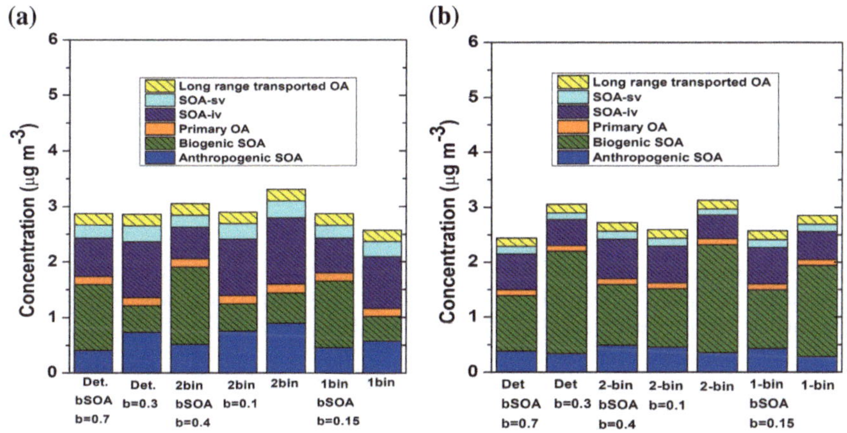

Fig. 59.3 Modeled OA composition for the seven schemes with satisfying performance in **a** San Pietro Capofiume, **b** Hyytiälä. The total average OA from all measurements was equal to 2.8 μg m^{-3} in San Pietro Capofiume and 2.1 μg m^{-3} in Hyytiälä

59.4 Conclusions

The evaluation of the simple functionalization scheme, both against stationary and airborne moving Zeppelin measurements, was satisfactory in both the PEGASOS campaigns. The modeled OA concentration and O:C showed a limited sensitivity to the assumed enthalpy of evaporation, for the OA levels both at the ground and in the 1 km of the atmosphere. Seven different aging schemes were successful in reproducing the observations. Their predictions about the contribution of the various OA sources were to a large extent consistent in both environments. This agreement is encouraging about our ability to constrain the SOA sources, despite uncertainties about the details of the corresponding processes.

QUESTIONER: Dr. Maria Kanakidou.

QUESTION: Among the mechanisms you have investigated in your study which one do you consider the most appropriate for use in global models?

ANSWER: The performance of the seven parameterizations is very similar leading to the conclusion that no parameterization is superior. However, the simple schemes (1-bin and 2-bin) have the advantage that they can also be implemented in the 1D-VBS, with much lower computational cost. In this case, the predictions of OA mass concentration and the volatility distributions remain the same, but one loses the ability to predict the O:C. The other schemes cannot be easily simplified for the 1D-VBS without introduction of errors. Regarding the bSOA aging mechanisms available findings suggest that there is significant second and later generation production of SOA from anthropogenic of SOA from anthropogenic precursors, while the corresponding later production of SOA from biogenic precursors may be a lot smaller and we should treat the chemical aging of anthropogenic and biogenic compounds independently following previous modeling efforts. Furthermore, fragmentation of organic compounds during chemical aging is an important reaction pathway that should be included as well. However, the robustness of these schemes, which were tuned to the measurements in two specific areas, will need to be tested in future work. For the time being, this work can be viewed as a parameter-fitting exercise demonstrating that there are multiple sets of 2D-VBS parameters that can result in predictions consistent with the measurements.

Acknowledgements This work was part of the PEGASOS project funded by the European Commission under the Framework Program 7 (FP7-ENV-2010-265148) and it was funded from the US National Science Foundation (US NSF) grant 1455244.

References

1. N.M. Donahue, J.H. Kroll, S.N. Pandis, A.L. Robinson, A two-dimensional volatility basis set: 1. Organic-aerosol mixing thermodynamics. Atmos. Chem. Phys. **11**, 3303–3318 (2011)
2. B.N. Murphy, N.M. Donahue, C. Fountoukis, S.N. Pandis, Simulating the oxygen content of ambient organic aerosol with the 2D volatility basis set. Atmos. Chem. Phys. **11**, 7859–7873 (2011)
3. B.N. Murphy, N.M. Donahue, C. Fountoukis, M. Dall'Osto, C.O. Dowd, A. Kiendler-Scharr, S.N. Pandis, Functionalization and fragmentation during ambient organic aerosol aging: application of the 2-D volatility basis set to field studies. Atmos. Chem. Phys. **12**, 10797–10816 (2012)

Chapter 60
A Parameterization of Heterogeneous Hydrolysis of N_2O_5 for 3-D Atmospheric Modelling

R. Wolke, Y. Chen, W. Schröder, G. Spindler and A. Wiedensohler

Abstract During night-time, the heterogeneous hydrolysis of N_2O_5 on the surface of deliquescent aerosol particles represents a major source for the formation of HNO_3 and leads to an important reduction of NO_x in the atmosphere. In Chen et al., Atmos. Chem. Phys. 18:673–689, 2018 [5], we investigate an improved parameterization of the heterogeneous N_2O_5 hydrolysis. This approach is based on laboratory experiments and takes into account the temperature, relative humidity, aerosol particle composition as well as the surface area concentration. The parametrization was implemented in the online coupled model system COSMO-MUSCAT (Consortium for Small-scale Modelling and Multi-Scale Chemistry Aerosol Transport, https://cosmo-muscat.tropos.de). In Chen et al., Atmos. Chem. Phys. 18:673–689, 2018 [5], the modified model was applied for the simulation of the HOPE-Melpitz campaign (10–25 September 2013) where especially the nitrate prediction over western and central Europe was analysed. The modelled particulate nitrate concentrations were compared with filter measurements over Germany. In this first study, the particulate nitrate results are significantly improved by using the developed N_2O_5 parametrization, particularly if the particulate nitrate was dominated by the local chemical formation (September 12, 17–18 and 25). The aim of the current study consists in an evaluation over a longer time period for different meteorological conditions and emission situations. For this reason, we have simulated the period from March to November 2010. The results were compared with other approaches and evaluated by filter measurements. The improvement was confirmed for the results in spring and autumn, but nitrate is strongly over-predicted also for the new parametrization during the summer time.

R. Wolke (✉) · Y. Chen · W. Schröder · G. Spindler · A. Wiedensohler
Leibniz Institute for Tropospheric Research (TROPOS), Permoserstrasse 15, 04303 Leipzig, Germany
e-mail: wolke@tropos.de

Present Address:
Y. Chen
Lancaster Environment Centre, Lancaster University, Lancaster LA1 4YQ, UK

C. Mensink et al. (eds.), *Air Pollution Modeling and its Application XXVI*, Springer Proceedings in Complexity,
https://doi.org/10.1007/978-3-030-22055-6_60

60.1 Introduction

The tropospheric chemistry is strongly influenced by the budget of nitrogen oxides (NO_x). The chemical mechanisms controlling this budget have also a large impact on ozone and the oxidizing capacity of the atmosphere. At day-time, nitrogen oxides are mainly removed from the atmosphere by the formation of nitric acid (HNO_3) via the reaction of NO_2 and OH. During night-time, the heterogeneous hydrolysis of N_2O_5 on the surface of deliquescent aerosol particles represents a major source for HNO_3 and acts as an important sink of NO_x in the atmosphere. The heterogeneous N_2O_5 hydrolysis was incorporated in global and regional chemistry transport models and its impact on atmospheric chemistry was investigated in several studies [4, 5, 10]. Usually, the heterogeneous N_2O_5 hydrolysis is implemented into chemical transport models as a first-order loss reaction.

$$N_2O_5 \overset{(aerosol)}{\rightarrow} 2HNO_3 \tag{60.1}$$

with a rate constant $k_{N_2O_5}$. In the originally version of COSMO-MUSCAT [17] this constant is calculated by an approach of [3] which depends only from the relative humidity (RH). Aiming to advance the representativeness of heterogeneous hydrolysis of N_2O_5, we propose a new parameterization depending on temperature, RH, and particle composition [5]. The influence of surface area concentration on the hydrolysis rate is also comprehensively considered. This new scheme was implemented in the model system COSMO-MUSCAT in order to investigate the impact of heterogeneous hydrolysis of N_2O_5 on the particulate nitrate formation. The model simulations were performed for different configurations and the period from March to November 2010. The results were compared and evaluated by filter measurements in Melpitz and two other Germany sites (Neuglobsow, Waldhof; www.umweltbundesamt.de).

60.2 Model Framework

Chemistry-transport model. COSMO-MUSCAT is a state-of-the-art multiscale model system developed by the modelling department of TROPOS, which is qualified for process studies in local and regional areas. The model system comprises the numerical weather forecast model of the German Weather Service (DWD) COSMO [11] and the chemistry transport model MUSCAT [17]. The model was applied to different investigations, e.g. air quality studies treating local as well as regional areas [7, 15]. Driven by the meteorological model, MUSCAT treats the atmospheric transport as well as chemical transformations for several gas phase species and particle populations. The transport processes include advection, turbulent diffusion, sedimentation, dry and wet deposition. In this study, the gas-phase reaction system is described by the chemical mechanism RACM-MIM2 [8, 16], which consists of 87 species and more than 200 reactions. The aerosol processes are represented by a mass-based approach.

The formation of secondary inorganic particulate matter is through reactions of ammonia with sulfuric acid and nitric acid, which are produced from gaseous precursors SO_2 and NO_x [7]. The applied particle–gas partitioning depends on temperature and humidity. By using the equilibrium approach of [9], the partitioning scheme was comparable to [6]. For the description of SOA an updated version of SORGAM [12] is implemented.

Parameterization of heterogeneous hydrolysis of N_2O_5. Based on the first-order reaction rate, Chang et al. [3] proposed

$$k_{N_2O_5} = \frac{1}{600\exp\left(-(\frac{RH}{28})^{2.8}\right) + a} \tag{60.2}$$

RH is the relative humidity in %, which was used as an indicator for the influence of hygroscopic growth of the particle surface. Chang et al. [3] used a = 5, that gives an asymptotic $k_{N_2O_5}$ of 0.2 min^{-1} if the relative humidity exceeds 60%. In the P2 approach discussed in [10], a = 17 was suggested which leads to a more suitable asymptotic $k_{N_2O_5}$ of nearly 0.06 min^{-1}. This approach was originally adopted in COSMO-MUSCAT to represent the heterogeneous hydrolysis of N_2O_5. In [5], we proposed a sophisticated parameterization with the full consideration of temperature, RH, aerosol compositions and mass concentration.

$$k_{N_2O_5} = \frac{1}{600\exp\left(-(\frac{RH}{28})^{2.8}\right) + 17} \cdot f_s \cdot f_{\gamma N_2O_5} \tag{60.3}$$

We adapted (60.2) with scaling factors f_s and f_γ which represent the impact of the particle surface S and the reaction probability γ, respectively. The factor f_γ depends on the particle mass composition. Additionally, an approach of [1] is incorporated that describes the organic coating suppression effect on the reaction probability γ. A detailed discussion of the developed parametrization is given in [5].

Emissions. The European anthropogenic emission inventory and the temporal-resolved emission factors are provided by TNO for the AQMEII project [17]. The inventory includes gaseous pollutants and primary emitted particulate matter. Based on the reasoning of other authors [2, 13], the mainly agricultural emissions of NH_3 were reduced by 50%. Biogenic emissions were generated online in dependence on land use and modelled meteorological conditions [14].

60.3 Results and Discussion

The model simulations were performed for the period from March to November 2010. The results are compared for four different parameterizations of the heterogeneous N_2O_5 hydrolysis: (1) Chen et al. [5] including SOA coating, (2) Chen et al. [5] without SOA coating, (3) approach P2 of [10], (4) without N_2O_5 hydrolysis.

The time series of the daily mean values of the nitrate concentration in Melpitz are presented in Fig. 60.1. The corresponding statistical values are summarized in Table 60.1. Monthly mean concentrations for Germany and some corresponding differences for March (top) and August (below) are shown in Fig. 60.2. Usually, COSMO-MUSCAT tends to over predict particulate nitrate in a reasonable range in long-term average. The results were comparable with other models in previous studies [15]. In the current study, this is also the case for the periods in spring and autumn.

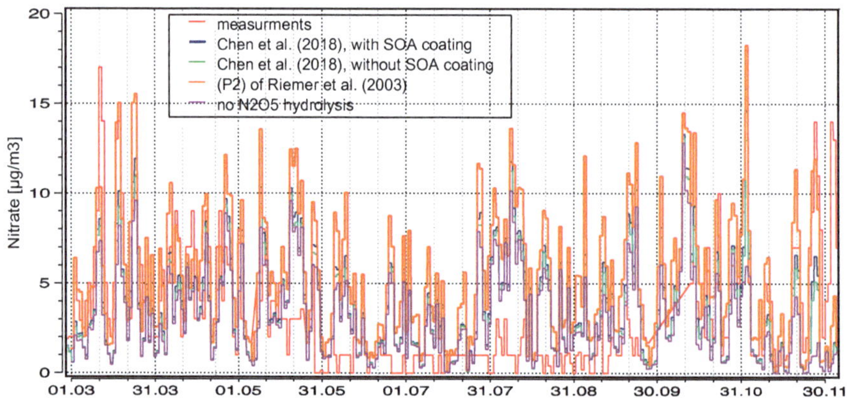

Fig. 60.1 Results of COSMO-MUSCAT for the TROPOS field site Melpitz for the period from March to November 2010: Daily impactor measurements and daily averaged model results of nitrate for different parameterizations of the N_2O_5 photolysis

Table 60.1 Comparison of mean nitrate concentrations [$\mu g/m^3$] for the Melpitz site and the corresponding correlation coefficients

Approach	March/April 2018	May–August 2018	September–November 2018
Measurements	4.62	1.46	3.51
Chen et al. [5], with SOA coating	3.88 (0.53)	3.50 (0.21)	3.36 (0.40)
Chen et al. [5], without SOA coating	4.09 (0.54)	3.65 (0.20)	3.60 (0.44)
(P2) in [10]	6.16 (0.50)	5.10 (0.18)	5.62 (0.49)
Without N_2O_5 hydrolysis	3.52 (0.51)	3.22 (0.20)	2.90 (0.18)

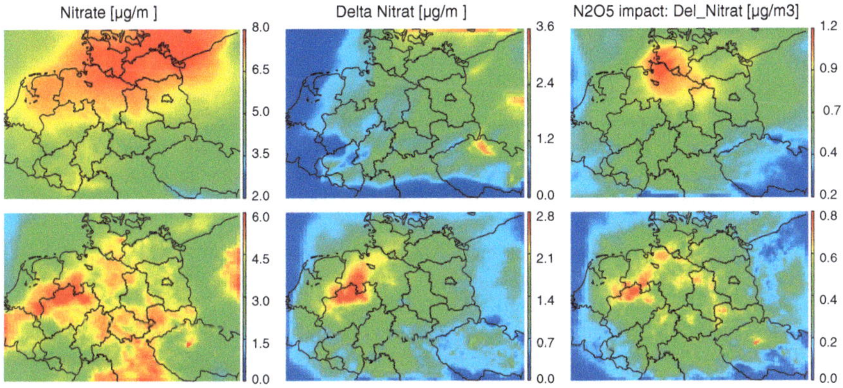

Fig. 60.2 Comparison of modelled monthly mean concentrations of nitrate [μg/m^3] at ground level for March (top) and August 2010 (bottom): nitrate concentration for parametrization (1) (left column), difference (3)-(1) (centre column), difference (1)-(4) (right column)

However, during the summer time particulate nitrate was significantly overestimated by the COSMO-MUSCAT model, despite a good agreement of meteorological simulation. Compared with filter measurements at Melpitz, it was overestimated by a factor of 2.4 for the parametrization of [5] with SOA coating and by a factor of 3.5 for the old approach.

Obviously, the strong overestimation cannot be caused by a deficient description of the N_2O_5 hydrolysis. Even the run without hydrolysis yields to an overproduction of more than 100%. Therefore, other causes should be originated this severe overprediction in summer. One possible reason could be a rigorous overestimation of NO soil fluxes due to manuring during the summer period. This issue needs further investigations.

In all cases, the (P2) approach produces more than 50% more nitrate than the parametrization of [5]. Furthermore, the nitrate concentration is significantly underestimated in the simulation without the consideration of the heterogeneous N_2O_5 hydrolysis in spring and autumn. The main results given for Melpitz in Table 60.1 were also confirmed by the analysis for the other two stations. In the difference plots in Fig. 60.2 the larges values occur in regions with high ammonia emission in north–western Germany and the Netherlands. In these areas, the ammonium nitrate formation is mainly restricted by the amount of available nitric acid. Therefore, the impact of changes in the hydrolysis rates is clearly to see here.

Additionally, the impact of the reduced N_2O_5 hydrolysis due to organic coating [1] on the particulate nitrate formation was investigated. Based on the new parametrization, the simulation results with and without organic coating were analyzed. In difference to the results in [5], we observed a considerable (more than 5% or 0.24 μg/m^3 on average) influence of coating on particulate nitrate in Melpitz.

Acknowledgements The work was supported by the ZIH Dresden and the JSC Jülich. Furthermore, we thank the German Weather Service for good cooperation.

References

1. T. Anttila, A. Kiendler-Scharr, R. Tillmann, T.F. Mentel, On the reactive uptake of gaseous compounds by organic-coated aqueous aerosols: theoretical analysis and application to the heterogeneous hydrolysis of N_2O_5. J. Phys. Chem. A **110**, 10435–10443 (2006)
2. A.M. Backes, A. Aulinger, J. Bieser, V. Matthias, M. Quante, Ammonia emissions in Europe, part II: how ammonia emission abatement strategies affect secondary aerosols. Atmos. Environ. **126**, 153–161 (2016)
3. J.S. Chang, R.A. Brost, I.S.A. Isaksen, S. Madronich, P. Middleton, W.R. Stockwell, C.J. Walcek, A threedimensional Eulerian acid deposition model: physical concepts and formulation. J. Geophys. Res.-Atmos. **92**, 14681–14700 (1987)
4. W.L. Chang, S.S. Brown, J. Stutz, A.M. Middlebrook, R. Bahreini, N.L. Wagner, W.P. Dubé, I.B. Pollack, T.B. Ryerson, N. Riemer, Evaluating N_2O_5 heterogeneous hydrolysis parameterizations for CalNex 2010. J. Geophys. Res.-Atmos. **121**, 5051–5070 (2016)
5. Y. Chen, R. Wolke, L. Ran, W. Birmil, G. Spindler, W. Schröder, H. Su, Y. Cheng, I. Tegen, A. Wiedensohler, A parameterization of the heterogeneous hydrolysis of N_2O_5 for mass-based aerosol models: improvement of particulate nitrate prediction. Atmos. Chem. Phys. **18**, 673–689 (2018)
6. M.V. Galperin, M.A. Sofiev, The long-range transport of ammonia and ammonium in the Northern Hemisphere. Atmos. Environ. **32**, 373–380 (1998)
7. D. Hinneburg, E. Renner, R. Wolke, Formation of secondary inorganic aerosols by power plant emissions exhausted through cooling towers in Saxony. Environ. Sci. Pollut. Res. **16**, 25–35 (2009)
8. M. Karl, H.P. Dorn, F. Holland, R. Koppmann, D. Poppe, L. Rupp, A. Schaub, A. Wahner, Product study of the reaction of oh radicals with isoprene in the atmosphere simulation chamber saphir. J. Atmos. Chem. **55**(2), 167–187 (2006)
9. M. Mozurkewich, The dissociation constant of ammonium nitrate and its dependence on temperature, relative humidity and particle size. Atmos. Environ. **27A**, 261–270 (1993)
10. N. Riemer, H. Vogel, B. Vogel, B. Schell, I. Ackermann, C. Kessler, H. Hass, Impact of the heterogeneous hydrolysis of N_2O_5 on chemistry and nitrate aerosol formation in the lower troposphere under photo smog conditions. J. Geophys. Res.-Atmos. **108** (2003)
11. U. Schättler, G. Doms, C. Schraff, *A Description of the Nonhydrostatic Regional Cosmo-Model. Part VII: User's Guide* (Deutscher Wetterdienst, Offenbach, 2013)
12. B. Schell, I.J. Ackermann, H. Hass, F.S. Binkowski, A. Ebel, Modeling the formation of secondary organic aerosol within a comprehensive air quality model system. J. Geophys. Res. Atmos. **106**(D22), 28275–28293 (2001)
13. J. Sintermann, A. Neftel, C. Ammann, C. Háni, A. Hensen, B. Loubet, C.R. Flechard, Are ammonia emissions from field-applied slurry substantially over-estimated in European emission inventories? Biogeosciences **9** (2012)
14. R. Steinbrecher, G. Smiatek, R. Koeble, G. Seufert, J. Theloke, K. Hauff, P. Ciccioli, R. Vautard, G. Curci, Intra- and inter-annual variability of VOC emissions from natural and semi-natural vegetation in Europe and neighbouring countries. Atmos. Environ. **43**(7), 1380–1391 (2009)
15. R. Stern, P. Builtjes, M. Schaap, R. Timmermans, R. Vautard, A. Hodzic, M. Memmesheimer, H. Feldmann, E. Renner, R. Wolke, A. Kerschbaumer, A model inter-comparison study focussing on episodes with elevated PM10 concentrations. Atmos. Environ. **42**, 4567–4588 (2008)
16. W.R. Stockwell, F. Kirchner, M. Kuhn, S. Seefeld, A new mechanism for regional atmospheric chemistry modeling. J. Geophys. Res. Atmos. **102**(D22), 25847–25879 (1997)
17. R. Wolke, W. Schröder, R. Schrödner, E. Renner, Influence of grid resolution and meteorological forcing on simulated european air quality: a sensitivity study with the modeling system COSMO-MUSCAT. Atmos. Environ. **53**, 110–130 (2012)

Chapter 61
Modelling Organic Aerosol in Europe: Improved CAMx and Contribution of Anthropogenic and Biogenic Sources

Jianhui Jiang, Sebnem Aksoyoglu, Imad El Haddad, Giancarlo Ciarelli, Emmanouil Oikonomakis, Hugo A. C. Denier van der Gon and André S. H. Prévôt

Abstract Chemical transport model (CTM) simulation of organic aerosol (OA) is always challenged by numerous sources and complicated formation processes of secondary organic aerosol. In this study, we conducted a source-specific, whole-year (2011) simulation of organic aerosol in Europe using the air quality model CAMx v6.3 with volatility basis set (VBS) scheme after implementing new findings from experimental studies. The VBS module was parameterized based on the latest data for gasoline and diesel vehicles and wood combustion from the smog-chamber experiments. The model performance was evaluated using OA measurements from the ACSM (Aerodyne Chemical Speciation Monitor) and AMS (Aerodyne Aerosol Mass Spectrometer) network and contributions from 6 different anthropogenic (gasoline and diesel vehicles with old or new technologies, biomass burning, and other sources (OP)) and biogenic sources were estimated. The

J. Jiang (✉) · S. Aksoyoglu · I. El Haddad · E. Oikonomakis · A. S. H. Prévôt
Laboratory of Atmospheric Chemistry, Paul Scherrer Institute, 5232 Villigen, Switzerland
e-mail: jianhui.jiang@psi.ch

S. Aksoyoglu
e-mail: sebnem.aksoyoglu@psi.ch

I. El Haddad
e-mail: imad.el-haddad@psi.ch

E. Oikonomakis
e-mail: emmanouil.oikonomakis@psi.ch

A. S. H. Prévôt
e-mail: andre.prevot@psi.ch

G. Ciarelli
Laboratoire Inter-Universitaire Des Systèmes Atmosphériques (LISA), Institut Pierre Simon Laplace, UMR CNRS 7583, Université Paris Est Créteil et Université Paris Diderot, Créteil, France
e-mail: giancarlo.ciarelli@lisa.u-pec.fr

H. A. C. Denier van der Gon
Department of Climate, Air and Sustainability, TNO, Utrecht, The Netherlands
e-mail: hugo.deniervandergon@tno.nl

C. Mensink et al. (eds.), *Air Pollution Modeling and its Application XXVI*, Springer Proceedings in Complexity,
https://doi.org/10.1007/978-3-030-22055-6_61

modified VBS scheme improved the model performance on OA simulation during the whole period by reducing the bias between model and measurements by up to 52%. The OA concentrations were dominated by biomass burning in winter, while biogenic emissions were the main sources in summer. The contribution of road traffic was relatively lower compared to studies in the USA. The contribution of new gasoline and diesel vehicles (after Euro IV emission standards or equipped with diesel particle filters) to the total OA was negligible.

61.1 Introduction

Organic aerosols (OA) contribute to 20–90% of atmospheric fine particulate matter pollution, which exerts significant impacts on human health and climate change. Among the numerous components of OA, secondary organic aerosol (SOA) generated from oxidation of organic gases is proved to be dominant [1]. Traditional chemical transport models (CTMs) tend to underestimate SOA substantially due to its numerous sources, composition and chemical-physical properties [2]. One of the most important reasons is the potentially high but unaccounted for contribution of semi-volatile organics (SVOCs) and intermediate-volatility organics (IVOCs). To improve the situation, the volatility basis set (VBS) scheme has been applied in CTMs like CAMx (Comprehensive Air quality Model with extensions) and CMAQ (Community Multiscale Air Quality) [3]. The VBS scheme characterizes organic species as surrogates by their volatilities [4], and therefore better simulate the oxidation of a wide range of organics. Despite its great potential, high uncertainties still remain. Recent studies found that precursors in traditional models account for only ~3–27% of the observed SOA from residential wood burning [5]; formation of SOA from diesel and gasoline vehicles were greatly influenced by temperature and vehicle condition (with or without diesel particle filter) [6]. Thus, in this study, we parameterized the regional model CAMx with VBS scheme based on recent findings from experimental studies, and simulated the contribution of major anthropogenic and biogenic sources on OA in Europe.

61.2 Method

The regional air quality model CAMx version 6.3 with VBS scheme [7] was used in this study. The model domain (15° W–35° E, 35° N–70° N) covered Europe with a horizontal resolution of 0.25° × 0.125°. The whole year of 2011 was modelled. Organic aerosol formation was estimated by the 1.5-D VBS organic aerosol chemistry/partitioning module [3]. We split the sources of gasoline vehicles (GV) and diesel vehicles (DV) in standard model to old and new (GO, GN, DO, DN), and the biomass/biogenic source (BB) to biomass burning (BB) and biogenic (BI). Parameters of yield for GO, GN, DO, DN were adjusted respectively according to chamber data with or without installation of diesel particle filter [6]; parameters for BB were updated by latest smog-chamber data by PSI-LAC.

The meteorological input was prepared by Weather Research and Forecasting Model WRF-ARW version 3.7.1 with the ECMWF (European Centre for Medium-Range Weather Forecasts) global atmospheric reanalysis ERA-Interim data. The MOZART global model data was adopted for initial and boundary conditions. Biogenic emissions were estimated by PSI-model developed by [8]. The anthropogenic emissions were obtained from the high-resolution European emission inventory TNO-MACC (Monitoring Atmospheric Composition and Climate)-III. The IVOC emissions for different sources were estimated as 25% and 20% of NMVOCs from gasoline and diesel vehicles, respectively [9], 4.5 times of POA from biomass burning [10] and 1.5 times of POA from other anthropogenic sources [11]. A factor of 3 was adopted for POA emission to compensate the influence of missing SVOCs in emission inventory. Along with the modified model (referred as NEW), a base case simulation (BASE) with default parameters of CAMx v6.3 was also conducted for comparison. Modelled OA was compared with measurements at 8 ACSM/AMS (Aerodyne Aerosol Chemical Speciation Monitor/Aerodyne aerosol mass spectrometer) stations.

61.3 Results and Discussion

61.3.1 Model Evaluation

Statistical metrics for OA concentrations by CAMx with modified (NEW) and default (BASE) parameterization are shown in Table 61.1. Zurich, Marseille and Bologna are urban stations, Paris is a suburban station, while Mace Head, Montsec, Finokalia and San Pietro Capofiume (SPC) are rural or remote stations. For all the urban stations, NEW parameterization effectively decreased the mean bias (by 24–52%) between model and measurements. For rural stations dominated by biogenic and biomass burning emissions, although mean bias decreased except for SPC, the mean error (ME) and root mean square error (RMSE) increased in Mace Head, Montsec and SPC. Improvement only occurred in Finokalia (MB by 21%, ME by 4%, RMSE by 4%).

61.3.2 Contribution of Anthropogenic and Biogenic Sources

Contribution of major anthropogenic and biogenic sources to surface OA concentrations varied with season and location. Biomass burning contributed most to winter OA at most of the sites, while biogenic sources dominated the summer OA (see example of Zurich in Fig. 61.1). In remote area such as Montsec (1570 m a.s.l.,

Table 61.1 Comparison between measured and modelled hourly OA concentration

Stations	MB ($\mu g\ m^{-3}$)		ME ($\mu g\ m^{-3}$)		RMSE ($\mu g\ m^{-3}$)		MFB		MFE	
	Base	New	Base	New	Base	New	Base	New	Base	New
Zurich	−2.06	−0.98	3.37	3.26	4.69	4.54	−0.36	−0.15	0.63	0.62
Marseille	−5.45	−4.16	5.58	4.57	7.64	6.83	−0.87	−0.60	0.92	0.74
Bologna	−11.84	−7.72	12.24	9.71	15.55	12.94	−0.85	−0.50	0.91	0.74
Paris	−6.02	−3.62	6.09	4.69	9.36	7.55	−0.98	−0.35	1.04	0.87
Mace Head	−0.05	−0.01	0.45	0.46	1.07	1.11	−0.77	−0.70	1.14	1.14
Montsec	−0.62	−0.42	2.05	2.09	2.68	2.71	−0.18	−0.07	0.80	0.84
Finokalia	−1.22	−0.96	1.53	1.47	2.17	2.08	−0.53	−0.43	0.69	0.66
SPC	−2.00	2.33	5.22	6.13	6.63	7.62	−0.06	0.46	0.57	0.78

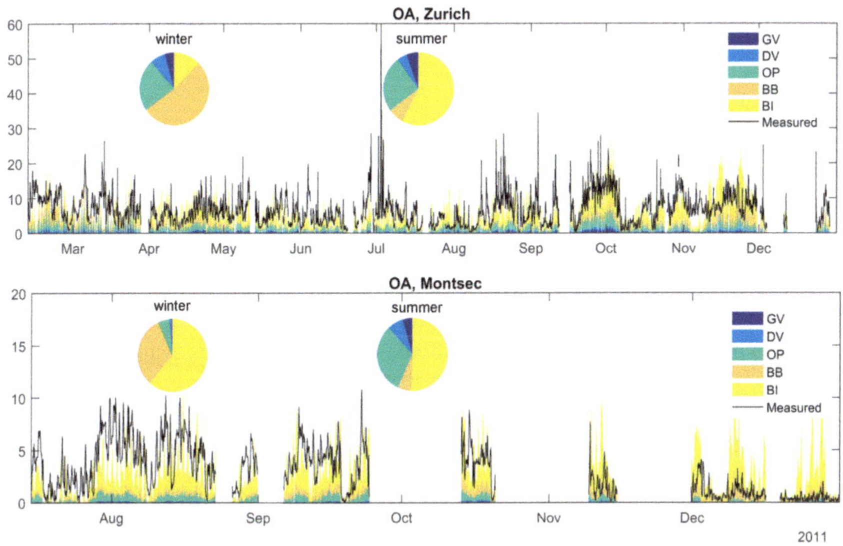

Fig. 61.1 Contribution of different sources to OA concentration s ($\mu g\ m^{-3}$). GV—Gasoline vehicles (old + new), DV—Diesel vehicles (old + new), OP—Other anthropogenic sources, BB—Biomass burning, BI—Biogenic

Fig. 61.1) and Mace Head (west coast of Ireland), the largest contribution was from biogenic sources throughout the whole year. The other anthropogenic sources (OP) including ships, industry and power plants were the third largest source for OA, with a contribution from 2 to 41% for the 8 stations over land. The highest contribution of OP was predicted to be about 99% in some grid cells over the ocean. The gasoline and diesel vehicle emissions accounted for 1–25% (Italy) of OA in Europe, which is relatively lower than in the USA (~35% for California according to [12]). The contribution of the new diesel vehicles equipped by diesel particle filters after Euro IV emission standard was negligible.

61.4 Conclusion

A whole-year simulation of OA in Europe was performed by the regional air quality model CAMx using the basic VBS and a modified VBS module with split sources and updated parameterization based on latest chamber experimental data. The validation by ACSM/AMS measurements indicated that modified VBS improved the model performance on OA simulation effectively, especially in urban areas. Biomass burning and biogenic sources were main source of OA in Europe in winter and summer, respectively. Contribution of gasoline and diesel vehicles was relatively lower compared with studies for the USA. With increasing strict regulations of vehicle emissions in Europe, the focus in modelling and chamber studies needs to be extended to OA from ships, industry and power plants, which is mostly estimated by default parameters in current CTMs.

QUESTIONER: Maria Kanakidou.

QUESTION: You have shown significant contribution of cooking organic aerosol (COA) to OA level found by the observations. Do you simulate the COA in one of the OA categories in your model?

ANSWER: No, the cooking organic aerosol is not simulated in this study. Although the COA is important, the cooking emission is not included in current emission inventory due to lack of data and very high uncertainty. The situation might be improved in future when well validated cooking emission data is available as model input.

Acknowledgements We thank Ramboll for help in CAMx modelling, ECMWF, UCAR, NASA for providing data for model input. We thank Canonaco F., O'Dowd C., Ovadnevaite J., Favez O., Gilardoni S., Marchand N., Minguillón MC., and Florou K. for providing ACSM/AMS data.

References

1. A. Hodzic, P.S. Kasibhatla, D.S. Jo et al., Rethinking the global secondary organic aerosol (SOA) budget: stronger production, faster removal, shorter lifetime. Atmos. Chem. Phys. **16**, 7917–7941 (2016)
2. A. Hodzic, J.L. Jimenez, S. Madronich et al., Modeling organic aerosols in a megacity: potential contribution of semi-volatile and intermediate volatility primary organic compounds to secondary organic aerosol formation. Atmos. Chem. Phys. **10**, 5491–5514 (2010)
3. B. Koo, E. Knipping, G. Yarwood, 1.5-Dimensional volatility basis set approach for modeling organic aerosol in CAMx and CMAQ. Atmos. Environ. **95**, 158–164 (2014)
4. N.M. Donahue, A.L. Robinson, C.O. Stanier et al., Coupled partitioning, dilution, and chemical aging of semivolatile organics. Environ. Sci. Technol. **40**, 2635–2643 (2006)
5. E.A. Bruns, I. El Haddad, J.G. Slowik et al., Identification of significant precursor gases of secondary organic aerosols from residential wood combustion. Sci. Rep. **6**, 27881 (2016)
6. S.M. Platt, I. El Haddad, S.M. Pieber et al., Gasoline cars produce more carbonaceous particulate matter than modern filter-equipped diesel cars. Sci. Rep. **7**, 4926 (2017)
7. Ramboll Environ, *User's Guide: The Comprehensive Air Quality Model with extensions (CAMx) Version 6.3*, California, USA (2016)
8. S. Andreani-Aksoyoglu, J. Keller, Estimates of monoterpene and isoprene emissions from the forests in Switzerland. J. Atmos. Chem. **20**, 71–87 (1995)
9. S.H. Jathar, T.D. Gordon, C.J. Hennigan et al., Unspeciated organic emissions from combustion sources and their influence on the secondary organic aerosol budget in the United States. Proc. Natl. Acad. Sci. **111**, 10473–10478 (2014)
10. G. Ciarelli, S. Aksoyoglu, I. El Haddad et al., Modelling winter organic aerosol at the European scale with CAMx: evaluation and source apportionment with a VBS parameterization based on novel wood burning smog chamber experiments. Atmos. Chem. Phys. **17**, 7653–7669 (2017)
11. A.L. Robinson, N.M. Donahue, M.K. Shrivastava et al., Rethinking organic aerosols: semivolatile emissions and photochemical aging. Science **315**, 1259–1262 (2007)
12. S.H. Jathar, M. Woody, H.O.T. Pye et al., Chemical transport model simulations of organic aerosol in Southern California: model evaluation and gasoline and diesel source contributions. Atmos. Chem. Phys. **17**, 4305–4318 (2017)

Chapter 62
Sea Spray Effects on Marine and Coastal Boundary Layer

George Kallos, C. Stathopoulos, P. Patlakas, G. Galanis, J. Al Qahtani and I. Alexiou

Abstract The air-water exchange processes are considered as complicated but very important for the boundary layer structure. The complex physical procedures related to momentum, moisture and thermal exchanges, render their numerical description into a quite challenging task. Ocean surface roughness and spray droplets have an impact on the lower atmospheric layers with profound effects. These involve alterations in the atmospheric stability profiles and microphysical processes such as the formation of sea salt particles. Water vapor produced by sea spray can alter humidity and temperature profiles close to the ocean surface, leading to modified stability conditions. This cascade of effects can be further extended to wind profiles. Sea salt particles derived from water droplet evaporation are a major source of CCN and GCCN that may lead in low cloud formation over the sea and/or the coastal boundary layer. To describe these processes numerically, an atmospheric-wave-spray coupled system is introduced. The RAMS/ICLAMS atmospheric and the WAM wave spectral model are interfaced, including schemes for ocean drag, water droplet thermodynamics and salt particles considered as predictive quantity. Model simulations performed in the Atlantic shoreline showed that droplet evaporation near the surface modifies the atmospheric stability and affects sea salt dispersion.

62.1 Introduction

In the air-sea interface zone, mechanical processes entail the transfer of momentum from the atmosphere to the ocean through the imposed drag of wind leading to wave generation. Likewise, a wave stress emerges, which is encountered by the surface

G. Kallos (✉) · C. Stathopoulos · P. Patlakas · G. Galanis
Department of Physics, National and Kapodistrian University of Athens, Athens, Greece
e-mail: kallos@mg.uoa.gr; g-kallos@otenet.gr

G. Galanis
Naval Academy of Greece, Section of Mathematics, Piraeus, Greece

J. Al Qahtani · I. Alexiou
Saudi Aramco, Dhahran, Saudi Arabia

C. Mensink et al. (eds.), *Air Pollution Modeling and its Application XXVI*, Springer Proceedings in Complexity,
https://doi.org/10.1007/978-3-030-22055-6_62

winds over waves. Different wave characteristics have an influence in surface winds, with an imposed oceanic drag expressed as sea surface roughness.

Thermodynamical processes, involve heat exchanges between the air and water surfaces and moisture transfers, expressing condensation and evaporation. Additionally, due to wave breaking and bubble bursting sea droplets and aerosols appear in the air surface above sea. Sea spray participates in the thermodynamic system of the air-sea interface zone through the exchange of heat and moisture between water droplets emerged in the lower atmosphere and the surrounding air. The release of sea droplets leads to a moistening of the near sea surface air and adds heat into the atmosphere. These thermodynamical interactions are expressed as heat fluxes. The effects influence temperature and moisture profiles near sea surface. This can alter buoyancy which affects non-linearly the near surface wind patterns [4].

Concurrently, the particles formed in the sea surface, noted as sea salt aerosols, have a significant impact on various processes of the marine atmosphere. This consists of interaction with short and long wave radiation, the role of CCN influencing marine clouds and further climate effects. The production, concentration, transportation and time scales in which sea salt particles remain in the atmosphere are highly depended on the prevailing atmospheric-wave characteristics.

For the description of the air-sea environment all the aforementioned processes are essential and necessary. The strength of links and feedback between atmosphere and ocean might deviate based on the meteorological conditions and the characteristics of the area where those occur. The air-sea interaction mechanisms lead to a cascade of effects in several meteorological features. This work focuses on investigating the role of mechanical and thermodynamical air-sea interaction processes in marine boundary layer and their feedback effects in sea salt aerosols.

62.2 Models and Methodology

To model efficiently the dynamical processes taking part in the air-sea regime, a combination of an atmospheric with a wave model is used. The RAMS/ICLAMS model [5] is interfaced with the WAM wave model under the OASIS-MCT coupler, used for communication. Parameterizations for sea surface roughness, sea spray fluxes and salt production are implemented. Precisely, sea surface conditions are considered in the atmospheric model by the dynamic sea surface roughness with Charnock parameter z_{ch} given by the fraction of wave induced stress τ_w with the total stress τ, as described in [3]. Sea spray fluxes derive from the difference in temperature and specific humidity, between the lifted water droplets and the surrounding air. In the current work, the sea spray model method described in [1], has been adopted. For sea salt production, a whitecap fraction parameterization is employed based on surface wind speed and relative humidity [2].

62.3 Case Study

In order to examine the aforementioned approaches, a sensitivity experiment in North Sea is considered. The case refers to 4–8 December, 2013. During this period, a winter storm took place in the area under study, accompanied with strong winds that favor wave breaking. The storm started from south Greenland on the 4th of December and moving east, entered North Sea. Two types of simulations have been performed. One with the impact of sea spray thermodynamics and a second one without. Model setup and initial conditions are the same for both cases.

Before the arrival of the storm in the main area of North Sea, in the first hours of the 5th of December, sea salt concentration at the surface is low and comparable for both runs (Fig. 62.1). The insertion of the storm in North Sea induced intensified surface winds and waves. This led to an increase in the produced sea salt aerosols, since concentration is related to surface wind speed magnitude and humidity. The amount of salt particles is higher when sea spray impact is considered (Fig. 62.2). In fact, the concentration increases in a range close to 300 μg/m^3.

The amplification in the concentration of salt particles is associated with the enhancement in surface wind speed and increase of water vapor in the lower atmosphere. This stems from the increase in the derived heat fluxes due to sea spray fluxes. More precisely, focusing on a specific location (56.01 N, 6.5 E) sensible heat fluxes are more enhanced close to 200 W/m^2 and latent heat fluxes up to 300 W/m^2 when sea spray is considered (Fig. 62.3).

The effects are evident in the profiles of wind speed and water vapor (Fig. 62.4). In wind speed, the deviation is increasing from the bottom till the first hundred meters. After this height differences reduced. Regarding water vapor, the profiles with height

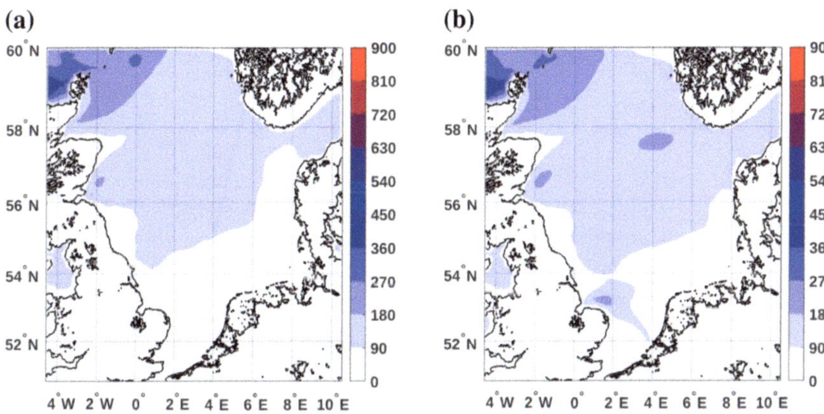

Fig. 62.1 Sea salt concentration in the surface (12 m) from **a** the run without the sea spray effects and **b** the run with sea spray effects, for 5 December 2013, 03:00 UTC

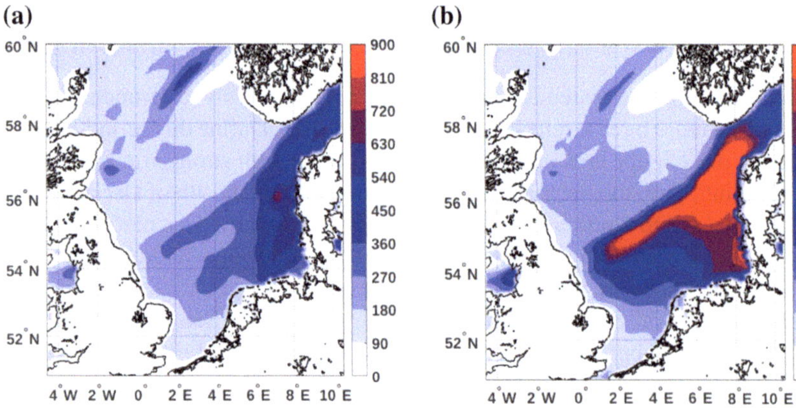

Fig. 62.2 Sea salt concentration in the surface (12 m) from **a** the run without the sea spray effects and **b** the run with sea spray effects, for 5 December 2013, 16:00 UTC

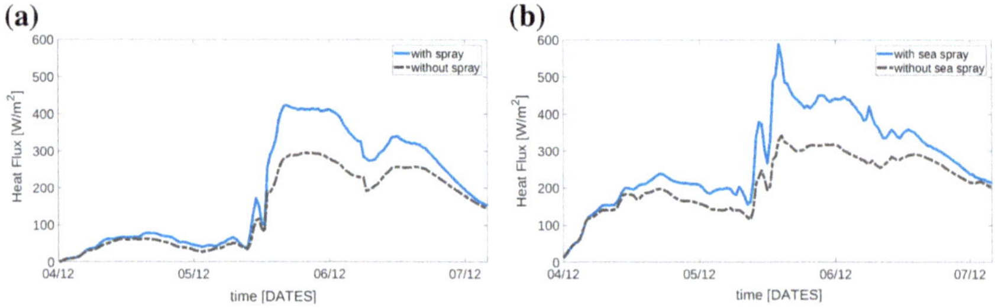

Fig. 62.3 **a** Sensible heat flux and **b** Latent heat flux evolution for the simulations with (continuous line) and without (dash line) the sea spray impact

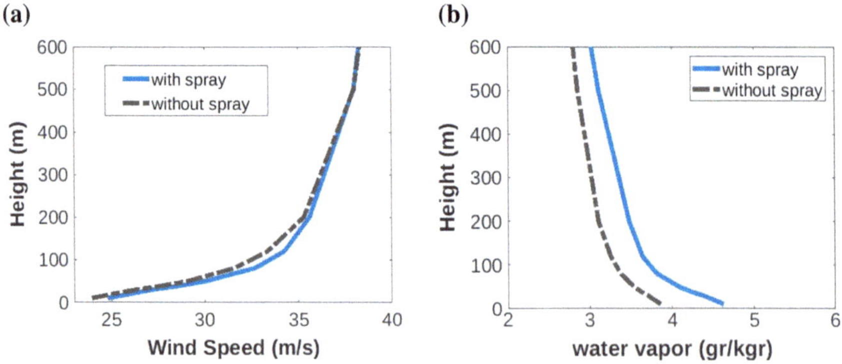

Fig. 62.4 **a** Vertical of differences with height for (**a**) wind speed and **b** water vapour, between the simulations with and without the sea spray impact

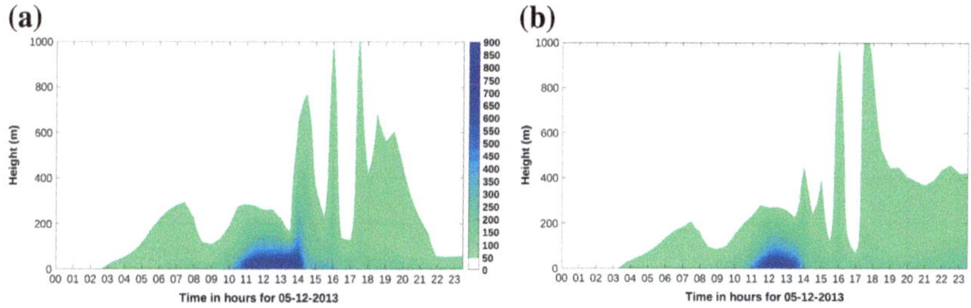

Fig. 62.5 Time-height cross section for salt concentration for simulations **a** with and **b** without the sea spray impact

presents similar behavior. The increase in the moistening of the near surface air due to sea spray is evident close to sea surface. The difference in the amount of water vapor decreases with the height and tends to minimize in altitudes above 500 m.

The concentrations with height in salt aerosols tend to be higher when the contribution of spray fluxes is significant (Fig. 62.5). It is also evident that salt particles disperse faster in the case where ocean spray is considered, as defined in the last hours of the demonstrated example.

62.4 Conclusions

The simulations showed that inclusion of sea state and sea spray effects lead to alterations in atmospheric stratification with intensified heat fluxes that affect the profiles of water vapour and wind speed. The impacts extended on sea salt production and concentration. Due to limited space, further discussion and evaluation of the modeling system is provided in [6].

References

1. E.L. Andreas, L. Mahrt, D. Vickers, An improved bulk air–sea surface flux algorithm, including spray-mediated transfer. Q. J. R. Meteorol. Soc. **141**, 642–654 (2015)
2. S.L. Gong, A parameterization of sea-salt aerosol source function for sub- and super-micron particles. Glob. Biogeochem. Cycles **17**, n/a–n/a (2003)
3. P. Janssen, *The Interaction of Ocean Waves and Wind* (Cambridge University Press, 2004)
4. D.H. Richter, F. Veron, Ocean spray: an outsized influence on weather and climate. Phys. Today **69**, 34–39 (2016)
5. S. Solomos, G. Kallos, J. Kushta, M. Astitha, C. Tremback, A. Nenes, Z. Levin, An integrated modeling study on the effects of mineral dust and sea salt particles on clouds and precipitation. Atmos. Chem. Phys. **11**, 873–892 (2011)
6. C. Stathopoulos, G. Galanis, G. Kallos, A study of mechanical and thermodynamical air-ocean interface processes with a coupled modeling approach (submitted to *Ocean Modeling*) (2018)

Chapter 63
On the Importance of Organic Mass for Global Cloud Condensation Nuclei Distributions

Georgios Fanourgakis, Nikos Kalivitis, Athanasios Nenes and Maria Kanakidou

Abstract Aerosol-cloud interactions constitute a major contributor of uncertainty in projections of anthropogenic climate change. The fraction of aerosol that activates to form cloud droplets (cloud condensation nuclei, CCN) is at the heart of aerosol cloud interactions. Towards this, we investigate the role of organic mass in the formation and evolution of CCN using the global 3-dimensional chemistry transport model TM4-ECPL coupled with the M7 aerosol microphysics module. The contribution of organics to the CCN levels is quantified by comparing the global surface distribution of aerosol particles and CCN computed with and without organic aerosol mass considerations, to the surface CCN observations. We also calculate the dynamical behavior of the CCN by computing their persistence times in atmosphere, i.e. the period over which the CCN concentrations show autocorrelation. It is found that organic species in aerosol modulate CCN concentrations by 50–90%—with a higher influence over land; furthermore, simulations compare better with observations when the impact of organics on CCN levels is taken into account.

G. Fanourgakis · N. Kalivitis · M. Kanakidou (✉)
Environmental Chemical Processes Laboratory, Department of Chemistry, University of Crete, Heraklion, Greece
e-mail: mariak@uoc.gr

G. Fanourgakis
e-mail: fanourg@uoc.gr

N. Kalivitis
National Observatory of Athens, P. Penteli 15236, Athens, Greece
e-mail: nkalivitis@uoc.gr

A. Nenes
Laboratory of Atmospheric Processes and their Impacts, School of Architecture, Civil & Environmental Engineering, École Polytechnique Federale de Lausanne, Lausanne 1015, Switzerland
e-mail: athanasios.nenes@epfl.ch

Institute of Chemical Engineering Sciences, Foundation for Research and Technology (ICEHT-FORTH), Hellas, Patras 26504, Greece

C. Mensink et al. (eds.), *Air Pollution Modeling and its Application XXVI*,
Springer Proceedings in Complexity,
https://doi.org/10.1007/978-3-030-22055-6_63

63.1 Introduction

The organic component represents a significant amount of the total submicron dry aerosol mass which, depending on the location and the atmospheric conditions, may reach the 30–70% of the total mass [1]. The contribution of the organic mass to the particles number concentrations and to the cloud condensation nuclei (CCN) are here quantified by performing simulations with the global 3-dimensional chemistry transport model TM4-ECPL [2, 6, 7] with and without emissions of primary organic particles and formation of secondary organic aerosols. The model is coupled to the aerosol microphysics module M7 [5]. All major aerosol components namely mineral dust, black carbon, sea-salt, organics, and sulfate are considered in the model as internally or externally mixed particles represented by log-normal mass and number distributions that evolve by coagulation, condensation and nucleation.

63.2 Results

63.2.1 Surface Distribution of CCN at Supersaturation Ratio 0.2%

The annual surface distribution of the CCN at 0.2% supersaturation ratio (denoted as $CCN_{0.2}$) obtained from the two simulations with the TM4-ECPL model, with and without organic mass, are shown in Fig. 63.1a, b respectively. Calculations of the CCN were performed using the κ-Köhler theory as developed by Petters et al. [3] that expresses the hygroscopicity κ of the aerosol particles as a function of the volume-weighted fraction hygroscopic parameter of each aerosol component. The hygroscopicity parameters used in the present study are $\kappa_{SO4} = 0.6$ (sulfate), $\kappa_{BC} = 0.0$ (black carbon), $\kappa_{DU} = 0.0$ (dust), $\kappa_{SS} = 1.0$ (sea-salt), $\kappa_{ORG} = 0.1$ (organics).

When organics are considered in the simulation (Fig. 63.1a, with organics), near-surface $CCN_{0.2}$ number concentrations are reaching ~3500 cm^{-3} at the polluted areas, which is close to the value of ~3000 cm^{-3} computed as the multi-model mean maximum during a model intercomparison based on the results of 15 global models [7]. Neglecting the contribution of organics to the formation of the $CCN_{0.2}$, the maximum of the near-surface global $CCN_{0.2}$ distribution drops to ~1200 cm^{-3} (Fig. 63.1b, without organics). In Fig. 63.1c is shown the percentage difference of the $CCN_{0.2}$ distribution between these two simulations. Over land, this difference is found to be >50% with the highest differences that reach 90% calculated in high polluted areas.

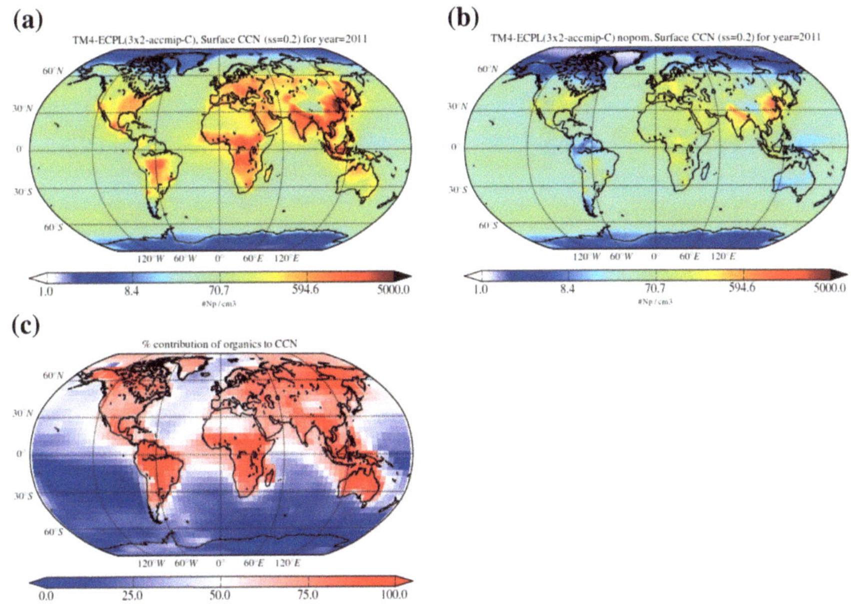

Fig. 63.1 Simulated annual surface concentration of $CCN_{0.2}$. Model simulations are **a** with organic mass and **b** without organic mass (units are number per cm^3), **c** percentage contribution of organics aerosol to $CCN_{0.2}$ (units are in %)

63.2.2 Comparison of CCN Number Concentration with Observations

Datasets of CCN measured at nine atmospheric observatories, one in Japan and eight from the Aerosols, Clouds, and Trace gases Research InfraStructure (ACTRIS) in Europe [4] were used to evaluate the model results. The observatories represent different environments typical for Atlantic, Pacific and Mediterranean maritime, boreal forest, and high alpine atmospheric conditions. In Fig. 63.2 the seasonal variations of the $CCN_{0.2}$ at the observational sites as computed by the two TM4-ECPL model simulations (with and without organics) are compared to the observations. At all stations but Jungfraujoch, the number of $CCN_{0.2}$ with organics that is larger than $CCN_{0.2}$ without organics is in a significantly better agreement with observations. The seasonal variation of $CCN_{0.2}$ is similar for most of the stations (except Puy de Dôme) in the presence or not of organics. For example, at Finokalia and Noto Peninsula the two simulations compute significantly higher $CCN_{0.2}$ levels during summer than during winter, while in Cabauw and Hyytiälä the number of $CCN_{0.2}$ remains invariant during the year.

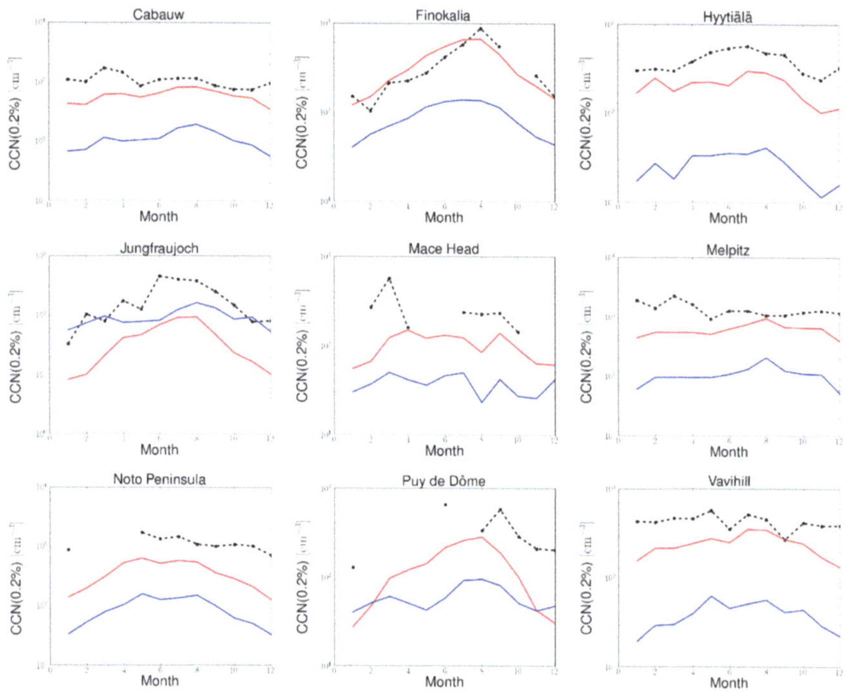

Fig. 63.2 Comparison of the seasonal variability of the $CCN_{0.2}$ from observational data (black lines with dots) and from simulations with organic mass (red line) and without organic mass (blue lines) consideration. Units are in number per cm^3

63.2.3 Persistence of $CCN_{0.2}$ in Atmosphere

The short-term dynamics of the CCN in the atmosphere is a complicated process that involves all CCN formation and removal processes. One way to quantify the dynamical behavior of CCN is by computing the autocorrelation function of the CCN population that can be used to determine the persistence time of CCN in atmosphere. Based on the measured $CCN_{0.2}$ and depending on the season (winter/summer) and other environmental conditions at the observational site, the $CCN_{0.2}$ persistence time varies from a few hours up to several days. In Fig. 63.3 the observational determined $CCN_{0.2}$ persistence times during winter and summer are shown along with the results of the simulations by the TM4-ECPL model, in the presence or not of organics.

It can be seen that in several cases the simulations are in agreement with observations, although in a few cases model results show a different order in the relative ratio of winter/summer persistence times. It is also seen that by neglecting the organic mass in the $CCN_{0.2}$ calculations the computed persistence times are drastically affected, only at one of the studied observational sites (Fig. 63.3) despite the fact that in the absence of organics the number of the $CCN_{0.2}$ is significantly smaller (Fig. 63.2). This result is not surprising since the persistence time depends only on the rate of

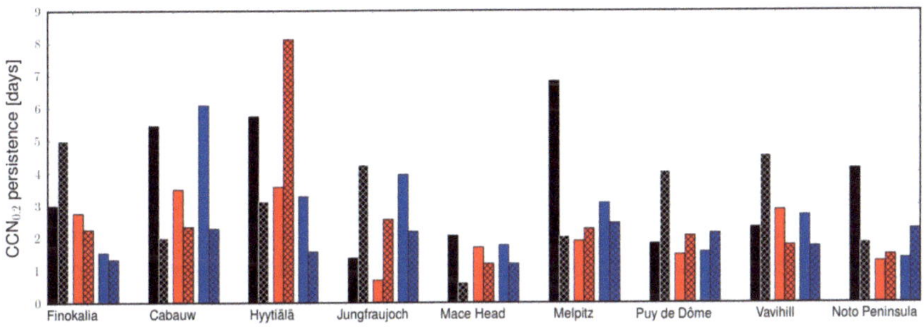

Fig. 63.3 Persistence times of $CCN_{0.2}$ in atmosphere at the observational sites. Black bars show the persistence times computed based on the observations of $CCN_{0.2}$, while the red and blue bars correspond to those calculated based on the TM4-ECPL model simulations with and without organics, respectively. For each pair of bars, the solid bar corresponds to winter and the dashed bar to summer

change of the CCN population and not on the CCN abundance. The effect of the presence of organics on the $CCN_{0.2}$ persistence is found to vary among sites with the most stricking changes in Cabauw, with higher persistence times without organics that are also closer to observations. In Hyytiälä also higher persistence times are computed when taking into account the contribution of organics and are significantly higher in summer than derived from observations. In addition, the summer-to-winter ratio of the persistence times is found to be affected by the presence of organics in particular at Hyytiälä, where, surprisingly, neglecting the organics the $CCN_{0.2}$ during winter is more persistent than during summer in agreement with observations, while at the high altitude Alpine station of Jungfraujoch the opposite trend is found. The larger persistence during summer than during winter found in the observations at this station is better captured by the model when considering the contribution of organics.

63.3 Conclusion

This global 3-dimensional modeling study has shown that the organic mass significantly affects the CCN number concentrations. The annual surface distribution of the $CCN_{0.2}$ number concentration is typically reduced by 50–90%, however, in environments without a significant amount of organics (i.e. Jungfraujoch) $CCN_{0.2}$ does not show noticeable sensitivity to organic mass. Finally, based on the calculated persistence times of the $CCN_{0.2}$ in the atmosphere, the inclusion of organics to the aerosol components does not significantly affect the dynamics of the CCN.

Question by Ulas Im:
What is the contribution of OA to the CCN over the tropospheric column?
Reply:

Organic aerosol concentrations are reduced with height despite the higher partitioning of the semi volatile organics into the aerosol phase at the lower temperatures found in the middle troposphere than near the surface. Therefore, the contribution of OA to CCN is the highest near the surface but remains important in the middle troposphere, although geographically restricted to the tropical continental regions and their outflow. There, convection brings from the surface to the middle and high troposphere significant amounts of OA originating from primary and secondary biomass burning sources or secondary formation from biogenic volatile organic compounds. For instance, according to our calculations, on an annual mean basis the contribution of OA to total CCN at 0.2% supersaturation at 500 hPa (around 5–6. Km height) in the continental tropics remains very high, above 50%.

Acknowledgements This work has been supported by the European FP7 collaborative project BACCHUS (Impact of Biogenic versus Anthropogenic emissions on Clouds and Climate: towards a Holistic UnderStanding. We acknowledge use of the ACTRIS database provided by Schmale, J. and co-workers.

References

1. M. Kanakidou, J.H. Seinfeld, Pandis et al., Organic aerosol and global climate modelling: a review. Atmos. Chem. Phys. **5**(4), 1053–1123 (2005)
2. M. Kanakidou, R.A. Duce, J.M. Prospero et al., Atmospheric fluxes of organic N and P to the global ocean. Global Biogeochem. Cycles **26**(3), 1–12 (2012)
3. M.D. Petters, S.M. Kreidenweis, A. Sandu et al., A single parameter representation of hygroscopic growth and cloud condensation nucleus activity. Atmos. Chem. Phys. **7**, 1961–1971 (2007)
4. J. Schmale, S. Henning, Henzing, et al., Collocated observations of cloud condensation nuclei, particle size distributions, and chemical composition. Sci. Data **4**, 170003 (2017)
5. E. Vignati, J. Wilson, P. Stier, M7: An efficient size-resolved aerosol microphysics module for large-scale aerosol transport models. J. Geophys. Res. Atmos. **109**(D22202) (2004)
6. N. Daskalakis, K. Tsigaridis, S. Myriokefalitakis, et al., Large gain in air quality compared to an alternative anthropogenic emissions scenario. Atmos. Chem. Phys. **16**, 9771–9784 (2016)
7. G.S. Fanourgakis, M. Kanakidou, A. Nenes, et al., Evaluation of global simulations of aerosol particle and cloud condensation nuclei number, with implications for cloud droplet formation. Atmos. Chem. Phys. **19**, 8591–8617 (2019)

Chapter 64
Biological Activity in Clouds: From the Laboratory to the Model

L. Deguillaume, H. Perroux, N. Wirgot, C. Mouchel-Vallon, N. Chaumerliac, M. Joly, V. Vinatier and A.-M. Delort

Abstract Microorganisms are present in the atmosphere and can survive in cloud droplets. They are able to transform organic compounds and can consequently compete with radical chemistry. Because the cloud system is a complex multiphase medium, the efficiency of biodegradation by cloud microorganisms has to be evaluated by numerical models of different complexity simulating multiphase chemical processes in clouds. However, only abiotic processes are taken into account in these numerical tools. The objective of this work was thus to integrate biological data in an atmospheric cloud chemistry model and to evaluate the effect of microorganisms in the transformation of chemical compounds. For this, experimental biodegradation rates of acetic and formic acids, formaldehyde and hydrogen peroxide by various bacterial representative of cloud biodiversity were experimentally measured. These values have been implemented in a cloud chemistry model describing aqueous phase chemistry with the explicit CLEPS (Cloud Explicit Physico-chemical Scheme) mechanism. Several simulations with and without biodegradation processes have been performed changing temperature (5 °C or 17 °C) and actinic flux to simulate summer or winter conditions. The chemical scenario (gas concentrations and emission/deposition) is representative of low-NOx emission with significant isoprene emissions.

64.1 Introduction

Numerical models of different complexity have been developed in the past to simulate cloud chemistry. Currently, only abiotic processes are implemented in these models.

L. Deguillaume (✉) · H. Perroux · C. Mouchel-Vallon · N. Chaumerliac
Université Clermont Auvergne, CNRS LaMP, 63000 Clermont-Ferrand, France
e-mail: L.Deguillaume@opgc.fr

N. Wirgot · M. Joly · V. Vinatier · A.-M. Delort
Université Clermont Auvergne, CNRS, SIGMA Clermont, ICCF, 63000 Clermont-Ferrand, France

C. Mouchel-Vallon
NCAR, National Center for Atmospheric Research, Boulder, CO, USA

C. Mensink et al. (eds.), *Air Pollution Modeling and its Application XXVI*,
Springer Proceedings in Complexity,
https://doi.org/10.1007/978-3-030-22055-6_64

But recently, the metabolic activity of microorganisms (bacteria, fungi, yeasts) has been shown to influence cloud chemistry [3]. In clouds, microorganism abilities to survive tend to activate their metabolism. They can resist and survive even they are exposed to harsh conditions (cold shock, presence of oxidants, osmotic shock, UV radiation,...) encountered in clouds. Cells in clouds can uptake molecular compounds used as nutrients, thus contributing to cloud chemistry. More specifically, microorganisms can degrade organic compounds (such as organic acids, formaldehyde or methanol) due to their carbon metabolism. They also modify the chemical budget of oxidants. For example, they can defend from the presence of H_2O_2 thanks to their oxidative stress metabolisms; this impacts the formation of $HO^{\bullet}$ radicals that are produced by H_2O_2 photolysis [6]. Rates of biotic and abiotic transformations determined in microcosms mimicking cloud environments and directly in cloud waters clearly highlighted that biodegradation are competitive with radical chemistry. However, those experiments have been done under bulk laboratory conditions. But the cloud system is a multiphase medium where chemical and biological transformations are impacted by microphysical processes and mass transfers between the various phases. The present study aims at integrating biological reactions studied in the lab into a cloud chemistry model reproducing microphysical and mass transfer processes.

64.2 Model Results

Four chemical compounds (formic and acetic acids, formaldehyde and H_2O_2) were selected and their biodegradation rates were experimentally evaluated. Three bacteria strains were chosen for their capacity to efficiently degrade these compounds; those strains have been isolated from cloud waters collected at the puy de Dôme station in France. These biodegradation transformations and the associated rates complete the CLEPS aqueous mechanism. The cloud chemistry model used in this study considers explicit chemistry in both gas and aqueous phases together with the dynamical mass transfer between these two phases. The aqueous mechanism is based on a new protocol that provides a representation of most probable oxidation pathways of organic compounds [5]. Radical versus microbiology efficiency were confronted modifying "key" environmental parameters (temperature and actinic flux).

The chemical scenario is representative of low NO_x conditions. In order to get a chemical photostationary state, a first simulation is performed using only the gas phase version of the model for a period of 31 days. Aqueous phase appears at noon the 31th day and lasts 12 h. An initial concentration of 1 μM of iron in the aqueous phase is considered, which is representative of values measured in continental cloud water [2]. This allows recycling oxidants resulting from the iron redox cycles. The size of the cloud droplets is fixed at 10 μM and the liquid water content is equal to $3\ 10^{-7}$ vol/vol. The initial acidity is set to pH = 4 and evolves afterwards.

Simulations are realized to represent two seasons (summer vs. winter). For the summer case, the actinic flux is at noon 1.5 higher than for a winter case. For the summer case, the temperature is prescribed to 17 °C and for wintertime conditions, the temperature is set at 5 °C. These two temperatures are similar to the temperature maintained in the lab to evaluate the biodegradation rates. For the wintertime, the isoprene emissions are decreased from 7.5×10^6 to 4.5×10^6 molec cm^{-3} s^{-1}, and the deposition rate of the secondary organic products is decreased from 5×10^{-5} to 1×10^{-5} s^{-1}.

Clouds are known to be oxidative reactors where chemicals are transformed by free radicals, mainly $HO^{\bullet}$ during the day. In this context, the dissolved organic matter is efficiently transformed where the molecules can be fragmented and/or more functionalized. Initially, these organic compounds results from the mass transfer from the gas phase or from the dissolution of the soluble fraction of the particle that acts as CCN. In this context, a major source of formic, acetic acids and formaldehyde in cloud droplets are their transfers from the gas phase. The oxidation in the aqueous phase of precursors, such as glyoxal, methyglyoxal, glycolaldehyde, pyruvic acid, etc., is also a significant source.

Results obtained with the CLEPS model for summer (Ref-sum) and winter (Ref-win) cases are presented in Fig. 64.1 with the time evolution of H_2O_2 and formic acid concentrations in the aqueous phase. Large differences are observed between the two seasons, due to temperature, photolysis and emission conditions in summer and winter, which leads to very different radical concentrations as shown with the $HO^{\bullet}$ time evolution also reported in Fig. 64.1. The $HO^{\bullet}$ radicals are mainly produced by the H_2O_2 reactivity (its photolysis and the Fenton reaction) and by its mass transfer from the gas phase during the first hours of the cloud.

Figure 64.2 summarizes the main sources and sinks of formic acid and H_2O_2. The contributions of reactions and of the mass transfer in the production/destruction of the studied species has been averaged over twelve hours (from 12:00AM–12:00PM) with the nighttime period starting after 7:30PM in summer and after 4:50PM in winter.

H_2O_2 increases during the first 4 h in Fig. 64.1 to reach a maximum concentration of 160 μM in summer (40 μM in winter) and then decreases until midnight to a value of 40 μM in summer (20 μM in winter). In Fig. 64.2, H_2O_2 is mainly produced by the reaction between Fe^{2+} and $HO_2^{\bullet}/O_2^{\bullet-}$ and by the self-reactions of $HO_2^{\bullet}/O_2^{\bullet-}$. The aqueous reactivity degrades H_2O_2 through the Fenton reaction ($Fe^{2+} + H_2O_2$) and

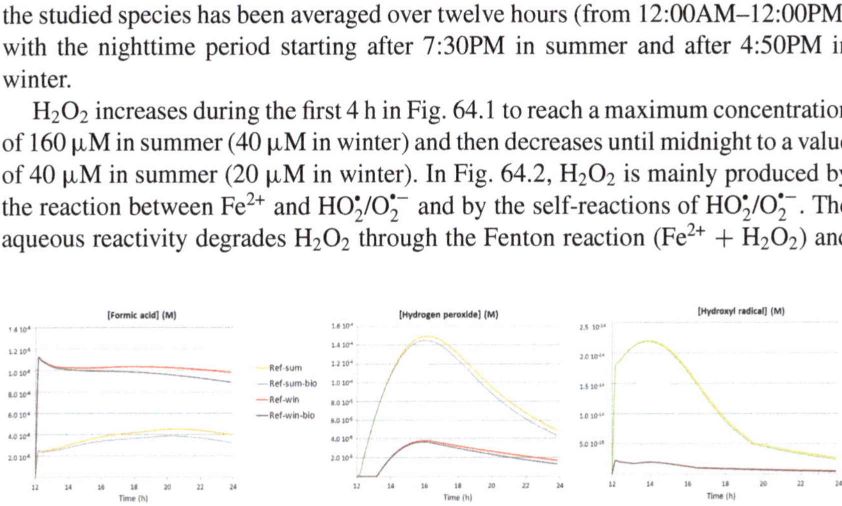

Fig. 64.1 Time evolution of the aqueous phase concentrations of H_2O_2, and formic acid for summer and winter cases without biodegradation (Ref-sum, Ref-win), with biodegradation (Ref-sum-bio, Ref-win-bio)

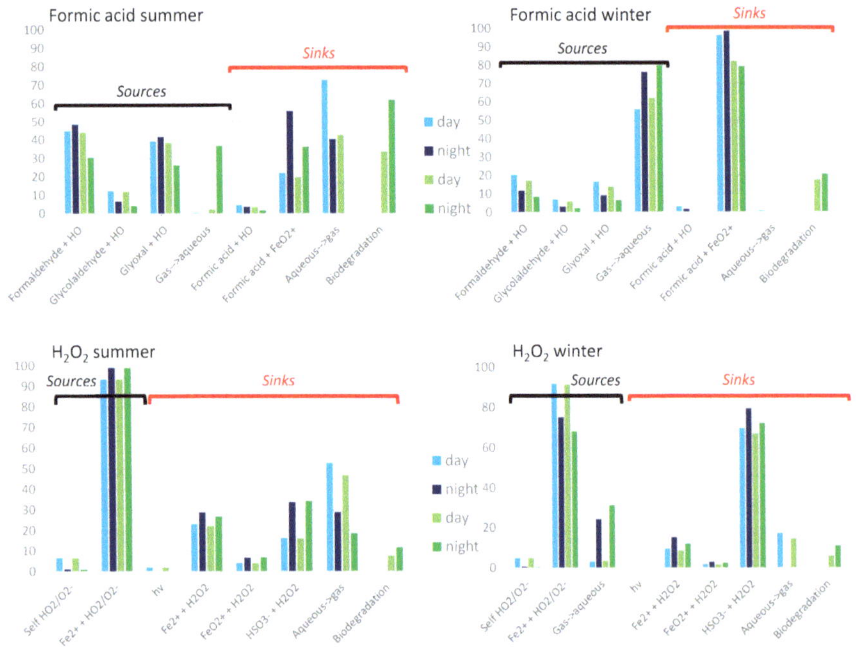

Fig. 64.2 Representation of sources and sinks of formic acid and hydrogen peroxide integrated over day and night for summer and winter cases. The relative contributions (%) of individual reactions contributing to the total destruction/production of the target species are indicated; Green: with biodegradation, in blue: without biodegradation

through the reaction of H_2O_2 with sulfite (HSO_3^-/SO_3^{2-}). For both seasons, for daytime cloud, reaction between Fe^{2+} and $HO_2^{\bullet}/O_2^{\bullet-}$ acts as the main source for H_2O_2 in the aqueous phase that is then transferred to the gas phase. Consequently the cloud acts as a source for H_2O_2 in the gas phase.

Formic acid is increasing in summer and decreasing in winter. During the night, its concentration decreases because the sources by oxidation become less efficient due to decrease of the $HO^{\bullet}$ concentration (Fig. 64.1). In the aqueous phase, formic acid is first transferred from the gas phase into the aqueous phase by the mass transfer and is thereafter produced (from 2 to 4 μM) in summer by the aqueous phase oxidation of formaldehyde (~45%), glyoxal (~40%) and glycolaldehyde (~12%) during daytime (Fig. 64.2). The aqueous formic acid is outgassed during the day that represents its main sink (around 70%). The other sinks are its oxidation by $HO^{\bullet}$ (less than 10%).

In winter, formic acid is rather destroyed; its concentration is larger than in summer and is reduced throughout the simulation (from 11 to 10 μM). The main sinks are its oxidation by the ferryl ion FeO^{2+} (~95%) and $HO^{\bullet}$ (~5%). In winter, the gas phase concentration of formic acid is higher and explains its efficient dissolution in the aqueous phase during the first time steps of the simulation. The sinks are totally different in winter. In this case, the transfer of formic acid to the gas phase disappears. During the night, the mass transfer is inverted and represents a major source of formic acid in the aqueous phase (~75%) that is now undersaturated in formic acid; aqueous formic acid is then degraded mainly by the ferryl ion and $HO^{\bullet}$.

Effect of Biodegradation Rates:

The time evolution of the 4 target species present similar behavior than what was observed for the Ref-sum and Ref-win cases (i.e. without considering biodegradation). But when looking more carefully in Fig. 64.2 at the details of their sources and sinks including biodegradation rates, the contribution of biodegradation appears clearly. In all cases, biodegradation sink is more important at nighttime. Biodegradation is globally more efficient during summer as explained by the biodegradation rates lower in the winter case (5 °C) in comparison to the summer case (17 °C).

In the Ref-sum-bio case, formic acid concentration decreases in comparison to the Ref-sum. At the end of the simulation time, the concentration is reduced by a 20%. H_2O_2 concentrations are not significantly modified by the implementation of biodegradation rates. In summer, formic acid during the day is degraded by the microorganisms significantly (~35%) and the mass transfer is reduced from 72 to 40%. During the night, this effect is amplified with the biodegradation process becoming dominant (60%) and the mass transfer becoming negligible. The biological activity degrades the H_2O_2 during the day and the night but in a lesser extent (7 and 12%, respectively).

Microorganisms are considered as biological catalysts that may compete with chemical reactivity and with photochemistry features in cloud waters, in particular. The results indicate that this biological activity has to be considered in cloud chemistry model since this modify the way the molecules are degraded both in the gas and the aqueous phases.

QUESTIONER: Paul Makar

QUESTION: The microorganisms degrade short chain organic acids? What do they produce? Do they make more molecular mass organic compounds?

ANSWER: Yes, they are also to degrade short chain organic compounds such as carboxylic acids because most of them can enter into the central metabolisms of several microorganisms, and these metabolites can be used to maintain energy levels through the production of ATP to synthesize larger molecules and produce biomass (proteins, nucleic acids, etc.).

QUESTIONER: Silvia Trini Castelli

QUESTION: What is the lifetime of the microorganisms, bacteria residing in the cloud?

ANSWER: Once in the cloud, the living bacteria are able to survive to this harsh medium. However, the viability of microorganisms in clouds could be altered by environmental factors such as exposition to H_2O_2, solar light, osmotic shocks and freeze-thaw cycles. Recent study from Joly et al. [4] have shown (https://doi.org/10.1016/j.atmosenv.2015.07.009) that the survival depends on the microorganism and on the stress; free-thaw appears to be the most stringent factor. See also the work from Amato et al. [1] (https://doi.org/10.5194/acp-15-6455-2015).

References

1. P. Amato et al., Survival and ice nucleation activity of bacteria as aerosols in a cloud simulation chamber. Atmos. Chem. Phys. **15**, 6455–6465 (2015)
2. L. Deguillaume et al., Classification of clouds sampled at the puy de Dôme (France) based on 10 yr of monitoring of their physicochemical properties. Atmos. Chem. Phys. **14**(3), 1485–1506 (2014)
3. H. Herrmann et al., Tropospheric aqueous-phase chemistry: kinetics, mechanisms, and its coupling to a changing gas phase. Chem. Rev. **115**(10), 4259–4334 (2015)
4. M. Joly et al., Survival of microbial isolates from clouds toward simulated atmospheric stress factors. Atmos. Environ. **117**, 92–97 (2015)
5. M. Mouchel-Vallon, et al., CLEPS 1.0: a new protocol for cloud aqueous phase oxidation of VOC mechanisms. Geosci. Model Dev. **10**(3), 1339–1362 (2017)
6. M. Vaïtilingom et al., Potential impact of microbial activity on the oxidant capacity and organic carbon budget in clouds. Proc. Natl. Acad. Sci. **110**(2), 559–564 (2013)

Chapter 65
Aerosol Indirect Effect on Air Pollution-Meteorology Interaction in an Urban Environment

Wanmin Gong, Ayodeji Akingunola, Shuzhan Ren, Stephen Beagley, Rodrigo Munoz-Alpizar, Paul Makar and Craig Stroud

Abstract Using a fully coupled air quality prediction model, simulations were carried out to investigate the impact of aerosol indirect effect on air pollution-meteorology interaction in an urban environment. We found that the aerosol indirect effect results in an increase in cloud droplet number concentration, a reduction in cloud droplet size, and an increase in cloud water. While, as a result, precipitation production is suppressed in low-level clouds, we found that, in a case of deep convective clouds, there is an enhancement of cloud ice and precipitation production at higher levels due to the increase in abundance of smaller drops being carried up in the updraft. There is also an indication of enhanced convective activity due to urban heating.

W. Gong (✉) · A. Akingunola · S. Ren · S. Beagley · P. Makar · C. Stroud
Air Quality Research Division, Science and Technology Branch, Environment and Climate Change Canada, Toronto, ON, Canada
e-mail: Wanmin.gong@canada.ca

A. Akingunola
e-mail: Ayodeji.akingunola@canada.ca

S. Ren
e-mail: Shuzhan.ren@canada.ca

S. Beagley
e-mail: Stephen.beagley@canada.ca

P. Makar
e-mail: Paul.makar@canada.ca

C. Stroud
e-mail: Craig.stroud@canada.ca

R. Munoz-Alpizar
Air Quality Modelling and Application, Meteorological Service of Canada, Environment and Climate Change Canada, Dorval, QC, Canada
e-mail: Rodrigo.munoz-alpiza@canada.ca

C. Mensink et al. (eds.), *Air Pollution Modeling and its Application XXVI*,
Springer Proceedings in Complexity, https://doi.org/10.1007/978-3-030-22055-6_65

65.1 Introduction

A previous study using a fully coupled version of the Environment and Climate Change Canada's (ECCC) air quality prediction model, GEM-MACH, has shown an increase in modelled cloud amount and liquid water content due to increased cloud droplet number concentration, a decrease in cloud droplet size and a reduction in warm precipitation, downwind of an urban-industrial area. The increased cloud amount due to the aerosol indirect effect reduced the downward shortwave radiative flux and air temperature at the surface, leading to a decrease in ozone over the region of enhanced cloud and an increase in particle sulphate from an increased capacity for aqueous-phase production [1]. In the meantime, observational evidence of urban-induced or urban-modified convective phenomena have been reported and linked to several possible attributing factors: urban roughness, urban heat island, and urban aerosols. Study of idealised cases suggests that enhanced urban aerosols can lead to a delay and suppression of precipitation from low-level clouds associated with urban induced convection but an invigoration of deep convection and enhancement of precipitation further downwind from the urban centre [2].

In this study, model simulations were carried out, using an updated version of the fully coupled GEM-MACH, for the 2015 Pan America Games period when a dense surface network for meteorological and air quality monitoring was set up. The model domain is centred over the Greater Toronto and Hamilton area in southern Ontario. Simulations were conducted with and without the aerosol indirect effect in combination with an urban module (Town Energy Balance) turned on or off, to investigate the impact of aerosol-cloud interaction on the lake-/land-breeze circulation and associated cloud and precipitation development and subsequent impact on modelled air quality in this urban environment.

65.2 Model and Simulation Setup

Version 2 of the Global Environmental Multiscale—Modelling Air-quality and CHemistry (GEM-MACH) model was used for this study. While a 2-bin configuration for aerosol size representation has been used for operational air quality forecast at ECCC [8], a 12-bin configuration was used here. This study also makes use of the interactive capability introduced previously within GEM-MACH to allow feedbacks between on-line aerosols and meteorology through aerosol radiative (direct) effect and aerosol microphysics (indirect) effect (see [1, 4] for a detailed description of the implementation). In addition, a parameterization for representing urban landscape, the heat and momentum exchange between urban surfaces and the atmosphere, based on a Town-Energy-Balance (TEB) model [3, 5], was also employed.

Simulations were carried out in a cascade fashion with model runs on a 10 km resolution North American domain which provide initial and boundary conditions for the runs on a 2.5 km resolution domain focused on southern Ontario. Figure 65.1

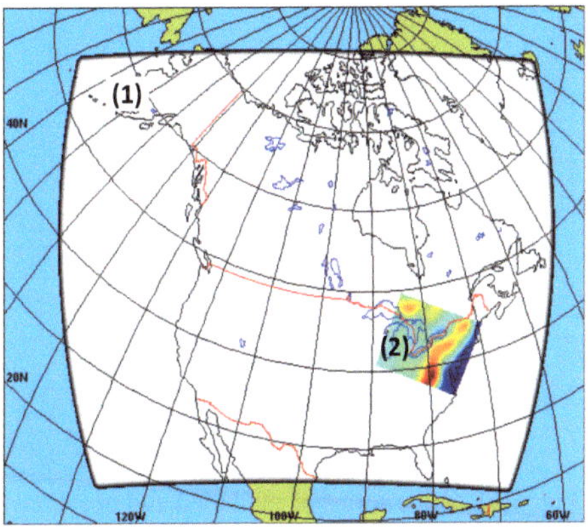

Fig. 65.1 Model domains: (1) at 10 km resolution and (2) nested domain at 2.5 km resolution

shows the model domain setup. A double-moment microphysics scheme [6, 7] was used for both 10 and 2.5 km resolution simulations, and Kain-Fritsch convective scheme was also employed at both scales. For a preliminary study, simulations were carried out for a 5 day period, July 25–29, 2015. The 10 km simulation starts at every 00Z from analysis (meteorology) and runs for 30 h while chemistry continues from the previous day forecast (at hour 24). The 2.5 km simulation starts at every 06Z and runs for 24 h, driven by the initial and boundary conditions provided by the 10 km simulation. Six experiments were conducted: (1) base case (no aerosol feedbacks), (2) aerosol direct effect was enabled, (3) both aerosol direct and indirect effects were enabled; experiment (4)–(6) were the repeat of experiment (1)–(3) but with TEB scheme enabled.

65.3 Results

Figure 65.2 shows the modelled air temperature at 1.5 m (AGL) from 4 experiments (1, 2, 3, and 6) compared with observations at "downtown" and "uptown" sites in the Greater Toronto Area. In general the model has a significant negative bias at night at these urban sites. The impact of the TEB scheme is mostly seen at night in raising the surface temperature (hence reducing the negative bias). This is consistent with the findings from Ren et al. [9] although the magnitude of the impact here is less in comparison. The smaller impact of TEB here is due to the fact that a number of TEB variables were diagnosed from 10 km fields (e.g., potential temperature and specific humidity) at the initialization of every 2.5 km runs which do not sufficiently resolve the variation in urban scale.

Figure 65.3 shows the modelled cloud water content (CWC), rain water content (RWC), and cloud droplet number concentration (N_d) at 18Z on July 25, 2015 at the model hybrid level 0.72 (roughly 3000 MASL) from three experiments: base run (1), with both aerosol direct and indirect effects (3), and with TEB in addition (6).

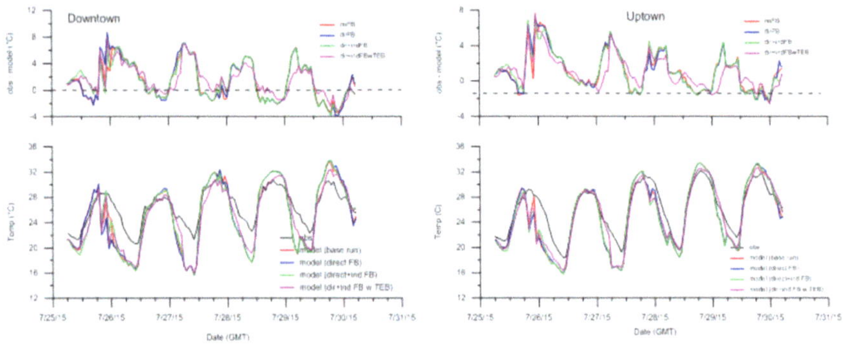

Fig. 65.2 Time series of modelled and observed surface air temperature at "downtown" and "uptown" locations in GTA. Top panels show model bias (observation—model)

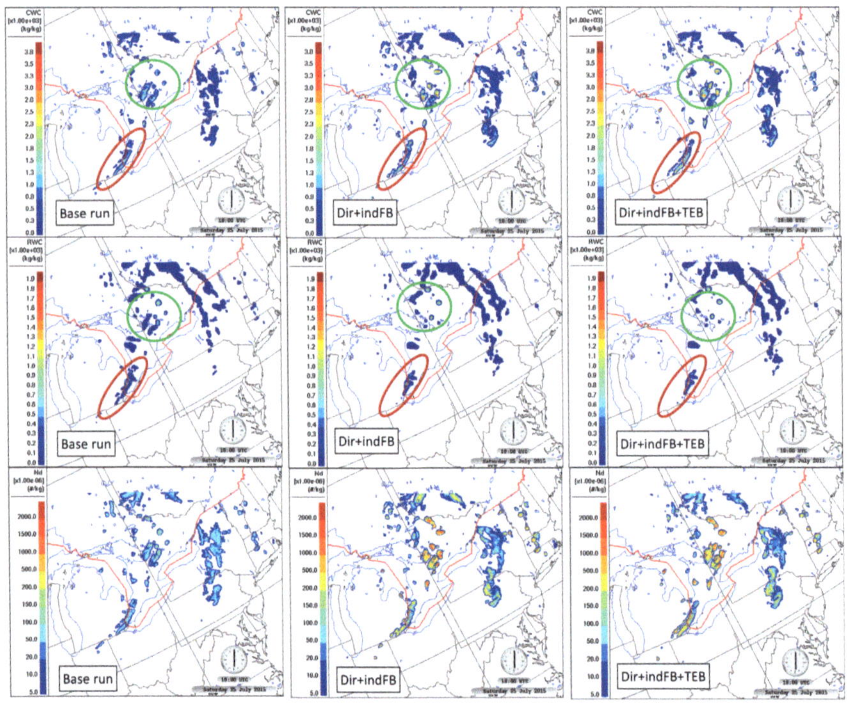

Fig. 65.3 Modelled cloud water content (top row), rain water content (middle row), and cloud droplet number concentration (bottom row) from three experiments: base case (left column), aerosol direct and indirect effects enabled (middle column), and with both aerosol effects and TEB enabled (right column)

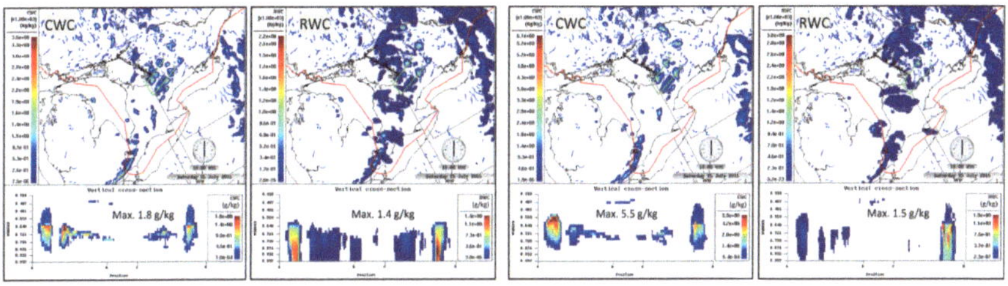

Fig. 65.4 Vertical cross-sections of modelled cloud water content and rain water content through convective cells north of lake Simcoe from two experiments: aerosol direct effect only (left two panels) and with both aerosol direct and indirect effects (right two panels)

This is a case with a cold front over eastern Great Lakes extending from a surface low (occluded) sitting over south end of Hudson Bay. Thunderstorm activities were observed over north of Lake Simcoe and east of Georgian Bay area. The modelled N_d is significantly enhanced when aerosol indirect effect was enabled, increasing from 50–100 kg^{-1} (base case) to over 1000 kg^{-1} in some areas (particularly over the area with thunderstorm activities). Correspondingly, there is an increase in modelled CWC with the aerosol indirect effect. The increased N_d results in decreased droplet sizes and hence reduced precipitation production. This is particularly evident over the tail end of the frontal system (circled in red). However there is also an increase in RWC within a few convective cells on the eastern side of Georgian Bay (circled in green). Vertical cross sections of CWC and RWC through two of the convective cells are shown in Fig. 65.4, comparing two of the experiments with and without the aerosol indirect effect. It can be seen that, with aerosol indirect effect, one of the convective cell is intensified with enhanced precipitation at higher levels. This is accompanied by an increase in cloud ice at upper levels in this cell (e.g., at ~5000 MASL; not shown), indicating that the increased precipitation in this case may be due to enhanced mixed phase process when more smaller droplets being made available at these levels through the updraft. Figure 65.3 also shows further increases in modelled RWC in these convective cells with TEB scheme enabled, indicating the impact of urban heating in intensifying the convective activities.

Question and Answer

Questioner: Alexander Baklanov
Questions:

(1) You demonstrated nice sensitivity studies of including different mechanisms of aerosol indirect effects. But, can you demonstrate that these implementations lead to improvement of model simulations vs. observational data?
(2) Your studies are realized for urban environment. In such conditions interaction of urban aerosols with urban effects (e.g., UHI) can be important and has non-linear character. Did you analyze such interaction?

Answer to (1): Our next step is to carry out simulations for the full month of July and to conduct statistical analysis including evaluation against observations in order to

determine if the representation of aerosol-cloud interaction will lead to an improvement in model predictions (e.g., in terms of forecast scores).
Answer to (2): Indeed, the investigation of interaction between urban aerosols and urban heat island effect is a main focus of the study. This initial analysis has been focused on the impact of aerosol-cloud interactions in this urbanized environment. We hope that the suite of simulations to be carried out over the full month of July period (see above), with and without the aerosol effects, in combination with the urban module (TEB), will shed some light on this aspect.

Questioner: Pieter De Meutter
Question: Did you look at the modelled outflow speed of convective system, since the downdraught depends on the microphysics? Observed wind speed/gust might be used to validate any finding on that.
Answer: We have not yet looked at the modeled outflow speed from convective cells. It is a good suggestion. We may be able to explore this by making use of weather radar products (e.g., radial velocities).

References

1. W. Gong et al., Modelling aerosol cloud meteorology interaction: a case study with a fully coupled air quality model GEM-MACH. Atmos. Environ. **115**, 695–715 (2015). https://doi.org/10.1016/j.atmosenv.2015.05.062
2. J.-Y. Han, J.-J. Baik, A.P. Khain, A numerical study of urban aerosol impacts on clouds and precipitation. J. Atmos. Sci. **69**, 504–520 (2012)
3. S. Leroyer, S. Blair, J. Mailhot, I.B. Strachan, Micro-scale numerical prediction over Montreal with the Canadian external urban modeling system. J. Appl. Meteor. Climate. **50**, 2410–2428 (2011)
4. P.A. Makar, et al., A process analysis of the impact of air-quality/weather feedbacks using GEMMACH. In: D. Steyn, R. Mathur (Eds.), Air Pollution Modeling and Its Application XXIII. Springer, pp. 229–235 (2014)
5. V. Masson, A physical based scheme for urban energy budget in atmospheric models. Bound.-Layer Meteor. **94**, 357–397 (2000)
6. J.A. Milbrandt, M.K. Yau, A multimoment bulk microphysics parameterization. Part I: analysis of the role of the spectral shape parameter. J. Atmos. Sci. **62**, 3051e3064 (2005a)
7. J.A. Milbrandt, M.K. Yau, A multimoment bulk microphysics parameterization. Part II: a proposed three-moment closure and scheme description. J. Atmos. Sci. **62**(9), 3065e3081 (2005b)
8. M.D. Moran, et al., Particulate-matter forecasting with GEM-MACH15, a new Canadian air-quality forecast model. In: Steyn DG, Rao ST (eds) Air Pollution Modelling and Its Application XX, Springer, Dordrecht, pp. 289–292 (2010)
9. S. Ren et al., Impact of urban land use and anthropogenic heat on air quality in urban environments. In: Proceedings 36th ITM on Air Pollution Modelling and Its Application, this volume (2018)

Chapter 66
Aerosol Intensive Optical Properties in the NMMB-MONARCH

Vincenzo Obiso, Oriol Jorba, Carlos Pérez García-Pando and Marco Pandolfi

Abstract The NMMB-MONARCH is a fully online integrated meteorology-chemistry model for short- and mid-term Chemical Weather Forecasts (CWF) developed at the Barcelona Supercomputing Center (Earth Sciences). The meteorological core is the Nonhydrostatic Multiscale Model on the B-grid (NMMB). The global aerosol module includes five natural and anthropogenic species: mineral dust, sea salt, organic matter, black carbon and sulfate. The full online coupling between the aerosol module and the model radiation scheme (RRTMG) has been recently implemented. We present in this work a first evaluation of the aerosol intensive optical properties required by the coupling mechanism: single scattering albedo (ω) and asymmetry factor (g). New aerosol refractive indexes from recent literature have also been tested. We found that the model performance improves when considering a less absorbing mineral dust. Moreover, in anthropogenic areas a purely scattering organic matter and a higher real index for sulfate partially reduce the model systematic biases.

66.1 Introduction

In literature very few studies report on model evaluations of intensive optical properties, even if these have been recognized to be crucial parameters controlling the radiative impact of the aerosols [7]. From the few available studies it seems clear that the performance of the state-of-the-art models is quite variable and in general poor [7, 10]. This general context suggests that still large uncertainties affect the aerosol microphysical and optical parameterizations and so that further investigation is needed to better assess and constrain the uncertainty sources. In this work aerosol optical properties for a period of 5 years (2012–2016) have been simulated over a

V. Obiso · O. Jorba (✉) · C. P. García-Pando
Earth Sciences Department, Barcelona Supercomputing Center (BSC), Barcelona, Spain
e-mail: oriol.jorba@bsc.es

M. Pandolfi
Institute of Environmental Assessment and Water Research, (IDAEA-CSIC), Barcelona, Spain

C. Mensink et al. (eds.), *Air Pollution Modeling and its Application XXVI*,
Springer Proceedings in Complexity,
https://doi.org/10.1007/978-3-030-22055-6_66

global domain with the NMMB-MONARCH [6, 12, 14, 15]. Simulated single scattering albedo (ω) and asymmetry factor (g) at 0.550 μm have been compared with AERONET Version 2.0 inversion data. The impact on these intensive properties of new aerosol refractive indexes, selected from recent literature, has also been evaluated. To the best knowledge of the authors, this kind of perturbed analysis represents a novelty in the context of the few evaluations of model intensive optical properties available in literature.

66.2 Observations

Observations taken from AERONET (Version 2.0) [4] have been used for our evaluation. It is well known [8] that there is a lack of Level 2 inversion data, mainly due to the optical depth (τ) filtering. Hence, we performed a manual screening of the Level 1.5 data by applying all the Level 2 quality-assurance criteria except filtering for $\tau_{440} > 0.4$. By requiring then at least 20 daily means per month over the 2012–2016 period, we selected 59 stations (full set) which we used for a global evaluation of the model performance (not shown here). A subset of 12 stations has been then selected for a more detailed analysis (including also the perturbed analysis), by applying the three following criteria: (i) monthly medians of observed τ above the threshold of 0.1 for more than 6 months (accuracy of inversion data); (ii) same number of stations dominated by natural and anthropogenic aerosols; (iii) homogeneous geographical distribution of the stations.

66.3 Model Simulations

The NMMB-MONARCH has been configured to run global meteorology-aerosol simulations over the period 2012–2016 (5-years period with 1 year of spinup) with initializing the meteorological fields every 24 h by means of NCEP FNL[1] analyses. The horizontal resolution for the global domain has been set to lon × lat = 1.4° × 1.0° and 48 vertical layers have been considered.

Emissions of organic carbon, black carbon and SO_2 from HTAP_v2.2[2] (anthropogenic), GFASv1.2[3] (biomass burning, including also dimethylsulfide: *DMS*) and from [16] (global open burning of domestic waste) have been used; also *DMS* ocean production fluxes from MOZART-4[4] simulations and online emission of monoterpenes and isoprene, for secondary organic aerosol (SOA) formation, from the MEGAN v2.04 model have been considered.

[1] https://doi.org/10.5065/D6M043C6.

[2] http://edgar.jrc.ec.europa.eu/htap_v2/.

[3] http://gmes-atmosphere.eu/oper_info/global_nrt_data_access/gfas_ftp/.

[4] http://cdp.ucar.edu.

The online aerosol-radiation coupling has been activated for all the 5 global species, using the default particle parameterization of the model (REF case) mainly based on the OPAC dataset [3]. The new refractive indexes used for the PTB case (real: n_r, and imaginary: n_i parts) are reported in Table 66.1 (references indicated in the last row). Also the percent variations with respect to the reference values and the wavelengths (λ) of the optical measurements are reported in the table (the acronym VIS stands for "visible range"). Note that a separate parameterization for primary organic matter (POM) and SOA has been adopted in the PTB case. Moreover, in the PTB case the new refractive indexes for dust and sea salt have been applied only in stations dominated by natural aerosols, while that for organic matter and sulfate only in stations dominated by anthropogenic aerosols.

66.4 Results

Simulated (both REF and PTB cases) and observed ω and g are plotted in Fig. 66.1 for Cairo_EMA_2 and La_Parguera (stations dominated by natural aerosols), Beijing and GSFC (stations dominated by anthropogenic aerosols).

Cairo_EMA_2 (top-left panel) is close to dust sources (dominated by dust in simulations). In the REF case the model strongly underestimates the observed ω and overestimates the observed g. These model biases seem to be related to a too high imaginary index for the reference dust. The new n_i value (PTB case), indeed, causes the simulated ω to significantly increase and the simulated g to slightly decrease, as expected, reducing thus both the mean biases. However a further correction is needed, above all for g, which could be generated for example by a higher fine to coarse ratio for mineral dust. La_Parguera (top-right panel) is a pure coastal station (with intrusions of transported dust during summer in simulations). In the REF case, we note a strong underestimation of the observed ω during summer (in conjunction with the dust intrusions) and a general overestimation of the observed g (above all during winter months). The need of a lower imaginary index for dust is confirmed, since the new n_i value (PTB case) causes the simulated ω to increase and the simulated g to slightly decrease in summer months. However, it seems that a missing of fine particles (for example fine sea salt or white organics) is the main contribution to the positive g mean bias in winter months.

Beijing (bottom-left panel) is a pure urban station (with very high fractions of carbonaceous aerosols in simulations). In the REF case, the model clearly underestimates the observed ω but shows a good agreement with the low observed g (coherently with the dominant presence of small species). The negative ω mean bias can be attributed to a too high imaginary index for the reference organic matter (0.006 at 0.550 μm) but also to the very high simulated fractions of black carbon (which also cause g to strongly decrease). GSFC (bottom-right panel) is characterized by an anthropogenic background not directly affected by urban emissions (higher sulfate and lower black carbon fractions in simulations). In the REF case, we note a less

Table 66.1 New real (n_r) and imaginary (n_i) indexes used for the PTB case

	Mineral Dust (a)	Sea Salt (b)	POM-urban (c)	SOA-biogenic (d, e)	Sulfate (f)
n_r λ	1.531 (~ref); 0.530 μm	1.557 (+3.8%); 0.500 μm	1.501 (−1.9%); 0.550 μm	1.486 (−2.9%); 0.532, 0.550 μm	1.546 (+8.1%); 0.532 μm
n_i λ	$2.5 \cdot 10^{-3}$ (−54.5%); 0.530 μm	$1.0 \cdot 10^{-8}$ (~ref); VIS	$1.0 \cdot 10^{-8}$ (−100%); 0.550 μm	$2.5 \cdot 10^{-5}$ (−99.6%); 0.532, 0.550 μm	$1.0 \cdot 10^{-8}$ (~ref); VIS
	(a) [1]; (b) [5]; (c) [13]; (d) [11]; (e) [9]; (f) [2].				

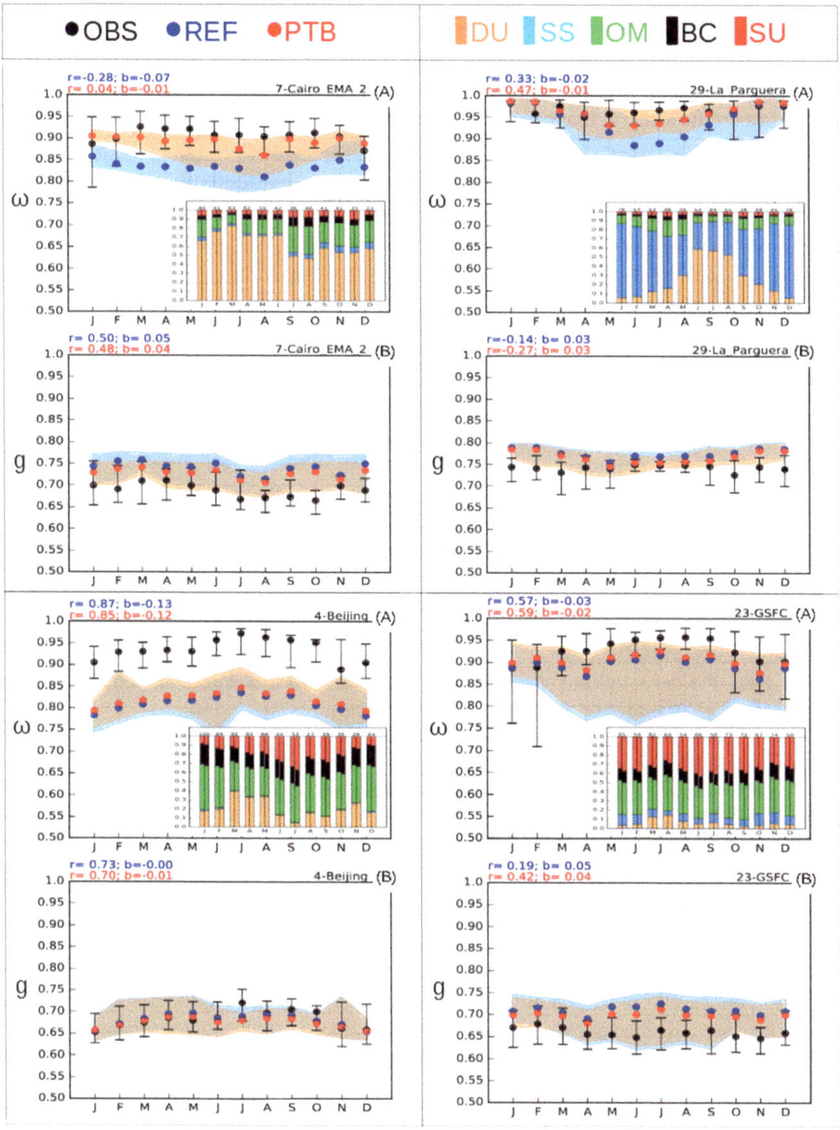

Fig. 66.1 Simulated (REF and PTB) and observed ω (plots A), g (plots B) and simulated τ-fractions (small sub plots inside the ω plot axes), in Cairo_EMA_2 (top-left panel), La_Parguera (top-right panel), Beijing (bottom-left panel) and GSFC (bottom-right panel). In τ-fractions sub-plots: for each month left and right columns refer to REF and PTB cases, respectively. In the legend: DU stands for mineral dust, SS for sea salt, OM for organic matter, BC for black carbon and SU for sulfate

pronounced underestimation of the observed ω (due to the higher fractions of sulfate, purely scattering species) but a strong overestimation of the observed g. Both these mean biases seem to be related to a too low real index for the reference sulfate. Applying the new refractive indexes (PTB case) in both stations causes the simulated ω to slightly increase and the simulated g to slightly decrease (more in GSFC, due to the higher presence of sulfate), as expected. However, evidently these microphysical variations are not enough to fully correct the model biases. In particular, in Beijing clearly lower fractions of black carbon are needed to fill the gap between simulated and observed ω. In GSFC, instead, higher fractions of fine (scattering) particles could cause the simulated g to decrease and so the strong positive mean bias to reduce.

References

1. C. Denjean, F. Cassola, A. Mazzino, S. Triquet, S. Chevaillier, N. Grand, T. Bourrianne, G. Momboisse, K. Sellegri, A. Schwarzenbock, E. Freney, M. Mallet, P. Formenti, Size distribution and optical properties of mineral dust aerosols transported in the western Mediterranean. Atmos. Chem. Phys. **16**, 1081–1104 (2016)
2. M.A. Freedman, C.A. Hasenkopf, M.R. Beaver, M.A. Tolbert, Optical properties of internally mixed aerosol particles composed of dicarboxylic acids and ammonium sulfate. J. Phys. Chem. A **113**, 13584–13592 (2009)
3. M. Hess, P. Köpke, I. Schult, Optical properties of aerosols and clouds: the software package OPAC. Bull. Am. Meteor. Soc. **79**(5), 831–844 (1998)
4. B.N. Holben, T.F. Eck, I. Slutsker, D. Tanré, J.P. Buis, A. Setzer, E. Vermote, J.A. Reagan, Y.J. Kaufman, T. Nakajima, F. Lavenu, I. Jankowiak, A. Smirnov, AERONET-A federated instrument network and data archive for aerosol characterization. Remote Sens. Environ. **66**, 1–16 (1998)
5. R. Irshad, R.G. Grainger, D.M. Peters, R.A. McPheat, K.M. Smith, G. Thomas, Laboratory measurements of the optical properties of sea salt aerosol. Atmos. Chem. Phys. **9**, 221–230 (2009)
6. O. Jorba, D. Dabdub, C. Blaszczak-Boxe, C. Pérez, Z. Janjic, J.M. Baldasano, M. Spada, A. Badia, M. Gonçalves, Potential significance of photoexcited NO2 on global air quality with the NMMB/BSC chemical transport model. J. Geophys. Res. **117**(D13) (2012)
7. C. Lacagnina, O.P. Hasekamp, H. Bian, G. Curci, G. Myhre, T. van Noije, M. Schulz, R.B. Skeie, T. Takemura, K. Zhang, Aerosol single-scattering albedo over the global oceans: comparing PARASOL retrievals with AERONET, OMI, and AeroCom models estimates. J. Geophys. Res.: Atmos. **120**, 9814–9836 (2015)
8. J. Li, B.E. Carlson, O. Dubovik, A.A. Lacis, Recent trends in aerosol optical properties derived from AERONET measurements. Atmos. Chem. Phys. **14**, 12271–12289 (2014)
9. P. Liu, Y. Zhang, S.T. Martin, Complex refractive indices of thin films of secondary organic materials by spectroscopic ellipsometry from 220 to 1200 nm. Environ. Sci. Technol. **47**, 13594–13601 (2013)
10. G. Myhre, B.H. Samset, M. Schulz, Y. Balkanski, S. Bauer, T.K. Berntsen, H. Bian, N. Bellouin, M. Chin, T. Diehl, R.C. Easter, J. Feichter, S.J. Ghan, D. Hauglustaine, T. Iversen, S. Kinne, A. Kirkevag, J.-F. Lamarque, G. Lin, X. Liu, M.T. Lund, G. Luo, X. Ma, T. van Noije, J.E. Penner, P.J. Rasch, A. Ruiz, O. Seland, R.B. Skeie, P. Stier, T. Takemura, K. Tsigaridis, P. Wang, Z. Wang, L. Xu, H. Yu, F. Yu, J.-H. Yoon, K. Zhang, H. Zhang, C. Zhou, Radiative forcing of the direct aerosol effect from AeroCom Phase II simulations. Atmos. Chem. Phys. **13**, 1853–1877 (2013)

11. T. Nakayama, K. Sato, Y. Matsumi, T. Imamura, A. Yamazaki, A. Uchiyama, Wavelength dependence of refractive index of secondary organic aerosols generated during the ozonolysis and photooxidation of α-pinene. Sci. Online Lett. Atmos. **8**, 119–123 (2012)
12. C. Pérez, K. Haustein, Z. Janjic, O. Jorba, N. Huneeus, J.M. Baldasano, T. Black, S. Basart, S. Nickovic, R.L. Miller, J.P. Perlwitz, M. Schulz, M. Thomson, Atmospheric dust modeling from meso to global scales with the online NMMB/BSC-Dust model—Part 1: model description, annual simulations and evaluation. Atmos. Chem. Phys. **11**, 13001–13027 (2011)
13. R.H. Shepherd, M.D. King, A.A. Marks, N. Brough, A.D. Ward, Determination of the refractive index of insoluble organic extracts from atmospheric aerosol over the visible wavelength range using optical tweezers. Atmos. Chem. Phys. **18**, 5235–5252 (2017)
14. M. Spada, Development and evaluation of an atmospheric aerosol module implemented within the NMMB/BSC-CTM. Polytechnic University of Catalonia (Ph.D. Thesis), Barcelona, Spain (2015)
15. M. Spada, O. Jorba, C. Pérez García-Pando, Z. Janjic, J.M. Baldasano, Modeling and evaluation of the global sea-salt aerosol distribution: sensitivity to size-resolved and sea-surface temperature dependent emission schemes. Atmos. Chem. Phys. **13**, 11735–11755 (2013)
16. C. Wiedinmyer, R.J. Yokelson, B.K. Gullett, Global emissions of trace gases, particulate matter, and hazardous air pollutants from open burning of domestic waste. Environ. Sci. Technol. **48**, 9523–9530 (2014)

Chapter 67
Characteristics and Source Contribution of Particulate Matters Acidity in City of Atlanta

Yu Qian and Armistead G. Russell

Abstract Aerosol acidity plays an important role by affecting aerosol formation, concentration, human health, and nutrient bioavailability. This study used thermodynamic model (ISORROPIA-II) and measurement data in Atlanta, GA, USA from 2009 to 2013. Aerosol pH in this research domain kept low (average 2.61) and did not change much over the past decade with the SO_2 emissions reduction. But the trend still shows significant seasonal patterns (high in the winter and low in the summer) and spatial variation (higher in urban and lower in rural area). Model performance were evaluated by comparing estimated and observed NH_3 gas phase partitioning. Also based on the source apportionment results for the same domain, multiple-linear-regression models were developed to show the source impacts on fine particle pH. With strong correlations (average $R^2 = 0.52$), in most cases, models indicate vehicles have the largest positive impact on pH and ammonia bisulfate contribute most negative impact on pH, even though specific results are different for different seasons and locations.

67.1 Introduction

Particles with fine diameter, such as $PM_{2.5}$, contain various components [10], as well as certain amount of liquid water [1]. Observation from the Southeastern US shows that ammonium, nitrate, and sulfate are major inorganic ions, which are closely related to particle acidity. Other species such as Na, Cl, Ca, Mg, etc. have relatively low proportion. Particle pH is calculated as the logarithmic scale of Hydronium ion (H_3O^+, simplified as H^+). H^+ substantially plays important roles in aerosol formation and chemical pathways. The SOA production rate from different precursors can

Y. Qian (✉) · A. G. Russell
School of Civil and Environmental Engineering, Georgia Institute of Technology, Atlanta, GA 30332, USA
e-mail: legendq@gatech.edu

A. G. Russell
e-mail: ted.russell@gatech.edu

C. Mensink et al. (eds.), *Air Pollution Modeling and its Application XXVI*,
Springer Proceedings in Complexity,
https://doi.org/10.1007/978-3-030-22055-6_67

be affected under strongly acidic conditions [7]. In addition, pH have impact on particle composition by control the partitioning of semi-volatile species (e.g., formic, acetic, ammonia and nitrate). Higher acidity will help these species more in the gas-phase. During the summer, this and other study showed that PM2.5 aerosol pH was always in low level, between 1.5 and 3 in the southeastern U.S. [2]. Because of the difficulties involved in the direct measurement of aerosol pH, different proxy methods were used and discussed in estimating aerosol acidity level in recent studies [6, 8]. Those methods include ion balances, cation/anion molar ratio, and thermodynamic models, but former two can be a poor reflect of H^+ as their relationship with pH is not significant [2, 4, 9]. In overall, particle pH can be an important reference for emission control strategies in the purpose of protecting human health [5]. Determine the relationship between pH and species or emission source can help evaluate the performance of current atmospheric model, help determine the control strategy, and help understand some chemical mechanisms related to aerosol acidity.

67.2 Methods

SEARCH Network (The Southeastern Aerosol Research and Characterization Study) [3] provided measurement data used in this study. 30 species and meteorological data were collected from SEARCH database from 2009 to 2013. The monitoring site is Jefferson Street (JST) (33.776°N, 84.413°W), represents urban Atlanta environment.

Aerosol pH represents the H^+ concentration in particle liquid water, as a log scale. In this study, thermodynamic model ISORROPIA-II was used to make estimation of aerosol pH value. ISORROPIA-II, a thermodynamic model, solve for the equilibrium concentrations of NH_4^-–SO_4^{2-}–NO_3^-–Na^+–Cl^-–Ca^{2+}–K^+–Mg^{2+} aerosol system to get H^+ concentration and liquid water content (LWC), under which aerosol pH can be calculated.

By the use of measurement data and Chemical Mass Balance (CMB) model CMB-GC, a widely used source apportionment model, daily $PM_{2.5}$ source contributions were obtained. 9 source categories were included in this study: light-duty gas vehicles (LDGV), heavy-duty diesel vehicles (HDDV), soil dust (SDUST), biomass burning (BURN), coal fire power plant (CFPP), ammonium sulfate (AMSULF), ammonium bisulfate (AMBSULF), ammonium nitrate (AMNITR), and secondary organic carbon (SOC).

67.3 Results

The daily pH time series from 2009 to 2013 for JST were presented in Fig. 67.1. Over the study period, pH is relatively stable over 5 years with significant seasonal pattern of lower in the summer (mean = 1.8) and higher in the winter (mean = 2.7). pH can be averagely differed by about 1 unit. This kind of pattern can be driven by variation

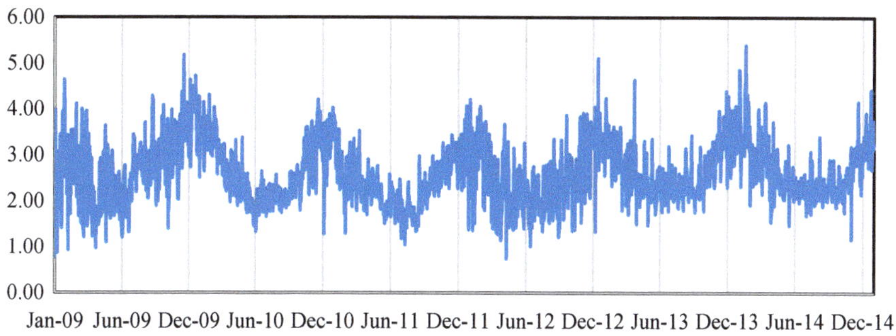

Fig. 67.1 Aerosol pH Trend for Jefferson Street, Atlanta, GA, USA from 2009 to 2014

of temperature, RH, and species concentration such as SO_4^{2-}, which potentially lower the pH during the summer by its higher level, although the presence of high RH, which can potentially rise the pH by adding LWC into aerosol. By comparing observed and model estimated NH_3–NH_4^{-} partitioning ratios, the model performance on estimating the equilibrium condition can be relatively reflected, and then tell the performance on aerosol pH estimation (67.1).

$$\text{Partitioning Ratio} = \varepsilon(NH_3) = \left[NH_{3\,(gas)}\right]/(\left[NH_{3\,(gas)}\right] + \left[NH_4^+{}_{(aerosol)}\right] \quad (67.1)$$

Those two set of ratios are in good agreement for JST ($R^2 = 0.81$, slope $= 0.92$) indicating that the estimation of aerosol pH through ISORROPIA-II is trustable (Fig. 67.2). 9 source categories were included in the source apportionment part to generate daily source contributions: light-duty gas vehicles (LDGV), heavy-duty diesel vehicles (HDDV), soil dust (SDUST), biomass burning (BURN), coal fire power plant (CFPP), ammonium sulfate (AMSULF), ammonium bisulfate (AMBSULF), ammonium nitrate (AMNITR), and secondary organic carbon (SOC). The impacts of those $PM_{2.5}$ source contributions on aerosol pH were investigated by multivariate linear regression method. A linear relationship between daily estimated aerosol pH and daily source contributions were assumed (67.2):

$$pH = b + \sum_{j=1}^{n} a_j^* S_j \quad (67.2)$$

where n is the number of source categories, a_j is the regression coefficient, and S_j is the source contribution for source j. Sensitivity of certain sources to pH can be reflected by corresponding regression coefficient. Also, the regression were conducted based on different season with the consideration of strong seasonal pattern of pH value. In addition to the regression work, model selections were also conducted to remove insignificant variables and trim the model. The results are summarized in Table 67.1.

Table 67.1 Result of Multivariate regression model between aerosol pH and $PM_{2.5}$ source contributions, separated by seasons and model estimated NH_3 partitioning ratio

		LDGV	HDDV	SDUST	BURN	CEPP	AMSVLF	AMBSLF	AMNTTR	SOC	R^2
Spring	***P_Value***			0.042			0.002	0.000	0.000	0.004	***0.52***
	Coef			−0.088			−0.052	−0.172	0.484	−0.059	
Summer	***P_Value***	0.000			0.008	0.000	0.042	0.000	0.000	0.007	***0.43***
	Coef	−0.294			−0.055	−1.743	−0.025	−0.102	0.646	−0.048	
Fall	***P_Value***		0.014			0.038	0.000	0.000	0.000		***0.6***
	Coef		0.082			−1.005	−0.105	−0.181	0.517		
Winter	***P_Value***	0.037	0.001		0.015	0.000	0.000	0.000	0.000		***0.51***
	Coef	0.100	0.174		−0.092	3.063	−0.127	−0.271	0.266		

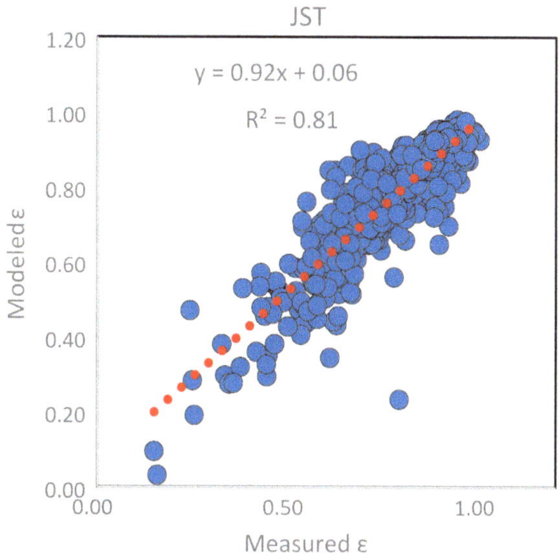

Fig. 67.2 Model evaluation—comparison between measure NH_3 partitioning ratio and model estimated NH_3 partitioning ratio

In general, all of those models showed relatively strong correlation between aerosol pH and $PM_{2.5}$ sources by examine R^2 values, which is 0.52 in average. In all four seasons, secondary sources such as ammonium sulfate and ammonium bisulfate are significantly correlated to pH regardless of season. Ammonium sulfate to pH sensitivity is about −0.08 in average and ammonium bisulfate to pH sensitivity is about −0.18 in average. Considering sulfuric acid is a strong acid and ammonia is a weak base, those species tend to contribute H^+ to lower the pH and make the aerosol more acidic. Other factors showed seasonable pattern, which is mainly related to the temperature and the variated chemical pathway under changed temperature.

References

1. A.G. Carlton, B.J. Turpin, Particle partitioning potential of organic compounds is highest in the Eastern US and driven by anthropogenic water. Atmos. Chem. Phys. **13**, 10203–10214 (2013). https://doi.org/10.5194/acp-13-10203-2013
2. H. Guo, L. Xu, A. Bougiatioti, K.M. Cerully, S.L. Capps, J.R. Hite, A.G. Carlton, S.-H. Lee, M.H. Bergin, N.L. Ng, A. Nenes, R.J. Weber, Fine-particle water and pH in the southeastern United States. Atmo. Chem. Phys. **15**, 5211–5228 (2015). https://doi.org/10.5194/acp-15-5211-2015
3. D.A. Hansen, E.S. Edgerton, B.E. Hartsell, J.J. Jansen, N. Kandasamy, G.M. Hidy, C.L. Blanchard, The southeastern aerosol research and characterization study: part 1—overview. J. Air Waste Manag. Assoc. **53**, 1460–1471 (2003). https://doi.org/10.1080/10473289.2003.10466318
4. C.J. Hennigan, J. Izumi, A.P. Sullivan, R.J. Weber, A. Nenes, A critical evaluation of proxy methods used to estimate the acidity of atmospheric particles. Atmos. Chem. Phys. **15**, 2775–2790 (2015). https://doi.org/10.5194/acp-15-2775-2015

5. J. Lelieveld, J.S. Evans, M. Fnais, D. Giannadaki, A. Pozzer, The contribution of outdoor air pollution sources to premature mortality on a global scale. Nature **525**, 367–371 (2015). https://doi.org/10.1038/nature15371
6. S. Song, M. Gao, W. Xu, J. Shao, G. Shi, S. Wang, Y. Wang, Y. Sun, M.B. McElroy, Fine particle pH for Beijing winter haze as inferred from different thermodynamic equilibrium models. Atmos. Chem. Phys. Discuss, 1–26 (2018). https://doi.org/10.5194/acp-2018-6
7. J.D. Surratt, A.W.H. Chan, N.C. Eddingsaas, M. Chan, C.L. Loza, A.J. Kwan, S.P. Hersey, R.C. Flagan, P.O. Wennberg, J.H. Seinfeld, Reactive intermediates revealed in secondary organic aerosol formation from isoprene. Proc. Natl. Acad. Sci. U. S. A. **107**, 6640–6645 (2010). https://doi.org/10.1073/pnas.0911114107
8. G. Wang, F. Zhang, J. Peng, L. Duan, Y. Ji, W. Marrero-Ortiz, J. Wang, J. Li, C. Wu, C. Cao, Y. Wang, J. Zheng, J. Secrest, Y. Li, Y. Wang, H. Li, N. Li, R. Zhang, Particle acidity and sulfate production during severe haze events in China cannot be reliably inferred by assuming a mixture of inorganic salts. Atmos. Chem. Phys. Discuss. 1–23 (2018). https://doi.org/10.5194/acp-2018-185
9. R.J. Weber, H. Guo, A.G. Russell, A. Nenes, High aerosol acidity despite declining atmospheric sulfate concentrations over the past 15 years. Nat. Geosci. **9**, 282–285 (2016). https://doi.org/10.1038/ngeo2665
10. Q. Zhang, J.L. Jimenez, M.R. Canagaratna, J.D. Allan, H. Coe, I. Ulbrich, M.R. Alfarra, A. Takami, A.M. Middlebrook, Y.L. Sun, K. Dzepina, E. Dunlea, K. Docherty, P.F. DeCarlo, D. Salcedo, T. Onasch, J.T. Jayne, T. Miyoshi, A. Shimono, S. Hatakeyama, N. Takegawa, Y. Kondo, J. Schneider, F. Drewnick, S. Borrmann, S. Weimer, K. Demerjian, P. Williams, K. Bower, R. Bahreini, L. Cottrell, R.J. Griffin, J., Sun, J.Y. Rautiainen, Y.M. Zhang, D.R. Worsnop, Ubiquity and dominance of oxygenated species in organic aerosols in anthropogenically-influenced Northern Hemisphere midlatitudes: UBIQUITY AND DOMINANCE OF OXYGENATED OA. Geophys. Res. Lett. **34**, n/a-n/a (2007). https://doi.org/10.1029/2007GL029979

Chapter 68
An Extreme Event of a Mesoscale Dust Front—A Case Study Over the Eastern Mediterranean

Nitsa Haikin and Pinhas Alpert

Abstract An extreme dust event occurred over southern Israel, where meteorological and air pollution observations suggested that the front was about 100 km wide and the extreme dust concentration formed an advancing "wall" of more than 1 km high. In this study, the atmospheric model RAMS was employed with nested grids over the event domain, where strong turbulence was demonstrated by the model. The simulations showed some turbulent features which supported the observations, yet weren't sufficient to explain the full development of the dust front.

68.1 Introduction

Dust or sand storms are rather frequent over the Eastern Mediterranean, where occasionally PM concentrations exceed background levels by an order of magnitude or more. Dust storms may build up gradually over a very short time, such was the case of the current study. A significant unpredicted dust and sand storm occurred on May 2nd 2007 over southern Israel, with a very strong wind shear at its front. The front was observed as an advancing brown wall. Numerous ground stations recorded an order of magnitude wind amplification accompanied by a sharp drop of temperature. The forecasters missed to predict this storm, however, an Israeli Air-Force pilot documented this event from the air and he estimated its advancing velocity as 60 km/hr, and its height as about 1300 m. RAMS was employed with high resolution grids, to study the dynamics of the boundary layer, which allowed and supported the development of this event.

N. Haikin (✉) · P. Alpert
Department of Geosciences, Tel-Aviv University, Tel Aviv, Israel
e-mail: nitsahai@tauex.tau.ac.il

P. Alpert
e-mail: pinhas@tauex.tau.ac.il

N. Haikin
Department of Physics, NRCN, Beer Sheva, Israel

C. Mensink et al. (eds.), *Air Pollution Modeling and its Application XXVI*,
Springer Proceedings in Complexity,
https://doi.org/10.1007/978-3-030-22055-6_68

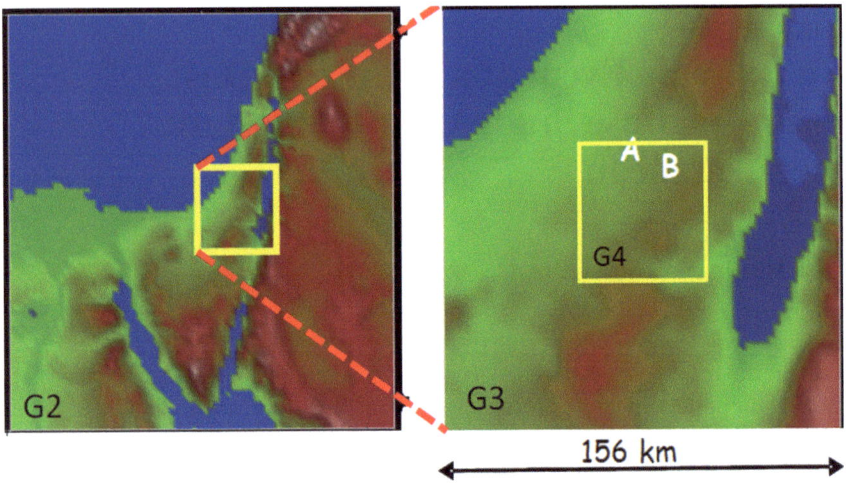

Fig. 68.1 The regional simulation domain, where the event occurred: grid 2 (left), grid 3 and grid 4 (right), with 8, 2 and 0.5 km horizontal resolution, respectively. 'A' and 'B' denote the study sites

68.2 Methods

Observations from 16 sites were analysed in order to identify the front timing and intensity. The main observed characteristics included an extreme wind amplification over a short time, with velocity twice to three times that of the sea-breeze front as found by Alpert and Rabinovich-Hadar [1], a sharp temperature drop and an estimated dust load of more than 1500 $\mu g/m^3$ at the peak. The synoptic pressure map showed a deep low over Egypt, forcing easterly winds over Israel. However, based on the observations, the evident dust front advanced from NW to SE. For further analysis of the event dynamics the three-dimensional, non-hydrostatic Regional Atmospheric Modelling System (RAMS; [2]) was employed with a configuration consisted of two-way nested grids with successive horizontal resolution of 32, 8, 2 and 0.5 km, centered over southern Israel (Fig. 68.1). The vertical grid interval started at 50 m agl. RAMS 18-hr simulation was initiated with NCEP 6 h reanalysis.

68.3 Results and Discussion

The simulation results showed a deep vertical circulation over the region, with a downwards flow over Israel. Although the simulated surface north-westerly flow only penetrated inland as far as site A, at that time the flow at 2 km high veered into north westerly over the whole G3 domain. A significant turbulence was demonstrated by the model, with significant TKE up to about 1500 m (Fig. 68.2), above which the vertical wind component *w* became negative. These results agree with the height

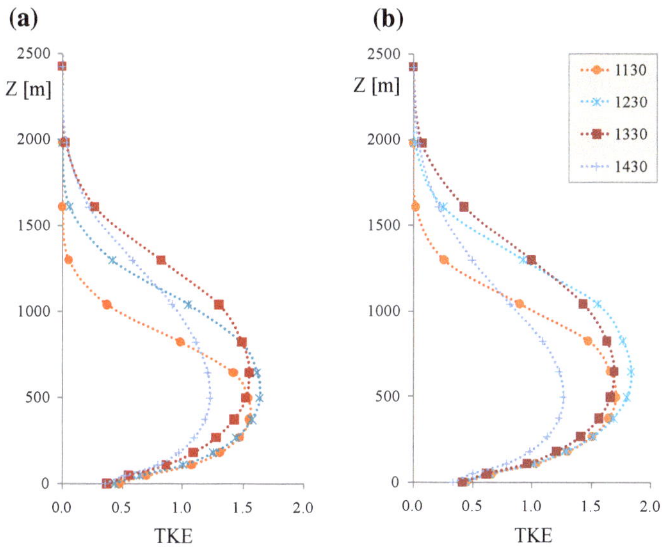

Fig. 68.2 Simulated hourly TKE along the event duration (**a**) over site A, (**b**) over site B, where the dust wall was documented (time in UTC). The peak intensity is at 1230–1330, where over site B the TKE is stronger than over site A

estimation of the observed dust wall, and might explain the vertical dimension of the dust front. A comparison with some ground meteorological observations showed a significant disagreement; however, the simulated coastal upper air profiles had a fair fit with the standard 12Z radiosonde, as shown in Fig. 68.3. The simulated wind over site A was much weaker than the observed, yet its direction was similar to the coastal one.

A further analysis based on available synoptic maps suggested that the model failure to identify the dust front development may be attributed to the synoptic miss-initiation of the deep low location and orientation. Local to mesoscale land-use features could have locally intensified the dust load and turbulence, while incorrect input values of these parameters may have caused a significant miscalculations by the model, as was also discussed by Gasch et al. [3].

68.4 Summary

The dust event in this study had unique characteristics by means of size, intensity and duration. The attempt to analyze the event by employing RAMS provided an insight of possible dynamic features and dimensions of the event, such as the intensity of the vertical turbulence, which might explain the vertical dimensions of the observed dust front. Under the current configuration, although fair agreement of upper-air wind

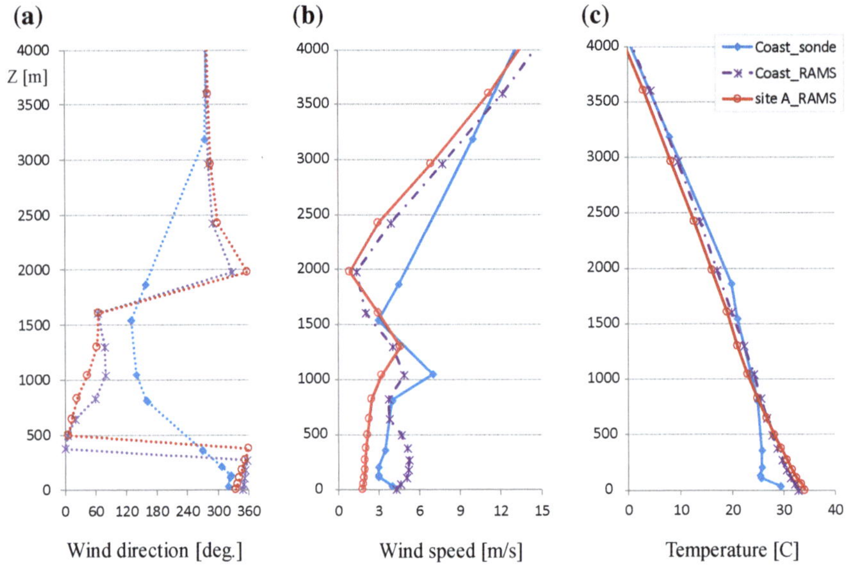

Fig. 68.3 Simulated and observed upper-air profiles of the wind (**a**, **b**) and temperature (**c**) over the coast and over site A

and temperature profiles was found over the coast, the simulation failed to identify the inland evolution of the apparently short-lived dust event. Further exploration is planned on this case.

Acknowledgements We wish to thank Roni Neumann from NRCN for his expert-forecaster insight on the case.

References

1. P. Alpert, M. Rabinovich-Hadar, Pre- and post-sea-breeze frontal lines—a meso-gamma-scale analysis over south Israel. J. Atmos. Sci. **30**, 2994–3008 (2003)
2. W.R. Cotton, R.A. Pielke Sr., R.L. Walko, G.E. Liston, C.J. Tremback, H. Jiang, R.L. McAnelly, J.Y. Harrington, M.E. Nicholls, G.G. Carrio, J.P. McFadden, RAMS 2001: current status and future directions. Meteorol. Atmos. Phys. **82**, 5–29 (2003)
3. P. Gasch, D. Rieger, C. Walter, P. Khain, Y. Levi, P. Knippertz, B. Vogel, Revealing the meteorological drivers of the September 2015 severe dust event in the Eastern Mediterranean. Atmos. Chem. Phys. **17**, 13573–13604 (2017)

Part VII
Modelling Air Pollution in a Changing Climate

Chapter 69
Climate Model Response Uncertainty in Projections of Climate Change Impacts on Air Quality

Fernando Garcia-Menendez, James East, Bret D. Pienkosz and Erwan Monier

Abstract Uncertainties in climate simulations can strongly propagate to estimates of climate change impacts, including its effects on air pollution. Here we use a coupled modeling framework to evaluate the role of climate model response in projections of climate-induced impacts on air quality. Within integrated economic, climate, and air pollution projections, climate model response is altered by modifying the climate sensitivity of the framework's Earth system component. We find that variations in climate sensitivity ranging from 2.0 to 4.5 °C per doubling of CO_2 can change projections of the climate penalty on O_3 and $PM_{2.5}$ pollution in the U.S. by more than 2 ppb and 0.5 $\mu g\ m^{-3}$. The impact of uncertainty due to climate model response can be as important as that related to greenhouse gas emissions scenario or natural variability.

69.1 Introduction

Projections of climate change under different greenhouse emissions scenarios have been widely used to investigate the potential impacts of a warmer climate on air quality. These analyses suggest that many locations may experience a "climate penalty" on air quality as climate change leads to meteorological conditions that enhance air pollution. Over the U.S., prior modeling studies consistently suggest a climate-induced increase in ozone (O_3) pollution over polluted regions, while there is less agreement on fine particulate matter ($PM_{2.5}$) impacts, which also vary across different $PM_{2.5}$ components [1].

F. Garcia-Menendez (✉) · J. East · B. D. Pienkosz
Department of Civil, Construction and Environmental Engineering, North Carolina State University, 27695 Raleigh, NC, USA
e-mail: f_garcia@ncsu.edu

E. Monier
Joint Program on the Science and Policy of Global Change, Massachusetts Institute of Technology, 02139 Cambridge, MA, USA

C. Mensink et al. (eds.), *Air Pollution Modeling and its Application XXVI*, Springer Proceedings in Complexity,
https://doi.org/10.1007/978-3-030-22055-6_69

The three principal sources of uncertainty in climate projections are greenhouse gas emissions scenario, natural variability, and climate model response [3]. While efforts to simulate climate change effects on air quality have explored the influence of emission scenarios and to a lesser degree natural variability, none have systematically assessed the impact of climate model response. Multi-model ensembles that project the effect of climate on pollutant concentrations inherently cover a range of climate responses across different climate-chemistry models. However, the influence of climate model response in these ensembles cannot be discerned from other sources of uncertainty including internal variability, parameter selection, and scenario representation. Analyses that compare the weights of major climate uncertainties on projections of temperature and precipitation have found that, although natural variability controls uncertainty initially and greenhouse gas emissions scenario eventually dominates, model response is consistently the second largest source of uncertainty [4]. It is critical that impact projections based on climate projections adequately consider their underlying uncertainties, including that related to climate model response. Here we specifically evaluate its role and compare it to the other major contributors to uncertainty in climate projections.

69.2 Methods

Our projections of climate-induced impacts on U.S. air quality are based on the integrated scenarios of economic activity, greenhouse gas emissions, and climate change from the U.S. EPA Climate Change Impacts and Risk Analysis project [4, 6]. Socioeconomic activity and climate projections are generated with the MIT IGSM-CAM integrated modeling framework [5]. Within the IGSM-CAM simulations, climate model response is altered using a cloud radiative adjustment method to modify climate sensitivity to greenhouse gas forcing [7]. We model changes in climate under climate sensitivities of 2.0, 3.0, or 4.5 °C mean surface temperature change per doubling of atmospheric CO_2. The climate projections are used to drive the CAM-Chem climate-chemistry model and simulate global air quality [2]. In the atmospheric chemistry simulations, the effect of climate is isolated by holding emissions of conventional air pollutants and precursors at start-of-the-century levels. In addition to variations in climate sensitivity, our simulation ensemble projects atmospheric chemistry at the start, middle, and end of the 21st century under a business-as-usual reference scenario (REF) and 2 climate policy scenarios with 2100 total radiative forcing targets of 4.5 W m^{-2} (P45) and 3.7 W m^{-2} (P37). Projections of U.S. air quality for each period and policy are based on 30-year atmospheric chemistry simulations under 5 unique sets of initial conditions in the IGSM-CAM to account for natural variability. The climate penalties on O_3 and $PM_{2.5}$ are estimated as the difference in concentrations under present and future climates.

69.3 Results

We explore the impacts of climate change on O_3 and $PM_{2.5}$ in the U.S. under 3 climate sensitivities representing lower (2.0 °C), middle (3.0 °C), and upper (4.5 °C) estimates of the high confidence range reported by Fifth Assessment Report of the United Nations Intergovernmental Panel on Climate Change (IPCC). Our ensemble-mean projections with a 3.0 °C sensitivity, across all 30-year simulations and 5 initializations, suggest that under the REF scenario, climate change can significantly impact U.S. air pollution (Fig. 69.1). For O_3 a substantial climate penalty is projected over the Northeast, Southeast, and Southern California, while a climate-driven reduction in concentrations is simulated over areas in the Midwest and West. The highest climate penalties on $PM_{2.5}$ are projected over the Eastern U.S., with lower increases in the West and some areas of the Midwest reflecting a climate-induced decrease in concentrations.

Variations in climate sensitivity strengthen or weaken the projected impacts on air quality. In the REF scenario, only a few locations are projected to experience a climate penalty on O_3 greater than 4 ppb by the end of the 21st century under a climate sensitivity of 2.0 °C. However, a 4.5 °C sensitivity leads to projections of impacts greater than 4 ppb over most of the Eastern U.S., with some locations experiencing penalties as large as 10 ppb. For $PM_{2.5}$, the largest projected climate-related increases, those over the Ohio River Valley, vary from 2 to 5 $\mu g\, m^{-3}$ across the climate sensitivities. In addition, the area projected to experience a climate-induced decrease in $PM_{2.5}$ concentration greatly shrinks as climate sensitivity is increased.

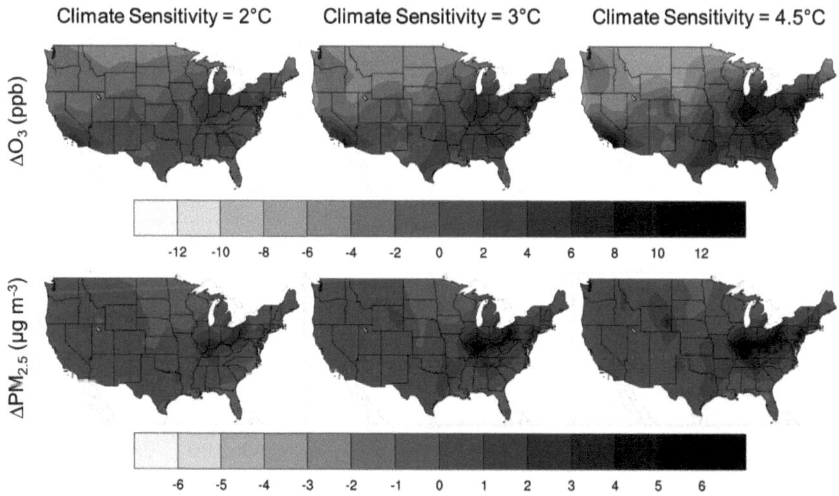

Fig. 69.1 Mean climate-induced change in annual-average ground-level 8-hour maximum O_3 and $PM_{2.5}$ at end of the 21st century relative to start of the century projected under REF scenario and 3 climate sensitivities

Relative to impacts at the end of the 21st century, the differences in projected climate penalties on U.S. population-weighted O_3 and $PM_{2.5}$ concentrations across the climate sensitivities considered are small at mid-century (Fig. 69.2). As climate sensitivity is increased, its impact on national annual average concentrations is not only reflected in the mean penalty values, but on the distribution of annual impacts as well. Figure 69.2 illustrates how a higher climate sensitivity may result in a wider distribution of impacts across the individual annual simulations included in our ensemble. The effect suggests greater risk of extreme air pollution episodes at a higher climate sensitivity, beyond that caused by a shift in distribution mean penalties.

Table 69.1 includes climate penalties projected across all policies and climate sensitivities considered. In the absence of global greenhouse gas mitigation policy (REF), our ensemble simulation projects climate-induced increases in U.S. population-weighted O_3 and $PM_{2.5}$ of 3.2 ± 0.3 ppb and 1.5 ± 0.1 μg m^{-3} by the end of

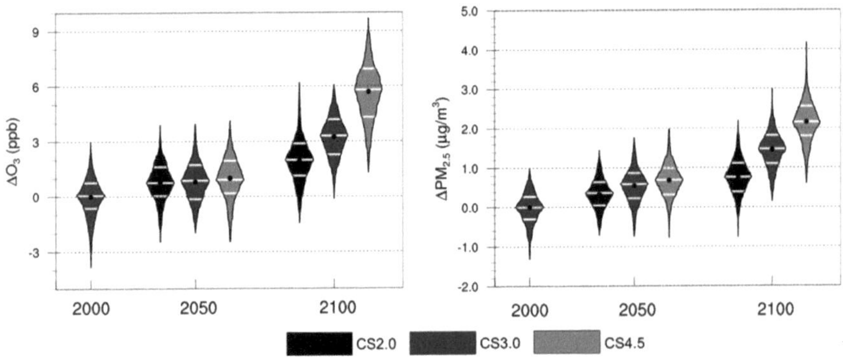

Fig. 69.2 Distribution of projected annual REF-scenario anomalies relative to 1981–2010 mean for U.S.-average annual population-weighted 8 h maximum O_3 and $PM_{2.5}$ under each climate sensitivity. Plots show distribution median and interquartile range (lines), and mean (dot), for each simulated period

Table 69.1 Mean climate-induced increases in U.S. population-weighted concentrations projected at the end of 21st century relative to start of the century for each climate sensitivity and policy scenario

Policy	Climate sensitivity (°C)	O_3 (ppb)	$PM_{2.5}$ (μg m^{-3})
P37	2.0	0.6 ± 0.3	0.1 ± 0.1
	3.0	0.6 ± 0.3	0.3 ± 0.1
	4.5	0.4 ± 0.3	0.5 ± 0.1
P45	2.0	0.3 ± 0.3	0.1 ± 0.1
	3.0	0.3 ± 0.3	0.4 ± 0.1
	4.5	0.6 ± 0.3	0.7 ± 0.1
REF	2.0	1.9 ± 0.3	0.8 ± 0.1
	3.0	3.2 ± 0.3	1.5 ± 0.1
	4.5	5.7 ± 0.4	2.1 ± 0.1

the century at the medium climate sensitivity. However, across the range of climate sensitivities considered, the O_3 penalty varies from 1.9 ± 0.3 to 5.7 ± 0.4 ppb. For $PM_{2.5}$, the range of climate sensitivities leads to population-weighted penalties spanning from 0.8 ± 0.1 to $2.1 \pm 0.1\ \mu g\ m^{-3}$. Policies leading to a stabilization of total radiative forcing by 2100 may significantly reduce these climate-induced increases in U.S. air pollution. Under the P45 and P37 scenarios, the range of O_3 penalties projected across the 3 climate sensitivities lowers to 0.3 ± 0.3 to 0.6 ± 0.3 ppb. The range of $PM_{2.5}$ penalties projected is similarly constrained to 0.1 ± 0.1 to $0.7 \pm 0.1\ \mu g\ m^{-3}$ under the stabilization scenarios.

69.4 Conclusions

Air quality and associated health effects are among the largest impacts reported in climate policy benefits assessments. Adequately characterizing the uncertainties that underlie climate projections and how these propagate to air quality simulations can add great value to these analyses. By perturbing climate sensitivity, our ensemble simulation of climate-induced changes to air quality shows that model response uncertainty can be as large as that caused by natural variability or emissions scenario. Our results also suggest that beyond reducing the climate penalty on air quality, climate policies could reduce the risks associated with higher-than-expected levels of response to climate forcing by the real climate system. Given the IPCC's latest conclusion that a best estimate for climate sensitivity can no longer be given due to lack of agreement across studies, it is critical that this source of uncertainty is carefully considered.

References

1. M. A. Fiore, V. Naik, E. M. Leibensperger, Air quality and climate connections. J. Air Waste Manage. Assoc. 65(6), 645–685 (2015)
2. F. Garcia-Menendez et al., U.S. air quality and health benefits from avoided climate change under greenhouse gas mitigation. Environ. Sci. Technol. **49**(13), 7580–7588 (2013)
3. E. Hawkins, R. Sutton, The potential to narrow uncertainty in regional climate predictions. Bull. Am. Meteorol. Soc. **90**(8), 1095 1108 (2009)
4. E. Monier et al., An integrated assessment modeling framework for uncertainty studies in global and regional climate change: the MIT IGSM-CAM (version 1.0). Geosci. Model Dev. **6**, 2063–2085 (2013)
5. E. Monier et al., A framework for modeling uncertainty in regional climate change. Clim. Change **131**(1), 51–66 (2015)
6. S. Paltsev et al., Integrated economic and climate projections for impact assessment. Clim. Change **131**(1), 21–33 (2015)
7. A.P. Sokolov, E. Monier, Changing the climate sensitivity of an atmospheric general circulation model through cloud radiative adjustment. J. Climate **25**, 6567–6584 (2012)

Chapter 70
Ozone Risk for Douro Vineyards in Present and Future Climates

Ana Isabel Miranda, Ana Ascenso, Carla Gama, Daniel Blanco-Ward, Alexandra Monteiro, Carlos Silveira, Carolina Viceto, Alfredo Rocha, Diogo Lopes, Myriam Lopes and Carlos Borrego

Abstract Tropospheric ozone (O_3) can damage vegetation, affecting productivity and quality of the crops. Vines, in particular, have an intermediate sensitivity to ozone. Moreover, an increase of ozone levels is foreseen under climate change scenarios. The Douro Demarcated Region is one of the most productive wine areas in Portugal; thus studying the ozone deposition over this region and assessing its potential effects is a nowadays concern. This work aims to evaluate the risk of Douro vineyards exposure to ozone in present and future climates. The chemical transport model CHIMERE, with a spatial resolution of 1 km^2, fed by meteorological data from the WRF model,

---

A. I. Miranda (✉) · A. Ascenso · C. Gama · D. Blanco-Ward · A. Monteiro · C. Silveira · D. Lopes · M. Lopes · C. Borrego
CESAM & Department of Environment and Planning, University of Aveiro, Aveiro, Portugal
e-mail: miranda@ua.pt

A. Ascenso
e-mail: ascenso.a@ua.pt

C. Gama
e-mail: carlagama@ua.pt

D. Blanco-Ward
e-mail: dblancoward@ua.pt

A. Monteiro
e-mail: alexandra.monteiro@ua.pt

C. Silveira
e-mail: carlos.silveira@ua.pt

M. Lopes
e-mail: myr@ua.pt

C. Borrego
e-mail: cborrego@ua.pt

C. Viceto · A. Rocha
CESAM & Department of Physics, University of Aveiro, Aveiro, Portugal
e-mail: carolinaviceto@ua.pt

A. Rocha
e-mail: alfredo.rocha@ua.pt

C. Mensink et al. (eds.), *Air Pollution Modeling and its Application XXVI*, Springer Proceedings in Complexity,
https://doi.org/10.1007/978-3-030-22055-6_70

was applied for the years 2003–2005 (present climate), for 2049 and 2064 (mid-term future) and for 2096 and 2097 (long-term future). The assessment of the potential damage in terms of productivity and quality was done through the analysis of ozone deposition and the application of concentration-response functions. The exposure indicator AOT40 (accumulated concentration of ozone above 40 ppb) for the period established in the Air Quality Framework Directive 2008/50/CE was also estimated. The model results show, for present and future climate, that the AOT40 levels in the entire Douro region are above the target value for the protection of vegetation. The results of the exposure-response functions suggest that the tropospheric ozone levels in the future, in the region, would influence the quality and productivity of the wine.

70.1 Introduction

Wine production is a major economic activity in Portugal; among the Portuguese wine production areas the Douro Demarcated Region (DDR) stands out by being where the famous Port Wine is produced. According to Soja et al. [5] vineyards exposure to high ozone levels can result in changes in their productivity and wine quality. Moreover, climate change (CC) is expected to have relevant impacts in Southern Europe, in particular in what concerns ozone levels. Lacressonnière et al. [4] estimated an increase of O_3 levels of approximately 1.5 ppb per decade in summer in Europe, under the RCP8.5 scenario.

Thus, the aim of this study is to assess the risk of Douro vineyards exposure to ozone in present and future climates. The O_3 concentration and deposition over the DDR were estimated by applying the air quality modelling system WRF-CHIMERE. To evaluate the potential damage in terms of productivity and quality, exposure-response functions were applied.

The present work is organized as follows: Sect. 70.2 describes the air quality modelling system setup and work methodology, the modelling results are presented in Sect. 70.3, and conclusions are drawn in Sect. 70.4.

70.2 Methodology

The air quality modeling system WRF-CHIMERE was applied over the DDR with a spatial resolution of 1 km^2. More details on the model input data and parameterizations can be found in [1].

To assess the ozone concentrations and deposition values over the DDR for the present and the future climates, a set of 5 scenarios and a total of 11 years were simulated and analyzed. The simulation periods were selected based on an assessment of the climatology of the region that identified the hottest years from an historical period (1986–2005) and from mid (2046–2065) and long-term (2081–2100) projected scenarios, which were based on the RCP8.5 pathway. The years 2003–2005 were selected to simulate the present climate, 2049 and 2064 for the mid-term future climate and the years 2096 and 2097 for the long-term future climate. The air quality simulations for the future climate were two-fold: one was done considering climate change only, i.e. emissions were kept the same as in present climate simulations, to assess the contribution of meteorological variables alone on future O_3 concentrations levels; the other also included future emissions assessing future O_3 concentrations resulting from both contributions (meteorology and emissions).

To evaluate the risk of Douro vineyard exposure to O_3 the legislated exposure indicator AOT40 (accumulated concentration of ozone above 40 ppb) was calculated as an average of AOT40 values for the simulated years, by scenario. The deposition level results were based on the total O_3 dry deposition accumulated over May to July, the same period established in the Air Quality Framework Directive 2008/50/CE for the calculation of the AOT40 indicator. The assessment of the potential damage in terms of productivity and quality was done through the application of the exposure-response functions calculated by Soja et al. [5] in their three-year experimental study on the effects of long-term ozone exposure in grapevines. For the purpose of this work, the function applied was the one calculated in the last year of the experiment and thus the one that represents the most damage in vines.

70.3 Modelling Results

A summary of the modelling results is presented in Fig. 70.1, which includes the results for present climate, and for long-term future climate, considering only the climate change and considering climate and emissions changes. The delimitated area in the maps is the DDR.

When considering the climate change impact on O_3 concentration, the modelling results show an averaged increase of 7% in the AOT40 levels, which is reflected in a 9 and 7% decrease in productivity and quality. When emission projections are applied, the AOT40 levels drop approximately 8%, reflecting an increase of 8% in productivity and 7% in quality. These results indicate that the reduction in AOT40 levels is mostly driven by emissions reduction, which is in agreement with Klingberg et al. [3] conclusions. Nevertheless, the target value for the protection of vegetation (18,000 $\mu g\ m^{-3}$ h) is still exceeded over all the DDR.

When discussing these outcomes it is important to remember that although AOT40 is an important indicator for assessing the risk of vegetation exposure to ozone, it is known that it has some limitations. For instance, it does not take into consideration the uptake of ozone by the plants, which depends on the stomatal flux, overestimating the potential damages. Furthermore, applying Soja et al. [5] functions brings also uncertainty to the results, since the variety and age of the vines can respond differently to ozone exposure [6].

Regarding dry O_3 deposition levels, both future simulation scenarios show a decrease in deposition of almost 50% when compared with the present climate simulations. This indicates that, as opposed to AOT40 levels, the dry deposition is mainly

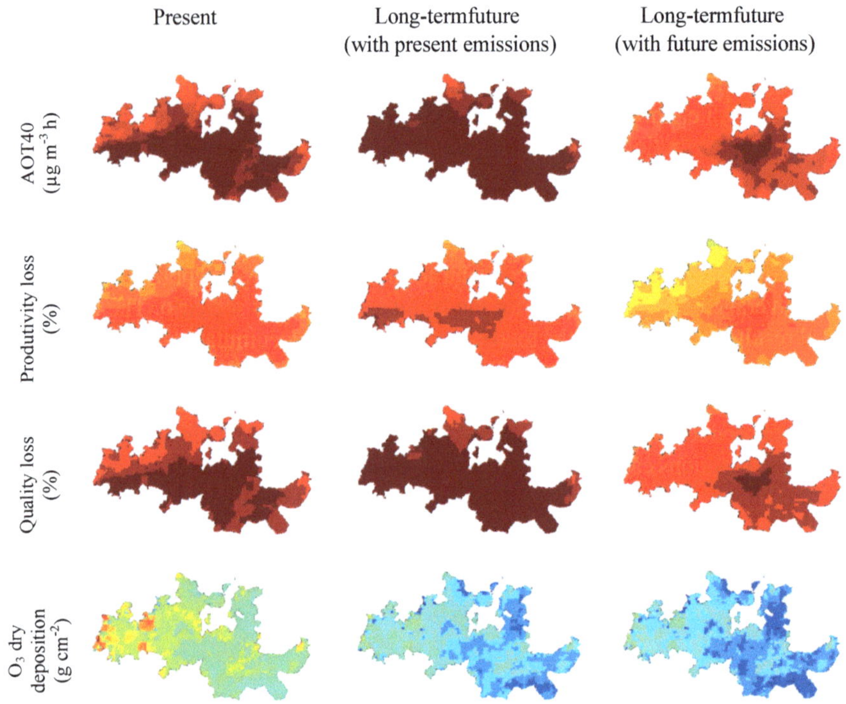

Fig. 70.1 Modelling results of AOT40, vines productivity and quality losses, and O_3 dry deposition, for present and long-term future climates, without and with future emissions

driven by meteorological factors. As Klingberg et al. [2] points out, even with the increased modelled future O_3 concentrations, in southern Europe, the risk for O_3 damage to vegetation is predicted to decrease at most sites, mostly as a result of drier conditions and higher temperature, as well as rising CO_2 concentrations.

70.4 Conclusions

The aim of this work was to estimate ozone concentrations and deposition over the DDR for present and future climates, in order to assess the risk of Douro vineyards exposure to O_3. Simulations for future climate were performed only considering the climate change impact in ozone concentrations and considering future emissions as well.

Modelling results show that emission reduction of ozone precursors can successfully decrease the AOT40 levels in the Douro Region. Nonetheless, the emission reduction projected by RCP8.5 is not enough to accomplish the target value for the protection of the vegetation established in the Air Quality Framework Directive since this value is exceeded over the entire region, suggesting a likely negative impact for the vegetation. If the emissions are maintained constant the exposure response functions indicate that, in long-term, the AOT40 levels can represent a potential damage to the vines causing over 35% loss in productivity and quality.

However, when ozone dry deposition levels are estimated a decrease of 50% is expected, for both future scenarios. Thus, the risk of exposure to ozone based on the AOT40 could have been overestimated. Therefore, it is important to define a threshold for the protection of vegetation that takes into account the potential uptake by the vegetation using dry deposition levels and developing particular dose-response functions.

Questions and answers

Question 1

QUESTIONER: Stefano Galmarini

QUESTION: Ozone uptake depends on the stomatal opening. How soil water content can influence stomata opening? How is the model considering this?

ANSWER: The soil water content (soil moisture) is an important factor for stomatal opening, since one of the functions of the stomata is the water regulation of the plant. The CHIMERE model considers this parameter when calculating the stomatal conductance as a factor comprised between 0 and 1, which represents the relative conductance determined by the soil water potential (SWP).

Question 2

QUESTIONER: Sebnem Aksoyoglu

QUESTION: How did you adjust EMEP emissions with 0.5° resolution to your high resolution domain of 1×1 km^2?

ANSWER: The emissions were disaggregated using an adaptation of the CHIMERE emiSURF pre-processor, which performs a spatial redistribution using the larger domain emissions data and allocates the emission values to each domain cell (1×1 km^2), based on a classification among four land use types: urban, crop, water and forest.

Question 3

QUESTIONER: Rostislav Kouznetsov

QUESTION: How are you measuring quality of the wine, or what kind of unit?

ANSWER: In this study the quality of the wine was related to the relative sugar content in the berries (total glucose + fructose), the units are normally grams of sugar per liter of wine.

Acknowledgements The authors wish to thank the financial support of FEDER through the COMPETE Program and the national funds from FCT—Science and Technology Portuguese Foundation for financing the DOUROZONE project (PTDC/AAG-MAA/3335/2014; POCI-01-0145-FEDER-016778). Thanks are also due, for the financial support to CESAM (UID/AMB/50017), to FCT/MEC through national funds, and the co-funding by the FEDER, within the PT2020 Partnership Agreement and Compete 2020.

References

1. A. Ascenso, Assessing Douro vineyards exposure to tropospheric ozone. Master thesis, University of Aveiro, 82 p, 2017
2. J. Klingberg, M. Engardt, J. Uddling, P.E. Karlsson, H. Pleijel, (2011), Ozone risk for vegetation in the future climate of Europe based on stomatal ozone uptake calculations. Tellus A **63**(1), 174–187
3. J. Klingberg, M. Engardt, P.E. Karlsson, J. Langner, H. Pleijel, Declining ozone exposure of European vegetation under climate change and reduced precursor emissions. Biogeosciences **11**(19), 5269–5283 (2014)
4. G. Lacressonnière, V.H. Peuch, R. Vautard, J. Arteta, M. Deque, M. Joly, B. Josse, V. Marecal, D. Saint-Martin, European air quality in the 2030s and 2050s: impacts of global and regional emission trends and of climate change. Atmos. Environ. **92**, 348–358 (2014)
5. G. Soja, T. Reichenauer, M. Eid, A.M. Soja, R. Schaber, H. Gangl, Long-term ozone exposure and ozone uptake of grapevines in open-top chambers. Atmos. Environ. **38**, 2313–2321 (2004)
6. U.S. EPA, Air quality criteria for ozone and related photochemical oxidants volume I of III (U.S. Environmental Protection Agency, Washington, DC, 2006), 821 p

Chapter 71
Linking North American Summer Ozone Pollution Episodes to Subseasonal Atmospheric Variability

E. Charles White, Dylan Jones and Paul Kushner

Abstract Extreme ozone pollution episodes can cause acute stress on the human respiratory system and damage vegetation. This study leverages long-term ozone measurement records from the United States Environmental Protection Agency Air Quality System (AQS) and the Environment and Climate Change Canada National Air Pollution Surveillance Program (NAPS) to identify ozone pollution episodes in summertime North America and assess meteorological patterns typical for those events. For this purpose, methods are adapted from studies of heat waves that similarly rely on station-based measurement records. Clustering methods are used to group ozone monitoring stations into large regions according to their likelihood of recording simultaneous ozone extremes. Multiday periods with abnormally high cluster-wide ozone concentrations are then identified as ozone episodes. Composite meteorological patterns associated with the ozone enhancement episodes are separately derived for each region. The composite patterns exhibit features commonly associated with elevated ozone, such as high temperatures and reduced cloud cover. In most regions, these features appear alongside anomalous synoptic-scale anticyclonic circulation in the mid-troposphere. These systems are themselves embedded in large-scale wave trains extending from the Pacific. The wave trains develop more than a week in advance of ozone episode onset and are plausibly related to sea surface temperature patterns with subseasonal persistence that also emerge in the composite meteorology. The persistence of these ozone episode circulation patterns could potentially be leveraged to improve forecasting and long-term projections of air quality.

71.1 Introduction

Ozone is an airborne pollutant that causes oxidative stress to the human respiratory system [10] and damages vegetation [5]. Short-term acute ozone exposure has been

E. C. White (✉) · D. Jones · P. Kushner
Department of Physics, University of Toronto, Toronto, ON M5S 1A7, Canada
e-mail: cwhite@physics.utoronto.ca

C. Mensink et al. (eds.), *Air Pollution Modeling and its Application XXVI*,
Springer Proceedings in Complexity,
https://doi.org/10.1007/978-3-030-22055-6_71

connected to the rate of incidence of acute respiratory illness [8], and to larger reductions in crop yields compared to equivalent long-term, lower concentration ozone exposure [6]. Ozone concentrations are positively correlated with temperature and stagnancy, and negatively correlated with cloud cover [4], owing to increased emissions of biogenic ozone precursor compounds at high temperatures and increased rates of photolysis, among other things. This suggests that ozone pollution episodes are more likely to happen during heat wave conditions. Previous studies have found that North American summertime heat waves may be preceded by particular large-scale meteorological patterns with subseasonal persistence [7, 9]. This study aims to determine whether summertime regional ozone pollution episodes in North America might similarly be preceded by such large-scale meteorological patterns.

71.2 Data

The ozone data used for this study are from the United States Environmental Protection Agency's Air Quality System (AQS) [11] and from Environment and Climate Change Canada's National Air Pollution Surveillance Program (NAPS) [2]. The ozone concentrations are maximum daily 8-hour average values (MDA-8) in June, July, and August (JJA) from 1990 to 2016. ERA-Interim reanalysis fields [1] are used to examine meteorology during ozone episodes.

71.3 Ozone Episode Identification

The AQS and NAPS stations span an area too large for any definition of an ozone pollution episode to be reasonably applied to all stations in aggregate. To remedy this, the AQS and NAPS stations are divided into smaller groups using a statistical clustering algorithm. Specifically, hierarchical agglomerative clustering [12] is used, with a similarity metric between each possible pairing of stations used as input. The similarity metric is a Jaccard distance [3] based on daily co-occurrences of extreme ozone concentrations (i.e. greater than historical 95th percentile at a location) between two stations. The results of the clustering are shown in Fig. 71.1.

Within each cluster, ozone pollution episodes are defined for all ozone monitoring stations in aggregate, with the daily 95th percentile ozone measurement, herein called O95, being the primary quantity of interest. An extreme threshold is defined as the mean plus standard deviation of O95 in a 5-year rolling window. The rolling window is applied to remove the influence of decreasing anthropogenic emissions of ozone precursor compounds over time. An ozone episode occurs when the extreme threshold is surpassed for 3 or more consecutive days. In the Northeast cluster, this definition identifies 51 ozone episodes over the 27-year period of interest.

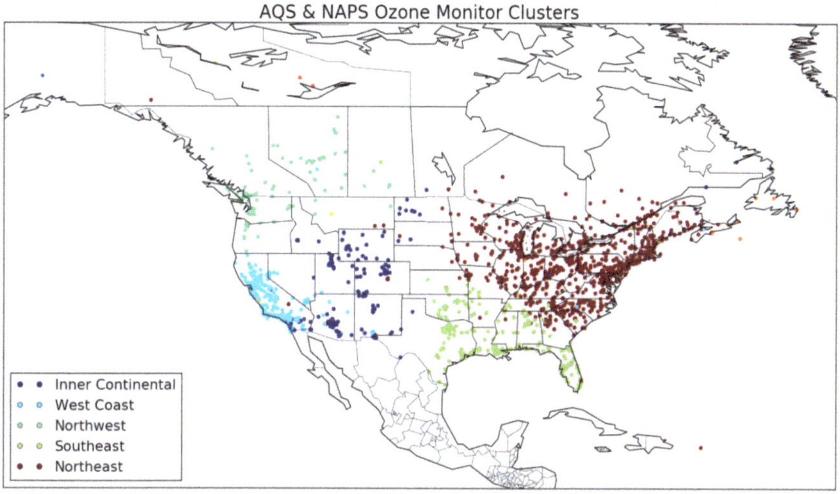

Fig. 71.1 AQS and NAPS stations clustered by co-occurrences of extreme daily ozone measurements

71.4 Composite Meteorological Patterns During Ozone Episodes

To gain an understanding of typical meteorological patterns that occur during ozone episodes, ERA-Interim reanalysis fields are averaged across the dates corresponding to the first day ("Day 0") of each ozone episode in a given region. Common meteorological features that emerge include increased temperatures in the afflicted region (Fig. 71.2a), with reduced cloud cover and a synoptic scale high pressure system aloft (Fig. 71.2b). These meteorological patterns are similar to those found during heat waves, which were identified by applying the same methodology described above to temperature measurements at AQS stations.

For ozone episodes in the Northeast region, the composite anticyclonic circulation anomaly aloft (Fig. 71.2b) appears to be part of a wave train extending from the North Pacific to the North Atlantic that forms one week in advance of the onset of the composite ozone episode (this also occurs in regions other than the Northeast). Further subcategorization of ozone episodes (e.g. those that overlap with heat waves vs. those that do not) reveals even longer-lived wave trains in some cases. Formation of these wave trains may be linked to sea surface temperature (SST) patterns. In particular, the composite Pacific SST pattern (Fig. 71.2c) for Northeast ozone episodes is markedly similar to a pattern observed to precede heat waves in the Eastern US [7], which were also preceded by an atmospheric wave train. Elements of this composite SST pattern form over a month in advance of the composite ozone episode. Though further work is required to establish the usefulness of such patterns for air quality prediction, this research suggests the presence of a link between regional-scale extreme ozone pollution episodes and subseasonal meteorological variability.

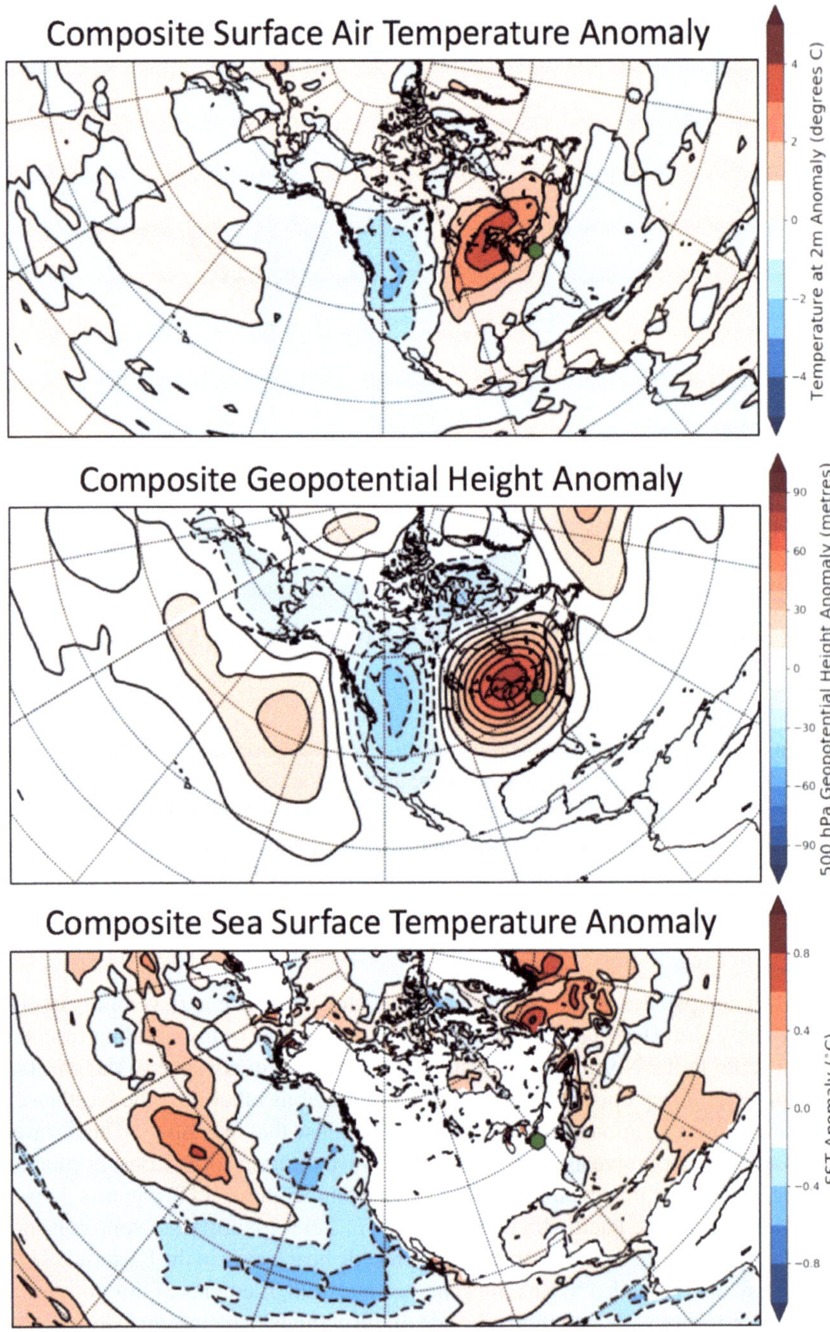

Fig. 71.2 ERA-interim (**a**, top) temperature, (**b**, middle) 500 hPa geopotential height, and (**c**, bottom) sea surface temperatures averaged over the first day of 51 ozone episodes in the Northeast region (hexagonal marker)

References

1. D.P. Dee, S.M. Uppala, A.J. Simmons, P. Berrisford, P. Poli, S. Kobayashi, U. Andrae, M.A. Balmaseda, G. Balsamo, D.P. Bauer, P. Bechtold, F. Vitart, The ERA-interim reanalysis: configuration and performance of the data assimilation system. Q. J. R. Meteorolog. Soc. **137**(656), 553–597 (2011). https://doi.org/10.1002/qj.828
2. Environment and Climate Change Canada, NAPS data products (2018), http://maps-cartes.ec.gc.ca/rnspa-naps/data.aspx?lang=en
3. P. Jaccard, The distribution of the flora in the alpine zone. New Phytol. **11**(2), 37–50 (1912). https://doi.org/10.1111/j.1469-8137.1912.tb05611.x
4. D.J. Jacob, D.A. Winner, Effect of climate change on air quality. Atmos. Environ. **43**(1), 51–63 (2009). https://doi.org/10.1016/j.atmosenv.2008.09.051
5. D.F. Karnosky, J.M. Skelly, K.E. Percy, A.H. Chappelka, Perspectives regarding 50 years of research on effects of tropospheric ozone air pollution on US forests. Environ. Pollut. **147**(3), 489–506 (2007). https://doi.org/10.1016/j.envpol.2006.08.043
6. J.M. McGrath, A.M. Betzelberger, S. Wang, E. Shook, X.-G. Zhu, S.P. Long, E.A. Ainsworth, An analysis of ozone damage to historical maize and soybean yields in the United States. Proc. Natl. Acad. Sci. **112**(46), 14390–14395 (2015). https://doi.org/10.1073/pnas.1509777112
7. K.A. McKinnon, A. Rhines, M.P. Tingley, P. Huybers, Long-lead predictions of eastern United States hot days from Pacific sea surface temperatures. Nat. Geosci. **9**(5), 389–394 (2016). https://doi.org/10.1038/ngeo2687
8. M. Szyszkowicz, B.H. Rowe, Respiratory health conditions and ambient ozone: a case-crossover study. Insights Chest Dis. **1**(1) (2016), http://insightsinchestdiseases.imedpub.com.myaccess.library.utoronto.ca/abstract/respiratory-health-conditions-and-ambient-ozone-a-casecrossover-study-8320.html
9. H. Teng, G. Branstator, H. Wang, G.A. Meehl, W.M. Washington, Probability of US heat waves affected by a subseasonal planetary wave pattern. Nat. Geosci. **6**(12), 1056–1061 (2013). https://doi.org/10.1038/ngeo1988
10. United States Environmental Protection Agency, National ambient air quality standards for ozone (Federal register) (2015)
11. United States Environmental Protection Agency, Air quality system data mart (2018), https://www.epa.gov/outdoor-air-quality-data
12. R. Xu, D.C. Wunsch, Hierarchical clustering, in *Clustering* (Wiley, 2008), pp. 31–62, http://onlinelibrary.wiley.com.myaccess.library.utoronto.ca/doi/10.1002/9780470382776.ch3/summary

Chapter 72
Assessing Potential Climate Change Impacts on Local Air Quality Using AERMOD

Jinliang Liu, Congtru Doan, Abby Salb, Yvonne Hall and Chris Charron

Abstract Model results in this study show that the modelled maximum ground level concentrations could vary significantly with the choice of meteorological data periods and source configurations, mainly due to changes in the climatology of wind speeds, their distribution and temperature. Modelled maximum ground-level concentrations could vary by as much as 30% during the historical period (1996–2016), and could decrease by over 50% for most low-level sources in the future (2051–2055) with projected climate change.

72.1 Introduction

Meteorological data is one of the critical inputs to an air dispersion model such as AERMOD [4]. It is recommended by the US EPA [5] that the most recent, readily available 5 consecutive years of meteorological data be used for regulatory purpose. Due to climate change, meteorological conditions have been experiencing dramatic changes in the past and could be in the future at most locations around the globe; therefore, using different time periods of meteorological data could result in significant differences in modelled concentrations, and thus impact regulatory conclusions. A previous similar study [1] focused on point sources using previous versions of AERMET/AERMOD; this study is an update by including more sources types/configurations, and using the latest versions of these models and meteorological data.

J. Liu (✉) · C. Doan · A. Salb · Y. Hall · C. Charron
Ontario Ministry of the Environment and Climate Change, Toronto, ON M9P 3V6, Canada
e-mail: Jinliang.Liu@Ontario.ca

C. Mensink et al. (eds.), *Air Pollution Modeling and its Application XXVI*,
Springer Proceedings in Complexity,
https://doi.org/10.1007/978-3-030-22055-6_72

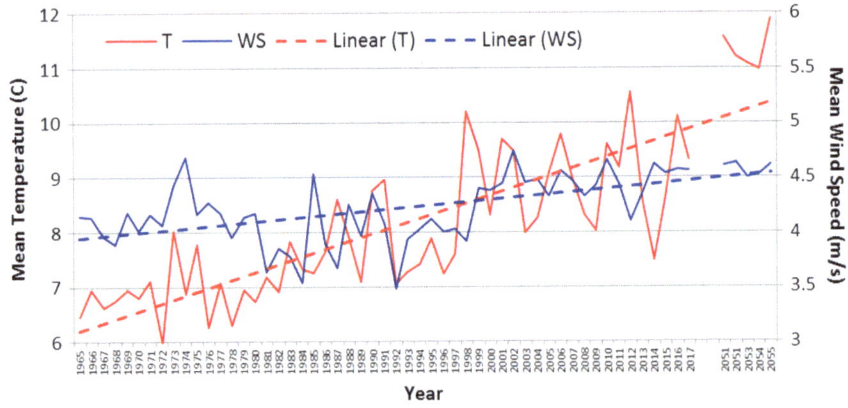

Fig. 72.1 Annual mean wind speed and temperature trends at the study location

72.2 Methodology

To quantify the impacts from the change of meteorological data periods on modelled concentrations, AERMOD was run for various hypothetical sources. Model results driven by the 1996–2000 historical meteorological data were used as benchmarks. Subsequently, AERMOD was run for a series of rolling 5-year periods of meteorological data up to year 2016 for the past, and years 2051–2055 for the future; then these model results were compared with the benchmarks (1996–2000) to understand the potential impacts from changes in climate. This study also tried to correlate these potential impacts to trends in key meteorological parameters such as wind speed (and distribution) and temperature over the past two decades and in the future.

72.2.1 Long-Term Climatological Trends

Long-term trends analysis (Fig. 72.1) shows that annual average temperature (T) and wind speed (WS) both have upward trends, while the upward trend in temperature is much more significant than that in the wind speed. It is obvious meteorological conditions can be very different among different 5-years to be used for air dispersion modelling, so do the subsequent modelled concentrations.

72.2.2 Aermet/Aermod

The latest versions of AERMET (16216)/AERMOD (16216r) [4] were used in this study. For simplicity, unified urban land use characteristics [2] were used around the hypothetical facility. The most representative surface and upper air stations (best to our knowledge) were used for AERMET. Modelled sources cover a wide arrange of source types in real practices, from point sources with different releasing height and exiting temperature/velocity, to area and volume sources usually releasing at/near surface level. Special source types like capped stacks and buoyant line sources were also included in this assessment. Receptors are allocated only along and outside of the property line out to a distance of 5 km from the dominant sources with spacing recommended by the MOECC [2]. Deposition was not considered in this study.

72.2.3 Future Climate Projections for 2051–2055

The 50th percentile projection of hourly wind speed and temperature [3] was used as potential future climate to produce the future 5-year (2051–2055) meteorological dataset; other variables (i.e. wind direction, cloud cover) are from the latest historical 5 years (2012–2016) and assumed not to change. As shown in Fig. 72.1, the projected 5-year average wind speeds and temperature are expected to increase by about 11% (3~15% year to year) and 2.8 °C (1~4.4 °C year to year), respectively, over those during 1996–2000, at the study location. It should be noted that future climate projections are subject to various uncertainties, especially surface wind speeds which are significantly influenced by small scale topographic conditions.

72.3 Results and Discussions

Along with the changes in average wind speeds (~10% increase starting from 1998–2002, with the maximum increase in 2051–2055), Figs. 72.2 and 72.3 respectively show the relative changes in the modelled maximum 1 and 24 h concentrations for different 5-year periods in comparison with the 1996–2000 model results, before meteorological anomaly removal. Figures 72.4 and 72.5 show the relative changes after meteorological anomaly removal.

Before meteorological anomaly removal, as much as 17% increases are observed in the modelled 1-hour maximum concentrations from low level releases (i.e. the two area sources, the central volume source), and up to ~33% increase from stacks with less plume rise (i.e. the non-buoyant stack 2) during the historical period. For these source types, maximum concentrations usually occur on or close to property line, increases of modelled concentration may have been caused by low winds at individual hours coinciding with other factors; whereas the modelled 24 h concentrations reduced from almost all sources, with most significant reductions (up to ~27%) for the two area sources; these reductions in 24 h concentrations are most possibly caused by the increased wind speed for most hours of the day, in other words, the average wind speed. The greatest reductions are found in the future (2051–2055), up to 45%

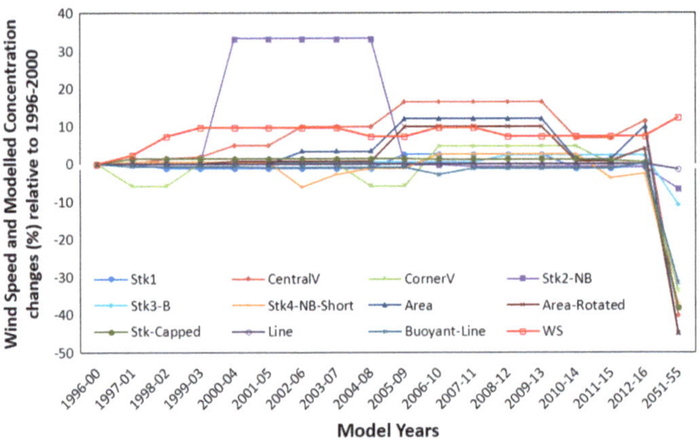

Fig. 72.2 Relative changes (%) in modelled maximum 1 h concentrations before meteorological anomalies removal

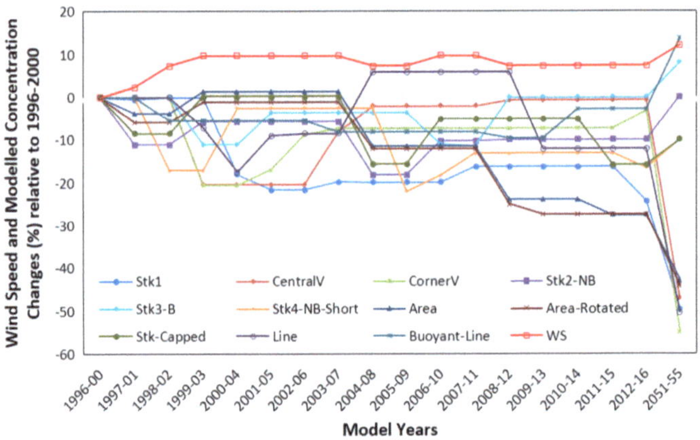

Fig. 72.3 Relative changes (%) in modelled maximum 24 h concentrations before meteorological anomalies removal

for the area sources in 1 h and up to 55% for the central volume sources in 24 h modelled concentrations.

General patterns in Figs. 72.2 and 72.3 remain after removing the meteorological anomalies (Figs. 72.4 and 72.5). By comparing Fig. 72.4 with Fig. 72.2, one can easily find out that almost all the increases in the modelled 1 h concentrations in Fig. 72.2 are results of hourly meteorological anomalies. Except two slight increases (as much as 3% for the capped stack and up to 5% for the central volume), all other sources see reductions in modelled concentrations by up to 13% in 1 h modelled concentration from the non-buoyant stack, and up to 26% in 24 h modelled concentration from the two area sources for the historical periods. The greatest reductions are again found to be in the future (2051–2055), as much as 69% for the buoyant line source

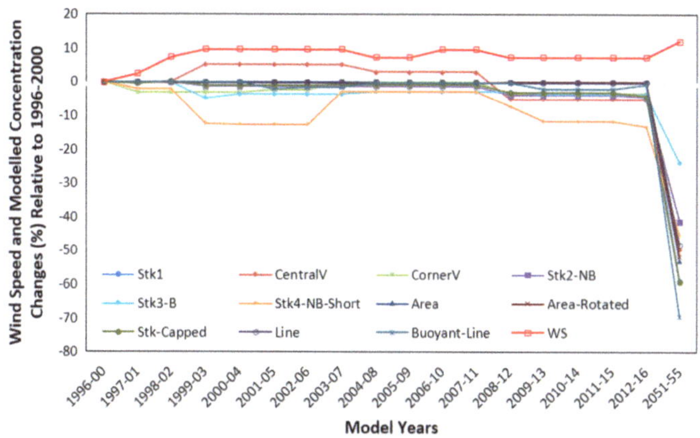

Fig. 72.4 Relative changes (%) in modelled maximum 1 h concentrations after meteorological anomalies removal

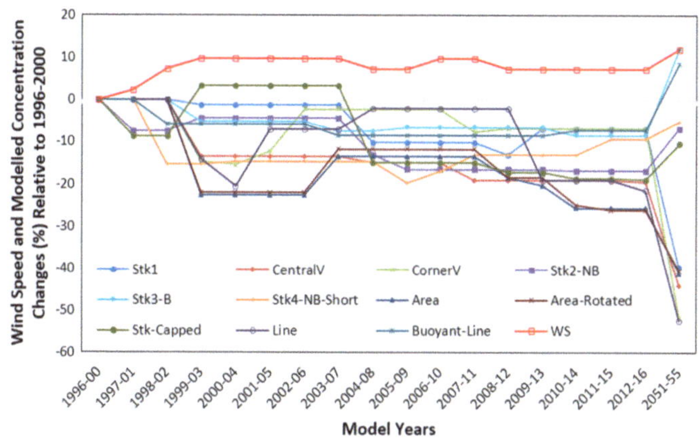

Fig. 72.5 Relative changes (%) in modelled maximum 24 h concentrations after meteorological anomalies removal

in 1 h and up to 52% for the central volume and the line sources in 24 h modelled concentrations.

Figure 72.6 illustrates that, the average wind speeds have been increasing because the frequency of low (high) wind speeds has decreased (increased) since 2001; mid-ranges of wind speeds are dominant in the future projected (50th percentile) wind speeds for 2051–2055, with much reduced low and high winds; these differences in the average wind speed and its distribution have definitely contributed to the changes in the modelled concentrations.

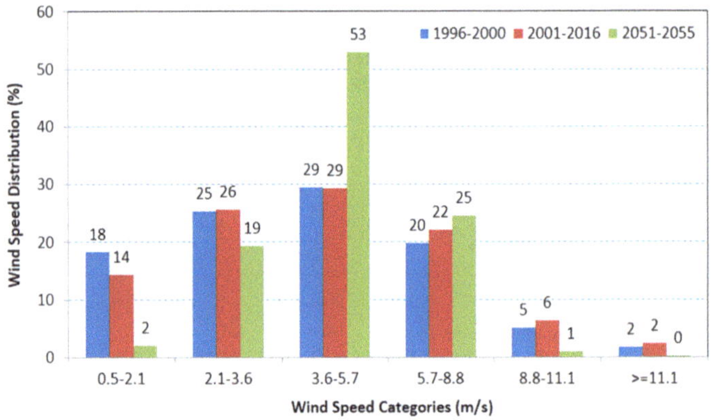

Fig. 72.6 Wind speed distributions during different time periods in the study area

72.4 Conclusion

- Both average wind speed and temperature has been increasing in the past 50 years and are expected to continue to increase in the future due to climate change; consequently, modelled concentrations are found very different when different 5-year periods are used. In other words, climate change does have significant impacts on local air quality. It is further found that not only the variation of average wind speed, but also its distribution has significant impacts on local air quality.
- For historical periods, modelled maximum ground-level concentrations could vary by as much as 30% depending on the choice of 5-year meteorological data periods, source configurations and averaging times.
- Modelled maximum ground-level concentrations for future (2051–2055) could be reduced by over 50% for most low-level (e.g. area and volume) sources.

While some physical explanations are provided in this study, more work is needed to further investigate the detailed mechanisms (i.e. changes in latent heat flux, stability and mixing heights) for the changes in modelled concentrations. Future projected meteorological conditions and associated impacts could be improved by introducing more projected variables, and using the 10th and 90th percentiles of the projected meteorological variables to produce a range of modelled concentrations to better assess uncertainties.

References

1. H. Liu, J. Liu, Assessing potential climate change impacts on small scale air dispersion modelling practices in Ontario—A case study with AERMOD, in *The 108th Air and Waste Management Association Annual Conference and Exhibition (ACE 2015) Proceedings*, 22–25 June 2015 (Raleigh, North Carolina, USA, 2015), pp. 11–23

2. MOECC, Air dispersion modelling guideline for Ontario, version 3.0, PIBs # 5165e03 (2017)
3. University of Regina, Ontario climate change data portal (2018), http://ontarioccdp.ca/
4. US EPA, AERMOD modelling system (2016), https://www.epa.gov/scram/air-quality-dispersion-modeling-preferred-and-recommended-models#aermod
5. US EPA, Revision to the guideline on air quality models: adoption of a preferred general purpose (flat and complex terrain) dispersion model and other revisions, final rule. Fed. Regist. **70**(216) (2005)

Part VIII
Air Quality Effects on Human Health and Ecology

Chapter 73
Multi-model Assessment of Air Pollution-Related Premature Mortality in Europe and U.S.: Domestic Versus Foreign Contributions

Ulas Im, Jørgen Brandt, Camilla Geels, Kaj Mantzius Hansen, Jesper Heile Christensen, Mikael Skou Andersen, Efisio Solazzo, Ioannis Kioutsioukis, Ummugulsum Alyuz, Alessandra Balzarini, Rocio Baro, Roberto Bellasio, Roberto Bianconi, Johannes Bieser, Augustin Colette, Gabriele Curci, Aidan Farrow, Johannes Flemming, Andrea Fraser, Pedro Jimenez-Guerrero, Nutthida Kitwiroon, Ciao-Kai Liang, Uarporn Nopmongcol, Guido Pirovano, Luca Pozzoli, Marje Prank, Rebecca Rose, Ranjeet Sokhi, Paolo Tuccella, Alper Unal, Marta Garcia Vivanco, Jason West, Greg Yarwood, Christian Hogrefe and Stefano Galmarini

Abstract The impact of air pollution on premature mortality in Europe and the United States (U.S.) in 2010 is modelled by a multi-model ensemble of regional models in the framework of the third phase of the Air Quality Model Evaluation International Initiative (AQMEII3). Introducing 20% emission reductions both globally and regionally in Europe, North America and East Asia were performed in order

U. Im (✉) · J. Brandt · C. Geels · K. M. Hansen · J. H. Christensen · M. S. Andersen
Department of Environmental Science, Aarhus University, Frederiksborgvej 399, 4000 Roskilde, Denmark
e-mail: ulas@envs.au.dk

J. Brandt
e-mail: jbr@envs.au.dk

C. Geels
e-mail: cag@envs.au.dk

K. M. Hansen
e-mail: kmh@envs.au.dk

J. H. Christensen
e-mail: jc@envs.au.dk

M. S. Andersen
e-mail: msa@envs.au.dk

E. Solazzo · L. Pozzoli · S. Galmarini
European Commission, Joint Research Centre (JRC), Ispra, VA, Italy
e-mail: efisio.solazzo@ec.europa.eu

C. Mensink et al. (eds.), *Air Pollution Modeling and its Application XXVI*, Springer Proceedings in Complexity,
https://doi.org/10.1007/978-3-030-22055-6_73

to calculate the domestic and foreign contributions to air quality and related premature mortality. Total premature mortality was estimated to be 414 000 in Europe and 160 000 in the U.S., using multi-model mean pollutant concentrations. The number of premature mortality cases calculated using concentration inputs from different air quality models can vary by up to a factor of three. Results show that the domestic emissions have the largest impacts on premature death, while foreign sources are a minor contributor to adverse impacts of air pollution.

73.1 Introduction

According to the World Health Organization (WHO), air pollution is now the world's largest single environmental health risk. Chemistry and transport models (CTMs) are useful tools to calculate the concentrations of health-related pollutants taking into account the non-linearities in the chemistry and the complex interactions between meteorology and chemistry. However, CTMs include different chemical and aerosol schemes that introduce differences in the representation of the processes. These uncertainties are introduced into the health impact estimates using output from these different CTMs. Multi-model (MM) ensembles can be useful to minimize these uncertainties introduced by the individual CTMs.

The AQMEII project attempts to bring together modellers from both sides of the Atlantic Ocean to perform joint regional model experiments using common boundary conditions, and emissions with a specific focus on regional modelling domains over Europe and North America [3]. AQMEII Phase 3 (AQMEII3) is devoted to performing joint modelling experiments with HTAP2.

L. Pozzoli
e-mail: pozzoli.ist@gmail.com

S. Galmarini
e-mail: Stefano.GALMARINI@ec.europa.eu

I. Kioutsioukis
Department of Physics, University of Patras, University Campus, 26504 Rio, Patras, Greece
e-mail: kioutio@upatras.gr

U. Alyuz · A. Unal
Eurasia Institute of Earth Sciences, Istanbul Technical University, Istanbul, Turkey
e-mail: ummugulsum.alyuz@bahcesehir.edu.tr

A. Unal
e-mail: alper.unal@itu.edu.tr

A. Balzarini · G. Pirovano
Ricerca sul Sistema Energetico (RSE SpA), Milan, Italy
e-mail: Alessandra.Balzarini@rse-web.it

G. Pirovano
e-mail: guido.pirovano@rse-web.it

73.2 Materials and Methods

In the present study, the simulated surface concentrations of health related air pollutants (CO, O_3, SO_2 and $PM_{2.5}$) for the year 2010 from twelve modelling groups from Europe and two groups from North America participating in the AQMEII (Air Quality Model Evaluation International Initiative) exercise, serve as input to the Economic Valuation of Air Pollution model (EVA: [1]), in order to calculate the impacts of these pollutants on human health, using exposure response functions from WHO, and the

R. Baro · P. Jimenez-Guerrero
Department of Physics, University of Murcia, Physics of the Earth, Campus de Espinardo, Ed. CIOyN, 30100 Murcia, Spain
e-mail: Rocio.Baro-Esteban@zamg.ac.at

P. Jimenez-Guerrero
e-mail: pedro.jimenezguerrero@um.es

R. Bellasio · R. Bianconi
Enviroware srl, Concorezzo, MB, Italy
e-mail: rbellasio@enviroware.com

R. Bianconi
e-mail: rbianconi@enviroware.com

J. Bieser
Chemistry Transport Modelling Group, Institute of Coastal Research, Helmholtz-Zentrum Geesthacht, Geesthacht, Germany
e-mail: Johannes.Bieser@hzg.de

A. Colette · M. G. Vivanco
INERIS, Institut National de l'Environnement Industriel et des Risques, Parc Alata, 60550 Verneuil-en-Halatte, France
e-mail: augustin.colette@ineris.fr

M. G. Vivanco
e-mail: m.garcia@ciemat.es

G. Curci · P. Tuccella
Department of Physical and Chemical Sciences, University of L'Aquila, L'Aquila, Italy
e-mail: gabriele.curci@aquila.infn.it

P. Tuccella
e-mail: paolo.tuccella@aquila.infn.it

Center of Excellence CETEMPS, University of L'Aquila, L'Aquila, Italy

A. Farrow · R. Sokhi
Centre for Atmospheric and Instrumentation Research (CAIR), University of Hertfordshire, Hatfield, UK
e-mail: a.farrow2@herts.ac.uk

R. Sokhi
e-mail: r.s.sokhi@herts.ac.uk

J. Flemming
European Centre for Medium Range Weather Forecast (ECMWF), Reading, UK
e-mail: Johannes.Flemming@ecmwf.int

associated external costs in Europe and North America. In addition, the impacts of a 20% global emission reduction scenario on the human health and associated costs have been calculated.

The EVA system is based on the impact-pathway chain [2], and calculates health impacts (morbidity and mortality for a range of health end-points based on exposure-response relations) and the associated external costs, due to short-term exposure to O_3, CO and SO_2 (soon also NO_2) and long-term exposure to PM2.5. The exposure is calculated based on gridded air pollution data and gridded population data. Exposure-response functions for air pollution and unit cost estimates are applied to calculate attributable cases and costs.

A. Fraser · R. Rose
Ricardo Energy & Environment, Gemini Building, Fermi Avenue, Harwell, Oxon OX11 0QR, UK
e-mail: andrea.fraser@gmail.com

R. Rose
e-mail: rebecca.rose@ricardo.com

N. Kitwiroon
Environmental Research Group, Kings' College London, London, UK
e-mail: nutthida.kitwiroon@kcl.ac.uk

C.-K. Liang · J. West
Department of Environmental Sciences and Engineering, University of North Carolina at Chapel Hill, Chapel Hill, NC, USA
e-mail: ckliang@live.unc.edu

J. West
e-mail: jjwest@email.unc.edu

U. Nopmongcol · G. Yarwood
Ramboll Environ, 773 San Marin Drive, Suite 2115, Novato, CA 94998, USA
e-mail: unopmongcol@ramboll.com

G. Yarwood
e-mail: gyarwood@environcorp.com

M. Prank
Atmospheric Composition Research Unit, Finnish Meteorological Institute, Helsinki, Finland
e-mail: marje.prank@fmi.fi

Department of Earth and Atmospheric Sciences, Cornell University, Ithaca, USA

M. G. Vivanco
CIEMAT, Avda. Complutense 40, 28040 Madrid, Spain

C. Hogrefe
National Exposure Research Laboratory, Computational Exposure Division, Office of Research and Development, United States Environmental Protection Agency, Research Triangle Park, NC, USA
e-mail: Hogrefe.Christian@epa.gov

73.3 Results and Conclusions

The MM mean and standard deviations from the twelve-model ensemble for different health outcomes are presented in Table 73.1. Results show that in Europe, the MM mean number of premature deaths due to air pollution is calculated to be ~400 000 ± 100 000, that agrees with the WHO estimate of 482 000. Estimated health impacts among different models can vary up to a factor of 3. $PM_{2.5}$ is calculated to be the major pollutant affecting the health impacts and the differences in models regarding the treatment of aerosol composition, physics and dynamics is a key factor. The geographical distribution of the total premature death in 2010 as calculated by the MM mean is presented in Fig. 73.1. In Europe, the largest number of cases are simulated to be in emission hot spots such as the Po Valley and the Benelux, as well as megacities such as London, Paris, Berlin and Athens. In North America, the premature death cases are mainly simulated over New York, Chicago, Detroit, Houston, Los Angeles and San Francisco.

The total MM mean costs due to health impacts of air pollution are estimated to be 400 billion € in Europe. Finally, the scenario with a 20% reduction in global anthropogenic emissions leads to a decrease of 18% of all health outcomes.

Results from the emission perturbation scenarios showed that a 20% reduction of global anthropogenic emissions avoids 54 000 (13%) and 27 500 (17%) premature deaths in Europe and the U.S., respectively. A 20% reduction of North American emissions avoids ~1 000 (0.2%) premature deaths in Europe and 25 000 (16%) premature deaths in the U.S., while a 20% decrease of emissions within the European source region avoids 47 000 (11%) premature deaths in Europe. Reducing the East Asian emission by 20% avoids ~2000 (1%) premature deaths in the U.S. These results show that the domestic emissions have the largest impacts on premature death, while foreign sources are a minor contributor to adverse impacts of air pollution.
QUESTIONER 1: Sebnem Aksoyoglu
QUESTION: In AQMEII, only anthropogenic emissions were harmonized, biogenic emissions were not. Is there any plan to check the effects of using different biogenic emissions on results or any future plans to harmonize biogenic emissions?

Table 73.1 MM mean health outcomes of air pollution in Europe and North America

Health outcomes	Europe	North America
Chronic bronchitis (CB)	360 ± 89	142 ± 74
Respiratory hospital admissions (RHA)	23 ± 5	10 ± 4
Cerebrovascular hospital admissions (CHA)	46 ± 11	19 ± 10
Congestive heart failure (CHF)	31 ± 6	13 ± 6
Lung cancer (LC)	55 ± 14	22 ± 11
Premature death (PD)	414 ± 98	165 ± 76
Infant mortality (IM)	403 ± 99	143 ± 75

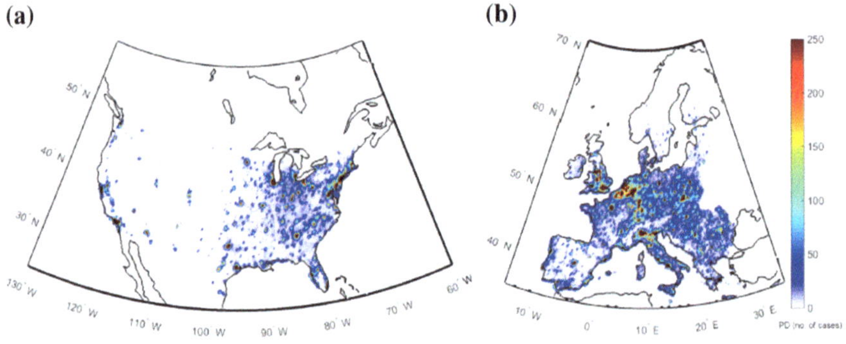

Fig. 73.1 Spatial distribution of premature deaths in 2010 in **a** Europe and **b** North America

ANSWER: Currently, there are no plans to harmonize biogenic emissions. Biogenic emissions are driven by meteorology in most model systems and the groups participating use their operational meteorological models to drive the chemistry and transport models.

QUESTIONER 2: SHUZAN REN
QUESTION: Given the fact that half of the population are living in urban regions and major air pollutants are from urban areas, urban population would have different exposure to pollutants. If the urban effects are taken into account, how would the conclusion on premature death rate be changed?
ANSWER: Taking into account urban exposure would require much higher spatial resolutions to reflect the fine distribution of emissions and population. This would maybe change the spatial distribution of the number of premature death cases over these areas. Regarding regional assessment studies focusing on e.g. Europe or North America, I do not know if the total number of premature death cases would change drastically anyway as majority of the studies estimating exposure response functions are based on measurements over urban background areas, which then we assume that it accounts for the urban vs. rural differences in premature death.

QUESTIONER 3: Moussiopoulos
QUESTION: I understand that you analyze emission scenarios with "frozen" meteorology. In view of the large uncertainties associated with the latter, and given the climate change issue, could you comment on the practical relevance of your scenario analysis results?
ANSWER: As the scenarios target the same year and only the anthropogenic emissions, natural emissions such as biogenic emissions or dust that are driven by meteorology, would not change the results from the perturbation scenarios. Different scenarios have to be designed to see the impact of meteorology or climate change on health impacts of air pollution.

Acknowledgements Aarhus University gratefully acknowledges the NordicWelfAir project funded by the NordForsk's Nordic Programme on Health and Welfare (grant agreement no. 75007), the REEEM project funded by the H2020-LCE Research and Innovation Action (grant agreement no.: 691739), and the Danish Centre for Environment and Energy (AU-DCE).

References

1. J. Brandt, J.D. Silver, J.H. Christensen, M.S. Andersen, J.H. Bønløkke, T. Sigsgaard, C. Geels, A. Gross, A.B. Hansen, K.M. Hansen, G.B. Hedegaard, E. Kaas, L.M. Frohn, Contribution from the ten major emission sectors in Europe and Denmark to the health-cost externalities of air pollution using the EVA model system—an integrated modelling approach. Atmos. Chem. Phys. **13**, 7725–7746 (2013). https://doi.org/10.5194/acp-13-7725-2013
2. R. Friedrich, P. Bickel, The impact pathway methodology, in *Environmental External Costs of Transport*, ed. by R. Friedrich, P. Bickel (Springer, Berlin, 2001), p. 326
3. S. Galmarini, B. Koffi, E. Solazzo, T. Keating, C. Hogrefe, M. Schulz, A. Benedictow, J.J. Griesfeller, G. Janssens-Maenhout, G. Carmichael, J. Fu, F. Dentener, Technical note: coordination and harmonization of the multi-scale, multi-model activities HTAP2, AQMEII3, and MICS-Asia3: simulations, emission inventories, boundary conditions, and model output formats. Atmos. Chem. Phys. **17**, 1543–1555 (2017). https://doi.org/10.5194/acp-17-1543-2017

Chapter 74
Using Multi-media Modeling to Investigate Conditions Leading to Harmful Algal Blooms

Valerie Garcia, Catherine Nowakowski, Christina Feng Chang, Penny Vlahos, Ellen Cooter, Chunling Tang and Marina Astitha

Abstract We used linked and coupled physical models to identify relationships among environmental variables across multiple sources and pathways to examine the impact of nitrogen loadings on chlorophyll α concentrations.

74.1 Introduction

Harmful Algal Blooms (HABs) degrade water quality (WQ) for drinking and recreational use, causing illness, and even death, in humans and other animals. Lake Erie supplies drinking water to over 11 million people, and HABs have threatened water quality in Lake Erie since the 1960's. David Schindler [3] conducted a series of precedent-setting field studies that pointed to phosphorous concentrations in the lake from point sources (such as sewage treatment plants), and non-point sources (such as agriculture), as the leading cause of nutrient enrichment. This phosphorous enrichment fueled the excessive growth of HABs. Removal of phosphorous in detergents and the implementation of no-till crop management practices has significantly reduced the amount of phosphorous entering the lake and led to recovery of the lake by 1990. However, since the mid-1990's, there has been a resurgence of HAB events, with the largest three events on record occurring in 2011, 2015 and 2017. In this study, we used a combination of multi-media modeling systems and chlorophyll α measurements to examine the drivers for this resurgence of HAB events in Lake Erie.

V. Garcia (✉) · E. Cooter · C. Tang
Computational Exposure Division, National Exposure Research Laboratory, U.S. Environmental Protection Agency, RTP, Washington, DC, NC, USA
e-mail: garcia.val@epa.gov

C. Nowakowski · C. F. Chang · M. Astitha
Civil and Environmental Engineering, University of Connecticut, Storrs, CT, USA

P. Vlahos
Marine Sciences, University of Connecticut, Groton, CT, USA

C. Mensink et al. (eds.), *Air Pollution Modeling and its Application XXVI*,
Springer Proceedings in Complexity, https://doi.org/10.1007/978-3-030-22055-6_74

74.2 Approach

Figure 74.1 shows the major components of the modeling system used in the study. The same land use (2011 National Land Cover Data; https://www.mrlc.gov/) and meteorology data (simulated by the Weather Research and Forecasting Model; http://www2.mmm.ucar.edu/wrf/) were used to drive all models. The United States Department of Agriculture (USDA) Environmental Policy Integrated Climate (EPIC) model (https://www.nfc.usda.gov/epic/) was coupled with the US Environmental Protection Agency (EPA) Community Multiscale Air Quality (CMAQ) model (https://www.epa.gov/cmaq). In this multi-media framework, applied fertilizer, depth and timing simulated by EPIC were provided as inputs to CMAQ, while deposition and evasion of ammonia (NH_3) concentrations were iteratively calculated in CMAQ to simulate the flux between the land surface and air [1]. Hydrological variables were simulated using the Variable Infiltration Capacity (VIC) Hydrological Model (http://www.hydro.washington.edu/Lettenmaier/Models/VIC/Overview).

Variables from this modeling system were used to examine associations with available chlorophyll α (chl-α) concentrations measured in Lake Erie and averaged seasonally (May–October) for 2002 through 2012. We used chl-α measurements as a surrogate for HABs as these measurements are readily available and have been shown to be significantly correlated with HABs (e.g., [4]).

Relevant variables from the coupled modeling system were used to regress on the chl-α measurements ($n = 187$). Depending on the variable (Table 74.1), measurements were paired with 12 km × 12 km gridded model output by summing or

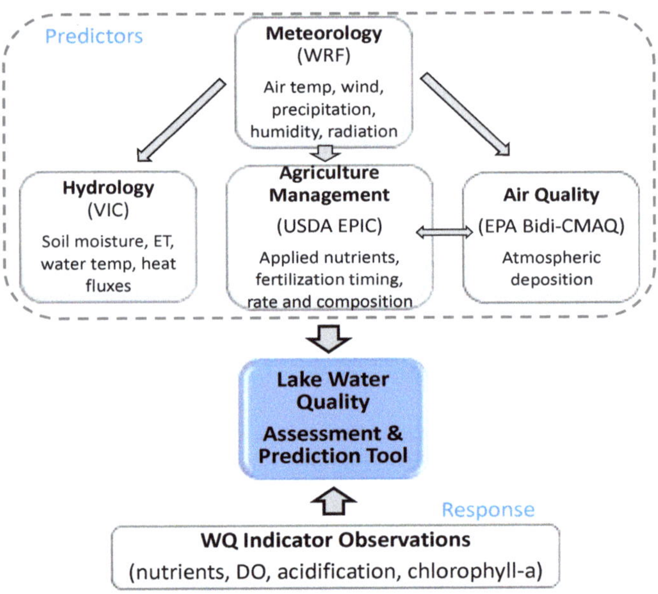

Fig. 74.1 Multi-media modeling systems used in study

Table 74.1 Explanatory variables from modeling system used in regression analysis (^ = grid-cell value, + = summed for watershed, * = averaged for watershed)

Meteorology (WRF)	Air quality (CMAQ)	Agriculture (EPIC)	Hydrology (VIC)
Humidity^	Total N Deposition+	Applied NH3 fertilizer+	Water temperature^
Air temperature (Min/Max)^	Dry Oxidized N Deposition+	Applied NO3 fertilizer+	Flow+
Precipitation	Dry Reduced N Deposition+	Applied Mineral P+	Evapotranspiration^
Radiation^	Wet Oxidized N Deposition+	Applied Organic P+	Soil moisture (3 layers)*
Wind speed^	Wet Reduced N Deposition+	Applied Organic N+	Snow water equivalent+
	Wet Organic N Deposition+	Runoff+	Sensible heat flux^
		Sediment+	Latent heat flux^
		N&P loss with sediment+	Net downward radiation^
		Labile P loss in surface runoff+	
		N in subsurface flow+	
		Soluble P loss - drainage+	
		N in drain tile flow+	

averaging across the watershed draining into the lake at the measurement location, or matching the measurement with the grid cell directly above its location. We used a Generalized Least Squares regression technique to account for spatial autocorrelation inherent in the measurements. Variable selection was based on forwards and backwards stepwise regression ($p \leq 0.05$) and 10-fold cross-validation. Plausible pathways were determined a priori and collinearity among explanatory variables was fully explored. The Variation Inflation Factor (VIF) was calculated to further check for the influence of collinearity. Variables were also lagged from 1 to 7 days to account for transport through the watershed and algal bloom growth.

In addition, we used a machine learning Random Forest technique to examine the most influential variables for predicting chl-α concentrations. This metric was calculated by permuting the variable multiple times and calculating the percent increase in the mean square error.

74.3 Results and Discussion

We found that evapotranspiration (3-day lag, grid-cell value), total ammonia fertilizer applied to crops (2-day lag, watershed), deep soil moisture (4-day lag, watershed), radiation (3-day lag, grid-cell value) and dry deposited oxidized nitrogen (5-day lag, watershed) were significantly associated with chl-α measurements. Significant variables from regression modeling fell within the top 20 most influential variables using the Random Forest technique. Figure 74.2 shows the results indicating robust prediction capability based on the 11-year timeseries.

The model variables allowed us to refine our analysis of nutrient loadings. For example, we were able to analyze dry, wet, reduced and oxidized N deposition (as opposed to total deposition normally used). In addition, we were able to examine different agricultural management approaches, such as till/no-till, irrigated/rainfed, and tile drainage, and fertilization type, rate and crop (instead of county fertilizer sales data normally used). It is interesting to note that phosphorous was not a significant explanatory variable in our regression model. This may be because the cyanobacteria community has changed from the phosphorous-limited Anabaena and Aphanizomenon species that are able to fix non-reactive N from the atmosphere, to Microcystis cyanobacteria communities. Microcystis requires bioavailable N (e.g., NH_3),

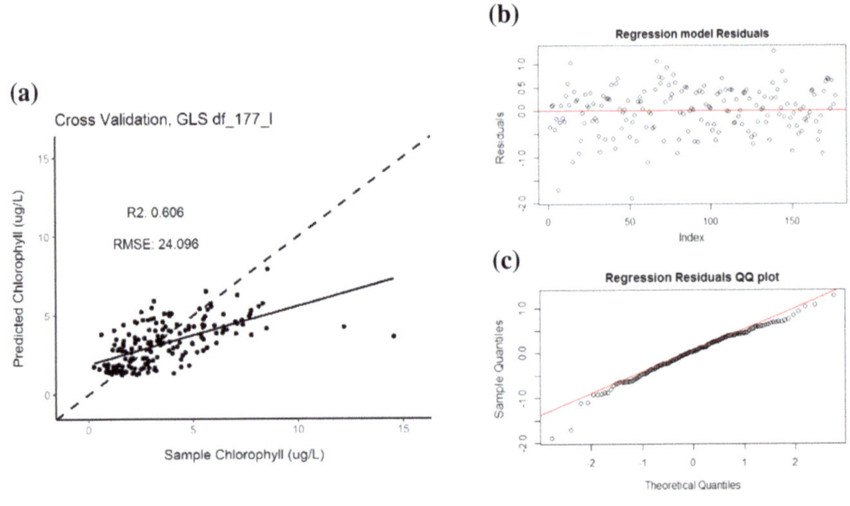

Fig. 74.2 Regression metrics **a** Coefficient of Determination (R^2); **b** residuals; and **c** QQ plot

but will dominate the cyanobacteria community with just small amounts available [2]. Microcystis blooms are more toxic, longer in duration and are more extensive than the earlier P-limited cyanobacteria communities experienced in the late 1960's and 1970's. A better understanding of the environmental conditions leading to the proliferation of these differing cyanobacteria communities may be gleaned through using physics-based, mass-balanced models that simulate the transport of nutrients through the land, air and water.

QUESTIONER: Clemens Mensink
QUESTION: To what extent do you think we understand the different processes? In other words, where are the biggest knowledge gaps to be found?
ANSWER: The cycling of nutrients throughout the environment is complex and extremely variable across time and space. Our understanding of these processes are limited, and data are sparse. Particular knowledge gaps include the biogeochemical cycling associated with nutrients (e.g., interactions involving microbes, organic N, nitrous oxide, carbon), and limitations in measurements (e.g., flux, dry deposition) and input data (e.g., soils, landcover, location/emissions of animal feeding operations).

QUESTIONER: Camilla Geels
QUESTION: You mention NH_3 is an imporant factor. It is well known that the application of manure and the following NH_3 emissions to the atmosphere is very dependent on the weather. To what degree do you include that in the atmospheric model? And is the spatial resolution of the atmospheric model high enough to describe the dry deposition of NH_3?
ANSWER: Temperature dependent processes, such as the increase in NH_3 volatilization as ambient temperature increases, is calculated for croplands (e.g., from applied fertilizer) and other non-agriculture vegetation. Emissions from animal feeding operations are based on animal density data and are seasonally adjusted. While CMAQ is run at a 12 km x 12 km resolution, relevant variables needed for calculating deposition and evasion are based on the varying landuse types within this grid structure (sub-grid level). This resolution is appropriate for the long-term objective of this study (national screening-level assessment of large waterbodies in the US to harmful algal blooms), however, it is recognized that NH_3 deposits locally, and running the model at a finer resolution would be valuable.

QUESTIONER: Paul Maker
QUESTION: What EPIC variables are being fed back into CMAQ as part of the 2-way coupling shown in one of your figures?
ANSWER: Fertilizer rates, depths and timing, and managed soil information are passed from EPIC to CMAQ on a daily time step. CMAQ uses this information to calculate the soil NH_4 pool, which in turn is used to calculate the NH_3 air-surface flux.

Disclaimer The views expressed in this paper are those of the authors and do not necessarily reflect the views and policies of U.S. Environmental Protection Agency.

References

1. J.O. Bash et al., Evaluation of a regional air-quality model with bidirectional NH3 exchange coupled to an agroecosystem model. Biogeosciences **10**, 1635–1645 (2013)
2. S. Levy, Microcystis Rising: Why Phosphorus Reduction Isn't Enough to Stop CyanoHABs. Env. Health Perspect. **125**(2), A34–A39 (2017)
3. D.W. Schindler, Evolution of phosphorus limitation in lakes. Science **195**, 260–262 (1977)
4. L.L. Yuan, A.I. Pollard, S. Pather, J.L. Oliver, L. D'Anglada, Managing microcystis: identifying national-scale thresholds for total nitrogen and chlorophyll a. Freshw. Biol. **59**, 1970–1981 (2014). https://doi.org/10.1111/fwb.12400

Chapter 75
Trying to Link Personal Exposure Measurement and Population Exposure Modelling: A Test Case in Liège, Belgium

Fabian Lenartz, Virginie Hutsemékers and Wouter Lefebvre

Abstract Commonly, population exposure is evaluated by crossing data of atmospheric pollution and population density maps. The former are usually actual measurements or simulated concentrations; depending on the approach or the model resolution, very different patterns may appear both in space and time, so that conclusions can vary significantly. The latter are usually based on residency information, and for many of us, do not reflect the typical wanderings, thus actual exposure. With the rise of portable devices, we are given the unprecedented opportunity to measure pollutant concentrations at a high time rate and to know the exact location of a subject. Moreover, the increase of computational capacities allows one to perform operational runs at spatial and temporal resolutions of about 10 m and 1 h respectively. Furthermore, if the subject writes an activity log, it is also possible to discriminate indoor and outdoor situations. In this ongoing work, we investigate the discrepancy in the evaluation of population exposure when using, on one hand different pollutant concentration maps e.g. yearly, daily or hourly average values, more or less sophisticated and/or refined models, different information related to the population e.g. static or dynamic and on the other hand actual data. Our region of interest for this test case is the city of Liège in Belgium.

75.1 Introduction

One of the two general methods to assess air pollution personal exposure is air monitoring, which depends either on direct measurements via personal monitors, or on indirect measurements via fixed-site analyzers combined with data on time

F. Lenartz (✉)
ISSeP, Rue du Chéra 200, 4000 Liège, Belgium
e-mail: f.lenartz@issep.be

V. Hutsemékers
AwAC, Avenue Prince de Liège, 5100 Jambes, Belgium

W. Lefebvre
VITO, Boeretang 200, 2400 Mol, Belgium

C. Mensink et al. (eds.), *Air Pollution Modeling and its Application XXVI*,
Springer Proceedings in Complexity,
https://doi.org/10.1007/978-3-030-22055-6_75

activity patterns [2]. Usefully, model output could be used instead of measurements made at fixed locations for they better catch or should better catch spatial variability than a closest station approach.

Should this technique prove to be advantageous, extrapolation to assess population exposure is far from direct. Usually this estimation is based on a static population density map that considers that everybody lives outdoor, therefore it obscures two important elements: (i) a population is dynamic and (ii) most of us spend most of their time indoor.

75.2 Experiment

In the framework of the OIE (*Outdoor and Indoor Exposure*) project, both a measurement campaigns and a modelling effort are done to evaluate personal exposure to black carbon (BC), nitrogen oxides (NO and NO_2), ozone (O_3) and fine dust ($PM_{2.5}$), and to relate to population exposure.

75.2.1 *Measurements*

The portable devices used for our measurement campaigns are on one hand the AethLabs AE51 and on the other hand the so-called Antilope, developed for ISSeP by the CECOTEPE. The first has already been used extensively in a series of studies, while the latter is a prototype based on three low-cost gas sensors from Alphasense and one low-cost particle sensor from Honeywell (see Figs. 75.1 and 75.2). In addition, a GPS sensor keeps track of the subject's movements, a weather sensor records temperature, relative humidity and barometric pressure in the direct vicinity of the printed circuit board, and another one checks when a pre-determined acceleration threshold is overrun.

75.2.2 *Modelling*

For our simulations, we have chosen the IFDM-OSPM model chain [1], recently re-baptized AtmoStreet. It's basically a two-level modelling system that consists of a Gaussian model for industry, shipping and major road emissions, a highly-parameterized box model for simulating pollution dispersion in street canyons, and a procedure to avoid double count of emissions, as well as to maintain a state of chemical equilibrium between NO_X and O_3. Receptor points are spread in a different way along line or around point sources for IFDM, and at the center of the street

Fig. 75.1 The portable mini-stations based on low-cost sensors

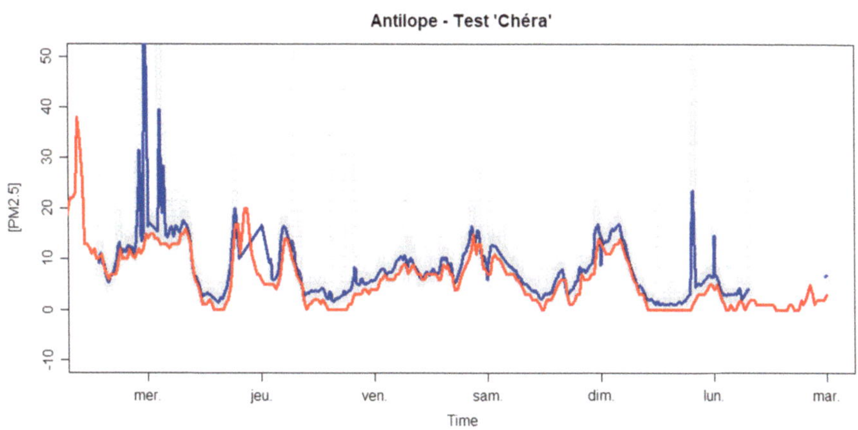

Fig. 75.2 Comparison of the raw (grey circles) and averaged (blue line) data from the mini-stations with a Grimm spectrometer (red line)

and both kerbsides for OSPM. Eventually, these scattered results are gridded at a 10-m resolution on an hourly basis. Additionally, other techniques are also used to provide pollutant ambient concentrations: a nearest-station approach, an optimal interpolation technique and the RIO model.

Since indoor activities usually account for more than 80% of a typical daily pattern, it is relevant to try and evaluate the concentration inside buildings or vehicles. A non-stationary node model has thus been developed to simulate indoor BC concentrations based on the outdoor temperature and pollution conditions. It takes into account

(i) sporadic exchange through a door, which opens according to a specific time pattern, (ii) continuous exchange due to a lack of impermeability either with the outside world or between the two considered rooms, and (iii) exchange through an air conditioning system.

75.2.3 Exposure

In order to have a sufficiently large pool of subjects we made an open call to volunteers. Such a sampling strategy prevents us from determining a confidence interval related to the estimated exposure level. Nevertheless, results can be extrapolated in some ways via the so-called "typical units" or "quota" methods, and then potentially be compared with modelling output. Each subject carries the instruments for a full week and is asked to fill in a questionnaire about his/her environment, as well as his/her habits in some domains relevant to air quality issues. Furthermore, the subject also has to fill in an activity and transport logbook to help us analyze the results.

75.3 Results

Very preliminary results of both outdoor and indoor models are shown in Figs. 75.3 and 75.4 respectively.

A one-week simulation of ambient NO_2 concentration was carried out using AtmoStreet for the area of Liège. Although it is based on dummy emissions and limited to a short time period, a typical pattern appears, highlighting both the valley and the narrowness of some streets.

Several one-day simulations of the indoor BC level at the ground level of a cloth store located along one of the city's main boulevards. For most runs, the model was able to represent adequately the concentrations observed in the main room—slightly underestimating them in average.

Both tools need now a lot of validation work before comparison with actual exposure measurements.

Questions

Questioner: Xuesong Zhang

Question: Once the calibration effort done, you could consider pushing these affordable instruments to other countries (e.g. developing ones) as their government and institutions might be interested in such devices to monitor their local air quality.

Answer: Presently, we are testing and using our mini-stations in Wallonia. This region has a temperate maritime climate and relatively low industry emissions, thus we know nothing about their performance under more "extreme" conditions. We'll

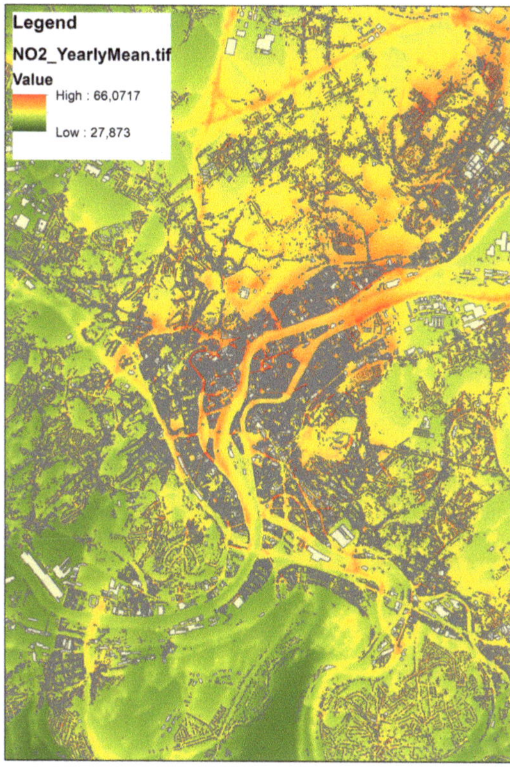

Fig. 75.3 One-week averaged NO_2 concentration

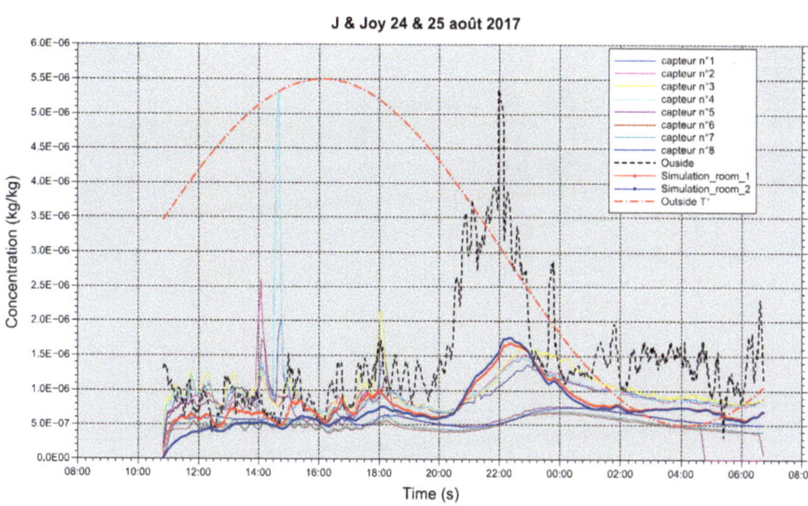

Fig. 75.4 One-day simulation of indoor BC concentration at the ground level of a building (solid blue and red lines) compared to actual measurements (other thinner solid lines)

decide in the near future if it's worth for us to carry a larger scale measurement campaign. Our original idea was to propose a service rather than an instrument but this point is also still under discussion.

Disclaimer All participants involved in the "Personal exposure" part of the OIE project have signed a consent allowing us to use their personal information as well as the measurements they made with our instruments.

References

1. W. Lefebvre, M. Van Poppel, B. Maiheu, S. Janssen, E. Dons, Evaluation of the RIO-IFDM-street canyon model chain. Atmos. Environ. **77**, 325–337 (2013). https://doi.org/10.1016/j.atmosenv.2013.05.026
2. A.Y. Watson, R.R. Bates, D. Kennedy (eds.), Assessment of human exposure to air pollution: methods, measurements, and models, in *Air Pollution, the Automobile, and Public Health* (National Academies Press (US), Washington (DC), 1988), https://www.ncbi.nlm.nih.gov/books/NBK218147/

Chapter 76
What Policy Makers and the Public at Large Should Know About Air Quality

Wouter Lefebvre

Abstract Poor air quality results in important health effects. However, the understanding of the problem by the public at large and by the policy makers is sometimes severely lacking. Therefore, it could be important to boil down the knowledge on air quality to some main points which then can be communicated to the stakeholders. The paper present such a list and is based on interviews and discussions with several air quality experts in the field. For the public at large, the main message is twofold. First of all, citizens have to acknowledge that every action they take has an influence on the quality of the air they breathe. Secondly, they have to understand the impact of air pollution on their own health and their neighbors, without blindly relying on rules like 'natural/green is good for air quality', which are often wrong. For the policy makers, a more heterogeneous set emerged from the discussions between the experts. First of all, policy makers have to know how much they (on their government level) can influence the air quality, and what effect certain actions can have on the population for which they are responsible. Secondly, they need to understand the uncertainties on the numbers as they exist now. Finally, stakeholders are encouraged to take action, from the local scale on, in order to get the actions at other levels moving (act local, think global).

76.1 Introduction

As an air pollution scientist, I regularly have to explain the work I do. Furthermore, I sometimes have to defend the actions taken by the government for the improvement of the air quality. However, when discussing this, one finds that what we think as basic knowledge is missing in the broader public. However, we cannot expect citizens to have a broad knowledge on air quality, next to all the other topics where other specialists surely would want the public at large to have more knowledge.

W. Lefebvre (✉)
VITO, Boeretang 200, 2400 Mol, Belgium
e-mail: wouter.lefebvre@vito.be

C. Mensink et al. (eds.), *Air Pollution Modeling and its Application XXVI*,
Springer Proceedings in Complexity,
https://doi.org/10.1007/978-3-030-22055-6_76

We can then ask ourselves the question: what exactly is the basic knowledge that we want the large public to know. And in extension, what should policy makers, who can have a large influence with their decisions on the air quality, know?

Thinking about this, I was able to compile a list of things that I would want that people know. In order to verify if my answers were representative for the scientific community at large, I used the last HARMO-conference to interview several air quality scientists (of different countries, backgrounds, …) on this topic. At first, the different answers that I received were wildly different, but in discussions with the different interviewees several common topics were found. This paper discusses these results.

76.2 Large Public

Two important messages were found for the public at large. The first one is that every citizen should know that air pollution has an adverse effect on their and their neighbors' health. The health effects of air pollution, in particulate particulate matter, are well-documented and are large. It is estimated that globally more people die due to ambient air pollution than due to transport accidents (by a factor of more than 4). The number of life years lost is also much larger for air pollution (by a factor of more than 2) than for transport injuries [1]. Nevertheless, the air pollution deaths are much less visible as air pollution is not a direct cause but a risk that increases a number of incidences of death causes. As a result, the health damage due to air pollution is not visible, but in order to create support for measures against air pollution, the knowledge of this existence is necessary (but not sufficient).

Secondly, people should acknowledge that every action they take can have an influence on air quality both close to them as far away. For instance, eating meat will increase ammonia emissions at the locations where the meat is produced, transport emissions in order to get the meat at the location, … Some of the emissions are unavoidable in the current system (food will have to be transported in many cases, leading to transport emissions). However, the realization of these linkages in the system are important.

When human beings are confronted with a complex system, they tend to apply simple and clear-cut rules in order to make decisions. However, this creates problems when these simple rules are incomplete or flat-out wrong. For instance, many people will consider that what is natural will be good for the air quality. When applied to for instance wood burning, this rule is wrong. A significant percentage of the particulate matter emissions, and thus concentrations, exposure and health effects is created by the combustion of wood.

76.3 Policy Makers

The policy makers add least need to understand the messages for the large public, from which they are part of. However, it would be good that they understand three extra points.

First of all, they should understand how much or how little (depending on pollutant, government level, …) they can influence air pollution. Linked to that is the question of how much effect certain actions will have on the population for which they are responsible. Linked to that question is also an understanding of the import–export balance of air pollution and thus the understanding that not only their constituents are influenced by the emissions abroad but also that the emissions in their area have an influence outside of their borders.

Secondly, the numbers concerning what they can do and what not, have uncertainties and it is important for the policy makers to have a clue of these uncertainties. In addition, temporal variability in meteorology creates an extra uncertainty in the measurements obscuring several trends in air pollution. Nevertheless, it is important that despite all uncertainties a certain knowledge base exists (such as the importance of air pollution on health) upon which no major disagreement exists within the scientific community albeit without an exact quantification. In other words, although we do not know the exact number of deaths due to air quality, we do know that this number is very high.

Therefore, as a third point, stakeholders are encouraged to take action, from the local scale on, which can help getting actions at higher levels moving (act local, think global).

76.4 Conclusion

This paper describes a basic set of knowledge that should be known by the public at large or more specifically by stakeholders. Once there is a consensus over these points, it would be interesting to define the best ways in communicating these points to the respective stakeholders and the large public.

Acknowledgements I want to thank the persons whom I collected interviews from: Silvia Trini Castelli, Goran Gašparac, Saravanan Arunachalam, Philippe Thunis, Kees Cuvelier and Stijn Janssen for their valuable input.

Reference

1. GBD, Global Burden of Disease (2017), Data: 2016. Source: https://vizhub.healthdata.org/gbd-compare/. Accessed 24 Jan 2018.

Part IX
Special Sessions

Chapter 77
An Atmospheric Scientist—The Contributions of Dr. Yitzhak Mahrer

Nitsa Haikin, George Kallos, Pinhas Alpert, Roni Avissar, Bob Bornstein and Roger Pielke Sr.

Abstract Dr. Yitzhak Mahrer, an Israeli atmospheric scientist, was one of the earliest contributors to the Regional Atmospheric Modeling System (RAMS), a leading model with abilities on a wide range of atmospheric scales. He was involved in many complex-terrain and coastal atmospheric dynamic studies, and was among the pioneers of air-pollution modeling, especially over the Eastern Mediterranean. Dr. Mahrer deceased on September 2017, and RAMS community has lost one of its founders, with his shy smile, funny remarks, and bright mind. While he also led multiple fields observational campaigns and graduated many students as a Professor at the Hebrew University of Jerusalem, we hereby present only a brief overview of his scientific contribution to the atmospheric modeling community.

N. Haikin (✉) · P. Alpert
Tel Aviv University, Tel Aviv, Israel
e-mail: nitsahai@tauex.tau.ac.il

P. Alpert
e-mail: pinhas@tauex.tau.ac.il

G. Kallos
University of Athens, Athens, Greece
e-mail: g-kallos@otenet.gr

R. Avissar
University of Miami, Florida, USA
e-mail: ravissar@rsmas.miami.edu

B. Bornstein
SJSU, San Jose, CA, USA
e-mail: pblmodel@hotmail.com

R. Pielke Sr.
University of Colorado, Colorado State University, Colorado, USA
e-mail: pielkesr@gmail.com

C. Mensink et al. (eds.), *Air Pollution Modeling and its Application XXVI*,
Springer Proceedings in Complexity,
https://doi.org/10.1007/978-3-030-22055-6_77

77.1 Introduction

Atmospheric modeling has enormously advanced over the last decades, due to the growing strength of computing, scientific development of numerical schemes which better represent physical behaviour of the atmosphere, newer capabilities of assimilation of observations into the models, and great minds on top of all. Dr. Yitzhak Mahrer was an Atmospheric modeller, who started his way with room-size computer and had gone a long way into the era of very strong computers of a notebook size. This paper is a brief presentation of his contribution to the scientific atmospheric modeling community.

77.2 Dr. Mahrer's Research—A Brief Overview

Yitzhak Mahrer started his atmospheric journey with his Msc. and PhD studies in meteorology, as Prof Neumann's student at the Hebrew University of Jerusalem. He passed away after four decades of atmospheric research covering a wide range of interests, including mesoscale meteorology, agricultural meteorology, microclimatology, and coastal and complex terrain air pollution modeling. In the 1970's he developed one of the first two-dimensional numerical model, and performed simulations of mesoscale atmospheric flow over mountainous slopes including sea-breeze dynamics over an island (e.g. [11, 13]).

The RAMS concept was created in the early 1980's at Colorado State University by merging three related models: the CSU cloud-mesoscale model [18], a hydrostatic version of the cloud model [17], and the sea breeze model Mahrer and Pielke [10], Cotton et al. [4]. A radiation scheme named after Mahrer-Pielke was integrated into the RAMS.

In the 1980's Dr. Mahrer published papers on a new numerical coordinate system; simulations over complex terrain including the need for a balance between the horizontal and vertical grid resolution, in order to represent the terrain features with reduced numerical error. To do this Mahrer [9] suggested an improved scheme for the terrain-following coordinate system which reduces numerical near-surface pressure errors, when calculating high vertical resolution near steep slopes. He was among the first scientists to introduce atmospheric dynamics into air-pollution modeling (e.g. [14]) Dr. Mahrer also used numerical simulations to explore atmospheric processes with important impacts on agriculture (e.g. [2]).

During his academic career as a professor at the Faculty of Agriculture of the Hebrew University of Jerusalem, Professor Mahrer led microclimate studies of flow within greenhouses and in the open field (e.g. [3, 12, 16]). Most of these studies combined observational field campaigns to support his modeling vision and developments.

In the 1990's Dr. Mahrer focused his research efforts on atmospheric air pollution aspects employing numerical modeling (e.g. [1, 7]), and agricultural applied research (e.g. [8]).

Dr. Mahrer worked with many top mesoscale modellers on variety of studies, such as the SKIRON system for Mediterranean dust forecast, a bilateral Israeli-Italian project on urban air pollution, a Middle East Regional Cooperation Program of regional transboundary pollutant transport. His latest publications treat aspects of topographic flow (e.g. [15]), atmospheric influence on agricultural phenomena (e.g. [21]), local and regional air pollution (e.g. [19, 20]), sea-breeze circulation (e.g. [6]) and patterns of inversion layers (e.g. [5]).

77.3 Conclusion

Dr. Yitzhak Mahrer contributed fundamental knowledge to the atmospheric science in many aspects of theory and applications, which is documented in his many papers. He educated many students, some of whom developed successful academic career, others reached senior positions in air pollution related professional occupations, and/or founded unique atmospheric applications. All of his students and colleagues have been tremendously influenced by his personality, his vision and knowledge and all deeply miss him. The scientific community has lost a great member of its community.

References

1. D. Alper Siman-Tov, M. Peleg, V. Matveev, Y. Mahrer, I. Seter, M. Luria, Recirculation of polluted air-masses over the Eastern Mediterranean coast. Atmos. Environ. **31**(10), 144–1448 (1997)
2. R. Avissar, Y. Mahrer, Mapping front-sensitive areas with a three dimensional local-scale numerical model, part 1: physical and numerical aspects. J. Appl. Meteorol. **27**, 400–413 (1988)
3. R. Avissar, Y. Mahrer Verification study of a numerical greenhouse microclimate model. Trans. Am. Soc. Agric. Eng. **25**(6), 1711–1720 (1982)
4. W.R. Cotton, R.A. Pielke Sr., R.L. Walko, G.E. Liston, C.J. Tremback, H. Jiang, R.L. McAnelly, J.Y. Harrington, M.E. Nicholls, G.G. Carrio, J.P. McFadden, RAMS 2001: Current status and future directions. Meteorol. Atmos. Phys. **82**, 5–29 (2003)
5. N. Haikin, E. Galanti, T.G. Reisin, Y. Mahrer, P. Alpert, Inner structure of atmospheric inversion layers over Haifa Bay in the eastern Mediterranean. Bound. Layer Meteorol. **156**, 471–487 (2015)
6. I. Levy, U. Dayan, Y. Mahrer, A five-year study of coastal recirculation and its effect on air pollutants over the East Mediterranean region. J. Geophys. Res. Atmos. **113**, D16121 (2008)
7. Y. Mahrer, Mesoscale models and their application to air quality studies. IL Nuovo Cimento C **14**(4), 347–361 (1991)
8. Y. Mahrer, G. Rytwo, Modeling and measuring evapotranspiration in a daily drip irrigated cotton field. Irrig. Sci. **12**(1), 13–20 (1991)

9. Y. Mahrer, An improved numerical approximation of the horizontal gradients in a terrain-following coordinate System. Mon. Weather. rev. **112**, 918–922 (1984)
10. Y. Mahrer, R.A. Pielke A numerical study of the airflow over irregular terrain. Beiträge zur Physik der Atmosphäre **50,** 98–113 (1977)
11. Y. Mahrer, R.A. Pielke, A numerical study of the air flow over mountains using the two-dimensional version of the University of Virginia mesoscale model. J. Atmos. Sci. **32**, 2144–2155 (1975)
12. O. Naot, Y. Mahrer, Modeling microclimate environments: a verification Study. Bound. Layer Meteorol. **46**(4), 333–354 (1989)
13. J. Neumann, Y. Mahrer, A theoretical study of the sea and land breezes of circular islands. J. Atmos. Sci. **31**, 2027–2039 (1974)
14. M. Segal, Y. Mahrer, R.A. Pielke, A numerical model study of plume fumigation during nocturnal inversion break-up. Atmos. Environ. **16**(3), 513–519 (1982)
15. E. Shilo, Y. Ashkenazy, A. Rimmer, S. Assouline, P. Katsafados, Y. Mahrer, Effect of wind variability on topographic waves: Lake Kinneret case. J. Geoph. Res. **112**, C12024 (2007)
16. E. Shilo, M. Teitel, Y. Mahrer, T. Boulard, Air-flow patterns and heat fluxes in a roof -ventilated multispan greenhouse with insect-proof screens over its openings. J. Agric. For. Meteorol. **122**(1–2), 3–20 (2004)
17. C.J. Tremback Numerical simulation of a mesoscale convective complex: model development and numerical results. Ph.D Dissertation,Atmos Sci Paper No 465, Colorado State University, Department of Atmospheric Science, Fort Collins, CO 80523, 1990
18. G.J. Tripoli, W.R. Cotton, The Colorado State University three-dimensional cloud/mesoscale model–1982. Part I: General theoretical framework and sensitivity experiments. J. Rech. Atmos. **16**, 185–220 (1982)
19. A. Wanger, M. Peleg, G. Sharf, Y. Mahrer, G. Kallos, V. Kortnoi, K. Lagouvardos, M. Varinou, A. Pappadopoulus, M. Luria, Some observational and modeling evidence of long-range transport of air pollutants from Europe towards the Israeli coast. J. Geoph. Res. **105**(D6), 7177–7186 (2000)
20. E. Weinroth, M. Luria, C. Emery, A. Ben-Nun, R. Bornstein, J. Kaplan, M. Peleg, Y. Mahrer, Simulations of Mideast transboundary ozone transport: A source apportionment case study. Atmos. Environ. **42**, 3700–3716 (2008)
21. E. Wineman, Y. Lenski, Y. Mahrer Solar heating of honey bee colonies (Apis mellifera L.) during the subtropical winter and its impact on hive temperature, worker population and honey production. Am. Bee J. **143**(7), 565–570 (2003)

MIX
Papier aus verantwortungsvollen Quellen
Paper from responsible sources
FSC® C105338

If you have any concerns about our products,
you can contact us on
ProductSafety@springernature.com

In case Publisher is established outside the EU,
the EU authorized representative is:
Springer Nature Customer Service Center GmbH
Europaplatz 3, 69115 Heidelberg, Germany

Printed by Libri Plureos GmbH
in Hamburg, Germany